Design Engineers Handbook

Bulletin 0224-B1

Design Engineers Handbook

Printed in the United States of America
ISBN 1-55769-018-9

preface

This handbook has been prepared by Parker-Hannifin to provide, in one document, technical data and reference material to the designer, builder and the user of equipment incorporating fluid power components. The information found in the following chapters is equally applicable to most industries including, Food Processing, Automotive, Forest Products, Woodworking, Machine Tool, Rubber, Plastic Molding, Construction, Packaging, Glass, White Goods and any others using automated equipment incorporating fluid power devices.

Since the information contained in this handbook is organized to assist the machine designer and manufacturer, as well as service and maintenance personnel, it should prove to be equally valuable to the college and vocational school student preparing to enter any of these fields.

The material includes pertinent information on the selection, installation and maintenance of fluid power products most commonly used in industry today. The table of contents and a complete index have been organized to allow for rapid identification and location of information in this book. As a further aid, the chapters have been organized into related topics and sub-topics, each of which is clearly labeled.

To the reader, the topical organization will readily lend itself to the analysis of any part of a fluid power system. We believe that the illustrations and circuits will be useful to improve product understanding, application, troubleshooting and service.

We would like to acknowledge the work done by the people who contributed to the writing and organization of this material in this handbook. Many deserve credit for our first effort to provide industry with a comprehensive fluid power handbook.

We sincerely hope you'll find the reference material valuable and useful and invite your constructive criticism.

Parker-Hannifin Corporation
Cleveland, Ohio 44112

table of contents

design engineers handbook

accumulators .... section a

cylinders .... section b

electrical devices .... section c

hydraulic filtration, fluids charts,
pressure drop tables .... section d

intensifiers .... section e

hydraulic motors .... section f

pneumatics .... section g

power units .... section h

pumps .... section i

brass connectors .... section j (a)

quick couplings .... section j (b)

hose .... section j (c)

thermoplastic tubing .... section j (d)

tube fittings .... section j (e)

glossary of terms .... section k (a)

general data .... section k (b)

symbols .... section k (c)

rotary actuators .... appendix

index

accumulators

accumulator sizing table a-2
hydraulic accumulators - operation .. a-2
applications a-4
helpful hints a-4
accumulator circuits a-5

hydraulic accumulators

An accumulator is a container in which fluid is stored under pressure as a source of fluid power. Storage of energy is accomplished by connecting the accumulator to a source of pressurized liquid to compress the volume of gas in the accumulator. The source pressure must exceed the pressure at which the liquid stored in the accumulator is to be used. As the system pressure drops below the original pressure, the gas expands, and pressurized liquid is forced out of the accumulator until it is empty or until the pressure in the accumulator is equal to the minimum operating pressure of the device being supplied with liquid.

Accumulators are used to supply oil to a system during temporary demands greater than the pump can supply. Or they can supply make-up oil for holding system pressure. They also are used as a prime source of hydraulic pressure in small systems of short operating life.

A hydro-pneumatic accumulator is a container in which compressed gas applies force to the stored liquid. They are, by far, the most popular type in use on industrial applications and are commercially available with a diaphragm, bladder or piston type separator between the gas and the oil. An inert gas, such as dry nitrogen, is used as the energy storage medium.

Since gas-loaded accumulators depend on the compression of gas for energy, ideal gas laws apply to their operation. The general equation is $P_1 V_1^n = P_2 V_2^n$. For isothermal compression or expansion, n = 1. Therefore the isothermal is: $P_1 V_1 = P_2 V_2$ = constant. For adiabatic compression or expansion, n = $k^{1.4}$ for nitrogen. Therefore the adiabatic equation is $P_1 V_1^{1.4} = P_2 V_2^{1.4}$ = constant.

Isothermal operation occurs when compression or expansion is slow enough to allow transfer of heat out of or into the accumulator. During compression, heat flows out of the accumulator, and during expansion heat flows into the accumulator. With standard 3000 psi accumulators the steel wall permits the isothermal equation to be used where compression or expansion occurs in about 3 minutes or longer.

Adiabatic operation occurs when compression or expansion is rapid so that there is no transfer of heat. The adiabatic equation is used where compression or expansion occurs in less than 1 minute.

Normally accumulators are sized using the ideal gas law and a value of $n^{1.4}$ for nitrogen. However, ideal gas law gives accurate results only at relatively low pressures. Gas change calculations can be made by assuming the proper form of the state of change and then applying one of the given formulas. This is tedious and can be done quickly, within engineering accuracy, using the Beachley graph of Table a-1. The graph allows the sizing of accumulators for isothermal or adiabatic operation or a combination of both. It will also allow interpretation of a process which is not completely isothermal or completely adiabatic.

Accumulators should be precharged to 100 psi less than the minimum operating pressure to prevent bottoming of the piston. Normally they are precharged at room temperature. However, in service the operating temperature of the unit will increase, depending on the system temperature of the oil. The Beachley chart can be used to determine room temperature accumulator precharge pressure for any system temperature.

The graph shows specific volume, and thus it can be used to determine the size of any accumulator.

The equation is:

$$\text{Accumulator size} = \frac{\text{in.}^3\text{ oil req'd.}}{\frac{V_3}{V_1} - \frac{V_2}{V_1}}$$

V_1 = Specific volume of gas at precharge pressure at operating temperature.

V_2 = Specific volume of gas at system pressure at operating temperature (or adiabatic temperature if adiabatic compression.)

V_3 = Specific volume of gas after discharge of oil (either isothermal or adiabatic).

Example: Let us assume 231 cubic inches of oil are required between pressures of 1000 and 2000 psi. Oil is charged slowly (isothermal) and discharged rapidly (adiabatic). Accumulator is precharged at 70° F. and will normally operate at 100° F.

Specific Volumes = V_1 = 405, V_2 = 190, V_3 = 285

$$\text{Accumulator size} = \frac{\text{in.}^3\text{ oil req'd.}}{\frac{V_3}{V_1} - \frac{V_2}{V_1}} = \frac{231}{\frac{285}{405} - \frac{190}{405}} =$$

$$\frac{231}{.234} = 987\text{ in.}^3\text{ or }4.27\text{ gallons}$$

In the above example, the precharge at operating temperature should be 900 psi. By following the pattern of constant volume lines, the graph shows the precharge at room temperature (70° F.) should be 840 psi.

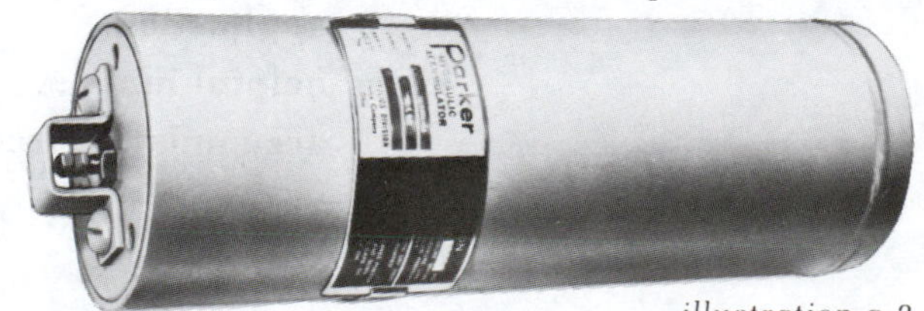

illustration a-2

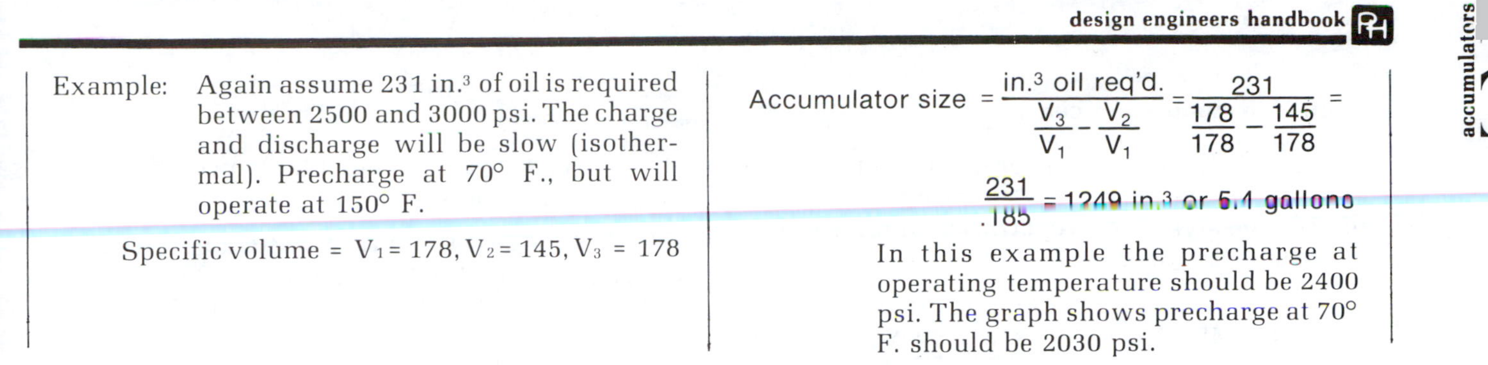

Example: Again assume 231 in.³ of oil is required between 2500 and 3000 psi. The charge and discharge will be slow (isothermal). Precharge at 70° F., but will operate at 150° F.

Specific volume = $V_1 = 178$, $V_2 = 145$, $V_3 = 178$

$$\text{Accumulator size} = \frac{\text{in.}^3 \text{ oil req'd.}}{\frac{V_3}{V_1} - \frac{V_2}{V_1}} = \frac{231}{\frac{178}{178} - \frac{145}{178}} =$$

$$\frac{231}{.185} = 1249 \text{ in.}^3 \text{ or } 5.4 \text{ gallons}$$

In this example the precharge at operating temperature should be 2400 psi. The graph shows precharge at 70° F. should be 2030 psi.

table a-1

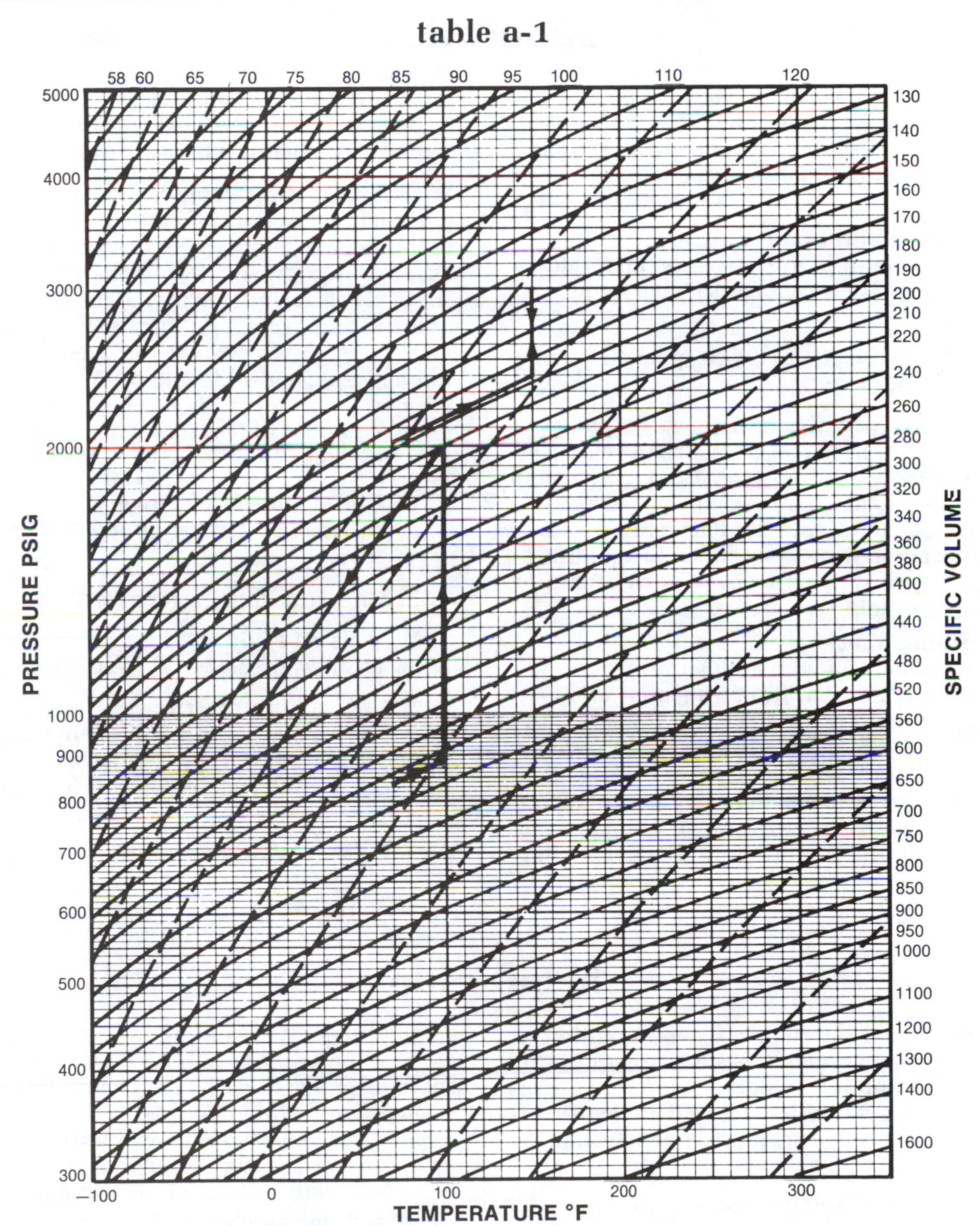

The vertical lines are constant temperature, and an isothermal process would follow these lines.

The solid curved lines are lines of constant gas volume.

The broken curved lines are lines of constant energy.

An adiabatic process would follow the pattern of the broken curved lines.

Courtesy of
Professor Norman H. Beachley
University of Wisconsin

hydraulic accumulators continued

This graphical technic has been adapted from the paper by Mr. Beachley (ref 1) which was based on the thermo dynamic properties of nitrogen published by Din (ref 2).

ref 1 N.H. Beachley "GRAFICAL DIMENSIONS OF ACCUMULATOR CHANGES USING REAL GAS DATA" proceeding, Fluidpower Systems and Controls Conference 1973.

ref 2 F. Din "THERMO DYNAMIC FUNCTIONS OF GASES VOLUME 3 - ETHANE, METHANE AND NITROGEN", Butterworth's, London.

applications

Piston-type accumulators can be used in hydraulic systems to perform many different functions. The following list indicates what a versatile, valuable tool the accumulator can be in your system:

1. **Maintaining Pressure**
   a. Compensating for pressure loss due to leakage, or pressure increase due to thermal expansion.

   b. For steady power while pump is doing other work.

   c. For emergency use in case of pump or power failure, such as pressurizing clamp on work piece in machine tool, or lift truck clamp carrying a load, or rubber curing and plastics presses, or supplying fluid power to withdraw tools to avoid work spoilage, or to maintain oil pressure on support bearing until generator can be brought to a stop.
2. **Absorbing Shock**
   a. Cushioning of valve closure in hydraulic power systems or fluid transmission lines.

   b. Cushioning load start, stop or reversal, as on hones, planers, etc.

   c. Cushioning loads being transported, as on lift trucks, tractor shovel loaders, and other material-handling and earthmoving equipment.

   d. Cushioning load reactions, as on head of rock crusher, or on pressure rolls of paper-making, calendaring or printing machines.
3. **Supplementing Pump Delivery**
   On machines of multiple operations where peak displacement requirements can be stored in the accumulator during low demand portions of the cycle or during idle period (on a continuously operating cycle, total demand cannot exceed total pump delivery over the cycle time. Operations requiring oil from accumulator must be of lower pressures than relief valve setting; volume of oil will be small if operation requires pressure close to relief setting.)
4. **Providing Source of Power**
   a. On electric circuit breakers and switch-gear.

   b. On hydraulic starters for diesel engines.
5. **Dispensing Fluid**
   Such as in a pressurized lubrication system.
6. **Acting as Fluid Barrier**
   Where two different fluids are being used in the same system.
   (Seal compound must be suitable for both fluids.)

helpful hints

Size selection first requires determination of needed oil displacement versus pressures. This information applied to performance charts selects smallest suitable size and indicates required gas precharge pressure. Size selection aid is available.

Fluids used in standard accumulators may be petroleum-base industrial oil or the water-base flame resistant fluids. Temperature range: minus 40°F. to plus 200°F. Seals can be furnished on special order for other fluids and for temperatures to minus 65°F. and plus 400°F.

Installation: Accumulators can be mounted in any position. Clamps must not prohibit thermal expansion. If shell mounting is necessary, avoid clamping that might distort shell. In multiple installation for additional oil volume, gas connections can be manifolded to level out combined action of accumulators and facilitates gas charging.

auxiliary gas bottles

Application: Where space will not permit installation of required accumulator, a smaller accumulator may often be used by connecting to it auxiliary gas bottle(s) which can be located in some nearby spot where space is available. In some cases an accumulator and gas bottle combination may be lower in total cost. Piston travel, confined to the accumulator, must be carefully calculated with ample margins.

illustration a-3

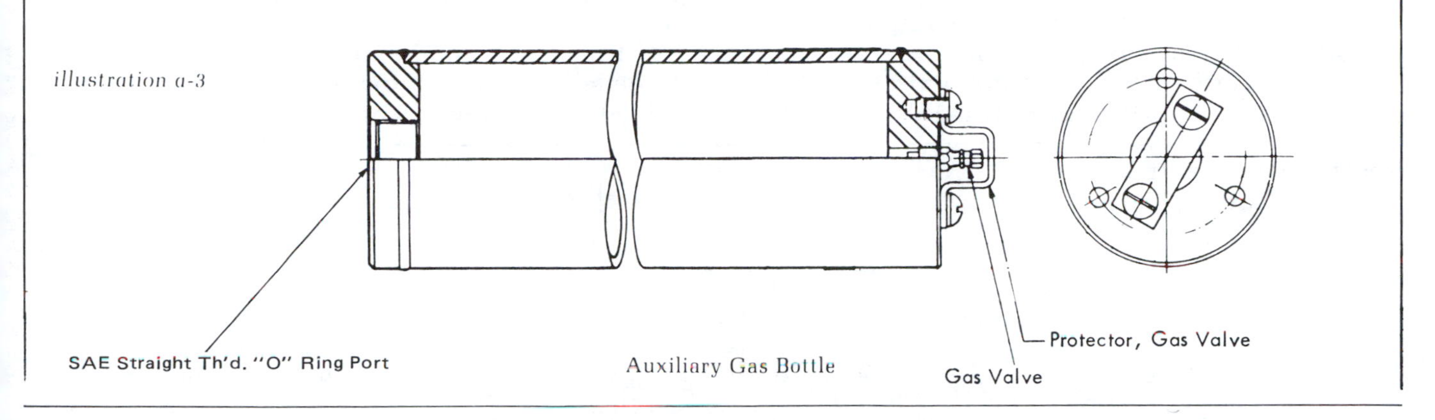

accumulator circuits

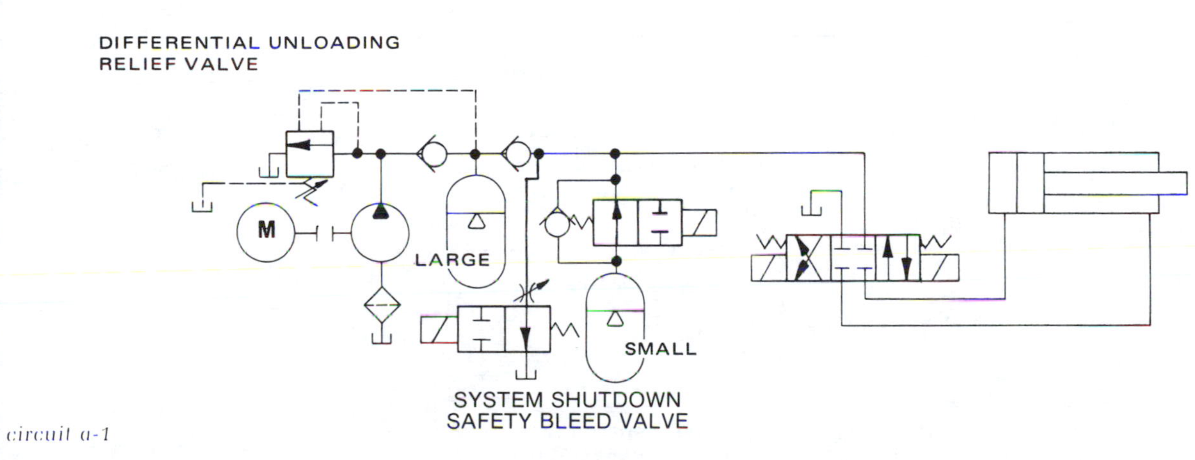

circuit a-1

Basic Accumulator Circuit
For circuits in which oil flow is required intermittently, with relatively long resting periods. A low volume pump, running continuously, can store pressurized oil in the accumulator—to be used in large volumes for short periods. Cylinder bore diameter must be large enough so that sufficient thrust will be obtained even at the low point of the pressure cycle, just before the pump is loaded to re-charge the accumulator. Recharging of the accumulator may take place at any part of the cycle — loading, clamping, unloading, etc. Small accumulator can be used, for reversal or two speed extension.

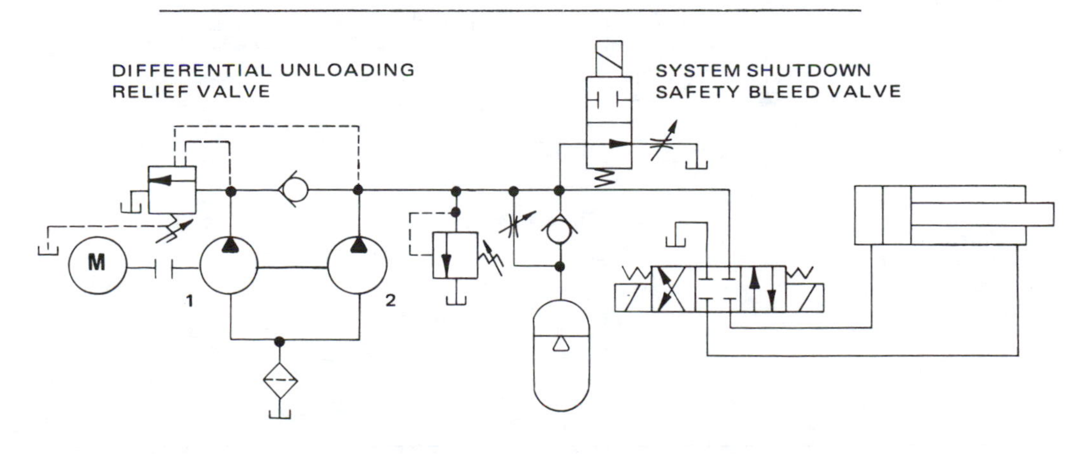

circuit a-2

accumulator circuits continued

Accumulator Unloading

Accumulator unloading, for low volume pumps. When accumulator has been charged up to the setting on the differential unload relief valve, pilot pressure (shown by dotted line) causes the D.U.R.V. to dump the full pump output to tank. This allows the pump to pump at a very low back pressure. When accumulator pressure falls 15% below the setting the D.U.R.V. loads the pump and starts charging the accumulator. Check valve is built into the differential unloading relief valve and prevents loss of accumulator oil through the D.U.R.V. when the pump is at a no load condition.

The differential unloading relief valve, is a valve designed specifically for this use. An ordinary unloading valve will not work.

Venting Relief with Small Unloading Valve

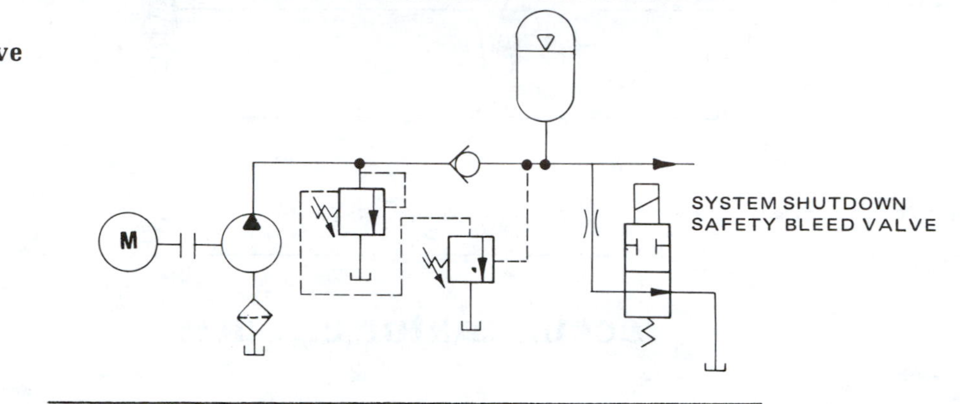

circuit a-3

Differential Unloading Relief Circuit

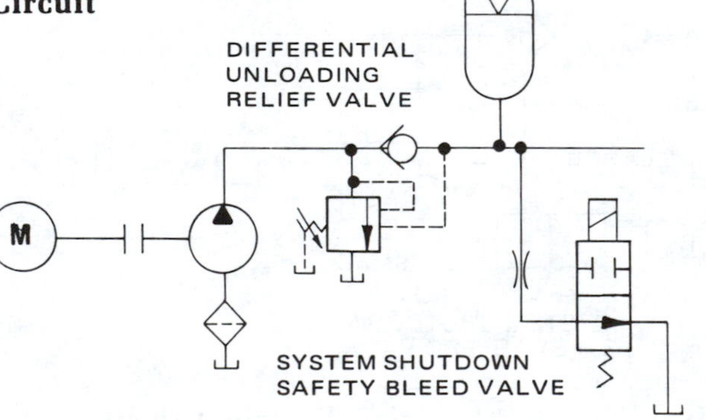

circuit a-4

Venting the Relief with Solenoid Controlled Two Way Valve

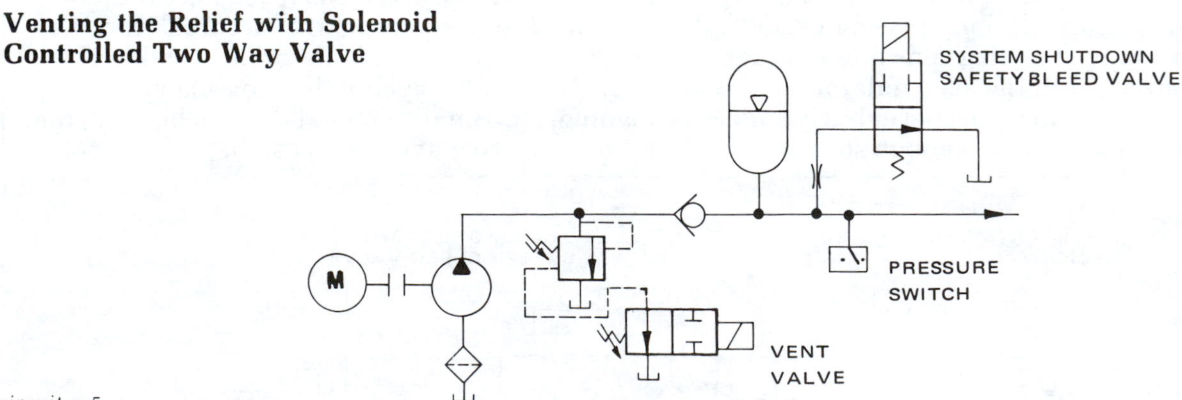

circuit a-5

Hi-Lo System Using an Accumulator

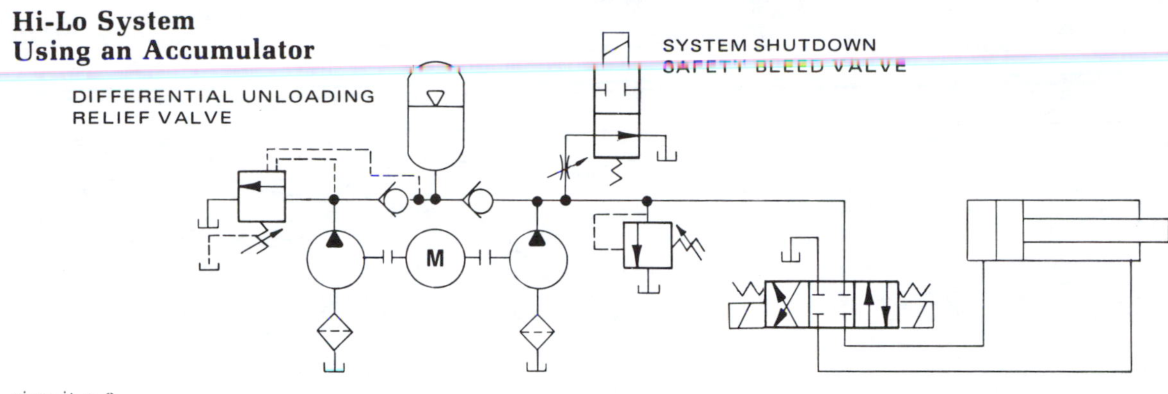

circuit a-6

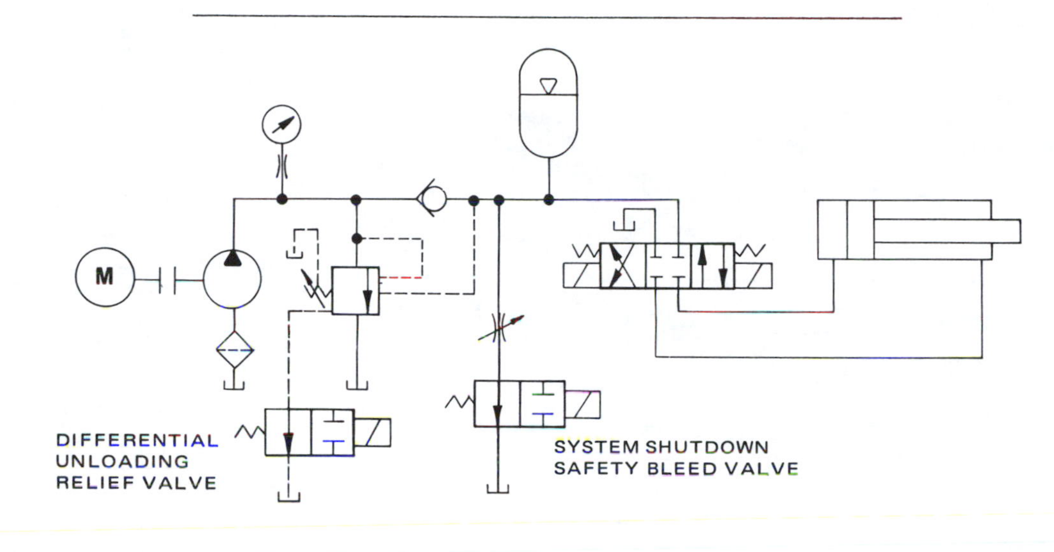

circuit a-7

cylinders

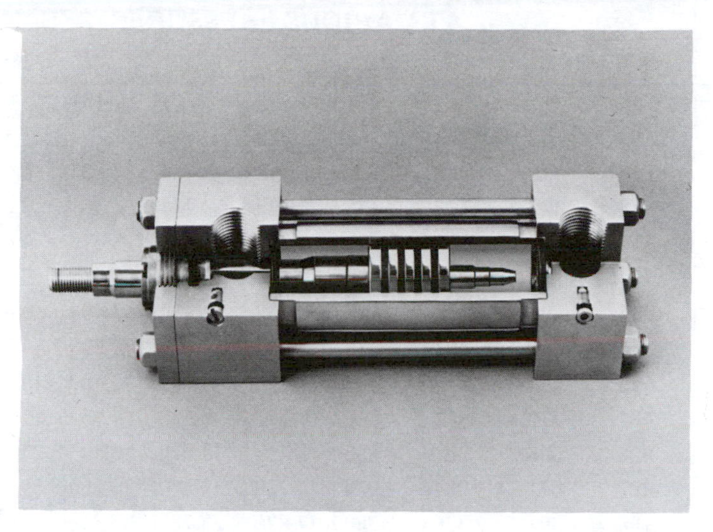

theoretical push and pull forces for cylinders b-2
mounting styles b-3
mounting accessories b-4
mounting classes b-4
cylinder stroke considerations b-4
determining acceleration and deceleration force for hydraulic cylinders b-6
hydraulic cylinder port sizes and piston speed b-8
determining deceleration force for air cylinder b-10
air requirement per inch of cylinder stroke b-11
cylinder ports b-14
seals b-15
cylinder options b-15
cushions b-16
cylinder mechanical motions b-18
cylinder circuits b-26
regenerative circuits b-29
operating principles and construction b-31
fundamental cylinders b-32
cylinder application b-33
trouble shooting cylinders b-38

theoretical push and pull forces for cylinders

CUSTOMARY U.S. UNITS

Cyl. Bore or Piston Rod. Dia. (in.)	Cyl. Bore Size (ϕ mm)	Area (sq. in.)	CYLINDER PUSH STROKE FORCE IN POUNDS AT VARIOUS PRESSURES (PSI)										Displacement per inch of Stroke (gallons)
			50	80	100	500	750	1000	1500	2000	2500	3000	
5/8	15.9	.307	15	25	31	154	230	307	461	614	768	921	.0013
1	25.4	.785	39	65	79	392	588	785	1,177	1,570	1,962	2,355	.0034
1-3/8	34.9	1.490	75	119	149	745	1,118	1,490	2,235	2,980	3,725	4,470	.0065
1-1/2	38.1	1.767	88	142	177	885	1,325	1,770	2,651	3,540	4,425	5,310	.00765
1-3/4	44.5	2.410	121	193	241	1,205	1,808	2,410	3,615	4,820	6,025	7,230	.0104
2	50.8	3.140	157	251	314	1,570	2,357	3,140	4,713	6,280	7,850	9,420	.0136
2-1/2	63.5	4.910	245	393	491	2,455	3,682	4,910	7,364	9,820	12,275	14,730	.0213
3	76.2	7.070	354	566	707	3,535	3,502	7,070	10,604	14,140	17,675	21,210	.0306
3-1/4	82.6	8.300	415	664	830	4,150	6,225	8,300	12,450	16,600	20,750	24,900	.0359
3-1/2	88.9	9.620	481	770	962	4,810	7,215	9,620	14,430	19,240	24,050	28,860	.0416
4	101.6	12.570	628	1,006	1,257	6,285	9,428	12,570	18,856	25,140	31,425	37,710	.0544
5	127.0	19.640	982	1,571	1,964	9,820	14,730	19,640	29,460	39,280	49,100	58,920	.0850
5-1/2	139.7	23.760	1,188	1,901	2,376	11,880	17,820	23,760	35,640	47,520	59,400	71,280	.1028
6	152.4	28.270	1,414	2,262	2,827	14,135	21,203	28,270	42,406	56,540	70,675	84,810	.1224
7	177.8	38.490	1,924	3,079	3,849	19,245	28,868	38,490	57,736	76,980	96,225	115,470	.1666
8	203.2	50.270	2,513	4,022	5,027	25,135	37,703	50,270	75,406	100,540	125,675	150,810	.2176
8-1/2	215.9	56.750	2,838	4,540	5,675	28,375	42,563	56,750	85,125	113,500	142,875	170,250	.2455
10	254.0	78.540	3,927	6,283	7,854	39,270	58,905	78,540	117,810	157,080	196,350	235,620	.3400
12	304.8	113.100	5,655	9,048	11,310	56,550	84,825	113,100	169,650	226,200	282,750	339,300	.4896

table b-1

NOTE: Deduct Force of Poston Rod Size from Bore Size for Pull Applications.

SI (METRIC) UNITS

Cyl. Bore or Piston Rod Dia. (in.)	Size in MM	Area in Sq. MM	CYLINDER PUSH FORCE IN NEWTONS AT VARIOUS PRESSURES IN BARS										Displacement for 1 MM of Stroke (Cu. MM)
			4	6.3	10	16	25	40	63	100	160	200	
5/8	15.87	197.9	79	125	198	317	495	792	1247	1979	3167	3959	197.9
1	25.40	506.7	203	319	507	811	1267	2027	3192	5067	8107	10134	506.7
1-3/8	34.93	958.0	383	604	958	1533	2395	3832	6035	9580	15328	19160	958.0
1-1/2	38.10	1140.1	456	718	1140	1824	2850	4560	7183	11401	18242	22802	1140.1
1-3/4	44.45	1551.8	621	978	1552	2483	3879	6207	9776	15518	24829	31036	1551.8
2	50.80	2026.9	811	1277	2027	3243	5067	8107	12769	20268	32429	40537	2026.9
2-1/2	63.50	3166.9	1267	1995	3167	5067	7917	12668	19952	31669	50671	63339	3166.9
3	76.20	4560.4	1824	2873	4560	7297	11401	18242	28730	45604	72966	91208	4560.4
3-1/4	82.55	5352.1	2141	3372	5352	8563	13380	21408	33718	53521	85634	107042	5352.1
3-1/2	88.90	6207.2	2483	3911	6207	9931	15518	24829	39105	62072	99315	124144	6207.2
4	101.60	8107.3	3243	5108	8107	12972	20268	32429	51076	81073	129717	162147	8107.3
5	127.00	12667.7	5067	7981	12668	20268	31669	50671	79807	126677	202683	253354	12667.7
5-1/2	139.70	15327.9	6131	9657	15328	24525	38320	61312	96566	153279	245247	306559	15327.9
6	152.40	18241.5	7297	11492	18242	29186	45604	72966	114922	182415	291864	364830	18241.5
7	177.80	24828.1	9931	15642	24829	39726	62072	99315	156421	248287	397260	496574	24828.7
8	203.20	32429.4	12972	20430	32429	51887	81073	129717	204305	324294	518870	648587	32429.4
8-1/2	215.90	36609.7	14644	23064	36610	58576	91524	146439	230641	366097	585755	732194	36609.7
10	254.00	50670.9	20268	31923	50671	81073	126677	202683	319226	506709	810734	1013417	50670.9
12	304.80	72966.0	29186	45968	72966	116746	182415	291864	459686	729660	1167457	1459321	72966.0

table b-2

REF. 1 $\#_f$ = 4.448 NEWTONS (N)
1 BAR = 14.504 PSI

mounting styles

In addition to the standard mountings shown the following information covers other mountings and mounting ideas that may prove helpful in your applications. When needed, special heads, caps, flanges or intermediate mountings can be provided. Sketches of your requirements, together with specifications relative to the application and forces involved should be submitted to the manufacturer.

Clevis Mountings — Cylinders should be pivoted at both ends, with the customer's pin in the piston rod knuckle parallel to the pivot pin supplied with the clevis.

Flange Mountings — Cylinders can be located by centering from the pilot diameter of the gland, or the alin-a-groove on the body. The flanges may be drilled for pins or dowels to prevent shifting after alignment has been obtained.

Lug and Side Tapped Mountings — Cylinders should be fixed at one end using fitted bolts, pins in the mounting lugs or shear keys so located as to resist the major load, whether push or pull.

Thrust Key Mountings — Thrust key mountings, of the integral key type eliminate the need of using fitted bolts or external keys on side mounted cylinders.

Tie Rod Mountings — Cylinders with tie rod mountings are recommended for applications where mounting space is limited.

Note: *If the tie rod nuts are removed during installation, be certain to retorque to manufacturer's specifications.*

In addition to the standard mountings shown the following information covers other mountings and mounting ideas that may prove helpful in your applications. When needed, special heads, caps, flanges or intermediate mountings can be provided. Sketches of your requirements, together with specifications relative to the application and forces involved should be submitted to the manufacturer.

Trunnion Mountings — Cylinders require lubricated pillow blocks with minimum bearing clearances. Pillow blocks should be carefully aligned and rigidly mounted so the trunnions will not be subjected to bending moments. The rod end connection should also be pivoted, with the customer's pin in the piston rod knuckle parallel to the trunnions. Trunnion pins are usually hard chrome plated.

Mounting Bolts — High tensile socket head screws are recommended for all mounting styles. Use 1/16" smaller than hole size.

TB	J	H	JB	HB	C
Tie Rods Extended, Styles TB, TC, TD. NFPA Styles MX3, MX2, MX1	Head Rectangular Flange, Style J. NFPA Style MF1	Cap Rectangular Flange, Style H. NFPA Style MF2	Head Square Flange, Style JB. NFPA Style MF5	Cap Square Flange, Style HB. NFPA Style MF6	Side Lugs, Style C. NFPA Style MS2
E	**F**	**CB**	**G**	**D**	**DB**
Centerline Lugs, Style E. NFPA Style MS3	Side Tapped, Style F. NFPA Style MS4	Side End Angles, Style CB. NFPA Style MS1	Side End Lugs, Style G. NFPA Style MS7	Head Trunnion, Style D. NFPA Style MT1	Cap Trunnion, Style DB. NFPA Style MT2
DD	**BB**	**BC**	**KTB**		
Intermediate Fixed Trunnion, Style DD. NFPA Style MT4	Cap Fixed Clevis, Style BB NFPA Style MP1	Cap Detachable Clevis, Style BC. NFPA Style MP2	Double Rod Cylinders	Intergral Key	

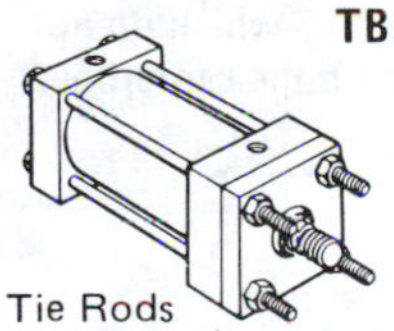

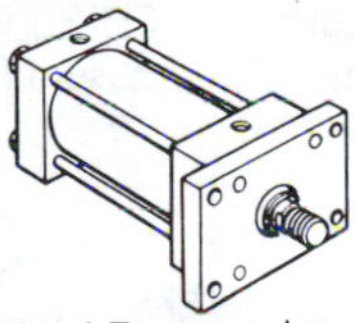

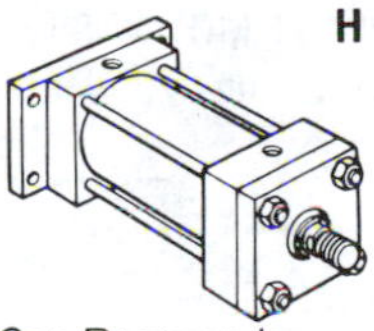

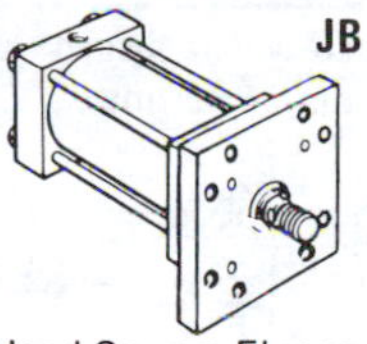

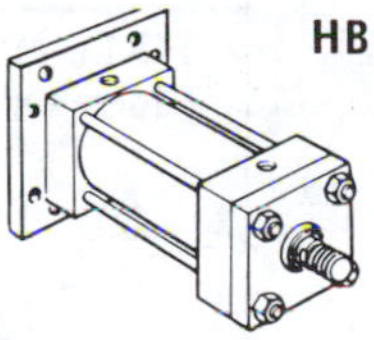

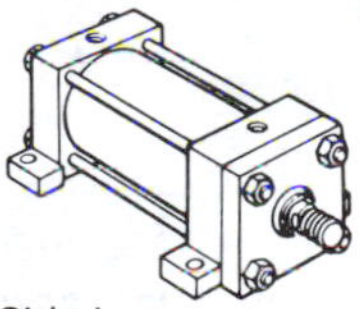

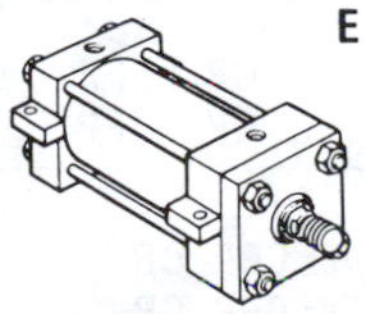

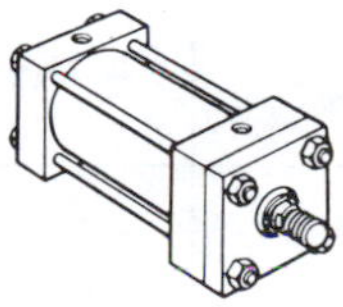

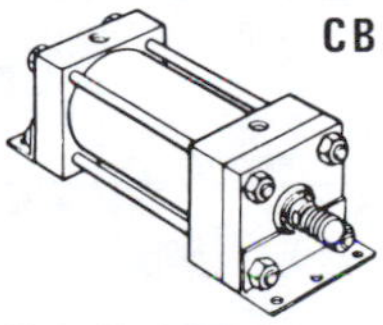

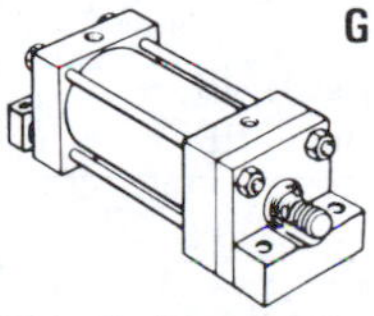

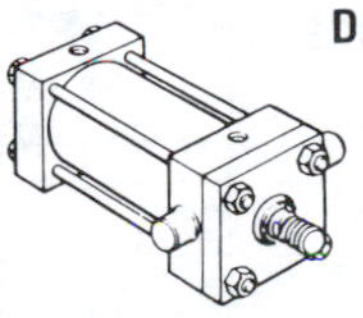

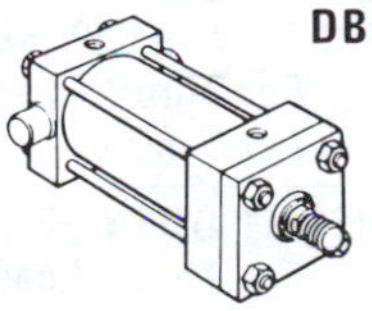

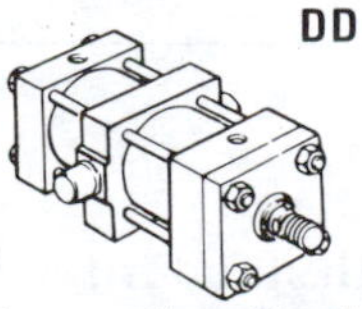

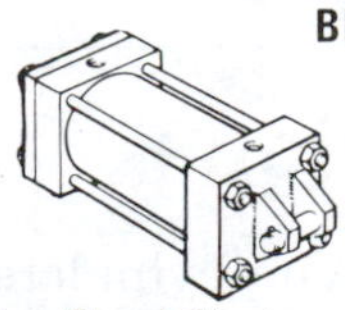

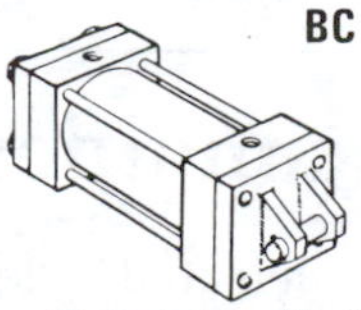

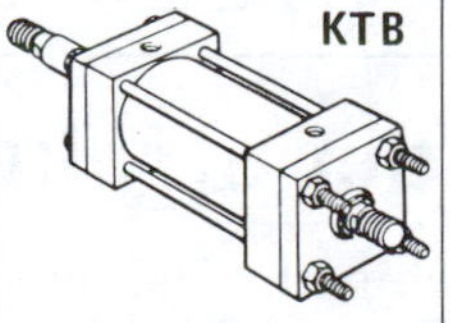

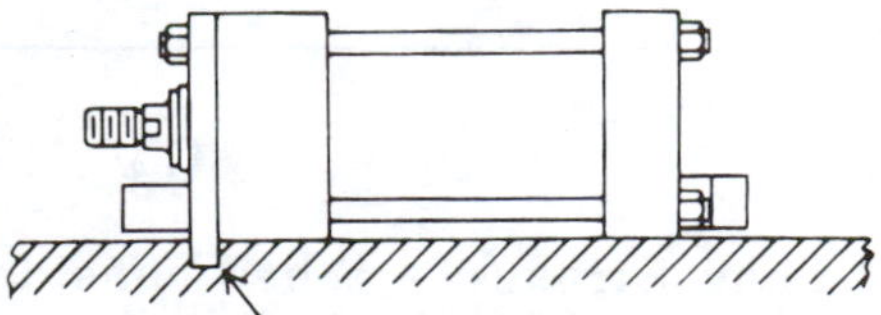

illustration b-1

*NFPA Syles conform to ANSI Standard, 1393.15-1971

mounting accessories

A complete range of cylinder accessories are available from most manufacturers to give versatility to present or future cylinder applications. These include ① Eye Bracket, ② Rod Clevis, ③ Pivot Pins, ④ Clevis Bracket and ⑤ Knuckle.

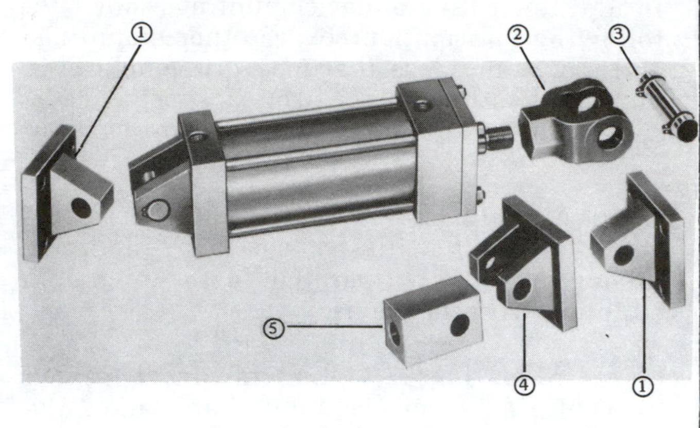

PH Industrial Cylinder with Accessories

illustration b-2

mounting classes

Standard mountings for power cylinders fall into two basic classes and three groups. The two classes can be summarized as follows:

Class 1 — Straight Line Force Transfer (Group 1 and 3).

Class 2 — Pivot Force Transfer (Group 2). Pivot mountings permit a cylinder to change its alignment in one plane.

Because a cylinder's mounting directly affects the maximum pressure at which the cylinder can be used, **table b-3** (page b-4) should be helpful in the selection of the proper mounting for your application. Stroke length, piston rod connection to load, extra piston rod length over standard, etc., should be considered for thrust loads. Alloy steel mounting bolts are recommended for all mounting styles, and thrust keys are recommended for Group 3.

CYLINDER MOUNTING CLASSES

FOR CYLINDERS THAT ARE RECOMMENDED TO 3000 PSI WORKING AND 5000 PSI NON SHOCK SERVICE

	CLASS 1 – GROUP 1	CLASS 2 – GROUP 2	CLASS 1 – GROUP 3
	FIXED MOUNTS which absorb force on cylinder centerline.	PIVOT MOUNTS which absorb force on cylinder centerline.	FIXED MOUNTS which do not absorb force on the centerline.
HEAVY-DUTY SERVICE			
For Thrust Loads –	Mtg. Styles HB, TC, E	Mtg. Styles DD, D, DB, BB	Mtg. Styles C, CP
For Tension Loads –	Mtg. Styles JB, TB, E	Mtg. Styles BB, DD, D, DB	Mtg. Styles C, CP
MEDIUM-DUTY SERVICE			
For Thrust Loads –	Mtg. Styles H, JB	–	Mtg. Styles G, GP, F, FP
For Tension Loads –	Mtg. Styles J, HB	–	Mtg. Styles G, GP, F, FP
LIGHT-DUTY SERVICE			
For Thrust Loads –	Mtg. Style J	–	Mtg. Styles CBP, CB *
For Tension Loads –	Mtg. Style H	–	Mtg. Styles CBP, CB *

* Mounting style CB recommended for maximum pressure of 500 p.s.i. in short stroke applications (to 5"). Longer strokes permit higher pressures. The use of a thrust key is recommended with this mounting. For more detailed information see manufacturer's product catalog.

table b-3

cylinder stroke considerations

Long Strokes — When considering the use of long stroke cylinders, it is necessary that the rod diameter be of such dimension so as to provide the necessary column strength. For tension (pull) loads, a correct rod size is easily selected by specifying cylinders with standard rod diameters, and using them at rated or lower pressures.

For compression (push) loads, the column

cylinder stroke considerations continued

strength must be carefully considered. This involves the stroke length, the length of the piston rod extension, the support received from the rod end connection and gland and piston bearings, the style of mounting and the mounting attitude. It is also necessary to consider the bearing loads on pistons and glands, and to keep bearing pressures within proper limits by increasing the distance between piston and gland bearings. This is economically accomplished by various means. Commonly, separation of the bearings is effected with a stop tube on the piston rod much like a large diameter spacer sleeve. Other designs are provided according to the application requirements. The **Stroke Selection Graph b-1, page b-6,** printed in this handbook will guide you where requirements call for unusually long strokes, used in push applications.

When specifying cylinders with long stroke and stop tube, be sure to call out the net stroke and the length of the stop tube. Machine design can be continued without delay by laying in a cylinder equivalent in length to the **Net Stroke Plus Stop Tube Length,** which is referred to as **Gross Stroke.**

piston rod compared to stroke

How to Use the Table

The selection of a piston rod for thrust (push) conditions requires the following steps:

1. Determine the types of cylinder mounting style and rod end connection to be used. Then consult the **table 4** [page b-5] and find the "stroke factor" that corresponds to the conditions used.

2. Using this stroke factor, determine the "basic length" from the equation:

$$\text{Basic Length} = \text{Actual Stroke} \times \text{Stroke Factor}$$

piston rod — stroke selection table

RECOMMENDED MOUNTING STYLES FOR MAXIMUM STROKE AND THRUST LOADS	ROD END CONNECTION	CASE	STROKE FACTOR
CLASS 1 – GROUPS 1 OR 3 Long stroke cylinders for thrust loads should be mounted using a heavy-duty mounting style at one end, firmly fixed and aligned to take the principle force. Additional mounting should be specified at the opposite end, which should be used for alignment and support. An intermediate support may also be desirable for long stroke cylinders mounted horizontally. Machine mounting pads can be adjustable for support mountings to achieve proper alignment.	FIXED AND RIGIDLY GUIDED.	I	.50
	PIVOTED AND RIGIDLY GUIDED	II	.70
	SUPPORTED BUT NOT RIGIDLY GUIDED	III	2.00
CLASS 2 – GROUP 2 Style – Trunnion on Head	PIVOTED AND RIGIDLY GUIDED	IV	1.00
Style – Intermediate Trunnion	PIVOTED AND RIGIDLY GUIDED	V	1.50
Style – Trunnion on Cap or Style – Clevis on Cap	PIVOTED AND RIGIDLY GUIDED	VI	2.00

table b-4

Graph b-1, page b-6, is prepared for standard rod extensions beyond the face of the gland retainers. For rod extensions greater than standard, add the increase to the actual stroke in arriving at the "basic length."

3. Find the load imposed for the thrust application by multiplying the full bore area of the cylinder by the system pressure.

4. Enter the graph along the values of "basic length" and "thrust" as found above and note the point of intersection:

a. The correct piston rod size is read from the diagonally curved line labeled "Rod Diameter" next **above** the point of intersection.

b. The required length of stop tube is read from the right of the graph by following the shaded band in which the point of intersection lies.

cylinder stroke considerations continued

5. If required length of stop tube is in the region labeled "consult factory," submit the following information for an individual analysis:

   a. Cylinder mounting style.

   b. Rod end connection and method of guiding load.

   c. Bore, required stroke, length of rod extension if greater than standard, and series of cylinder used.

   d. Mounting position of cylinder.

   Note: *If at an angle or vertical, specify direction of piston rod.*

   e. Operating pressure of cylinder.

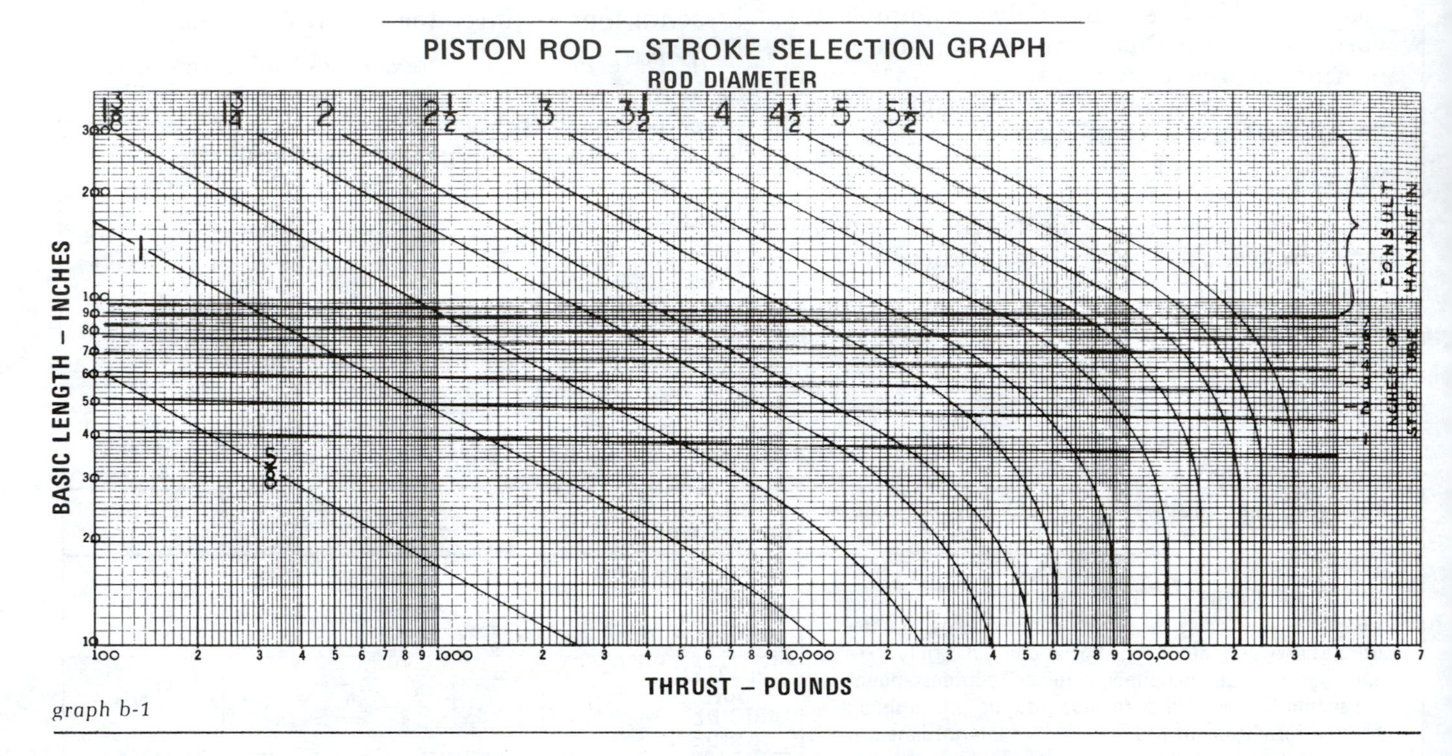

graph b-1

determining acceleration and deceleration force for hydraulic cylinders

The **Uniform Acceleration Force Factor Graph** and the accompanying formula can be used to rapidly determine the forces required to accelerate and decelerate a cylinder load. To determine these forces, the following factors must be known; total weight to be moved, maximum piston speed, distance available to start or stop the weight (load), direction of movement i.e. horizontal or vertical, and load friction. By use of the known factors and the "g" factor from **graph 2**, the force necessary to accelerate or decelerate a cylinder load may be found by solving the formula (as shown in graph on page b-7) applicable to given set of conditions.

Nomenclature

V = Velocity in feet per minute
S = Distance in inches
F = Force in pounds
W = Weight of load in pounds
g = Force factor
f = Friction of load on machine ways in pounds

To determine the force factor "g" from the graph, locate the intersection of the maximum piston velocity line and the line representing the available distance. Project downward to locate "g" on the horizontal axis. To calculate the "g" factor for distances and velocities exceeding those shown on the chart, this formula can be used:

$$g = \frac{V^2}{S} \times .0000517$$

Example: *Horizontal motion of a free moving 6,000 pound load is required with a distance of 1/2" to a maximum speed of 120 feet per minute. Formula (1) F = Wg should be used.*

F = 6,000 pounds X1.50 (from table)= 9,000 pounds (page b-7)

Assuming a maximum available pump pressure of 1,000 p.s.i., a 4" bore cylinder should be selected, operating on push stroke at approximately 750 p.s.i. pressure at the cylinder to allow for pressure losses from the pump to the cylinder.

determining forces continued

Assume the same load to be sliding on ways with a coefficient of friction of 0.15. The resultant friction load would be 6,000 × 0.15 = 900 pounds. Formula (2) $F = Wg + f$ should be used.

$F = 6{,}000$ lbs. × 1.5 (from table) + 900 = 9,900 lbs. (page b-7)

Again allowing 750 p.s.i. pressure at the cylinder, a 5" bore cylinder is indicated.

Example: *Horizontal deceleration of a 6,000 pound load is required by using a 1" long cushion in a 5" bore cylinder having a 2" diameter piston rod. Cylinder bore area (19.64 Sq. In.) minus the rod area (3.14 Sq. In.) results in a minor area of 16.5 Sq. In. at head end of cylinder. A 1,000 p.s.i. pump delivering 750 p.s.i. at the cylinder is being used to push the load at 120 feet per minute. Friction coefficient is 0.15 or 900 pounds.*

In this example, the total deceleration force is the sum of the force needed to decelerate 6,000 pound load, and the force required to counteract the thrust produced by the pump.

W = Load in pounds = 6,000
S = Deceleration distance in inches = 1"
V = Maximum piston speed in feet per minute = 120
g = .74 (from table)
f = 900 pounds

Use formula (3) $F = Wg - f$

$(F = Wg - f) = (F = 6{,}000 \times .74 - 900) = 3{,}540$ lbs.

The pump is delivering 750 p.s.i. acting on the 19.64 Sq. In. piston area producing a force (F_2) of 14,730 pounds. This force must be included in our calculations. Thus $F + F_2 = 3{,}540 + 14{,}730 = 18{,}270$ pounds total force to be decelerated.

The total deceleration force is developed by the fluid trapped between the piston and the head. The fluid pressure is equal to the force (18,270 pounds) divided by the minor area (16.5 Sq. In.) equals 1107 p.s.i. This pressure should not exceed the non-shock rating of the cylinder.

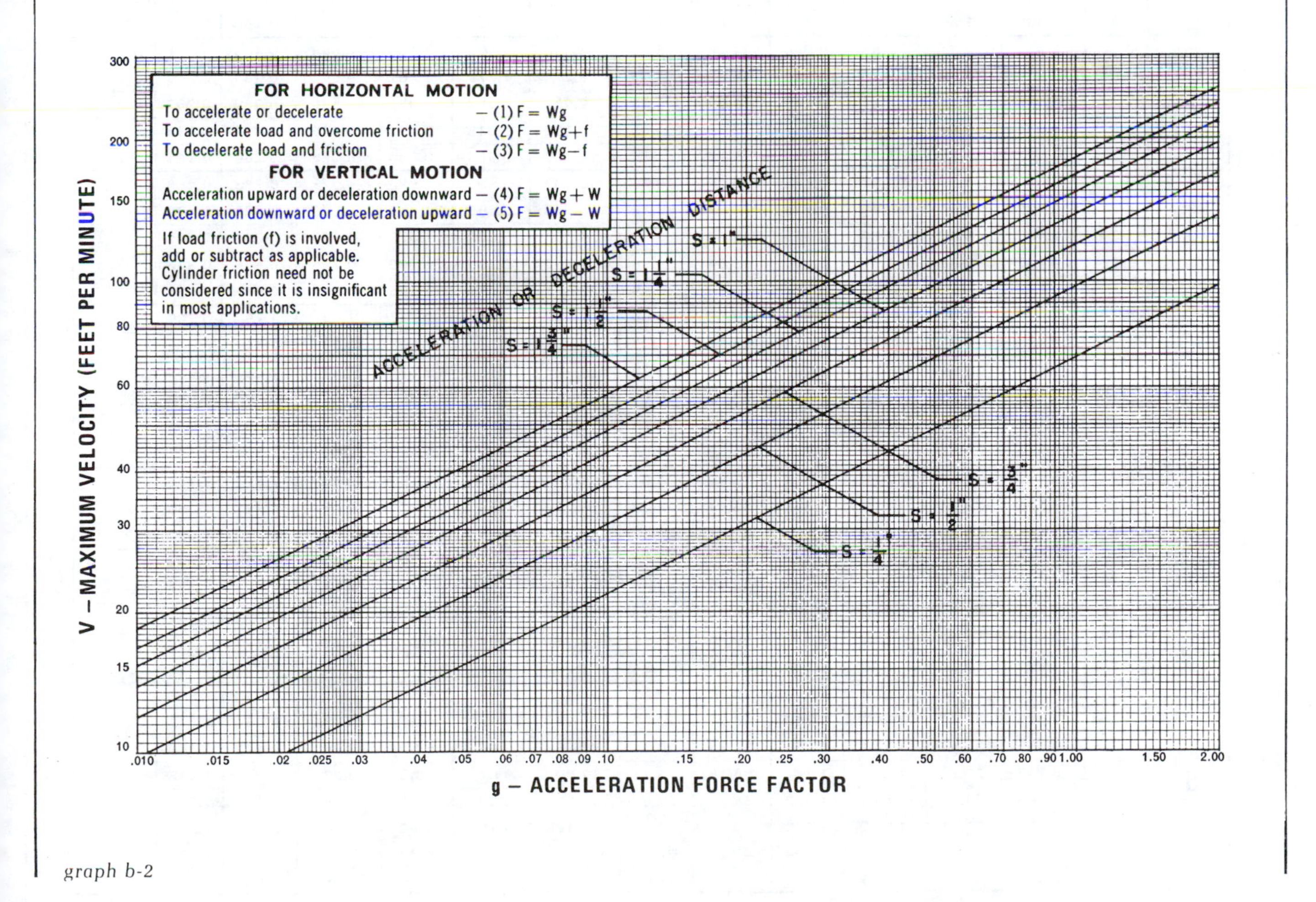

graph b-2

hydraulic cylinder port sizes and piston speed

One of the factors involved in determining the speed of a hydraulic cylinder piston is fluid flow in connecting lines, generally measured in gallons per minute, introduced to, or expelled from, cap end cylinder port. (Due to piston rod displacement, the flow at head end port will be less than at cap end.) Fluid velocity, however, is measured in feet per second. In connecting lines this velocity should generally be limited to 15 feet per second to minimize fluid turbulence, pressure loss and hydraulic shock.

Piston speed for cylinders can be calculated from data shown in **table b-5** and **b-6.** The table shows fluid velocity flow fo major cylinder area as well as for the net area at the rod end for cylinders 1" through 14" bore size.

If desired piston speed results in fluid flow in excess of 15 feet per second in connecting lines, consider the use of larger lines up to cylinder port, using either oversized ports or two ports per cap.

CYLINDER BORE-INCHES	PISTON ROD		CYLINDER NET AREA SQ. IN.	FLUID DISPLACEMENT AT 10 FT. PER MINUTE PISTON VELOCITY		FLUID VELOCITY (IN FEET PER SECOND) THROUGH EXTRA HEAVY PIPE AT 10 F.P.M. PISTON SPEED.								
	DIA.-INCHES	AREA SQ. IN.		G.P.M.	C.F.M.	1/4	3/8	1/2	3/4	1	1-1/4	1-1/2	2	2-1/2
1	0	0	0.785	0.41	0.054	1.82	0.92	0.56	0.30	0.183	0.102	0.074	0.045	.....
	1/2	0.196	0.589	0.30	0.041	1.33	0.68	0.41	0.21	0.134	0.075	0.055	0.033	.....
	5/8	0.307	0.478	0.16	0.033	0.71	0.36	0.22	0.12	0.071	0.040	0.029	0.017	.....
1½	0	0	1.77	0.92	0.123	4.09	2.09	1.259	0.680	0.410	0.230	0.167	0.100	.....
	5/8	0.307	1.46	0.76	0.101	3.38	1.73	1.040	0.562	0.338	0.190	0.138	0.082	.....
	1	0.785	0.98	0.51	0.068	2.27	1.16	0.699	0.378	0.228	0.128	0.093	0.055	.....
2	0	0	3.14	1.63	0.218	7.27	3.71	2.238	1.209	0.728	0.408	0.296	0.177	.....
	5/8	0.307	2.84	1.48	0.197	6.56	3.35	2.019	1.091	0.657	0.368	0.267	0.160	.....
	1	0.785	2.36	1.23	0.164	5.45	2.79	1.678	0.907	0.546	0.306	0.222	0.133	.....
	1-3/8	1.485	1.66	0.86	0.115	3.84	1.96	1.180	0.638	0.384	0.215	0.156	0.094	.....
2½	0	0	4.91	2.55	0.341	11.36	5.80	3.496	1.890	1.138	0.638	0.463	0.277	.....
	5/8	0.307	4.60	2.39	0.319	10.65	5.44	3.278	1.771	1.067	0.598	0.434	0.260	.....
	1	0.785	4.12	2.14	0.286	9.54	4.87	2.937	1.587	0.956	0.536	0.389	0.233	.....
	1-3/8	1.485	3.42	1.78	0.237	7.93	4.05	2.439	1.318	0.794	0.445	0.323	0.193	.....
	1-3/4	2.405	2.50	1.30	0.174	5.96	2.96	1.783	0.963	0.580	0.325	0.236	0.141	.....
3¼	0	0	8.30	4.31	0.576	19.20	9.81	5.909	3.193	1.923	1.078	0.783	0.468	.....
	1	0.785	7.51	3.90	0.521	17.38	8.88	5.349	2.891	1.741	0.976	0.708	0.424	.....
	1-3/8	1.485	6.81	3.54	0.473	15.77	8.05	4.851	2.622	1.579	0.885	0.642	0.384	.....
	1-3/4	2.405	5.89	3.06	0.409	13.64	6.96	4.196	2.268	1.366	0.765	0.556	0.333	.....
	2	3.142	5.15	2.68	0.357	11.93	6.09	3.671	1.984	1.195	0.670	0.486	0.291	.....
4	0	0	12.57	6.53	0.872	29.09	14.85	8.95	4.84	2.91	1.63	1.19	0.709	.....
	1	0.785	11.78	6.12	0.818	27.27	13.93	8.39	4.54	2.73	1.53	1.11	0.665	.....
	1-3/8	1.485	11.08	5.76	0.769	25.65	13.10	7.89	4.27	2.57	1.44	1.05	0.625	.....
	1-3/4	2.405	10.16	5.28	0.705	23.52	12.01	7.24	3.91	2.36	1.32	0.96	0.574	.....
	2	3.142	9.42	4.89	0.654	21.82	11.14	6.71	3.63	2.19	1.22	0.89	0.532	.....
	2-1/2	4.909	7.66	3.98	0.532	17.73	9.05	5.45	2.95	1.78	1.00	0.72	0.432	.....
5	0	0	19.64	10.20	1.363	45.45	23.21	13.99	7.56	4.55	2.55	1.85	1.108	.....
	1	0.785	18.85	9.79	1.308	43.64	22.28	13.43	7.26	4.37	2.45	1.78	1.064	.....
	1-3/8	1.485	18.15	9.43	1.260	42.01	21.45	12.93	6.99	4.21	2.36	1.71	1.024	.....
	1-3/4	2.405	17.23	8.95	1.196	39.88	20.37	12.27	6.63	3.99	2.24	1.63	0.973	.....
	2	3.142	16.49	8.57	1.144	38.18	19.50	11.75	6.35	3.82	2.14	1.56	0.931	.....
	2-1/2	4.909	14.73	7.65	1.022	34.09	17.41	10.49	5.67	3.41	1.91	1.39	0.831	.....
	3	7.069	12.57	6.53	0.872	29.09	14.85	8.95	4.84	2.91	1.63	1.19	0.709	.....
	3-1/2	9.621	10.01	5.21	0.695	23.18	11.84	7.13	3.86	2.32	1.30	0.95	0.565	.....
6	0	0	28.27	14.69	1.962	65.45	33.42	20.14	10.88	6.55	3.67	2.67	1.596	.....
	1-3/8	1.485	26.79	13.92	1.859	62.01	31.67	19.08	10.31	6.21	3.48	2.53	1.512	.....
	1-3/4	2.405	25.87	13.44	1.795	59.88	30.58	18.43	9.96	5.60	3.36	2.44	1.460	.....
	2	3.142	25.13	13.06	1.744	58.18	29.71	17.90	9.67	5.83	3.27	2.37	1.418	.....
	2-1/2	4.909	23.37	12.14	1.622	54.1	27.6	16.64	8.99	5.42	3.04	2.20	1.32	.....
	3	7.069	21.21	11.02	1.472	49.1	25.1	15.10	8.16	4.92	2.76	2.00	1.20	.....
	3-1/2	9.621	18.65	9.69	1.294	43.2	22.1	13.29	7.18	4.32	2.42	1.76	1.05	.....
	4	12.566	15.71	8.16	1.090	36.4	18.6	11.19	6.05	3.64	2.04	1.48	0.89	.....

table b-5

hydraulic cylinder port sizes and piston speed

7	0	0	38.49	20.00	2.671	89.1	45.5	27.41	14.81	8.92	5.00	3.63	2.17	. . . . .
	1-3/8	1.485	37.00	19.22	2.568	85.7	43.7	[illegible]	14.24	8.58	4.81	3.49	2.09	. . . . .
	1-3/4	2.405	36.08	18.74	2.504	83.5	42.7	25.70	13.89	8.36	4.69	3.40	2.04	. . . . .
	2	3.142	35.34	18.36	2.453	81.8	41.8	25.17	13.60	8.19	4.59	3.33	2.00	. . . . .
	2-1/2	4.909	33.58	17.44	2.330	77.7	39.7	23.92	12.92	7.78	4.36	3.17	1.90	. . . . .
	3	7.069	31.42	16.32	2.181	72.7	37.1	22.38	12.09	7.28	4.08	2.96	1.77	. . . . .
	3-1/2	9.621	28.86	14.99	2.003	66.8	34.1	20.56	11.11	6.69	3.75	2.72	1.63	. . . . .
	4	12.566	25.92	13.47	1.799	60.0	30.6	18.46	9.98	6.01	3.37	2.45	1.46	. . . . .
	4-1/2	15.904	22.58	11.73	1.567	52.3	26.7	16.08	8.69	5.23	2.93	2.12	1.28	. . . . .
	5	19.635	18.85	9.79	1.308	43.6	22.3	13.43	7.26	4.37	2.45	1.78	1.06	. . . . .
8	0	0	50.27	26.12	3.489	116.4	59.4	35.80	19.35	11.65	6.53	4.74	2.84	1.977
	1-3/8	1.485	48.78	25.34	3.385	112.9	57.7	34.74	18.78	11.31	6.34	4.60	2.75	1.918
	1-3/4	2.405	47.86	24.86	3.321	110.8	56.6	34.09	18.42	11.09	6.22	4.51	2.70	1.882
	2	3.142	47.12	24.48	3.270	109.1	55.7	33.56	18.14	10.92	6.12	4.45	2.66	1.853
	2-1/2	4.909	45.36	23.57	3.149	105.0	53.61	32.31	17.46	10.51	5.892	4.278	2.560	1.784
	3	7.069	43.20	22.44	2.998	100.0	51.06	30.77	16.63	10.01	5.612	4.074	2.438	1.699
	3-1/2	9.621	40.65	21.12	2.821	94.1	48.04	28.95	15.65	9.42	5.279	3.834	2.294	1.598
	4	12.566	37.70	19.59	2.616	87.3	44.56	26.85	14.51	8.74	4.897	3.556	2.128	1.483
	4-1/2	15.904	34.36	17.85	2.385	79.5	40.62	24.47	13.23	8.20	4.464	3.241	1.939	1.351
	5	19.635	30.63	15.91	2.126	70.9	36.21	21.82	11.79	7.10	3.979	2.889	1.729	1.205
	5-1/2	23.758	26.51	13.77	1.840	61.4	31.33	18.88	10.20	6.15	3.444	2.500	1.496	1.043
10	0	0	78.54	40.80	5.451	181.8	92.84	55.94	30.23	18.21	10.203	7.408	4.433	3.089
	1-3/4	2.405	76.14	39.56	5.284	176.2	89.99	54.23	29.31	17.65	9.890	7.181	4.297	2.994
	2	3.142	75.40	39.17	5.233	174.5	89.12	53.70	29.02	17.48	9.795	7.112	4.255	2.965
	2-1/2	4.909	73.63	38.25	5.110	170.4	87.03	52.44	28.34	17.07	9.565	6.945	4.156	2.896
	3	7.069	71.47	37.13	4.960	165.4	84.48	50.91	27.51	16.57	9.284	6.741	4.034	2.811
	3-1/2	9.621	68.92	35.80	4.783	159.5	81.47	49.09	26.53	15.98	8.953	6.501	3.890	2.710
	4	12.566	65.97	34.27	4.578	152.7	77.98	46.99	25.39	15.29	8.570	6.223	3.724	2.595
	4-1/2	15.904	62.64	32.54	4.347	145.0	74.04	44.61	24.11	14.52	8.137	5.908	3.535	2.463
	5	19.635	58.91	30.60	4.088	136.4	69.63	41.96	22.67	13.65	7.652	5.556	3.325	2.317
	5-1/2	23.758	54.78	28.46	3.802	126.8	64.75	39.02	21.09	12.70	7.116	5.167	3.092	2.154
	6	28.274	50.27	26.12	3.489	116.4	59.42	35.80	19.35	11.65	6.530	4.741	2.837	1.977
	6-1/2	33.183	45.36	23.57	3.148	105.0	53.6	32.31	17.46	10.52	5.89	4.278	2.560	1.784
	7	38.485	40.06	20.81	2.780	92.7	47.4	28.53	15.42	9.29	5.20	3.778	2.261	1.575
12	0	0	113.10	58.76	7.849	261.8	133.7	80.55	43.53	26.22	14.69	10.668	6.383	4.448
	2	3.142	109.96	57.12	7.631	254.5	130.0	78.32	42.32	25.49	14.28	10.371	6.206	4.324
	2-1/2	4.909	108.19	56.21	7.508	250.4	127.9	77.06	41.64	25.08	14.05	10.205	6.106	4.255
	3	7.069	106.03	55.08	7.359	245.4	125.3	75.52	40.81	24.58	13.77	10.001	5.984	4.170
	3-1/2	9.621	103.48	53.76	7.182	239.5	122.3	73.70	39.83	23.99	13.44	9.760	5.840	4.069
	4	12.566	100.53	52.23	6.977	232.7	118.8	71.60	38.70	23.30	13.06	9.482	5.674	3.954
	4-1/2	15.904	97.19	50.49	6.745	225.0	114.9	69.23	37.41	22.53	12.63	9.168	5.486	3.822
	5	19.635	93.46	48.55	6.486	216.4	110.5	66.57	35.98	21.67	12.14	8.816	5.275	3.676
	5-1/2	23.758	89.34	46.41	6.200	206.8	105.6	63.63	34.39	20.71	11.61	8.427	5.042	3.513
	6	28.274	84.82	44.06	5.887	196.4	100.3	60.42	32.65	19.66	11.02	8.001	4.787	3.336
	6-1/2	33.183	79.92	41.52	5.547	185.0	94.5	56.92	30.76	18.53	10.38	7.538	4.510	3.143
	7	38.485	74.61	38.77	5.179	172.7	88.2	53.14	28.72	17.30	9.69	7.038	4.211	2.934
	7-1/2	44.179	68.92	35.80	4.783	159.5	81.5	49.09	26.53	15.98	8.95	6.501	3.890	2.710
	8	50.266	62.83	32.64	4.360	145.4	74.3	44.75	24.19	14.57	8.16	5.926	3.546	2.471
	8-1/2	56.745	56.35	29.27	3.911	130.5	66.6	40.14	21.69	13.06	7.32	5.315	3.181	2.216
14	0	0	153.94	79.97	10.683	356.3	182.0	109.6	59.25	35.68	20.00	14.52	8.688	6.054
	2-1/2	4.909	149.03	77.42	10.343	345.0	176.2	106.2	57.36	34.55	19.36	14.06	8.411	5.861
	3	7.069	146.87	76.30	10.193	340.0	173.6	104.6	56.53	34.05	19.08	13.85	8.289	5.776
	3-1/2	9.621	144.32	74.97	10.016	334.1	170.6	102.8	55.55	33.45	18.75	13.61	8.145	5.676
	4	12.566	141.37	73.44	9.811	327.3	167.1	100.7	54.42	32.77	18.37	13.33	7.979	5.560
	4-1/2	15.904	138.03	71.71	9.579	319.5	163.2	98.3	53.13	32.00	17.93	13.02	7.791	5.428
	5	19.635	134.30	69.77	9.320	310.9	158.8	95.7	51.70	31.13	17.45	12.67	7.580	5.282
	5-1/2	23.758	130.18	67.63	9.035	301.3	153.9	92.7	50.11	30.18	16.91	12.28	7.347	5.120

table b-5 (cont.)

determining deceleration force for air cylinder

Cushion ratings for **Air Cylinders Only** are described in **table b-6** and **graph b-3.** To determine whether a cylinder will adequately stop a load without damage to the cylinder, the wieght of the load (including the weight of the piston and the piston rod from **table b-6)** and the maximum speed of the piston rod must first be determined. Once these two factors are known, the **Kinetic Energy Graph** may be used. Enter the graph at its base for the value of weight determined, and project vertically to the required speed value. The point of intersection of these two lines will be the cushion rating number required for the application.

To determine the total load to be moved, the weight of the piston and rod must be included.

Total Weight = weight of the piston and non-stroke rod length (Column 1) + weight of the rod per inch of stroke × the inches of stroke (Column 2) + the load to be moved.

WEIGHT TABLE

Bore Diameter	Column 1 Basic Wgt. (lbs.) for Piston & Non-Stroke Rod	Rod Diameter	Column 2 Basic Wgt. (lbs.) for 1" Stroke
1-1/2	1.5	5/8	.087
2	3.0	1	.223
2-1/2	5.4	1-3/8	.421
3-1/4	8.3	1-3/4	.682
4	14.2	2	.89
5	29	2-1/2	1.39
6	41	3	2.0
8	89	3-1/2	2.73
10	115	4	3.56
12	161	5	5.56
14	207	5-1/2	6.73

table b-6

Example: *a 3-1/4" bore cylinder, having a 1" diameter rod and 25" stroke; load to be moved is 85 pounds. Total load to be moved is then 8.3 lbs. + .223 lbs./in. × 25 in. + 85 lbs. or a total of 99 lbs.*

KINETIC ENERGY GRAPH — AIR CYLINDERS

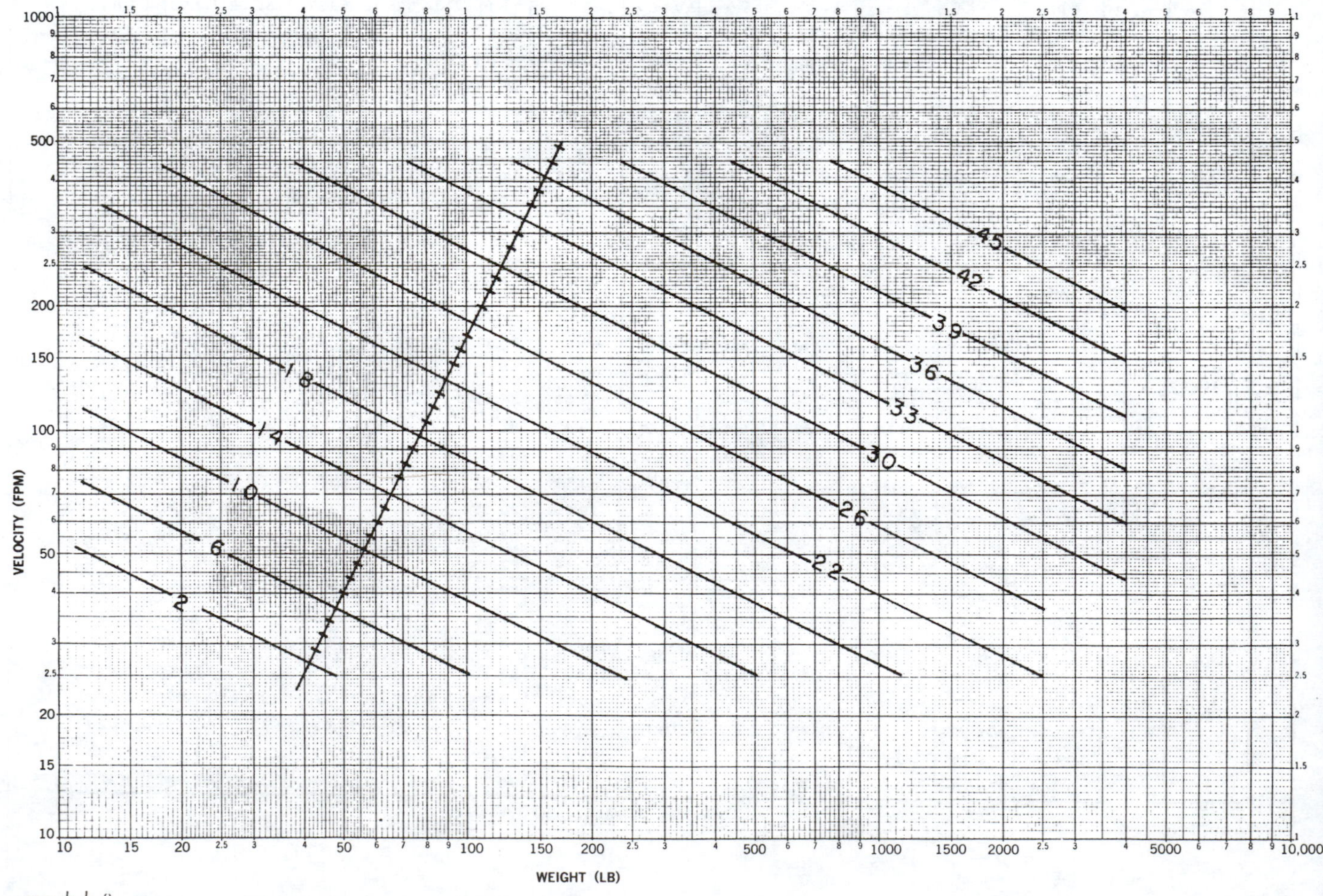

graph b-3

deceleration force continued

Now refer to **table b-7** and find the cushion ratings, using bore size and rod diameter of the cylinder selected. If a simple circuit is used, with no meter out or speed control, use the "no back pressure, Column A" values. If a meter out or speed control is to be used, use the back pressure column values. If the cushion rating found in **table b-7, page b-11,** is **greater** than the number determined in **graph b-3,** then the cylinder will stop the load adequately. If the cushion rating in **table b-7** is **smaller** than the number found in **graph b-3,** then a larger bore cylinder should be used. In those applications where back pressures exist in the exhaust lines, it is possible to exceed the cushion ratings shown in **table b-7.** In these cases, consult the factory and advise the amount of back pressure.

AIR CYLINDER CUSHION RATINGS TABLE

Bore Diameter	Rod Diameter	Rating With No Back Pressure	Rating w/Back Pressure
1-1/2	Cap End	12	17
	5/8	8	14
	1	3	8
2	Cap End	14	20
	5/8	12	18
	1	9	15
	1-3/8	6	11
2-1/2	Cap End	17	23
	5/8	14	20
	1	14	19
	1-3/8	12	18
	1-3/4	8	13
3-1/4	Cap End	21	26
	1	18	24
	1-3/8	17	23
	1-3/4	16	22
	2	13	19
4	Cap End	23	28
	1	20	27
	1-3/8	20	26
	1-3/4	19	25
	2	17	23
	2-1/2	17	22
5	Cap End	26	31
	1	23	28
	1-3/8	23	28
	1-3/4	22	28
	2	20	26
	2-1/2	19	25
	3	18	24
	3-1/2	15	20
6	Cap End	26	31
	1-3/8	26	31
	1-3/4	26	31
	2	24	29
	2-1/2	24	29
	3	22	28
	3-1/2	21	27
	4	20	26
8	Cap End	29	35
	1-3/8	29	35
	1-3/4	29	34
	2	27	33
	2-1/2	26	32
	3	26	32
	3-1/2	26	32
	4	25	31
	5	23	29
	5-1/2	22	28
10	Cap End	33	39
	1-3/4	32	38
	2	31	37
	2-1/2	31	36
	3	30	36
	3-1/2	30	36
	4	30	36
	5	28	34
	5-1/2	27	33
12	Cap End	35	41
	2	33	39
	2-1/2	33	38
	3	33	38
	3-1/2	32	38
	4	32	38
	5	31	36
	5-1/2	31	36
14	Cap End	38	43
	2-1/2	37	42
	3	36	42
	3-1/2	36	41
	4	36	41
	5	35	40
	5-1/2	34	40

table b-7

air requirement per inch of cylinder stroke

The amount of air required to operate a cylinder is determined from the volume of the cylinder and its cycle in strokes per minute. This may be determined by use of the following formulae which apply to a single-acting cylinder.

$$V = \frac{3.1416\ L\ D^2}{4} \qquad C = \frac{f\ V}{1728}$$

Where:

- V = Cylinder volume, cu. in.
- L = Cylinder stroke length, in.
- D = Internal diameter of cylinder, in.
- C = Air required, cfm
- f = Number of strokes per minute

The air requirements for a double-acting cylinder is almost double that of a single-acting cylinder, except for the volume of the piston rod.

air requirements continued

The air flow requirements of a cylinder in terms of cfm should not be confused with compressor ratings which are given in terms of free air. If compressor capacity is involved in the consideration of cylinder air requirements it will be necessary to convert cfm values to free air values. This relationship varies for different gauge pressures.

Thrust (pounds) = operating pressure × area of cylinder bore.

Note: *That on the "out" stroke the air pressure is working on the entire piston area but on the "in" stroke the air pressure works on the piston area less the rod area.*

Graph b-4 and **b-5** offer a simple means to select pneumatic components for dynamic cylinder applications. It is only necessary to know the force required, the desired speed and the pressure which can be maintained at the inlet to the F-R-L "Combo." The graphs assume average conditions relative to air line sizes, system layout, friction, etc. At higher speeds, consider appropriate cushioning of cylinders.

The general procedure to follow when using these graphs is:

1. Select the appropriate graph depending upon the pressure which can be maintained to the system—**graph b-4** for 100 psig and **graph b-5** for 80 psig.

2. Determine appropriate cylinder bore. Values underneath the diagonal cylinder bore lines indicate the maximum recommended dynamic thrust developed while the cylinder is in motion. The data in the table at the bottom of each graph indicates available static force for applications in which clamping force

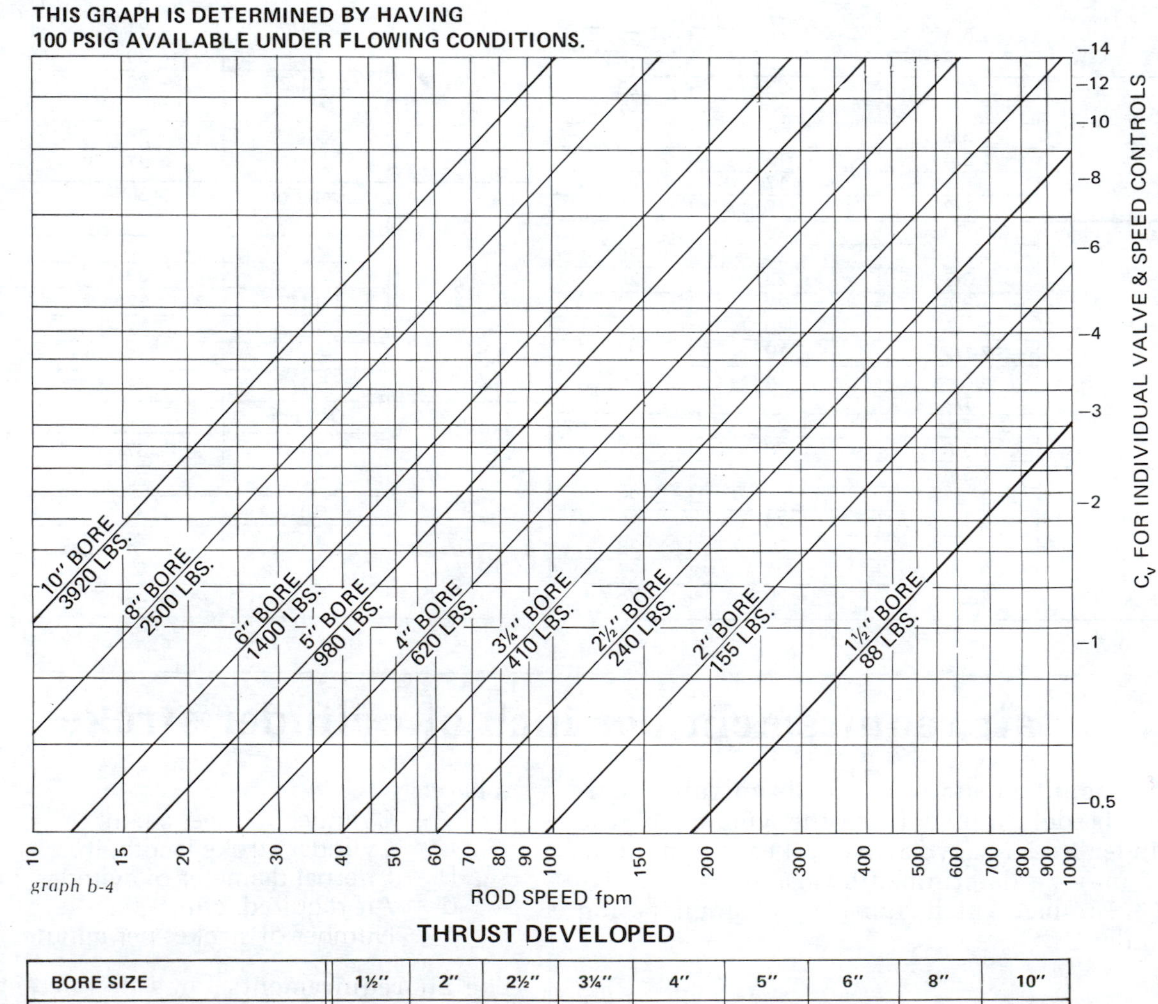

graph b-4

THRUST DEVELOPED

BORE SIZE	1½"	2"	2½"	3¼"	4"	5"	6"	8"	10"
DYNAMIC THRUST (lbs.)	88	155	240	410	620	980	1400	2500	3920
STATIC THRUST (lbs.)	177	314	491	830	1250	1960	2820	5020	7850

table b-8

air requirements continued

is a prime consideration in determining cylinder bore.

3. Read upward on appropriate rod speed line to intersection with diagonal cylinder bore line. Read right from intersection point to determine the required C_V of the valve and the speed controls. Both the valve and speed controls must have this C_V.

The following examples illustrate use of the graphs:

Example 1: *Assume it is necessary to raise a 900-pound load 24 inches in two seconds. With 100 psig maintained at the inlet to the F-R-L, use* **graph b-4.** *The 5-inch bore cylinder is capable of developing the required thrust while in motion. Since 24 inches in two seconds is equal to 60 fpm, read upward on the 60 fpm line to the intersection of the 5-inch bore diagonal line. Reading to the right indicates that the required valve and speed controls must each have a* C_V *of over 1.9.*

Example 2: *Assume similar conditions to Example 1 except that only 80 psig will be available under flowing conditions. Using* **graph b-5,** *a 6-inch bore cylinder is indicated. Read upward on the 60 fpm line to the intersection point. Interpolation of the right-hand scale indicates a required valve and speed control* C_V *of over 2.8.*

Example 3: *Assume similar conditions to Example 1 except that the load is being moved in a horizontal plane with a coefficient of sliding friction of 0.2. Only a 180-pound thrust is now required (900 lb. × 0.2). Consult* **graph b-4.** *The 2-1/2-inch bore cylinder will develop sufficient thrust, and at 60 fpm requires a valve and speed control* C_V *of about 0.5.*

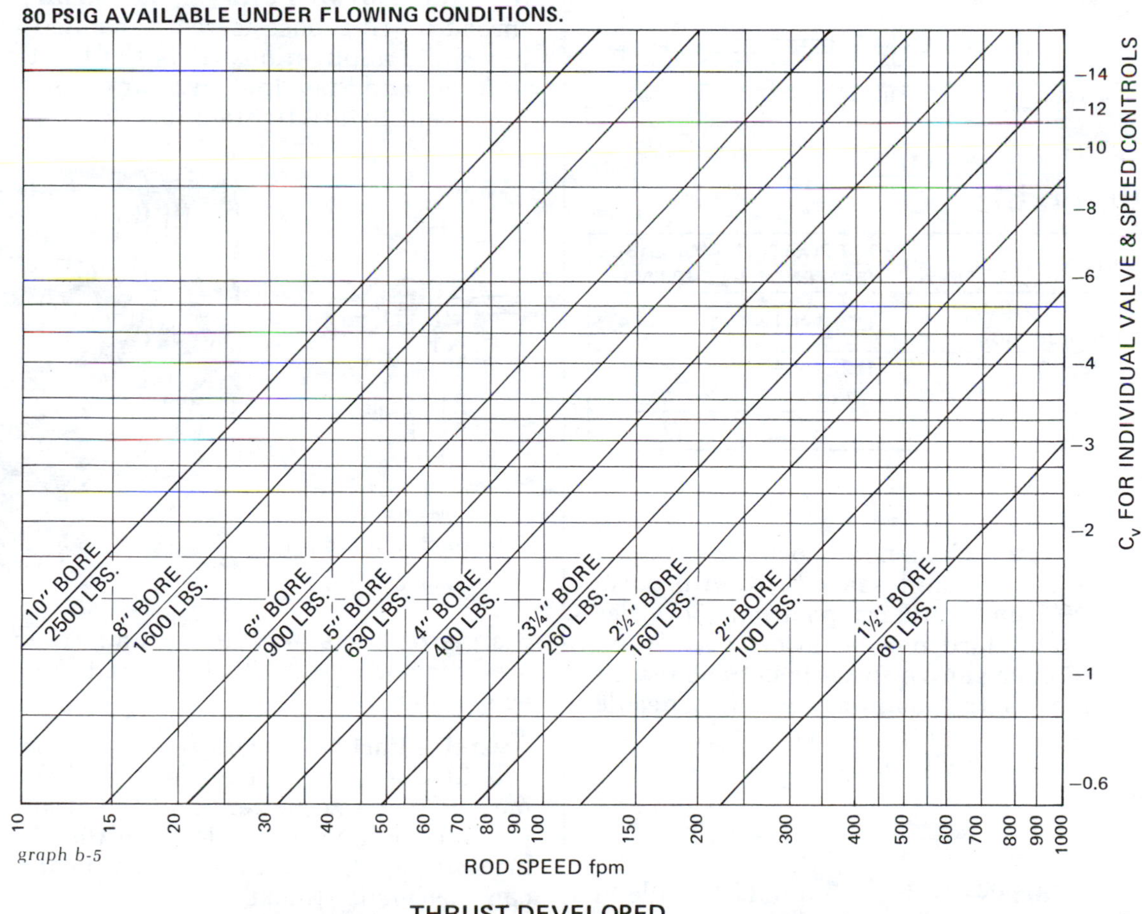

graph b-5

THRUST DEVELOPED

BORE SIZE	1½"	2"	2½"	3¼"	4"	5"	6"	8"	10"
DYNAMIC THRUST (lbs.)	60	100	160	260	400	630	900	1600	2500
STATIC THRUST (lbs.)	141	251	393	663	1000	1570	2260	4010	6280

table b-9

cylinder ports

port sizes

One of the factors involved in determining the speed of a hydraulic cylinder piston is fluid flow in connecting lines, generally measured in gallons per minute, introduced to, or expelled from, cap end cylinder port. (Due to piston rod displacement, the flow at head end port will be less than at cap end.) Fluid velocity, however, is measured in feet per second. In connecting lines this velocity should generally be limited to 15 feet per second to minimize fluid turbulence, pressure loss and hydraulic shock.

port location

Standard port location is position 1, as shown in **illustration b-3** below. Cushion adjustment needle and check valve are at positions 2 and 4 (or 3), depending on mounting style.

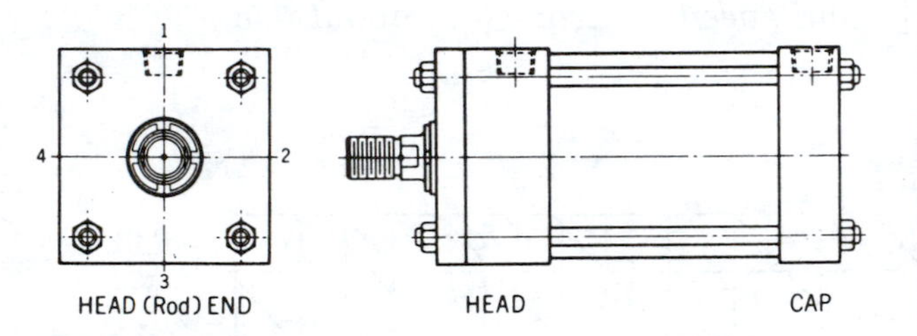

illustration b-3

position table

MOUNTING STYLE	PORT POSITION AVAILABLE	
	HEAD END	CAP END
T, TB, TC, TD, BC, H, HB, J, JB, DD	1, 2, 3 or 4	1, 2, 3 or 4
BB, BD	1, 2, 3 or 4	1 or 3
D	1 or 3	1, 2, 3 or 4
C, E, F, G	1	1

table b-10

Heads or caps which do not incorporate mounting can be rotated and assembled with ports 90° or 180° from standard position. To order other than standard port location, specify by position number shown in **table b-10** above. In such assemblies, the cushion adjustment needle and check valve rotate accordingly, since their relationship with port position does not change.

Manifold Ports — Side mounted cylinders, can be furnished with the cylinder ports arranged for mounting and sealing to a manifold surface. The ports are drilled and counterbored for O-ring seals which are provided.

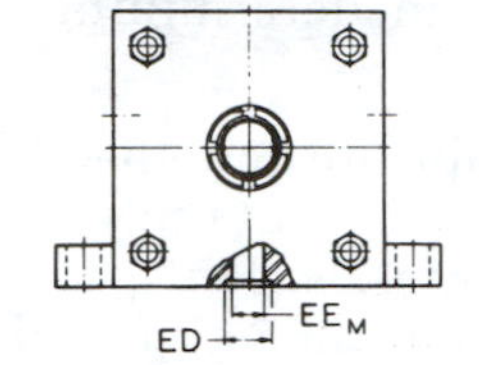

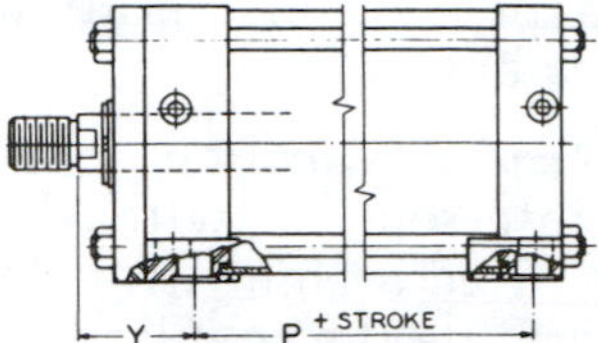

illustration b-4

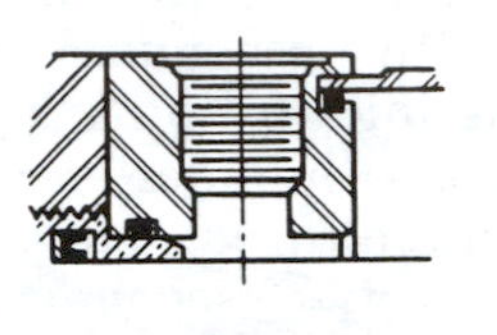

S.A.E. straight thread.
illustration b-5

Straight Thread Ports — The S.A.E. straight thread. O-ring boss is recommended for most hydraulic applications. It is the least prone to leakage and has the advantage of direction alignment before tightening.

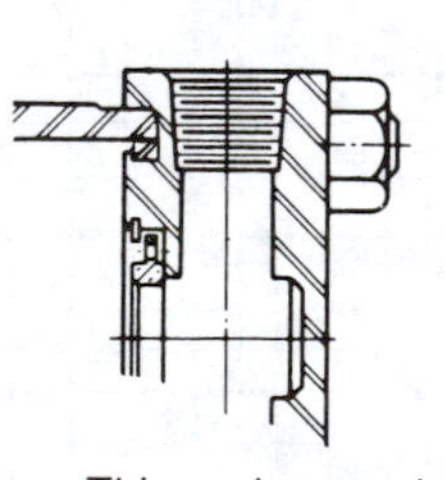

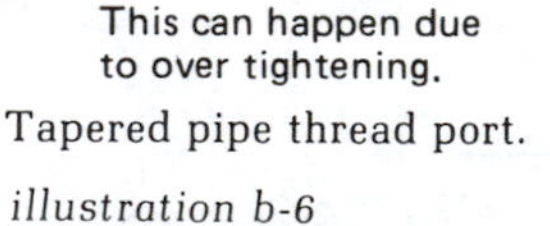

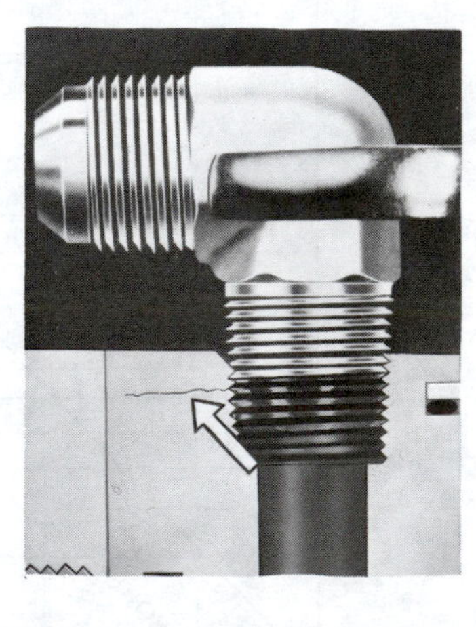

Tapered pipe thread port.
illustration b-6

Standard Ports — NFPA standard cylinders are furnished with NPTF tapered pipe threads as standard.

Oversize Ports — Oversize NPTF ports can be provided on more bore sizes. Welded port bosses, one size larger than the standard, are provided which protrude from the side of the head or cap. In addition, special heads or caps can be made thicker to accomodate larger ports. In these cases, application details should be supplied to the factory relative to speed, load conditions, pressure and circuitry details.

seals

Combination rod sealing arrangements perform dual functions. In (a), "V" packing reduces leakage of pressurized fluid from cylinder. Wiper prevents foreign materials from being drawn into cylinder during return stroke of piston rod. Arrangement at (b) includes two synthetic-type seals in an assembly that can be removed from cylinder as a unit. Inner seal performs primary sealing function. Oil getting past primary seal is trapped in cavity formed by inner lip of outer seal and deposited on rod for return to cylinder during return stroke. Outer lip of outer seal performs wiping function to exclude foreign materials.

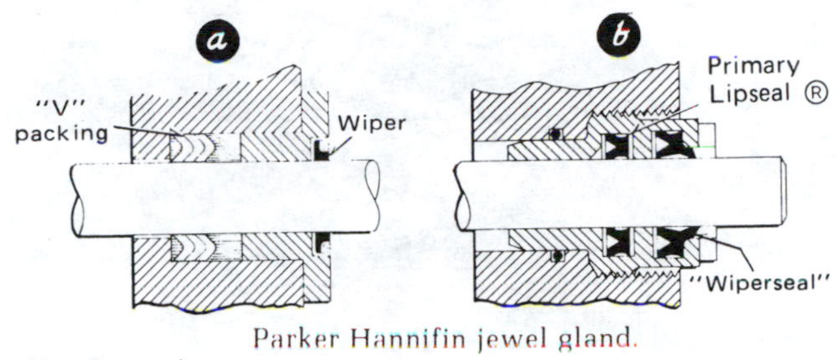

Parker Hannifin jewel gland.

illustration b-8

Buna N, fluorocarbon and polyurethane are the more popular materials used for cylinder seals. Of these, polyurethane has highest resistance to abrasion and a longest wear life than does Buna N. However, Buna N and polyurethane materials have temperature limitations of about 200°F. Other materials are required for applications where heat is a problem.

Viton® has good heat-resistant characteristics, to about 400°F. If higher heat resistance is necessary, Teflon® can be used. Teflon will withstand heat of approximately 500°F. However, before using these seal materials, check with the individual manufacturer for his recommendations.

Standard Seals — Class 1 Service Kits are standard, and contain seals of Nitrile (Buna N) elastomers for standard fluid service. These seals are suitable for use when air, hydraulic (mineral-type) oil, water-glycol fluid or water-in-oil emulsions are the operating medium.

The recommended operating temperature range for Class 1 seals is +10°F to +165°F. These seals will function at temperatures up to 200°F with reduced life.

Special Seals — Class 5 Service Kits contain seals of fluorocarbon elastomers (Viton®*) for special fluid service. These seals are especially suitable for most straight synthetic phosphate ester and phosphate ester base (fire-resistant) fluids. They can also be used when air, hydraulic oil, water glycol or water-in-oil emulsions are the operating medium.

The recommended operating temperature range for Class 5 seals is +10°F to +350°F. These seals will function at temperatures up to +400°F with reduced life.

To order Class 1 or 5, specify operating medium.

* Registered tradename of E.I. duPont de Nemours and Co., Inc.

cylinder options

the piston

The piston, which is permanently locked to the piston rod, is of one-piece construction. It is pilot-fitted to the piston rod. This pilot ensures piston concentricity with the centerline of the piston rod. It has a wide surface contacting the cylinder body; this reduces bearing loads during mechanical deflection. It also has a long thread engagement to the piston rod to provide greater shock absorption.

Iron piston rings are standard for hydraulic service and are furnished unless otherwise specified. (Clearance flow of 1 to 3 cubic inches per minute can be expected.) Parker Lipseal® piston construction is optional. It provides virtually zero leakage under static conditions, because the seals are fully dynamic and self-compensating even with variations in pressure, mechanical deflections and wear. The Lipseals have backup washers to prevent extrusion, and can be used in applications up to 3000 p.s.i. hydraulic pressure.

stroke adjusters

For the requirement where adjusting the stroke is specified some manufacturers have several designs to offer, one of which is illustrated below **(b-9)**. This is suitable for infrequent adjustment and is economical.

cushions

stepped floating cushions combine the best features of known cushion technology.

Deceleration devices or built-in "cushions" are optional and can be supplied at head end, cap end, or both ends without change in envelope or mounting dimensions. Parker cylinder cushions are a stepped design and combine the best features of known cushion technology.

Standard straight or tapered cushions have been used in industrial cylinders over a very broad range of applications, research has found that both designs have their limitations.

As a result, Parker has taken a new approach in cushioning of industrial hydraulic cylinders and for specific load and velocity conditions have been able to obtain deceleration curves that come very close to the ideal. The success lies in a stepped sleeve or spear concept where the steps are calculated to approximate theoretical orifice areas curves.

In the cushion performance chart, pressure traces show the results of typical orifice flow conditions. Tests of a three-step sleeve or spear shown three pressure pulses coinciding with the steps. The deceleration cushion plunger curves shape comes very close to being theoretical, with the exception of the last 1/2 inch of travel. This is a constant shape in order to have some flexibility in application. The stepped cushion design shows reduced pressure peaks for most load and speed conditions, with comparable reduction of objectionable stopping forces being transmitted to the load and the support structure.

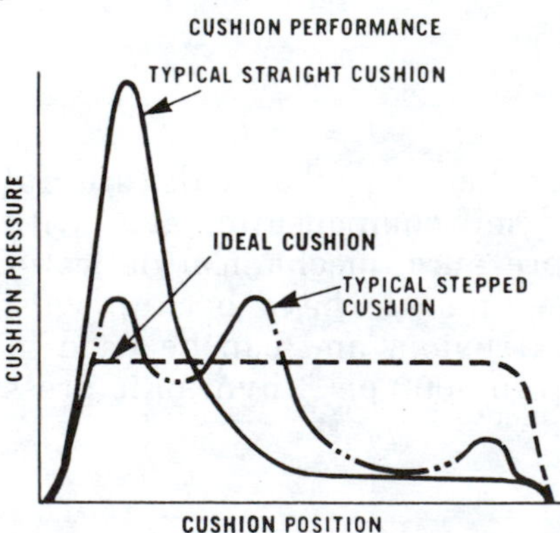

All Parker Hannifin cushions are adjustable except those at head end of 1-1/2", 2" and 2-1/2" bore size cylinders equipped with Code 2 diameter piston rods. On these three sizes a cushion at head end is supplied without cushion adjustment needle or check valve, and is nonadjustable. The Series 2H cylinder design incorporates the longest

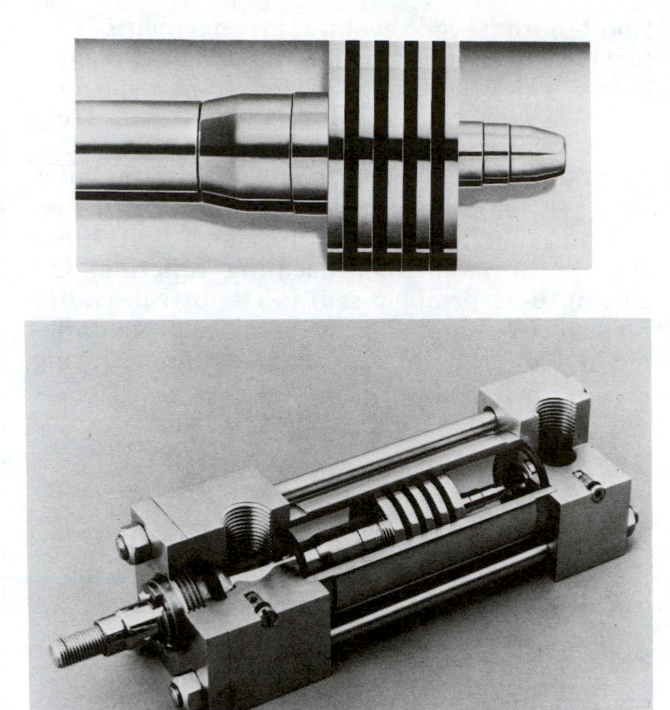

cushion sleeve and cushion spear that can be provided in the standard envelope without decreasing the rod bearing and piston bearing lengths.

1. When a cushion is specified at the head end:
   a. A self-centering stepped sleeve is furnished on the piston rod assembly.
   b. A needle valve is provided that is flush with the side of the head even when wide open, except in 10" and 12" bore models. It may be identified by the fact that it is socket-keyed. It is located on side number 2, in all mounting styles except D, DB, DD, JJ, HH and E. In these styles it is located on side number 3.
   c. A springless check valve is provided that is also flush with the side of the head and is mounted on the face opposite the needle valve except on mounting styles D, DB, DD, JJ, HH and E, where it is mounted on side number 3, next to the needle valve.
   It may be identified by the fact that it is slotted.
   d. The check and needle valves are interchangeable in the head.
2. When a cushion is specified at the cap end:
   a. A cushion stepped spear is provided on the piston rod.
   b. A "float check" self-centering bushing is provided which incorporates a large flow check valve for fast "out-stroke" action.
   c. A socket-keyed needle valve is provided that is flush with the side of the cap when wide open. It is located on side number 2 in all mounting styles except D, DB, DD, JJ, HH and E. In these it is located on side number 3.

cylinder options continued

PH stroke adjuster.

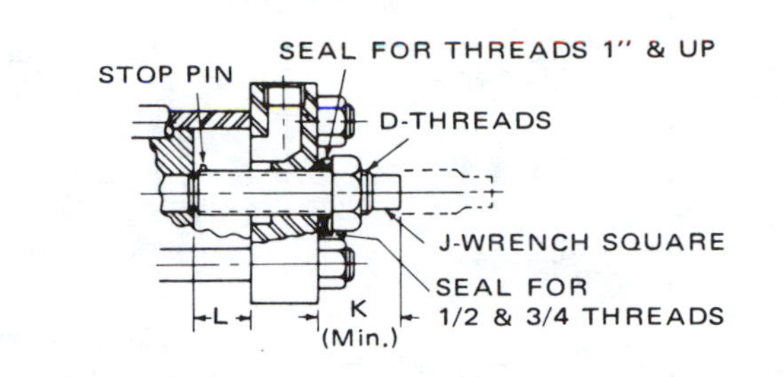

illustration b-9

gland drain port

A gland drain may be desirable on cylinders with exceptionally long stroke, or with restricted outlet port flow at the head end, or with constant back-pressure as in differential circuit applications. When specified, a 1/8" NPTF gland drain port between Lipseal® and Wiperseal can be supplied.

air bleed

When cylinders are mounted horizontally with inlet ports on top, or mounted vertically and cycled their full stroke, they are usually self-bleeding. On Hannifin 2H cylinders, vent screws for bleeding air can be provided when specified. They are added to both ends of the cylinder. To ensure proper location, the cylinder side (designated by side number reference) should be specified on the order.

When it is impractical to utilize vent screws in the cylinder body, vent ports can be provided for remote air bleeding through tubing to a remote location. In this case, 1/8" NPTF ports are added to both ends of the cylinder body. Cylinder side (designated by side number reference) should be specified on the order.

water service modifications

Standard — When requested, Parker-Hannifin can supply Series 2H cylinders with standard modifications that make the cylinders more nearly suitable for use with water as the fluid medium. The modifications include chrome-plated, stainless steel piston rod. On orders for water-service cylinders, be sure to specify the maximum operating pressure or the load and speed conditions. (These factors must be taken into account because of the lower tensile strength of stainless steels available for use in piston rods.)

Special — If required, special materials for any of the cylinder elements can be supplied. In these cases, consult the factory.

Warranty — Parker-Hannifin will warrant Series 2H cylinders modified for water service to be free of defects in materials or workmanship. On the other hand, Parker-Hannifin cannot accept responsibility for premature failure of cylinder function, where failure is caused by corrosion, electrolysis or mineral deposits within the cylinder.

cylinder formulas

Cylinder Force

$$F = P \times A$$

$$P = \frac{F}{A}$$

$$A = \frac{F}{P}$$

$$A = .7854 \times D^2$$

F = Force (lbs)
P = Pressure (psi)
A = Area (inches squared)

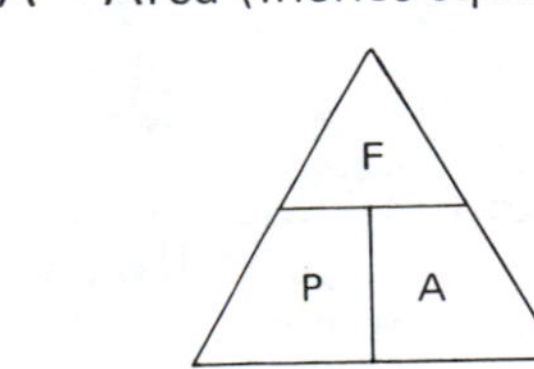

Cylinder Speed

$$\text{cylinder speed in ft./min.} = \frac{\text{G.P.M.} \times 19.25}{A}$$

$$\text{cylinder velocity (in./min.)} = \frac{924 \times \text{G.P.M.}}{\pi \times D^2}$$

$$\text{pump volume in G.P.M.} = \frac{A \times L \times 60}{231 \times \text{time in seconds (cylinder stroke)}}$$

$$\text{Piston V ft./sec.} = \frac{231 \times \text{gpm}}{720 \times A}$$

applications of cylinders for providing a variety of fundamental mechanical motions.

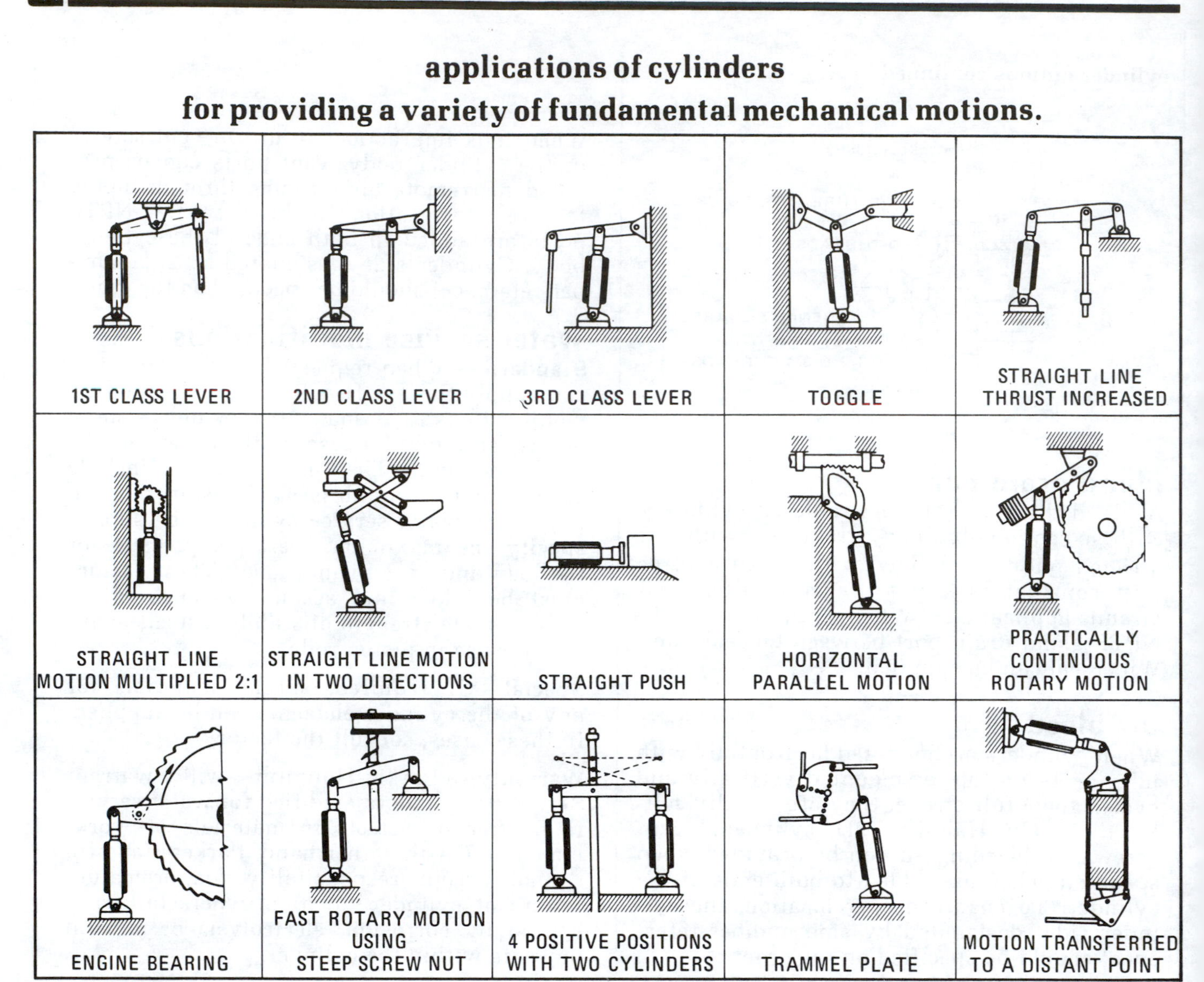

illustration b-10

Toggle Mechanism

For operations such as coining and marking requiring exact depth control, and requiring extremely high force for a very short distance, the toggle lever system can be useful.

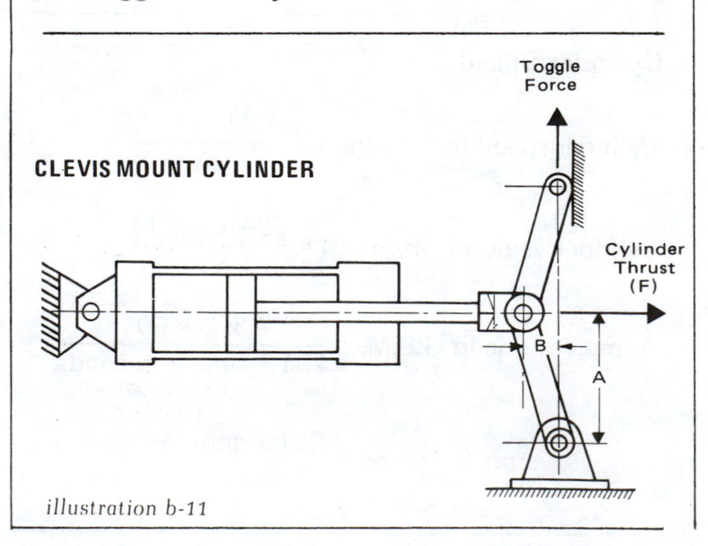

illustration b-11

In **illustration b-11,** cylinder thrust is horizontal and toggle force is taken off vertically. Bearings at each end of the toggle lever must be closely fitted and heavy enough to carry the full toggle thrust.

A calculation of toggle force can be made with the following formula, with T and F in the same units, and A and B in the same units. Note that dimension A is not the lever length, but for high leverage toggle calculations it can be used for lever length, with only small error, since the lever is nearly vertical.

$$T \text{ (Toggle Force)} = \frac{F \text{ (Cylinder Thrust)} \times A}{2B}$$

Example: *Find the toggle force from a cylinder thrust of 8300 lbs., if the toggle lever is 14 inches long and is 1/2 inch from vertical (Distance B).*

mechanical motion continued

Solution: *T = 8300 × 14 ÷ 2 × 1/2 = 116,200 lbs. This is a multiplication of 14 times the direct cylinder thrust. The remaining travel distance of the toggle arm at any point in the cylinder stroke is twice the difference between distance A and the true length, pin-to-pin, of the toggle arm. Distance A can be found by geometry or from a scale layout.*

Thrust Exerted by a Cylinder Working at an Angle

Cylinder thrust, F, is horizontal in this figure. Only that vector force, T, which is at right angles to the lever axis is effective for turning the lever. The value of T varies with the acute angle "A" between cylinder and lever axes.

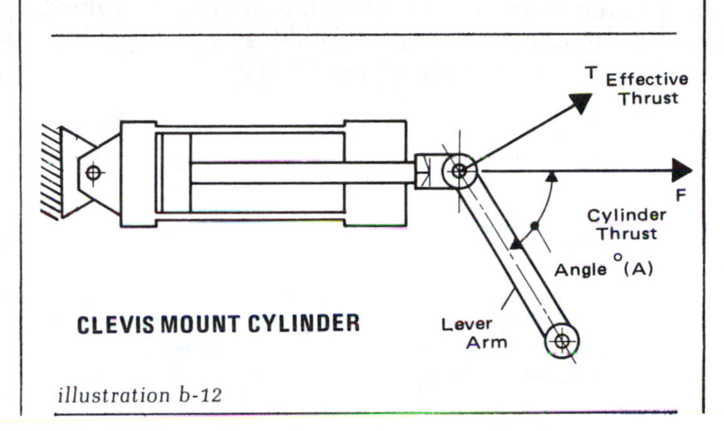

illustration b-12

To calculate T, multiply cylinder thrust times the power factor taken from **table b-11** below.

Example: *A 3.25-inch bore cylinder working 1000 p.s.i. gauge pressure will develop a 8300 lb. thrust (8.294 in. area × 1000). Effective thrust when working at a 65° angle is: 8300 × 0.906 (from table 10) = 7519.8.*

power factor table

ANGLE A DEGREES	PWR. FACTOR (SIN A)	ANGLE A DEGREES	PWR. FACTOR (SIN A)
5	0.087	50	0.766
10	0.174	55	0.819
15	0.259	60	0.867
20	0.342	65	0.906
25	0.423	70	0.940
30	0.500	75	0.966
35	0.573	80	0.985
40	0.643	85	0.996
45	0.707	90	1.000

table b-11

Determining Cylinder Stroke for Operating a Hinged Lever by the Chord Factor Method

If the cylinder is rotating the lever to an equal angle each side of the perpendicular as in **illustration b-13,** the length of stroke can very easily be determined by multiplying lever length (pin-to-pin) times the chord factor from **table b-12.** If the movement is not equal on each side of the perpendicular, the stroke may be determined by using another example shown in this section.

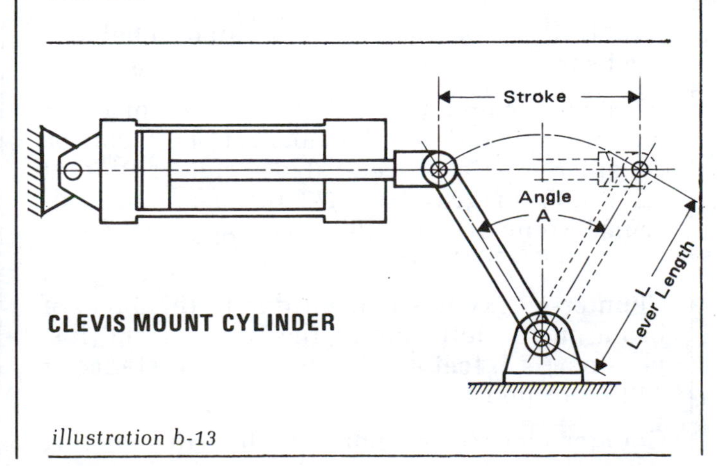

illustration b-13

Example: *The cylinder stroke needed to swing a 15-inch lever through a 120 degree arc, when mounted as in* **illustration b-12,** *is found by taking the factor 1.732 (from* **table b-12**) *times 15 (lever length) = 25.98" (stroke length).* Many times a stock cylinder with standardized stroke length can be used by lengthening or shortening the lever arm for the desired travel.

ANGLE A DEGREES	CHORD FACTOR	ANGLE A DEGREES	CHORD FACTOR
5	0.087	85	1.351
10	0.174	90	1.414
15	0.261	95	1.475
20	0.347	100	1.532
25	0.433	105	1.587
30	0.518	110	1.638
35	0.601	115	1.687
40	0.684	120	1.732
45	0.765	125	1.774
50	0.845	130	1.813
55	0.923	135	1.848
60	1.000	140	1.879
65	1.075	145	1.907
70	1.147	150	1.932
75	1.217	155	1.953
80	1.286	160	1.970

table b-12

mechanical motion continued

Cranes and Beams

Because crane working angles constantly change, constructing a rough model on paper is often necessary to show the point at which the greatest cylinder thrust is needed. An exact calculation can then be made.

Only the portion of cylinder thrust at right angles to the beam axis is effective for turning the beam. This can be calculated by the method shown below.

Example: *Find the maximum load that can be lifted by the crane when the angles are as shown, and the cylinder thrust, F, is 19635 lbs. Translate the 19635 lbs. cylinder thrust into* F_1*, 9817.5 lbs. at right angles to the beam, using the power factor of 0.500 lbs. for a 30° angle from* **table b-12, page b-19.** *Next, translate this to* F_2*, 3272.5 lbs. thrust at the weighted end of the beam. This is done with simple proportion by the length of each arm from the base pivot point.* F_2 *is 1/3* F_1*, since the lever is 3 times as long. Then, using the power factor* **table b-12,** *find the maximum hanging load that can be lifted at a 45° angle between beam and load weight.*

Calculations for Heavy Beam

With heavy beams, it is necessary to take into account the weight of the beam itself. If the beam is distributed uniform in weight across its length, the calculation is relatively easy. In **illustration b-15** the beam has a uniform weight of 90 pounds per foot. It is partially counter balanced by a load weight of 400 pounds on the left side of the fulcrum, and must be raised by the cylinder force applied at a point 9 feet from the right side of the fulcrum.

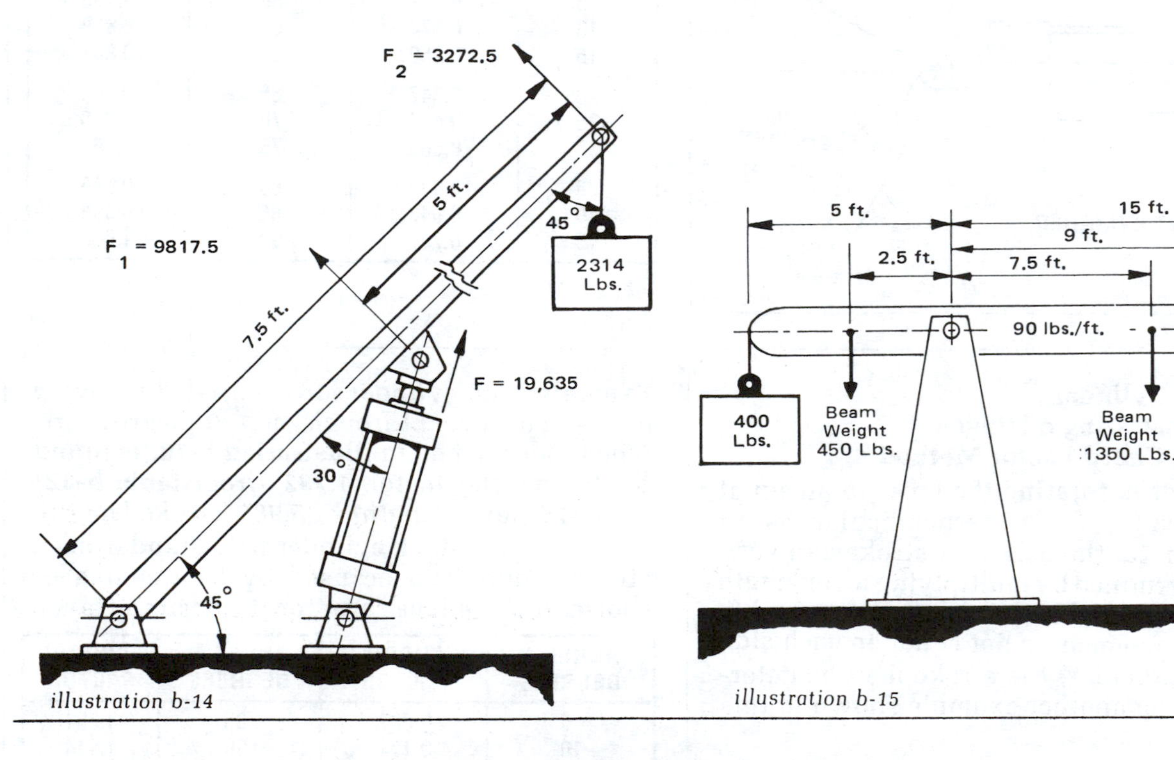

illustration b-14

illustration b-15

A method of solution is to use the principle of moments. A moment is a torque force consisting of (so many) pounds applied at a lever distance of (so many) feet or inches. The solution here is to find how much cylinder thrust is needed to just balance the beam. Then by increasing the hydraulic cylinder thrust about 10 to 15% to take care of friction losses, the cylinder would be able to raise the beam.

Using the principle of moments, it is necessary to calculate all of the moment forces which are trying to turn the beam clockwise, then calculate all the moment forces trying to turn the beam counter-clockwise, then subtract the two. In this case they must be equal to balance the beam.

Clockwise moment due to the 15 feet of beam on the right side of the fulcrum: This can be considered as a concentrated weight acting at its center of gravity 7-1/2 feet from the fulcrum. Moment = 90 (lbs. per foot) × 15 feet × 7-1/2 feet = 10125 foot pounds.

Counter-clockwise moment due to the 5 feet of beam on the left side of the fulcrum: 90 (lbs. per foot) × 5 feet × 2-1/2 feet (CG distance) = 1125 foot pounds.

Counter-clockwise moment due to hanging

continued on page b-24

mechanical motion continued

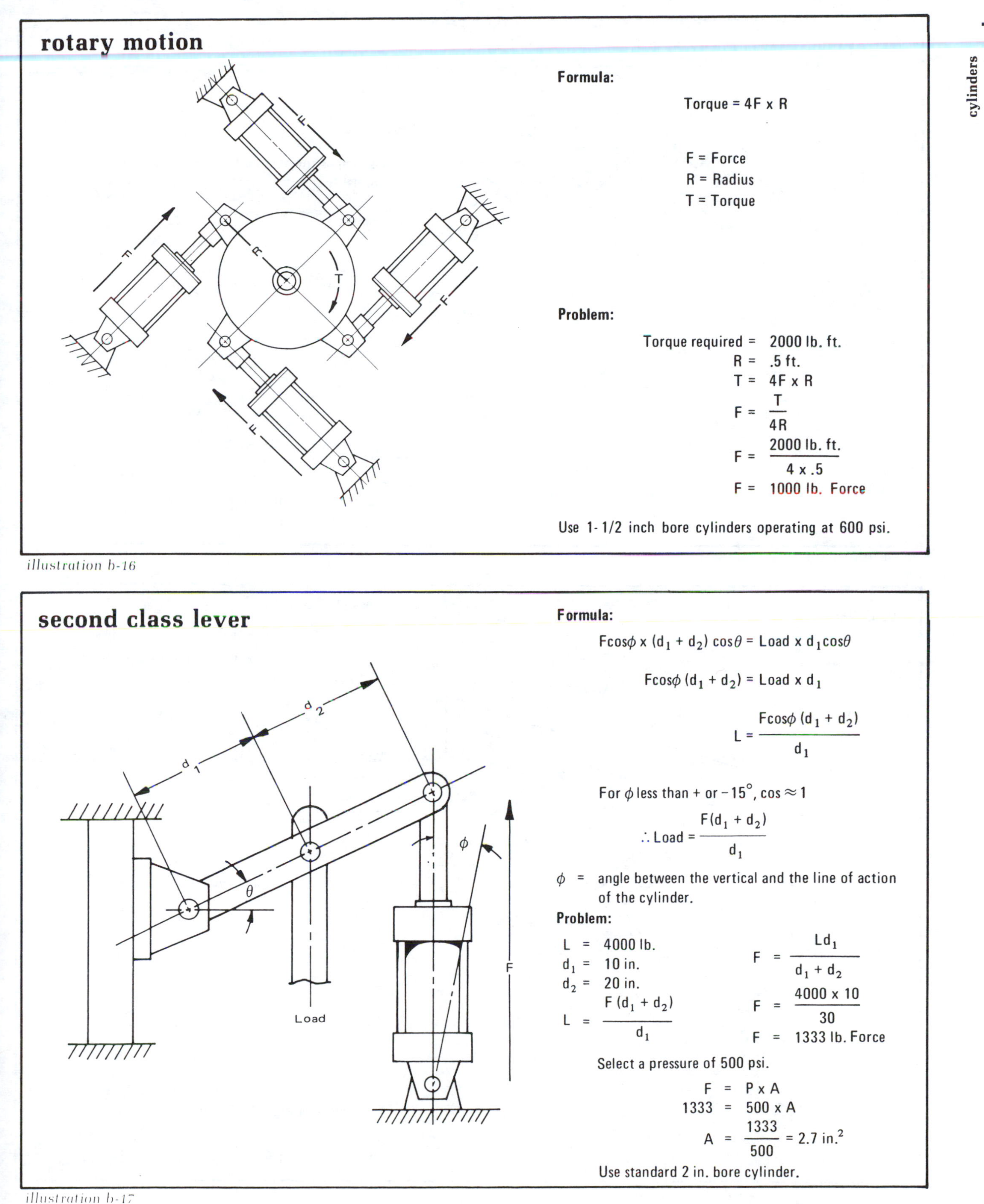

rotary motion

Formula:

$$\text{Torque} = 4F \times R$$

F = Force
R = Radius
T = Torque

Problem:

Torque required = 2000 lb. ft.

R = .5 ft.

$$T = 4F \times R$$

$$F = \frac{T}{4R}$$

$$F = \frac{2000 \text{ lb. ft.}}{4 \times .5}$$

F = 1000 lb. Force

Use 1-1/2 inch bore cylinders operating at 600 psi.

illustration b-16

second class lever

Formula:

$$F\cos\phi \times (d_1 + d_2)\cos\theta = \text{Load} \times d_1\cos\theta$$

$$F\cos\phi\,(d_1 + d_2) = \text{Load} \times d_1$$

$$L = \frac{F\cos\phi\,(d_1 + d_2)}{d_1}$$

For ϕ less than + or − 15°, cos ≈ 1

$$\therefore \text{Load} = \frac{F(d_1 + d_2)}{d_1}$$

ϕ = angle between the vertical and the line of action of the cylinder.

Problem:

L = 4000 lb.
d_1 = 10 in.
d_2 = 20 in.

$$L = \frac{F\,(d_1 + d_2)}{d_1}$$

$$F = \frac{Ld_1}{d_1 + d_2}$$

$$F = \frac{4000 \times 10}{30}$$

F = 1333 lb. Force

Select a pressure of 500 psi.

$$F = P \times A$$

$$1333 = 500 \times A$$

$$A = \frac{1333}{500} = 2.7 \text{ in.}^2$$

Use standard 2 in. bore cylinder.

illustration b-17

mechanical motion continued

lever

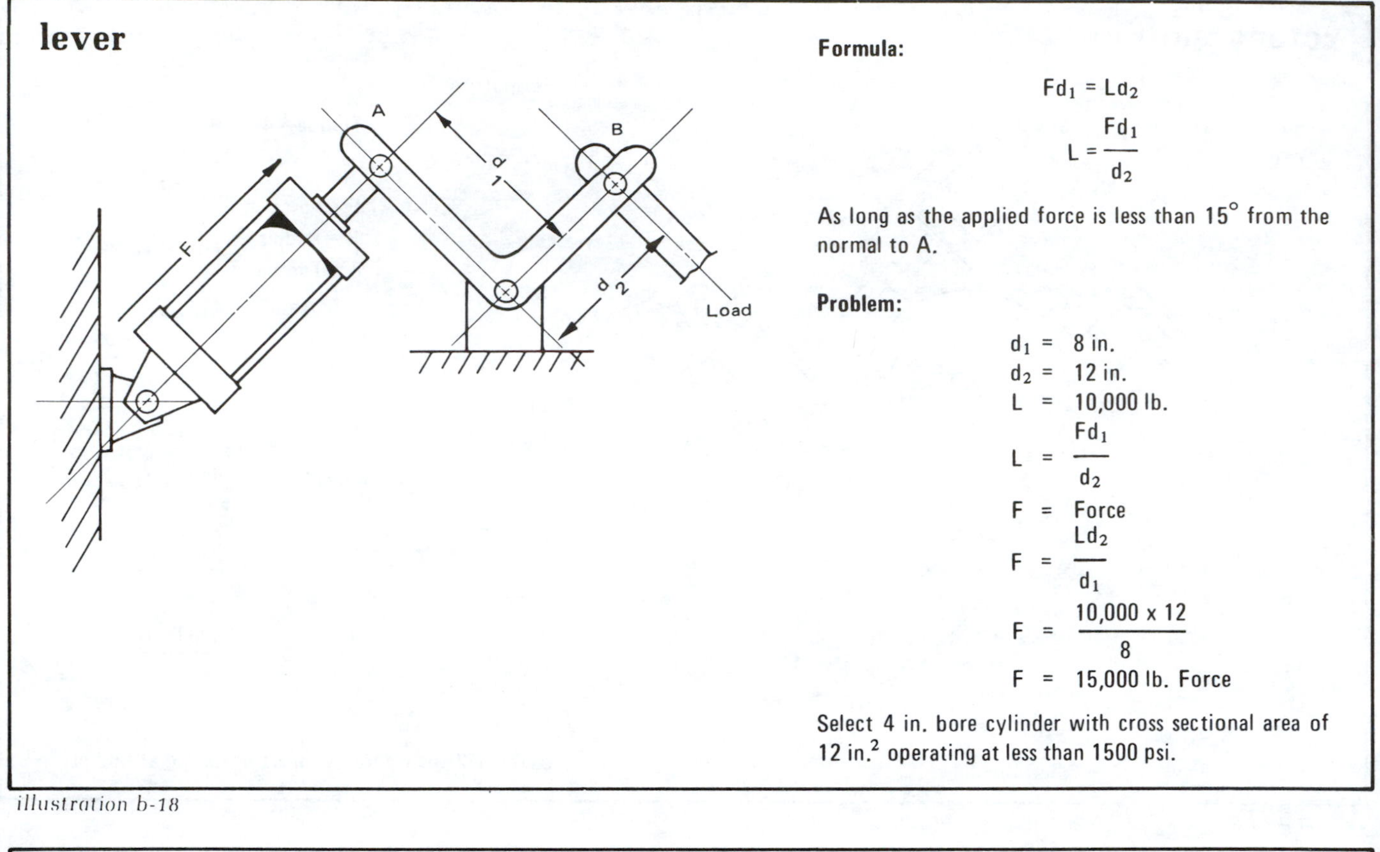

Formula:

$$Fd_1 = Ld_2$$

$$L = \frac{Fd_1}{d_2}$$

As long as the applied force is less than 15° from the normal to A.

Problem:

d_1 = 8 in.
d_2 = 12 in.
L = 10,000 lb.

$$L = \frac{Fd_1}{d_2}$$

F = Force

$$F = \frac{Ld_2}{d_1}$$

$$F = \frac{10{,}000 \times 12}{8}$$

F = 15,000 lb. Force

Select 4 in. bore cylinder with cross sectional area of 12 in.^{2} operating at less than 1500 psi.

illustration b-18

crane

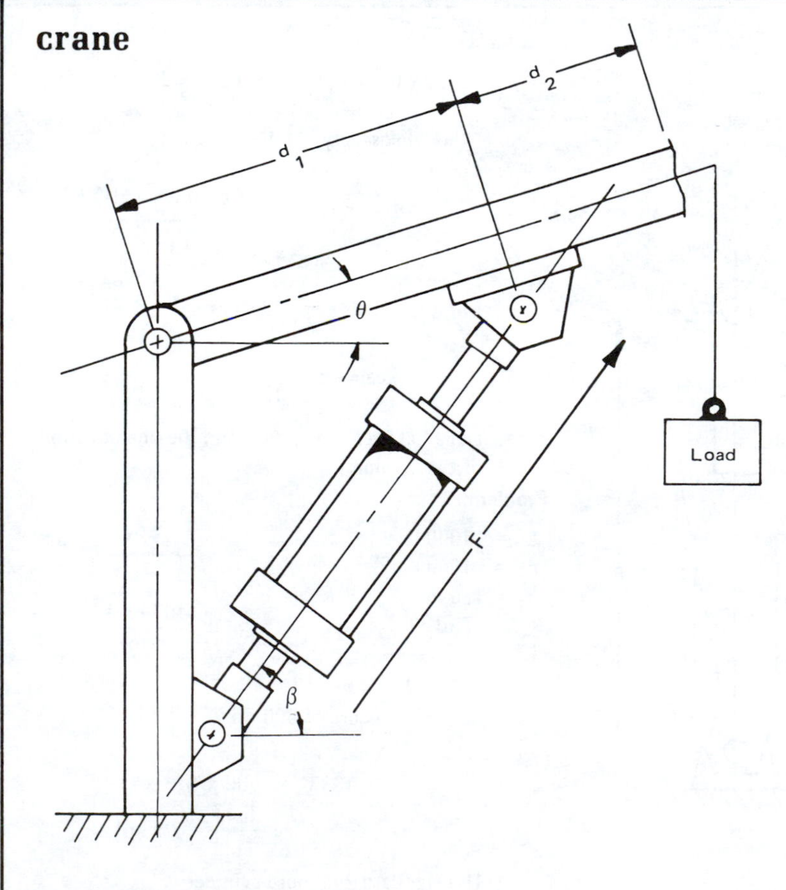

Formula:

$$F\sin\beta \times d_1\cos\theta = L\,(d_1 + d_2)\cos\theta$$

$$F\sin\beta \times d_1 = L(d_1 + d_2)$$

$$L = \frac{F\sin\beta \times d_1}{d_1 + d_2}$$

Where β in minimum angle between center line of cylinder and the horizontal.

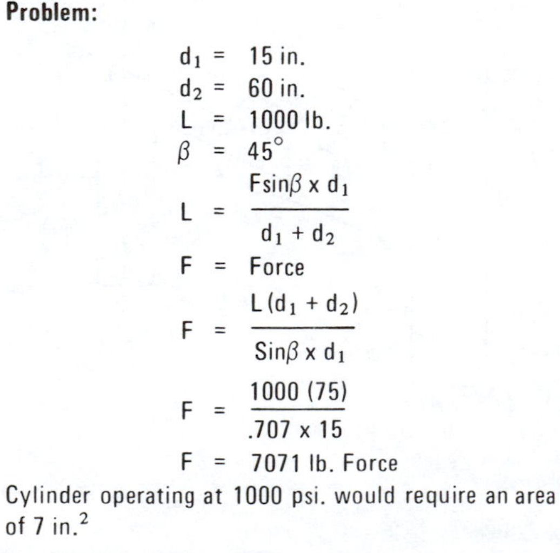

Problem:

d_1 = 15 in.
d_2 = 60 in.
L = 1000 lb.
β = 45°

$$L = \frac{F\sin\beta \times d_1}{d_1 + d_2}$$

F = Force

$$F = \frac{L\,(d_1 + d_2)}{\sin\beta \times d_1}$$

$$F = \frac{1000\ (75)}{.707 \times 15}$$

F = 7071 lb. Force

Cylinder operating at 1000 psi. would require an area of 7 in.^{2}

Use standard 3 1/4 in. bore cylinder.

illustration b-19

mechanical motion continued

first class lever

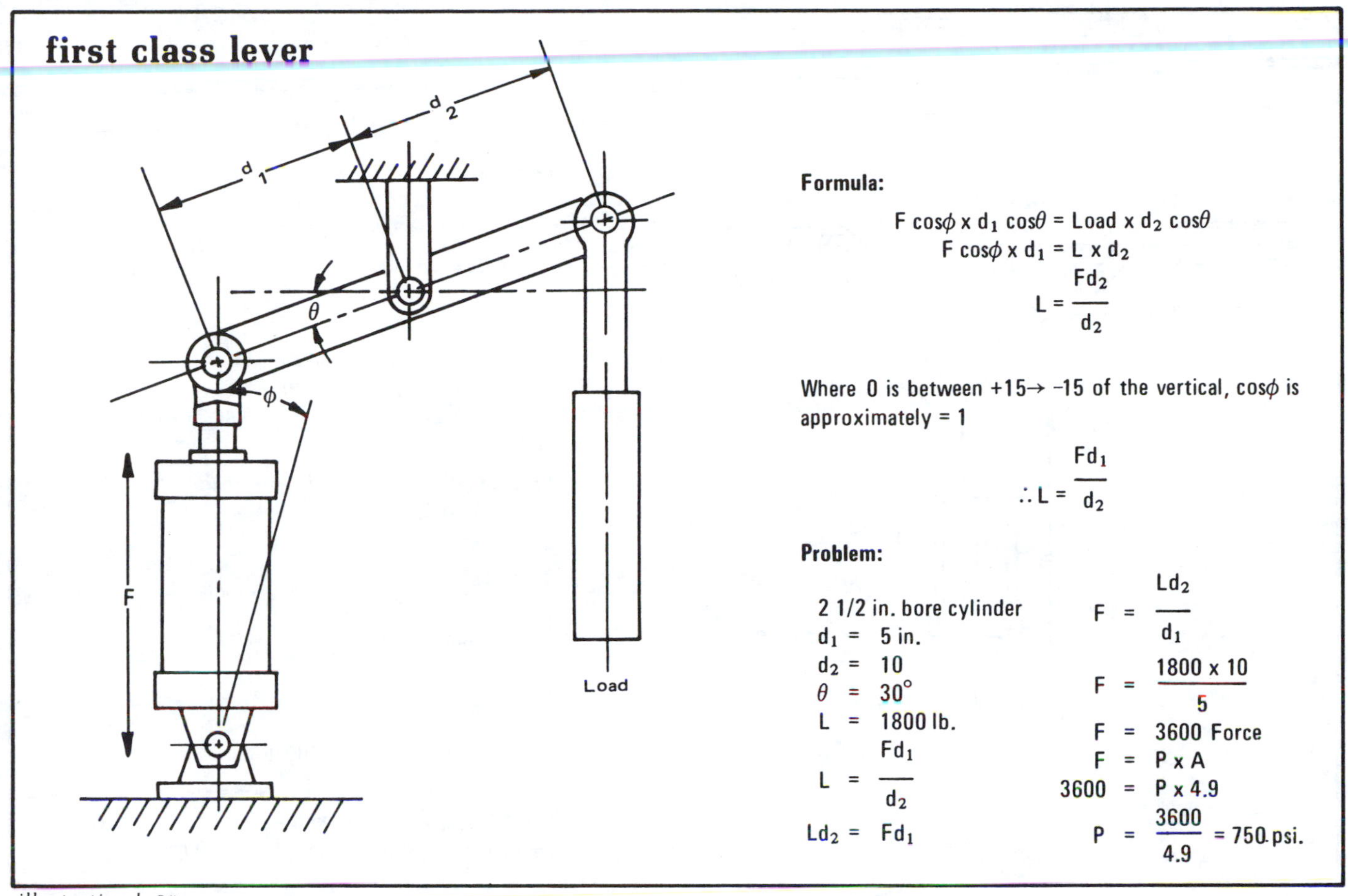

Formula:

$$F \cos\phi \times d_1 \cos\theta = \text{Load} \times d_2 \cos\theta$$

$$F \cos\phi \times d_1 = L \times d_2$$

$$L = \frac{Fd_2}{d_2}$$

Where 0 is between +15→ −15 of the vertical, $\cos\phi$ is approximately = 1

$$\therefore L = \frac{Fd_1}{d_2}$$

Problem:

2 1/2 in. bore cylinder

d_1 = 5 in.

d_2 = 10

θ = 30°

L = 1800 lb.

$$L = \frac{Fd_1}{d_2}$$

$$Ld_2 = Fd_1$$

$$F = \frac{Ld_2}{d_1}$$

$$F = \frac{1800 \times 10}{5}$$

F = 3600 Force

F = P x A

3600 = P x 4.9

$$P = \frac{3600}{4.9} = 750 \text{ psi.}$$

illustration b-20

first class lever variation

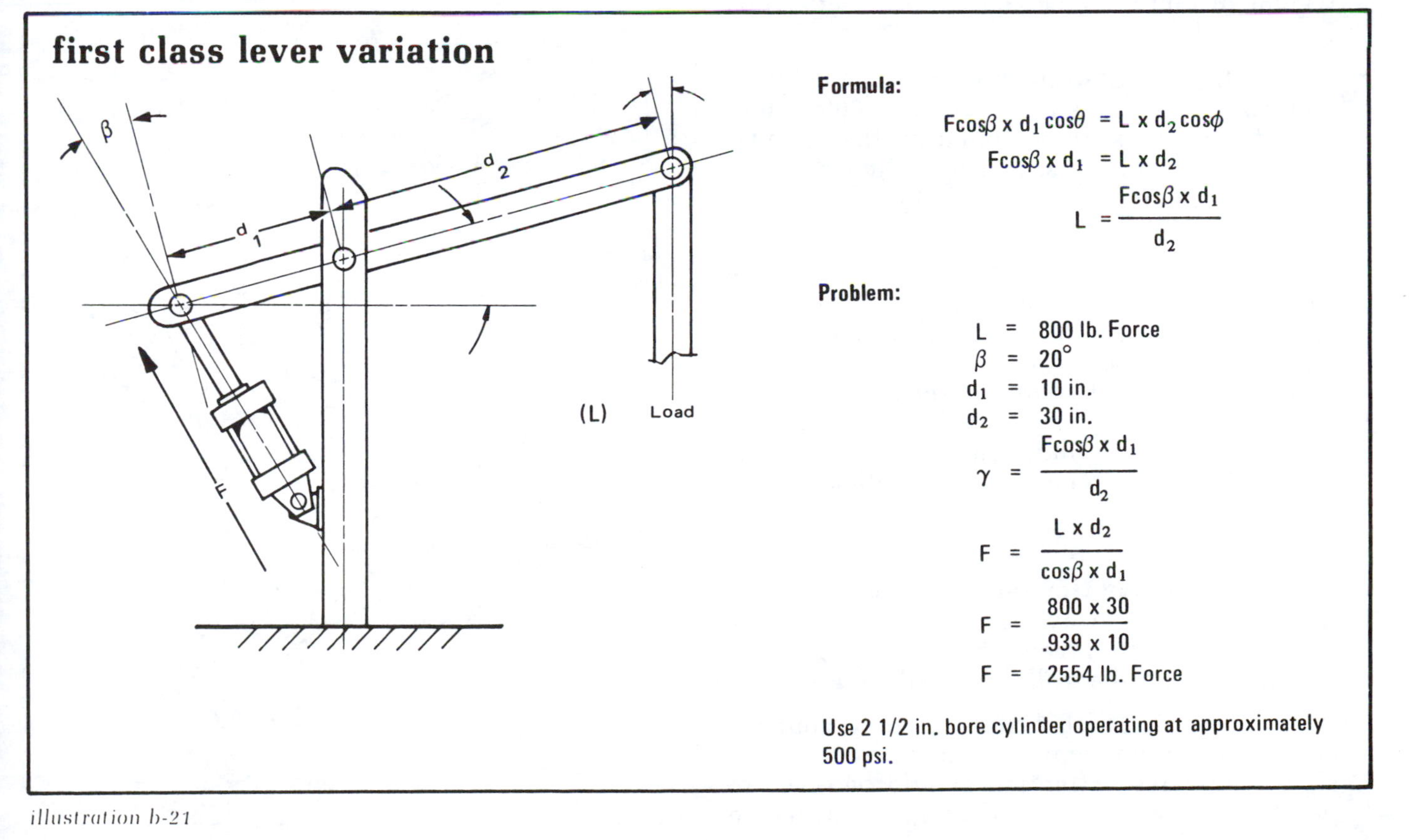

Formula:

$$F\cos\beta \times d_1 \cos\theta = L \times d_2 \cos\phi$$

$$F\cos\beta \times d_1 = L \times d_2$$

$$L = \frac{F\cos\beta \times d_1}{d_2}$$

Problem:

L = 800 lb. Force

β = 20°

d_1 = 10 in.

d_2 = 30 in.

$$\gamma = \frac{F\cos\beta \times d_1}{d_2}$$

$$F = \frac{L \times d_2}{\cos\beta \times d_1}$$

$$F = \frac{800 \times 30}{.939 \times 10}$$

F = 2554 lb. Force

Use 2 1/2 in. bore cylinder operating at approximately 500 psi.

illustration b-21

mechanical motion continued

toggle variation

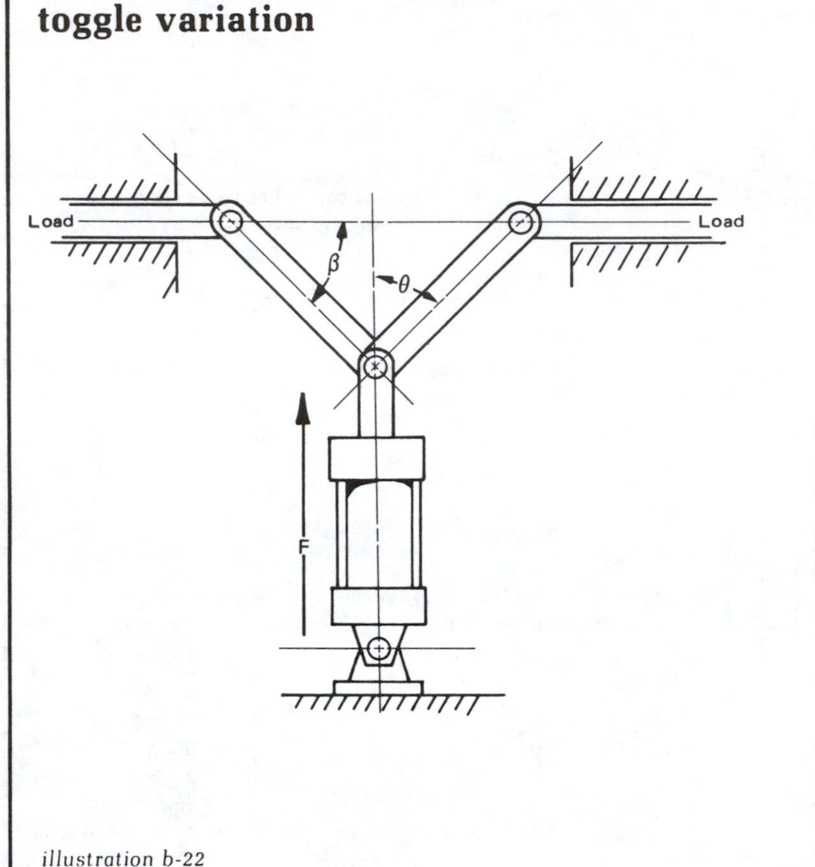

illustration b-22

Formula:

$$\text{Load} = \frac{F}{2\cos\theta}\sin\beta$$

$$\text{Load}_1 = \text{Load}_2$$

θ between 30° -- 60°
β between 30° -- 60°

Problem:

Load is 500 pounds on both sides, therefore, total load is 1000 pounds.

$\theta = 45°$
$\beta = 45°$

$$L = \frac{F\sin\beta}{2\cos\theta}$$

$$F = \frac{L2\cos\theta}{\sin\beta}$$

$$F = \frac{2 \times 1000 \times .7}{.7} = 2000 \text{ lb. force}$$

Use 1 in. bore cylinder with a cross sectional area of about 1 in.^{2} operating at less than 1000 psi.

weight of 400 pounds: 400 lbs. × 5 feet = 2000 foot pounds.

Subtracting counter-clockwise from clockwise moments: 10125 - 1125 - 2000 = 7000 foot pounds that must be supplied by the cylinder for balance conditions. To find cylinder thrust: 7000 foot pounds ÷ 9 feet (distance from fulcrum) = 777.77 pounds.

Note: *When working with moments, that only the portion of the total force which is at right angles to the beam is effective as a moment force. If the beam is at an angle to the cylinder or to the horizontal, then the effective portion of the concentrated or distributed weight, and the cylinder thrust, can be calculated with power factors by the method shown in this section.*

Cylinders Moving Horizontal Loads

Rolling Loads — A cylinder thrust of 1/10 the load weight will move loads which operate on low-friction needle, roller, or ball bearings.

An air cylinder with meter-out flow controls can be used on some applications, even at slow feed rates. Attention should be given to deceleration at the end of cylinder stroke to

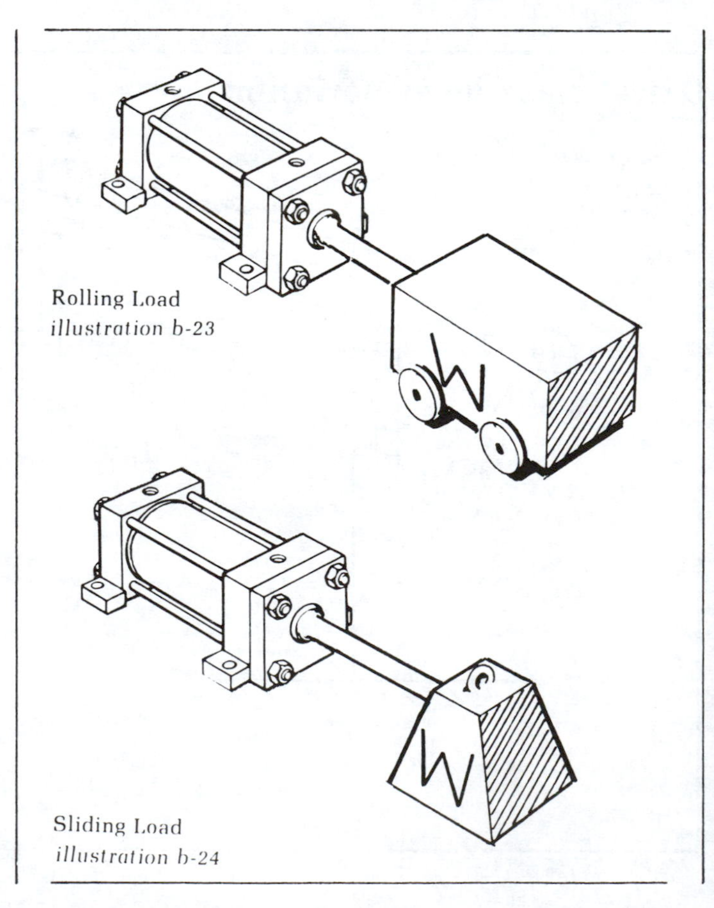

Rolling Load
illustration b-23

Sliding Load
illustration b-24

mechanical motion continued

prevent momentum of the load from damaging the cylinder and/or the machine.

Sliding Loads — Either air or hydraulic cylinders can be used for moving high friction sliding loads. (**illustration b-24, page b-24**).

On applications that require rapid indexing from one positive stop to another, an air cylinder will give more rapid action than hydraulics if the load is within its capacity. Air cylinders should **not** be used for slow speed or controlled feeding of a sliding load with a large area of surface friction, as a chattering motion will result. Hydraulic cylinders with a meter-out flow control should be used in these applications. In some instances an air/oil system will also give acceptable performance.

The force needed to push a sliding load varies with surface material, lubrication, unit loading and other factors. For lightly lubricated machined slides, the cylinder thrust should be equal to 1/2 to 3/4 of the load weight to get the load started. A thrust of 1/5 to 1/6 load weight will keep it moving.

To operate an air cylinder at high speeds, the cylinder should be sized to develope twice the thrust needed to balance the load.

Cylinders for Lifting

Differential Lift — Where overhead clearance is not sufficient for a direct lift, a differential lift is sometimes an ideal solution. A shorter length, larger diameter cylinder is usually best in this application.

The arrangement illustrated will give a 2:1 mechanical reduction. Size the cylinder with twice the piston area and half the stroke needed for a straight lift of the same load.

Possible side thrust from the cylinder is prevented by running the pulley attached to the cylinder rod in horizontal guides.

In addition to needing less head room, this arrangement allows the cylinder to work on full piston area, and the rod packings are not subjected to high pressure.

Vertical Lifting — Air cylinders used to lift a load must always be sized to exert a force greater than the weight of the load. An air cylinder which exerts a 500 lb. force can support a 500 lb. load, but cannot move it. For normal applications, an air cylinder should develop 25% more thrust than needed to support the load. Twice the force needed to support the load is required for fast operation.

A hydraulic or air/oil system must be used if the cylinder is to be stopped at an intermediate point for loading or unloading.

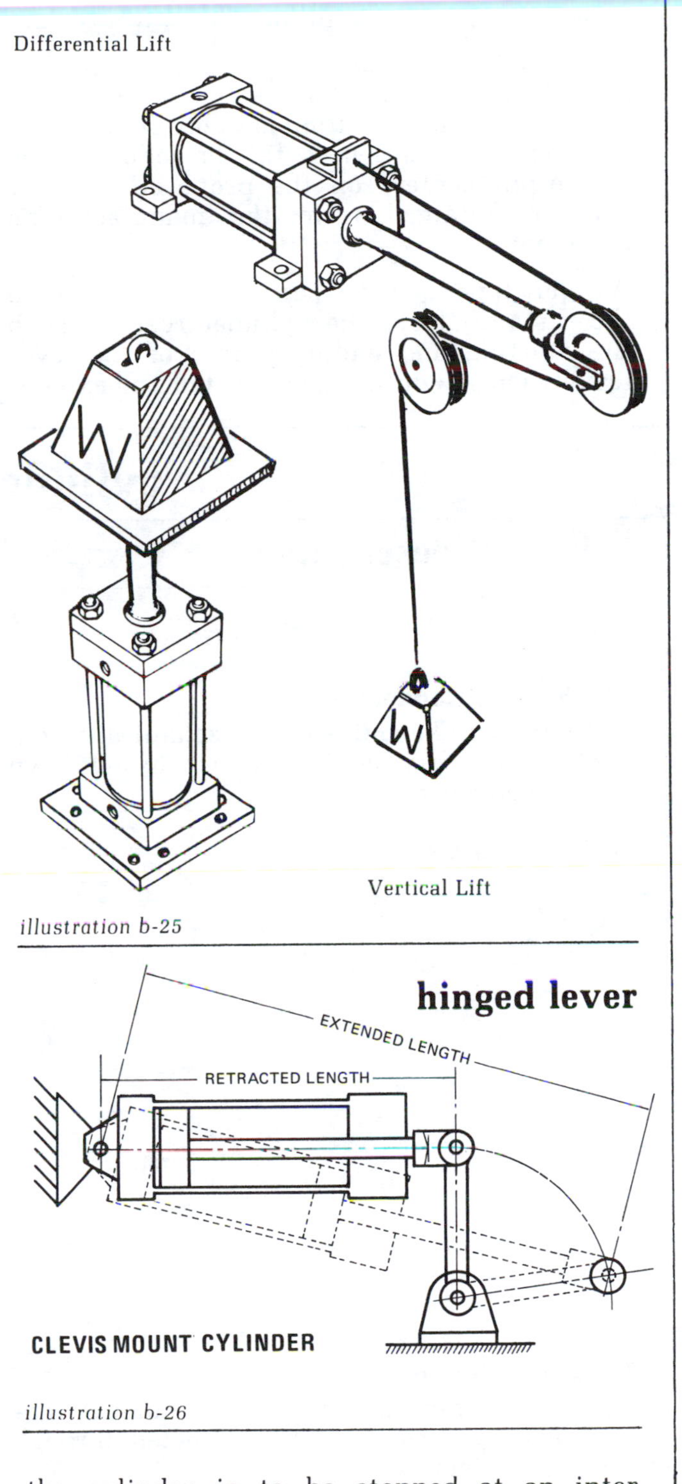

illustration b-25

hinged lever

illustration b-26

Determining Cylinder Stroke for Operating a Hinged Lever Pushing at an Angle by the Scale Layout Method

In all cases a sketch should be made, showing the length and angular travel of the lever, and

mechanical motion continued

showing the mounting position of the cylinder.

If desired, an exact solution can be worked out by mathematics.

For those not familiar with mathematical methods, an easy solution is to lay out all parts to exact scale, either to full or reduced size. Pin-to-pin centers on the proposed cylinder can be obtained from the manufacturer's drawings.

A ruler, tape, or scale can be used to measure the distance from the cylinder rear hinge to the starting and ending points of the lever travel. These will be the retracted and extended cylinder lengths. The travel of the cylinder piston (or stroke) will be the difference between these two measurements.

It may be necessary to experiment with different hinge locations until the best mounting position for the cylinder can be determined.

For a given amount of angular travel, the longest cylinder stroke is required when the cylinder is mounted at right angles to the lever center position as illustrated in this hand book. All other cylinder mounting locations will need a shorter stroke.

cylinder circuits

description	circuit diagram
Mechanical Connection Actuator will keep in step if mechanically connected via cross head (1) which has its own guides or bearings.	*circuit b-1*
Rack and Pinion By using a rack and pinion mechanical connection any error will be corrected through the mechanism. **Note:** *The cross shaft must be of sufficient strength to transmit maximum load.*	*circuit b-2*

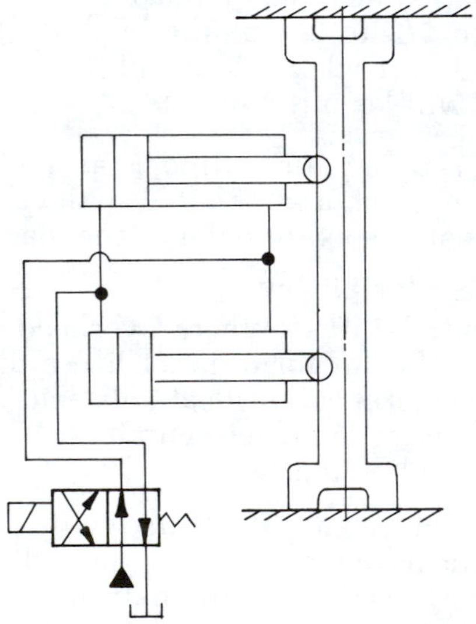

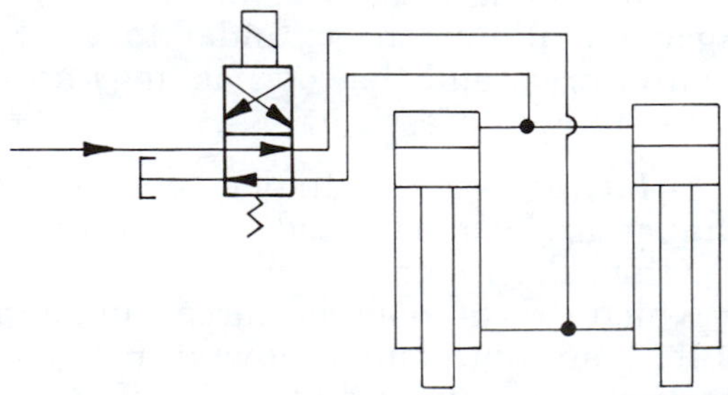

cylinder circuits continued

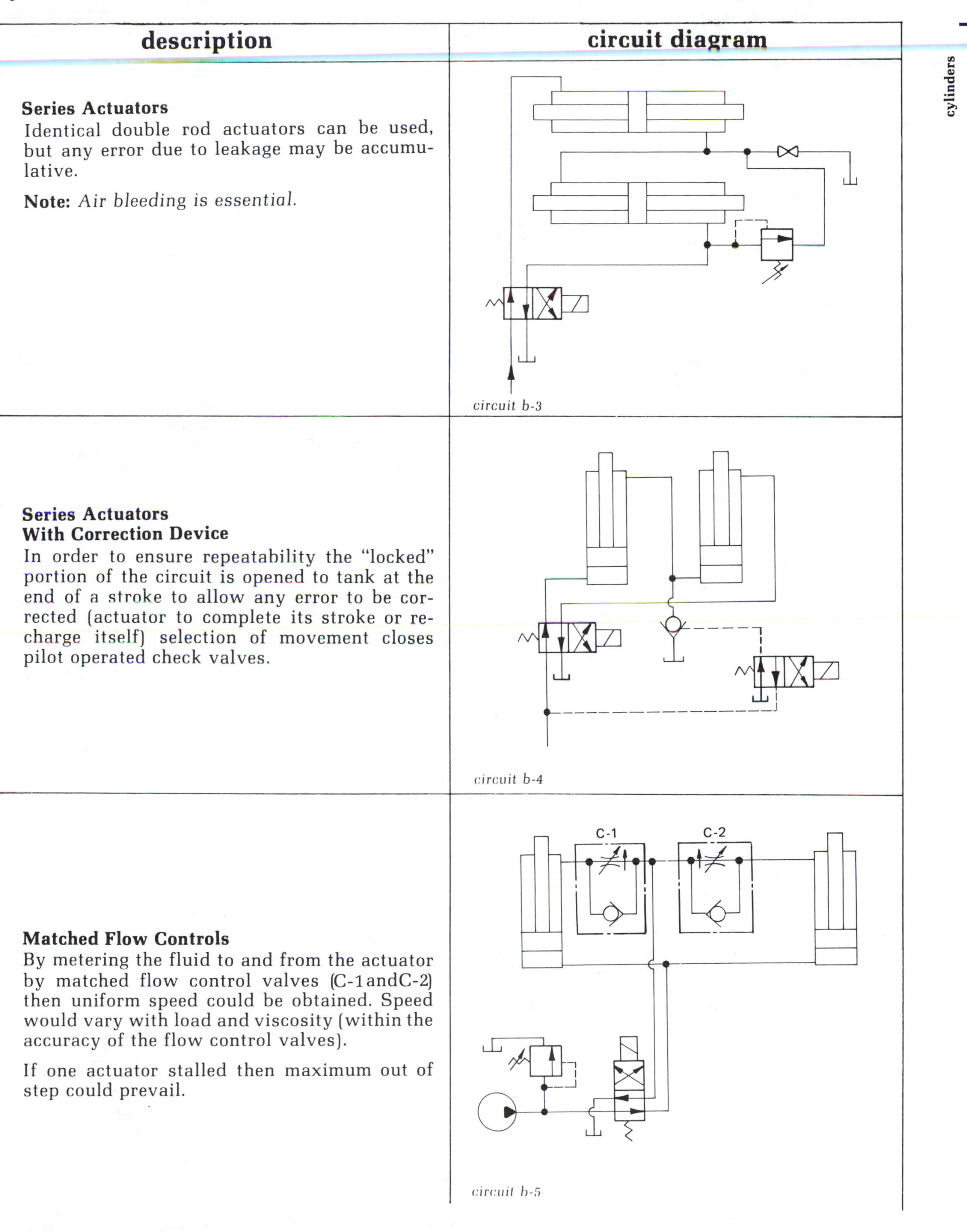

Series Actuators
Identical double rod actuators can be used, but any error due to leakage may be accumulative.

Note: *Air bleeding is essential.*

circuit b-3

Series Actuators
With Correction Device
In order to ensure repeatability the "locked" portion of the circuit is opened to tank at the end of a stroke to allow any error to be corrected (actuator to complete its stroke or recharge itself) selection of movement closes pilot operated check valves.

circuit b-4

Matched Flow Controls
By metering the fluid to and from the actuator by matched flow control valves (C-1andC-2) then uniform speed could be obtained. Speed would vary with load and viscosity (within the accuracy of the flow control valves).

If one actuator stalled then maximum out of step could prevail.

circuit b-5

cylinder circuits continued

description	circuit diagram

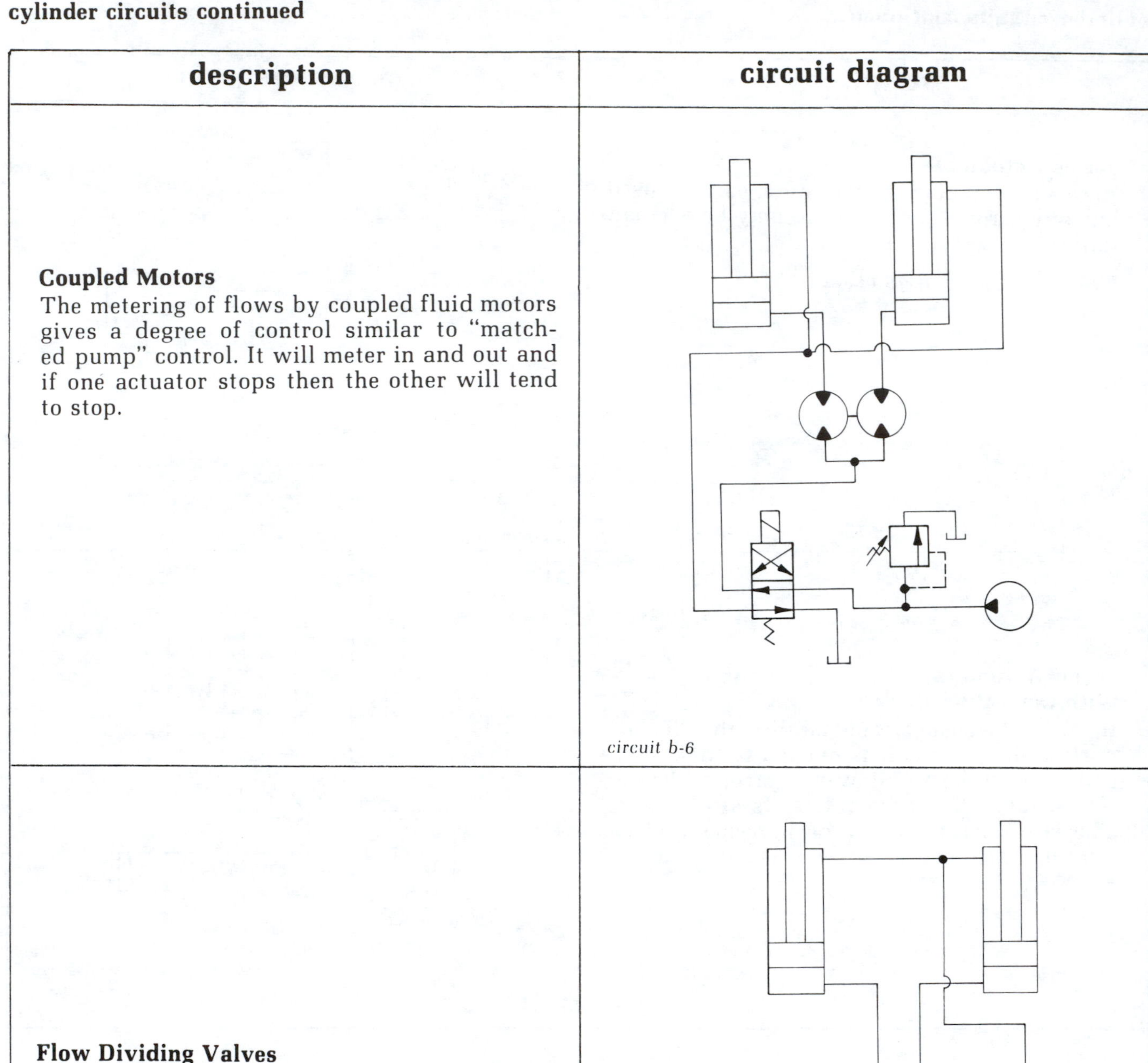

Coupled Motors
The metering of flows by coupled fluid motors gives a degree of control similar to "matched pump" control. It will meter in and out and if one actuator stops then the other will tend to stop.

circuit b-6

Flow Dividing Valves
These valves can be used to select any "combination" or division. (i.e. 1:1, 2:1, 3:1). It requires check valves around it to give free flow in opposite direction.

It can only be used on flows equal to or above the valve setting. No positive control is possible if flow drops. Internal leakage permits actuator to complete stroke if necessary. This means that maximum force out of balance can prevail.

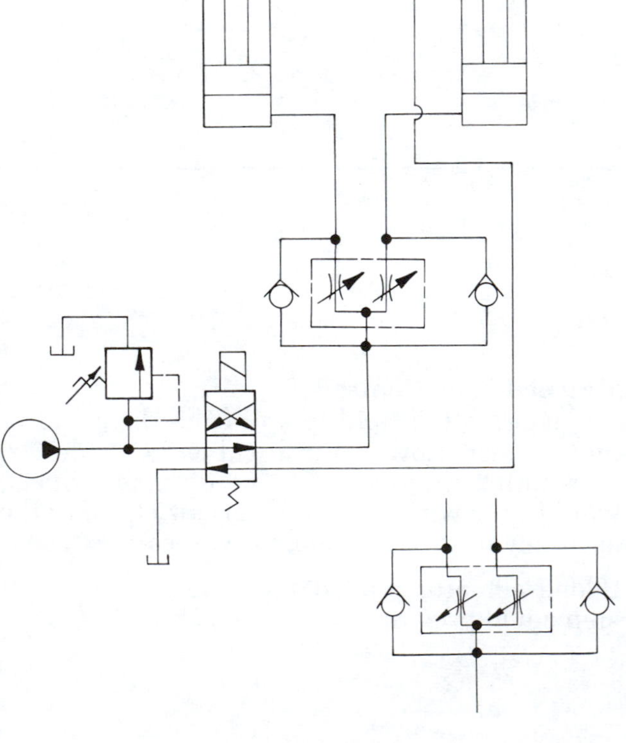

circuit b-7

regenerative circuits

The basic principle of regneration is, by use of suitable valving, connect the rod end of the cylinder with the blind end, so the oil which normally would flow to tank from the rod end will join the pump oil causing the cylinder to advance at an increased rate of speed.

During regeneration equal pressure is applied to both sides of the cylinder piston, the net thrust delivered by the rod will be the same as if the effective pressure were applied only to the rod area. So, force during regeneration, calculations are: thrust = system pressure × rod area (square inches).

The return oil from the rod end fills up an equivalent volume on the cap side of the piston, so the pump volume need only fill up a space equivalent to the volume of the rod. To calculate rod speed, take the pump volume, in cubic inches per minute, and divide the rod area (in square inches) into the cubic inches per minute. This will give you speed in inches per minute.

Calculate rod speed as in cylinder speed. To find oil flow at Point "A" in **circuit b-10,** calculate how much oil will have to flow to make the piston travel at this calculated speed. This will be speed (inches per minute) × piston area (square inches). Convert to GPM by dividing by 231 (cubic inches to the gallon). To find the oil flow at Point "B" in **circuit b-10,** take the result and subtract the pump volume from it.

EXAMPLE: *Assume a system pressure of 1500 PSI, pump volume of 9 GPM, piston diameter 8", rod diameter 6".* Force = 28.27 square inches (rod area)× *1500 PSI = 42,405 lbs. Speed = 9 GPM × 231 (cu. ins./gal.) ÷ 28.27 square inches = 74" per minute. Oil flow at "A" = 74 × 50.3 (piston area) ÷ 231 = 16.1 GPM. Oil flow at "B" = 16.1 - 9 GPM (from pump) = 7.1 gpm.*

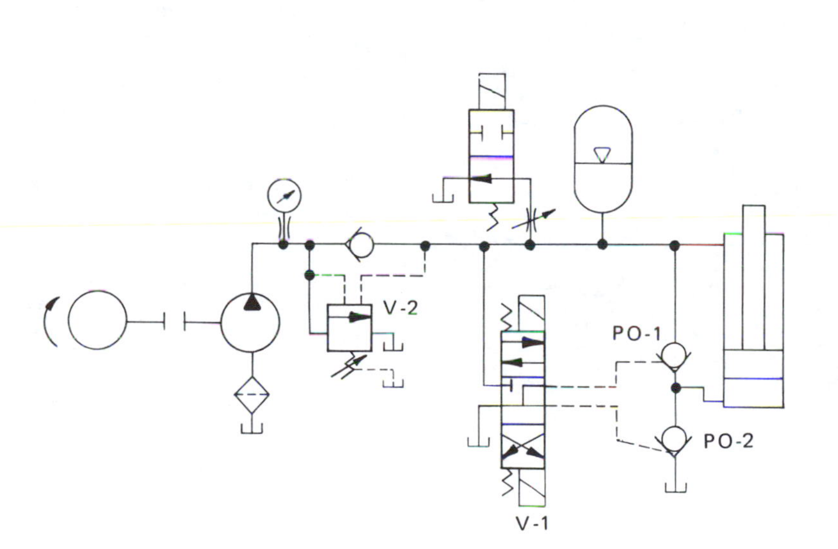

circuit b-8

The regeneration in this circuit is accomplished by energizing (V^1) which opens pilot operated check valve (PO1) or returns the cylinder by opening (PO2) when the accumulator is charged the system is unloaded through the differential unloading relief valve. (V^2)

Consideration When Using Regenerative Circuits

1. In a regeneration circuit, the force generated will be that of system pressure acting only on the rod area. The remainder of the piston area is concelled out by an equal and opposing pressure acting on the piston area on the rod side.

2. When the full thrust of the cylinder is required, the pressure on the rod end must be valved and connected to tank.

3. Regeneration is mainly used with large rod cylinders, 2:1 (piston to rod area ratio). If used with small rod cylinders, the extending speed is sometimes too great, the thrust too small, and the return speed is too slow.

4. When a 2:1 ratio cylinder is retracting, dis-

regenerative circuits continued

charge oil velocity from the blind end of the cylinder is twice the velocity of the oil from the pump. Select pipe and valving large enough to handle this volume.

5. With 2:1 ratio cylinders (usually used in regenerative circuits), pressure intensification occurs in the rod end during the forward stroke if the discharge oil is restricted or blocked. A safety relief valve should be installed in the rod end if intensification could endanger the cylinder or plumbing.

6. The regenerative portion of the cycle is usually for moving the machine member rapidly into working position. The actual thrust required is slight.

7. When large rod cylinders are used in regenerative circuits, the oil level in the reservoir will fluctuate more than for small rod cylinders. Make sure the reservoir is large enough so the oil level will not drop dangerously low as to cavitate the pump as the cylinder extends.

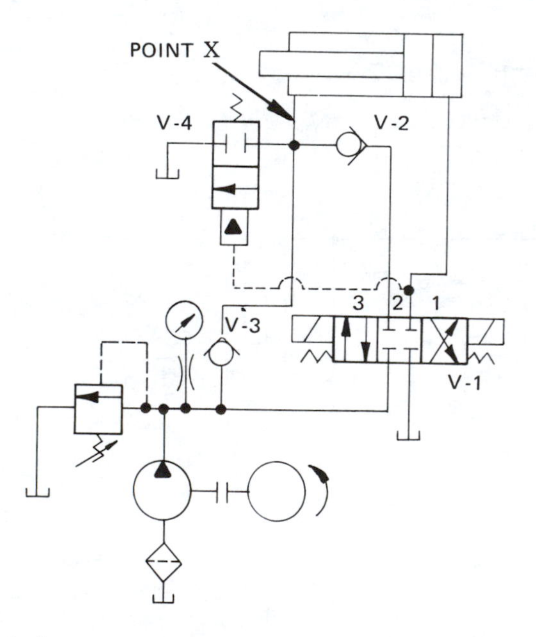

circuit b-9

Shifting the 4-way to Position 1 causes the cylinder to start extending. A regenerative circuit is one which discharge oil from rod end passes through check valve V-3 and joins system oil to the cap end of the cylinder. Circuit regenerates until or unless work resistance builds up on pilot of V-4 causing it to shift. Rod oil then goes directly to tank and circuit becomes nonregenerative and capable of developing full tonnage.

Shifting the 4-way to Position 3 causes the cylinder to retract, and for pump oil to pass through check valve V-2. Valve V-4, with no pilot pressure, closes, preventing pump oil from by-passing to tank.

If it is desired to keep the cylinder from moving by gravity, a counterbalance valve should be added at Point X. The pilot valve must be a spool-type having better throttling character than a poppet type.

The Directional Valve has a closed center spool. It could be any other spool center which isolates cylinder ports from the pump in the neutral position such as a tandem center spool for unloading.

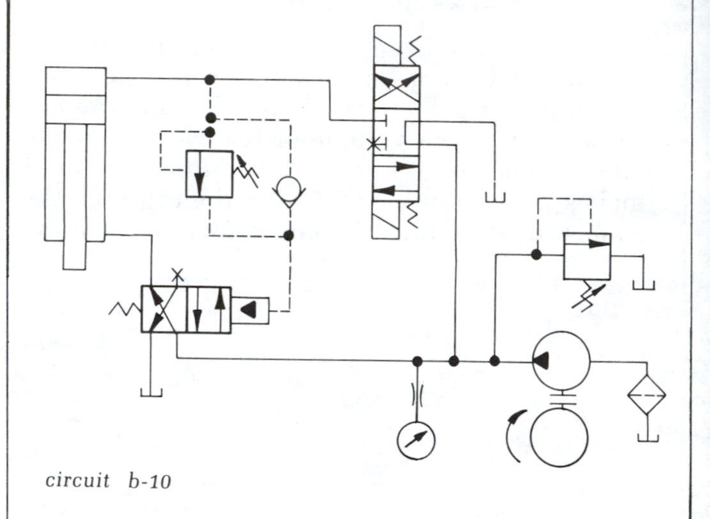

circuit b-10

In b-10, the 4 way valve is standard 4-way with "CYL 2" port plugged and cylinder connected to "CYL 1" port. A1/4" sequence valve with internal free flow check is connected for internal pilot, external drain. A standard 4-way, pilot operated, with "CYL 2" plugged, can have about half the capacity of the 3 position 4-way valve.

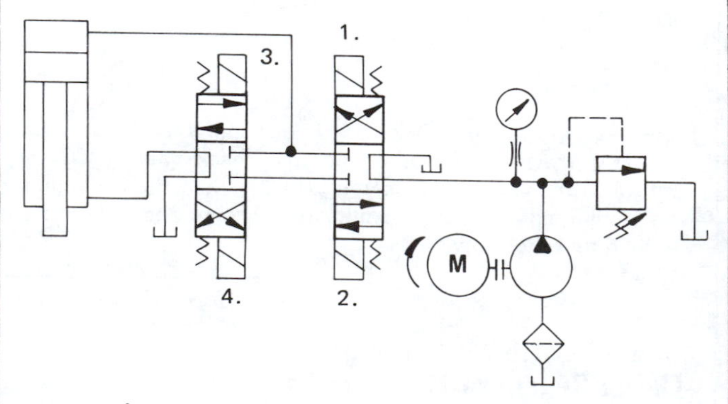

circuit b-11

In b-11 solenoid control of regenerative forward, normal forward, reverse, or stop, in the cylinder stroke.

Energize Solenoids 1 and 3 for regenerative forward, Solenoid 1 only for normal forward,

regenerative circuits continued

Solenoids 2 and 4 for retract. De-energize all solenoids for stop.

In b-12 the spool of the directional valve has both cylinder ports connected to the pressure port when it is in center position. This is the regenerative position. The side position of the directional valve are normal extend, and retract.

A regenerative circuit is used to cause a cylinder to advance faster than it could with the pump volume alone. It can **only** be used to extend a cylinder — never to retract it.

circuit b-12

operating principles and construction

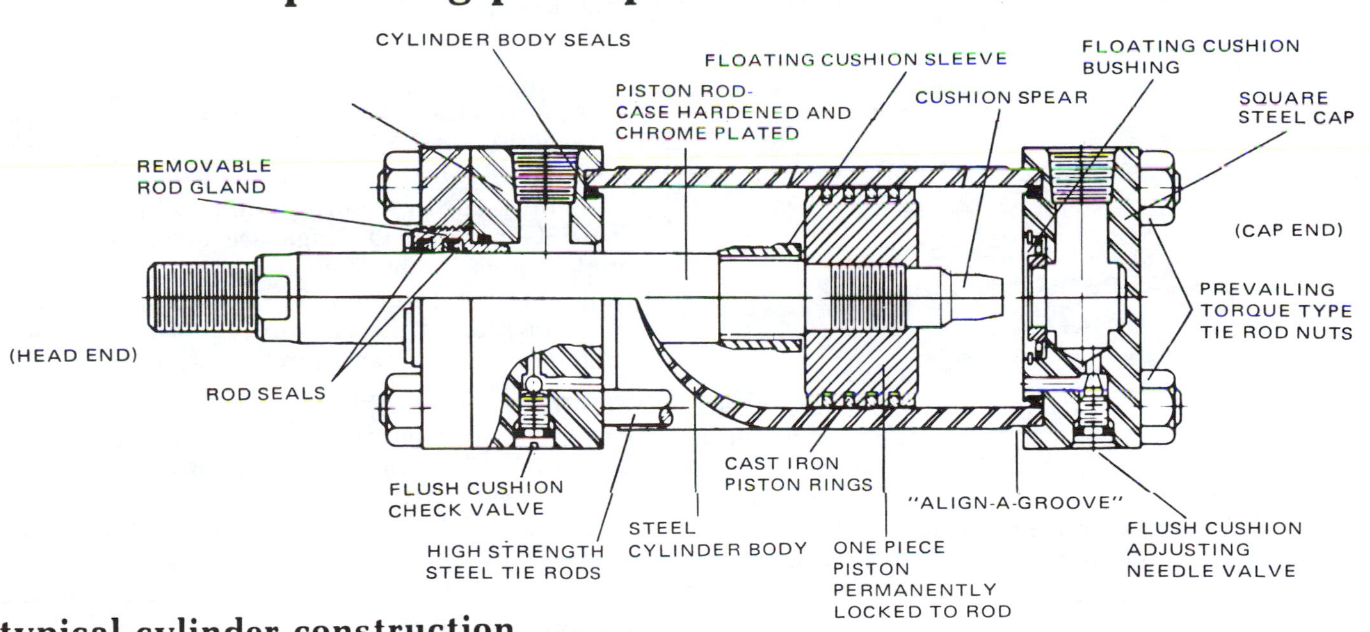

typical cylinder construction

illustration b-27

cylinder operation

Cylinders are used in the majority of applications to convert fluid energy into straight line motion. For this reason, they are often called linear actuators.

Cylinders are manufactured in a variety of diameters, stroke lengths, and mounting styles. They may be classified, according to construction, into four types; tie-rod, threaded, welded, and flanged. Cylinders are also made using retaining rings.

$$\text{Area} = \frac{\pi D^2}{4} \text{ or Area} = .7854 \times D^2$$

When calculating force developed on the return stroke, pressure does not act on the rod area of the piston, therefore the rod area must be subtracted from the total piston area.

operating principles and construction continued

basic construction

The major components of a cylinder are the head, cap, tube tie rods, piston, piston rod, rod bearing and seals.

Cylinder Heads and **Caps** are usually made from rolled steel or cast iron. Some are also from aluminum or bronze.

Cylinder Tubes are usually brass, steel or aluminum. The inside, and sometimes the outside, is plated or anodized to improve wear characteristics and reduce corrosion. In some applications, cylinder tubes can also be made of fibreglass.

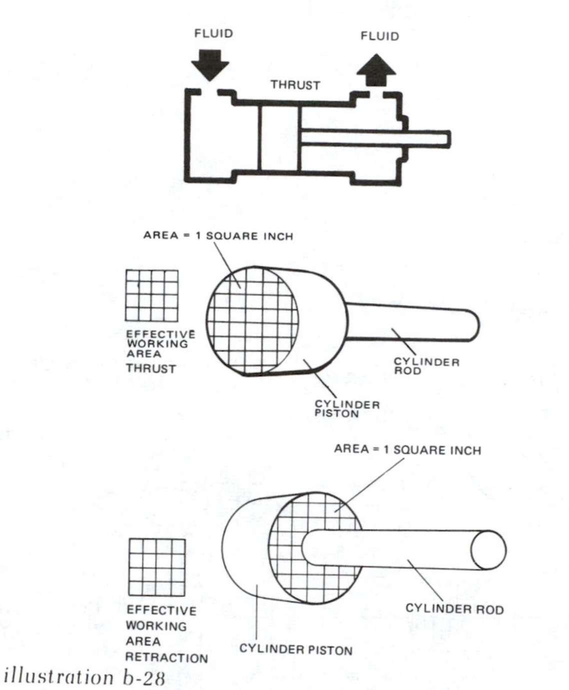

illustration b-28

Pistons vary in design and materials used. Most are made of cast iron or steel. Several methods of attaching the piston to the rod are used. Cushions, are an available option on most cylinders and most often, can be added with no change in envelope dimensions.

Piston Rods are generally high strength steel, case-hardened, ground, polished and hard chrome plated for wear and corrosion resistance. Corrosive atmosphere conditions usually require rods of stainless steel, which may be chrome plated for wear resistance.

Rod Glands or **Bearings** are used on the head end of most industrial cylinders to support the piston rod as it travels back and forth. The gland also acts as a retainer for the rod packing or seals. Most are made of ductile iron or bronze and usually are removable without disassembling the entire cylinder.

The gland usually contains a piston rod wiper or scraper on the outboard side to remove dirt and contamination from the rod, and prevent foreign material from being drawn into the packings. A primary seal is used to seal the cylinder pressure.

The gland usually contains a piston rod wiper or scraper on the outboard side to remove dirt and contamination from the rod, and prevent foreign material from being drawn into the packings. A primary seal is used to seal the cylinder pressure.

Seals are generally made from Nitrile or fluoro carbon elastomers, polyurethane, lether or Teflon®. Commonly used seal shapes are shown in **illustration b-8.** The Lipseal® shape is commonly used for both piston and piston rod seals. Generally, O-Rings are used for static applications such as head to tube, piston to rod, and head to gland. Cup or V-packings are used for sealing piston and piston rod. Piston rings are usually cast iron.

Tie-Rods are usually high tensile steel with either cut or rolled threads, prestressed during assembly. Prestressing with proper torque prevents separation of parts when subjected to pressure and reduces the need for locknuts, although locknuts are sometimes used.

fundamental cylinders

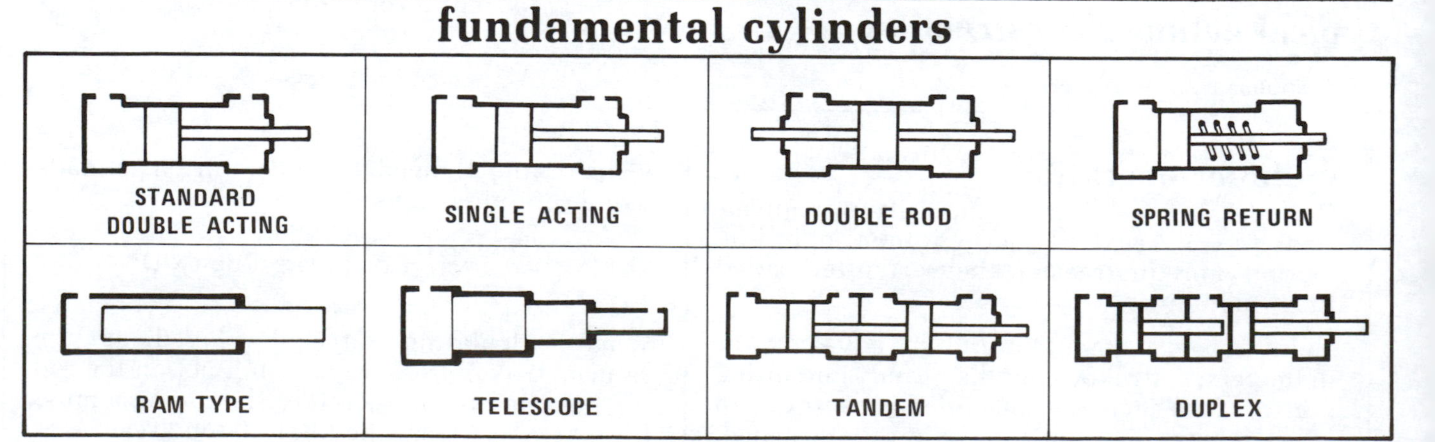

illustration b-29

fundamental cylinders continued

Standard Double-Acting. Power stroke is in both directions and is used in the majority of applications.

Single-Acting Cylinder. When thrust is needed in only one direction, a single-acting cylinder may be used. The inactive end is vented to atmosphere through a breather/filter for pneumatic applications, or vented to reservoir below the oil level in hydraulic application.

Double-Rod Cylinders. Used when equal displacement is needed on both sides of the piston, or when it is mechanically advantageous to couple a load to each end. The extra end can be used to mount cams for operating limit switches, etc.

Spring Return, Single-Acting Cylinders. Usually limited to very small, short stroke cylinders used for holding and clamping. The length needed to contain the return spring makes them undesirable when a long stroke is needed.

Ram Type, Single-Acting Cylinders. Containing only one fluid chamber, this type of cylinder is usually mounted vertically. The weight of the load retracts the cylinder. They are sometimes known as "displacement cylinders", and are practical for long strokes.

Telescoping Cylinders. Available with up to 4 or 5 sleeves; collapsed length is shorter than standard cylinders. Available either single or double-acting, they are relatively expensive compared to standard cylinders.

Tandem Cylinders. A tandem cylinder is made up of two cylinders mounted in line with pistons connected by a common piston rod and rod seals installed between the cylinders to permit double acting operation of each. Tandem cylinders allow increased output force when mounting width or height are restricted.

Duplex Cylinders. A duplex cylinder is made up of two cylinders mounted in line with pistons not connected and with rod seals installed between the cylinders to permit double acting operation of each. Cylinders may be mounted with piston rod to piston (as shown) or back to back and are generally used to provide three position operation.

cylinder application

causes of cylinder failure

Standard cylinders are not designed to take piston rod side loading. They must be carefully and accurately mounted so the rod is not placed in a bind at any part of the stroke. In many cases the cylinder must have a clevis or trunnion mount to allow it to swing as the direction of the load changes. Use guides on the load mechanism, if necessary, to assure that no side load is transmitted to the cylinder rod.

Rod Buckling

Column failure, or the buckling of the rod, may occur if the cylinder stroke is too long in relation to the rod length to rod diameter. The **piston rod-stroke selection graph 1** [page b-5] will give the minimum safe rod size for normal applications.

Rod Bearing Failure

Rod bearing failures usually occur when the cylinder is at maximum extension. Failure occurs most often on hinge or trunnion mount cylinders, in which the rear support point is located considerably behind the rod bearing. Where space permits, order cylinders with longer stroke than actually needed. Do not permit the piston to approach close to the front end under load.

Stop Tube

On applications where it is necessary to allow the piston to "bottom out" on the front end, cylinders may be ordered with a stop tube. The stop tube should especially be considered on long strokes if the length between supports exceeds 10 times the rod diameter.

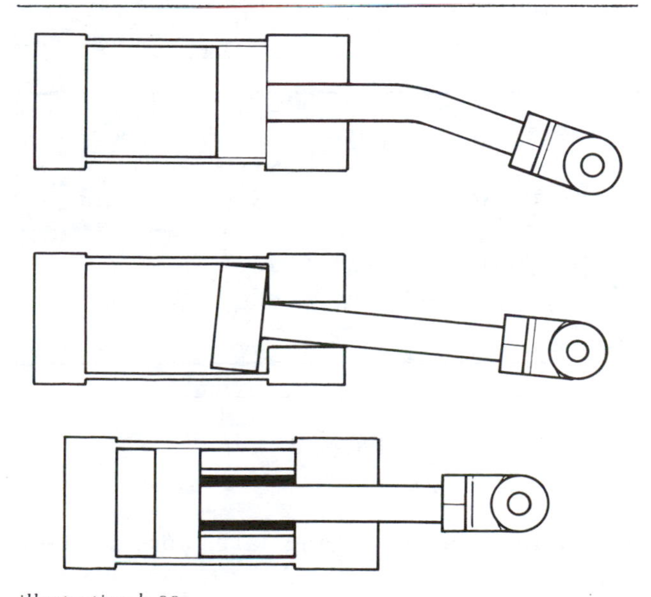

illustration b-30

cylinder application continued

Tension and Compression Failures
Standard cylinders are designed with sufficiently large piston rods. They will never fail either in compression or tension, if the cylinder is operated within the pressure rating of the manufacturer.

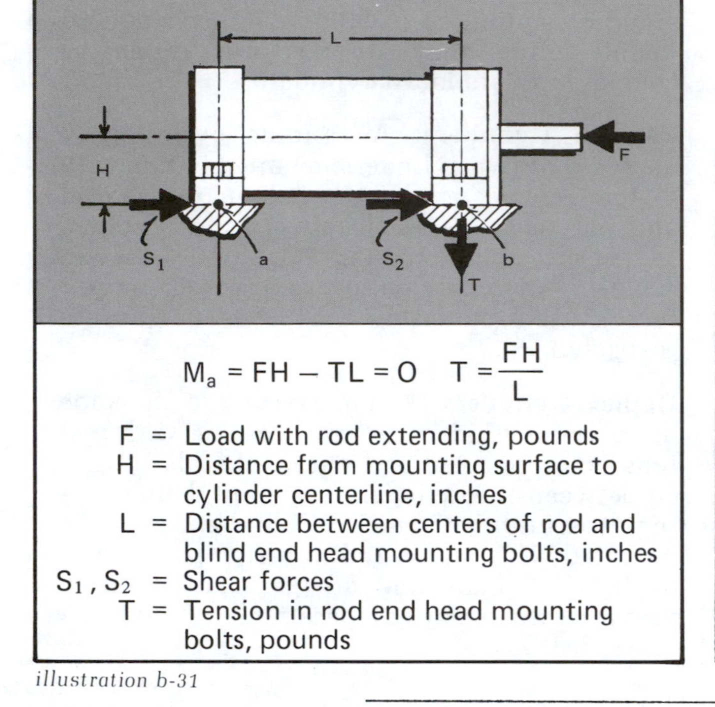

illustration b-31

short noncenterline mounted cylinders

Relatively short, fixed noncenterline mounted cylinders can subject mounting bolts to large tension forces which, in combination with shear forces, overstress the bolts. Condition shown for a cylinder rod extending against a load illustrates manner in which tension is developed in rod end head mounting bolts. Note from formula that, for a constant load, tension increases with decrease in distance (L) between rod and blind end head mounting bolt centers. Similar analysis can be applied to determine stressing of blind end head mounting bolts by substituting for load (F) the rod tension force developed during retraction, and by equating moments about point (B) to zero.

noncenterline cylinders

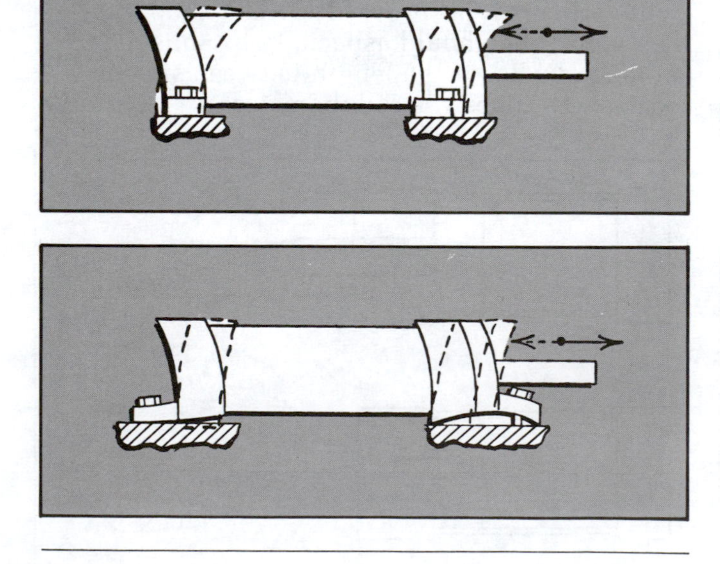

illustration b-32

In applications involving large forces, cylinders with noncenterline type mountings tend to sway under load.

noncenterline cylinder/support

ORIGINAL SUPPORT

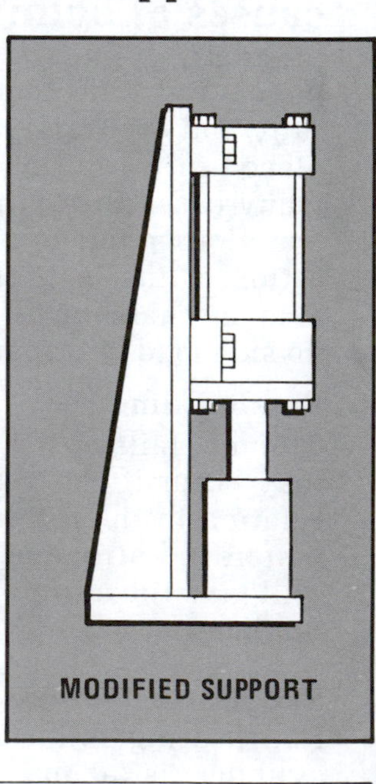

illustration b-33

Use of noncenterline type cylinder mountings may require strengthening of machine members to resist bending under load.

cylinder application continued

long cylinders/spreader type tie rods

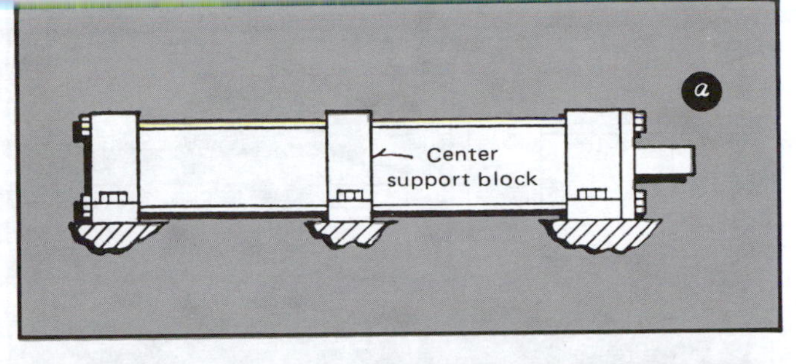

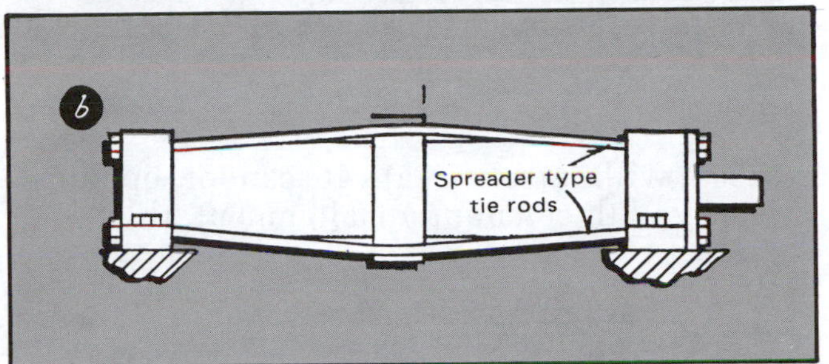

Relatively long cylinders with fixed mountings often require additional supports to prevent excessive sag or vibration. Arrangement at (a) uses an extra mounting block located midway along cylinder body. At (b), spreader-type tie rods are used to increase rigidity in center portion of cylinder. Where one end of a cylinder must be overhung, as at (c), an additional supporting member can be provided.

illustration b-34

cylinders using dowel pins

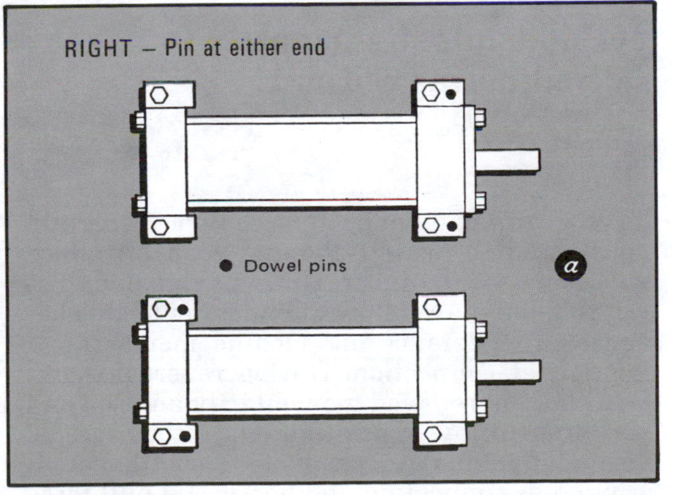

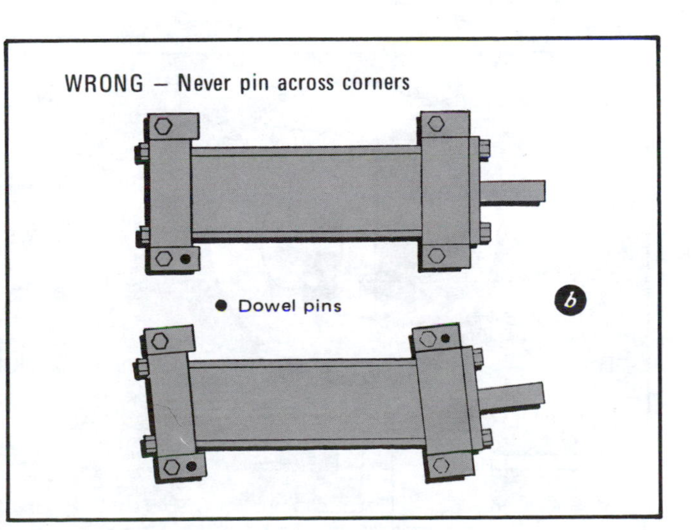

illustration b-35

Cylinders that are pinned in place to help secure alignment and resist shock loads should be pinned at either end, (a). Choice of end depends upon direction of major shock load. If dowel pins are used across corners, (b), cylinder may be warped by operating temperatures and pressures or shock loads.

cylinder application continued

cylinder misalignment

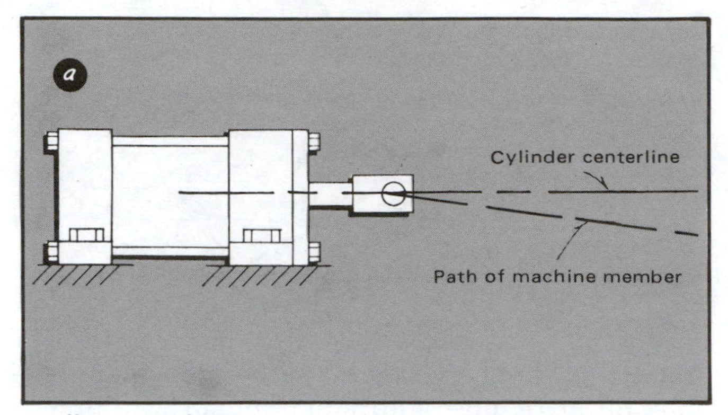

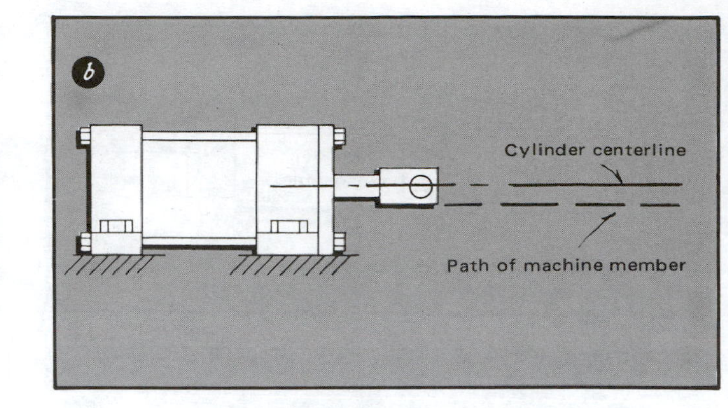

illustration b-36

Misalignment of fixed-mounted cylinders with work slides can be of two types. The cylinder can tolerate slight misalignment that increases with stroke, (a). It cannot operate properly with constant misalignment, (b).

cylinder misalignment correction

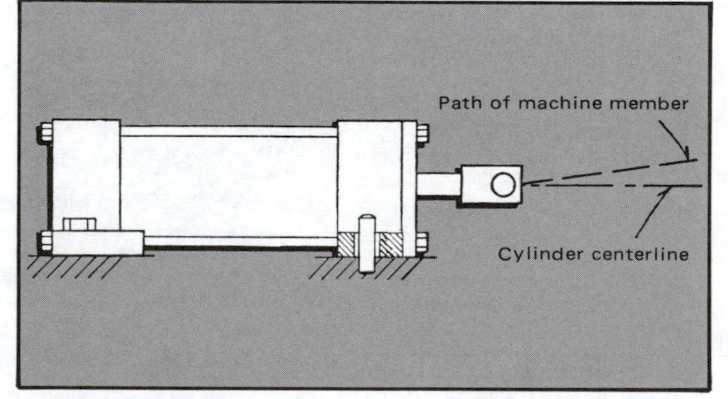

illustration b-37

Sometimes a relatively long-stroke cylinder can be made somewhat self-aligning by allowing the rod end head to float. In arrangement shown, holes in the side lugs at the rod end cylinder head permit some movement of front of cylinder with respect to dowel pins. Cylinder body flexes slightly about fixed rear mounting.

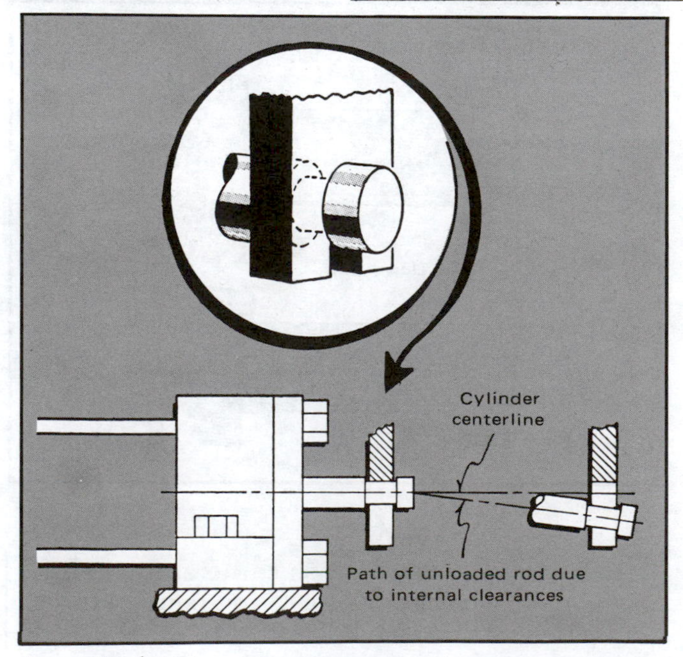

illustration b-38

cylinder misalignment/ self-adjusting rod end

Type of rod end connection shown—generally considered to be self-aligning—can introduce side load on a cylinder. During extension, piston rod end may follow cylinder centerline because of resistance of machine member and friction at connection. During reversal, however, load is relieved momentarily and rod end can drop to lower position. On return stroke, side load—equal to product of coefficient of friction at connection and horizontal pull force —is set up.

cylinder application continued

shear key mounting

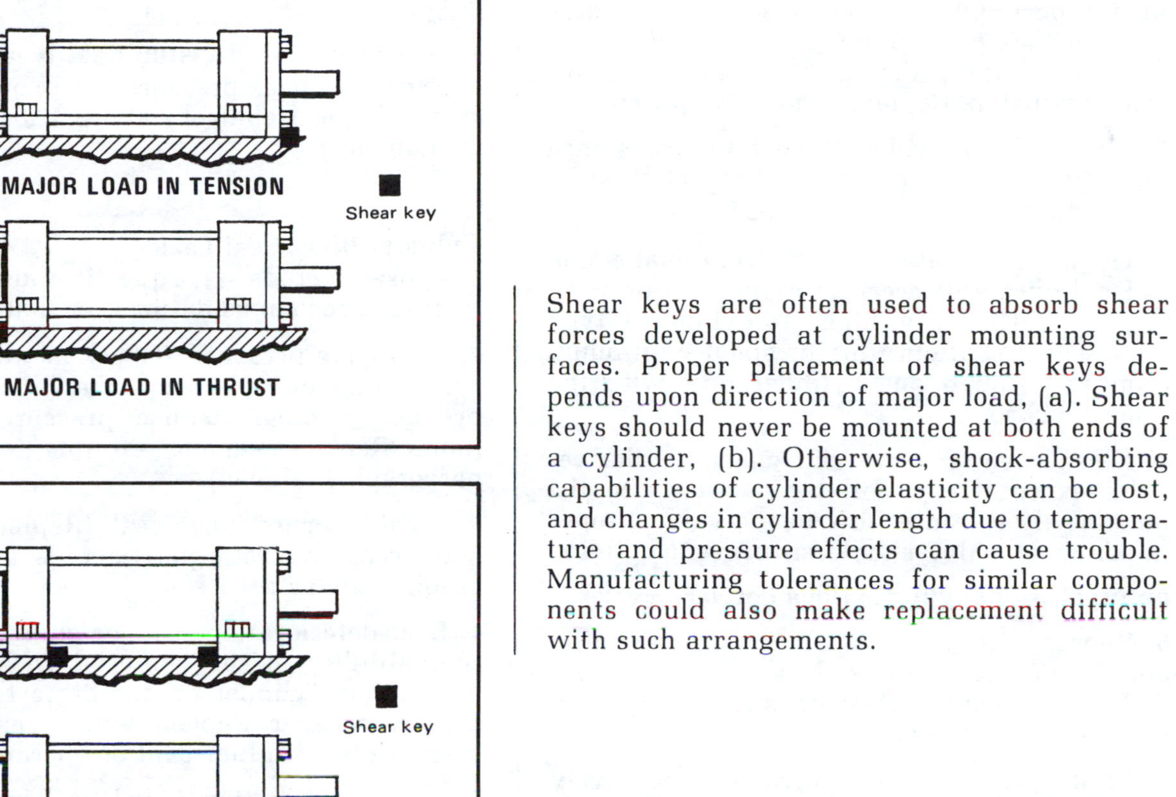

illustration b-39

Shear keys are often used to absorb shear forces developed at cylinder mounting surfaces. Proper placement of shear keys depends upon direction of major load, (a). Shear keys should never be mounted at both ends of a cylinder, (b). Otherwise, shock-absorbing capabilities of cylinder elasticity can be lost, and changes in cylinder length due to temperature and pressure effects can cause trouble. Manufacturing tolerances for similar components could also make replacement difficult with such arrangements.

trunnion mounting

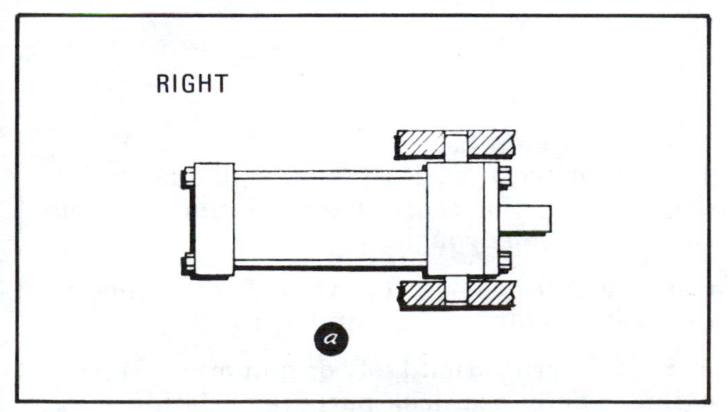

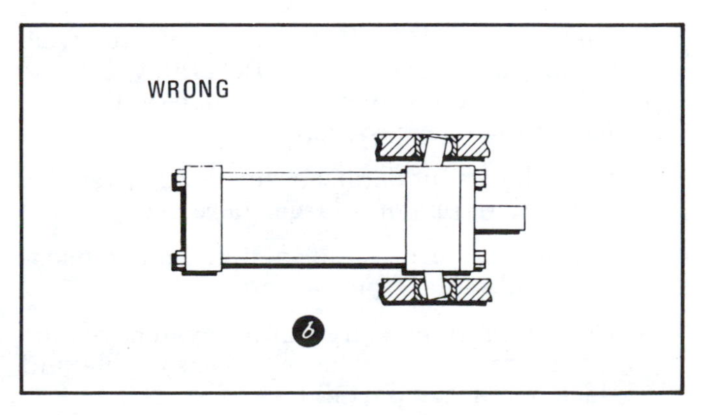

illustration b-40

Trunnions for pivot-mounting of cylinders are generally designed to resist shear loads only, (a). Use of self-aligning bearings that have small bearing areas acting at a distance from the trunnions and the cylinder head introduce bending forces that can overstress the trunnions, (b).

trouble shooting cylinders

Cylinder Drifts

1. Piston seal leak. (Pressurize one side of cylinder piston and disconnect fluid line at opposite port. Observe leakage. One to three cubic inches per minute is considered normal for piston rings. Virtually no leak with soft seals on piston. Replace seals as required.)

2. Other circuit leaks. (Check for leaks thru operating valve and correct. Correct leaks in connecting lines.)

3. Open center valve with conventional single rod cylinder will creep if restriction on tank port is sufficiently high. Use tandem type valve spool configuration or spool with pump dumped through one cylinder port with the other blocked.

4. Closed center valve can cause similar results except creep will be according to the amount of clearance flow in the valve. **Proper notching of valve spool can prevent building up pressure in cylinder lines between cycles.**

5. Spools with pressure blocked and cylinder ports completely relaxed will also prevent drift if the moving element is not affected by gravity or vibration.

6. Pilot operated check valves can positively lock fluid in cylinder lines. Care must be exercised to insure adequate pilot pressure when rod differential may cause intensification.

Cylinder Fails to Move the Load When Valve is Actuated

1. Pressure too low. (Check pressure at cylinder to make sure it is to circuit requirements.)

2. Piston seal leak. (Operate valve to cycle cylinder and observe fluid flow at valve exhaust ports at end of cylinder stroke. Replace seals if flow is excessive.)

3. Cylinder is undersized for load. (Replace cylinder with one of a larger bore size.)

4. Piston rod broken at piston end. (Disassemble and replace piston rod.)

5. Contamination in hydraulic system resulting in scored cylinder bore. (Disassemble and replace necessary parts.)

Erratic or Chatter Operation

1. Excessive friction due to load misalignment. (Correct cylinder to load alignment.)

2. Cylinder sized too close to load requirements. (Reduce load or install larger cylinder.)

3. Large difference between static and kinetic friction. (Install speed control valves to provide back pressure to control stroke.)

Excessive or Rapid Piston Seal Wear

1. Excessive back pressure due to over-adjustment of speed control valves. (Correct valve adjustment.)

Cylinder Body Seal Leak

1. Loose tie rods. (Torque tie rods to manufacturer's recommendations for that bore size.)

2. Excessive pressure. (Check maximum pressure rating on cylinder nameplate or vendor product catalog. Reduce pressure to rated limits. Replace seal and retorque tie rods as in paragraph 1. above.)

3. Pinched or extruded seal. (Replace cylinder body seal and retorque tie rods as in paragraph 1. above.)

4. Seal deterioration [soft or gummy]. (Check compatibility of seal material with lubricant used if air cylinder or operating fluid if hydraulic cylinder. Replace with a seal which is compatible with lubricant or operating fluid.)

5. Seal deterioration [hard or loss of elasticity]. (Usually due to exposure to elevated temperature. Shield by shielding cylinder from heat source. Replace seal as in paragraph 1. above.)

6. Seal deterioration [loss of radial squeeze due to flat spots or wear on O.D. or I.D.]. (Can occur as normal wear due to high cycle rate or length of service. Replace seal as in paragraph 1.)

Rod Gland Seal Leak

1. Torn or worn seal. (Examine piston rod for dents, gouges or score marks. Replace piston rod if surface is rough.)

Check gland bearing for wear. If clearance is excessive, replace gland and seals.

2. Seal deterioration [soft or gummy]. (Repeat cylinder body seal leak paragraph listing 4.)

3. Seal deterioration [hard or loss of elasticity]. (Repeat cylinder body seal leak paragraph listing 5.)

4. Seal deterioration [flat spots on I.D.]. (Repeat cylinder body seal leak paragraph listing 6.)

electrical devices

electric motor horsepower c-2
dimensions for foot-mounted a-c motors
with single straight shaft extension . c-3
standard enclosures for electric motors c-4
three-phase motor design c-4
motor starter, conduit and wire size . c-5
solenoid cycle rate c-6
wet armature solenoid c-6
50 and 60 cycle solenoids c-7
solenoid force and voltage c-8
trouble shooting solenoid valves c-9

electric motor horsepower

ELECTRIC MOTOR HORSEPOWER REQUIRED TO DRIVE A HYDRAULIC PUMP

GPM	100 PSI	200 PSI	250 PSI	300 PSI	400 PSI	500 PSI	750 PSI	1000 PSI	1250 PSI	1500 PSI	2000 PSI	2500 PSI	3000 PSI
1/2	.04	.07	.09	.11	.14	.18	.26	.35	.44	.53	.70	.88	1.10
1	.07	.14	.18	.21	.28	.35	.52	.70	.88	1.05	1.40	1.76	1.92
1-1/2	.10	.21	.26	.31	.41	.52	.77	1.03	1.29	1.55	2.06	2.58	3.09
2	.14	.28	.35	.42	.56	.70	1.04	1.40	1.76	2.10	2.80	3.53	4.20
2-1/2	.17	.34	.43	.51	.69	.86	1.29	1.72	2.15	2.58	3.44	4.30	5.14
3	.21	.42	.53	.63	.84	1.05	1.56	2.10	2.64	3.15	4.20	5.28	6.30
3-1/2	.24	.48	.60	.72	.96	1.20	1.80	2.40	3.00	3.60	4.80	6.00	7.20
4	.28	.56	.70	.84	1.12	1.40	2.08	2.80	3.52	4.20	5.60	7.04	8.40
5	.35	.70	.88	1.05	1.40	1.75	2.60	3.50	4.40	5.25	7.00	8.80	10.50
6	.42	.84	1.05	1.26	1.68	2.10	3.12	4.20	5.28	6.30	8.40	10.56	12.60
7	.49	.98	1.23	1.47	1.96	2.45	3.64	4.90	6.16	7.35	9.80	12.32	14.70
8	.56	1.12	1.40	1.68	2.24	2.80	4.16	5.60	7.04	8.40	11.20	14.08	16.80
9	.62	1.24	1.55	1.86	2.48	3.10	4.65	6.18	7.73	9.28	12.40	15.56	18.58
10	.70	1.40	1.75	2.10	2.80	3.50	5.20	7.00	8.80	10.50	14.00	17.60	21.00
11	.77	1.54	1.93	2.31	3.08	3.85	5.72	7.70	9.68	11.50	15.40	19.36	23.10
12	.84	1.68	2.10	2.52	3.36	4.20	6.24	8.40	10.50	12.60	16.80	21.00	25.20
13	.89	1.78	2.23	2.67	3.56	4.45	6.68	8.92	11.20	13.40	17.80	22.40	26.72
14	.96	1.92	2.40	2.88	3.84	4.80	7.20	9.60	12.00	14.40	19.20	24.00	28.80
15	1.05	2.10	2.63	3.15	4.20	5.25	7.80	10.50	13.20	15.70	21.00	26.40	31.50
16	1.10	2.20	2.75	3.30	4.40	5.50	8.25	11.00	13.80	16.50	22.00	27.60	33.00
17	1.17	2.34	2.93	3.51	4.68	5.85	8.78	11.70	14.60	17.60	23.40	29.20	35.10
18	1.26	2.52	3.15	3.78	5.04	6.30	9.35	12.60	15.80	18.90	25.20	31.60	37.80
19	1.30	2.60	3.25	3.90	5.20	6.50	9.75	13.00	16.30	19.50	26.00	32.60	39.00
20	1.40	2.80	3.50	4.20	5.60	7.00	10.40	14.00	17.60	21.00	28.00	35.20	42.00
25	1.75	3.50	4.38	5.25	7.00	8.75	13.10	17.50	21.90	26.20	35.00	43.80	52.50
30	2.10	4.20	5.25	6.30	8.40	10.50	15.60	21.00	26.40	31.50	42.00	52.80	63.00
35	2.45	4.90	6.13	7.35	9.80	12.20	18.40	24.50	30.60	36.70	49.00	61.20	73.50
40	2.80	5.60	7.00	8.40	11.20	14.00	20.80	28.00	35.20	42.00	56.00	70.40	84.00
45	3.15	6.30	7.87	9.45	12.60	15.80	23.60	31.50	39.40	47.30	63.00	78.80	94.50
50	3.50	7.00	8.75	10.50	14.00	17.50	26.00	35.00	44.00	52.50	70.00	88.00	105.00
55	3.85	7.70	9.63	11.60	15.40	19.30	28.60	38.50	48.40	57.80	77.00	96.80	115.50
60	4.20	8.40	10.50	12.60	16.80	21.00	31.20	42.00	52.80	63.00	84.00	105.60	126.00
65	4.55	9.10	11.40	13.60	18.20	22.80	33.80	45.50	57.20	68.20	90.00	114.40	136.50

table c-1

The **table c-1,** above is based on a pump efficiency of 85%, and is calculated from the formula:

$$HP = GPM \times PSI \div (1714 \times .85)$$

As horsepower varies directly with flow or pressure, multiply proportionately to determine values not shown. For example, at 4000 PSI, multiply 2000 PSI values by 2.

DIMENSIONS FOR FOOT-MOUNTED A-C MOTORS WITH SINGLE STRAIGHT SHAFT EXTENSION

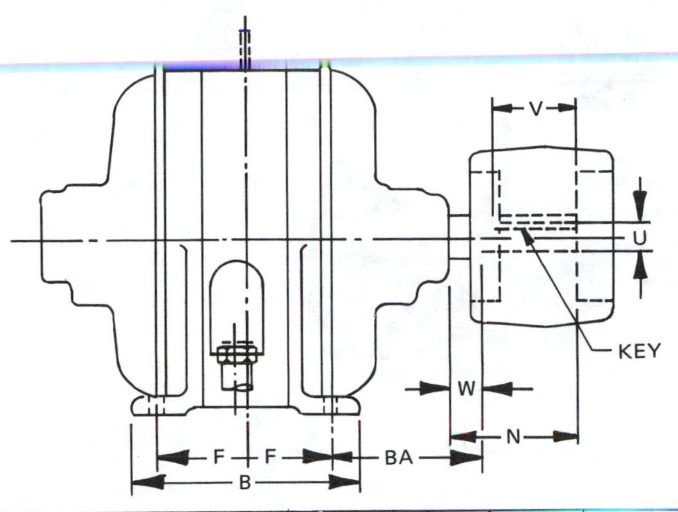

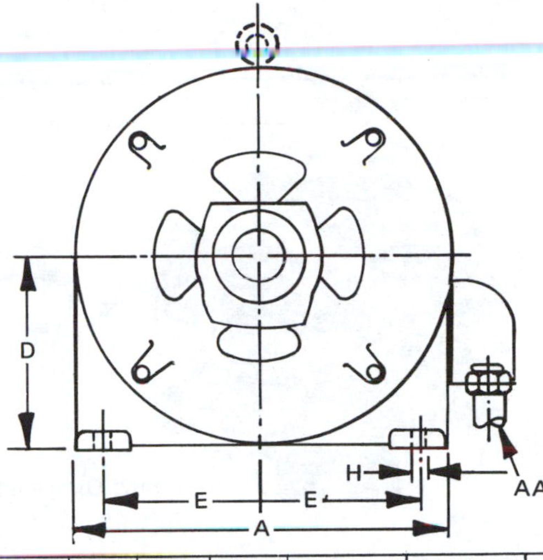

Frame Number	A Max.	B Max.	D*	E	F	BA	H	N-W	U	V Min.	Key Width	Key Thickness	Key Length	AA, Min. Conduit Size
42	....	....	2-5/8	1-3/4	27/32	2-1/16	9/32 slot	1-1/8	3/8	....	....	3/64 flat	....	....
48	....	....	3	2-1/8	1-3/8	2-1/2	11/32 slot	1-1/2	1/2	....	....	3/64 flat	....	....
56	....	....	3-1/2	2-7/16	1-1/2	2-3/4	11/32 slot	1-7/8	5/8	....	3/16	3/16	1-3/8†	....
66	....	....	4-1/8	2-15/16	2-1/2	3-1/8	13/32 slot	2-1/4	3/4	....	3/16	3/16	1-7/8†	....
182	9	6-1/2	4-1/2	3-3/4	2-1/4	2-3/4	13/32	2-1/4	7/8	2	3/16	3/16	1-3/8	3/4
184	9	7-1/2	4-1/2	3-3/4	2-3/4	2-3/4	13/32	2-1/4	7/8	2	3/16	3/16	1-3/8	3/4
213	10-1/2	7-1/2	5-1/4	4-1/4	2-3/4	3-1/2	13/32	3	1-1/8	2-3/4	1/4	1/4	2	3/4
215	10-1/2	9	5-1/4	4-1/4	3-1/4	3-1/2	13/32	3	1-1/8	2-3/4	1/4	1/4	2	3/4
254U	12-1/2	10-3/4	6-1/4	5	4-1/8	4-1/4	17/32	3-3/4	1-3/8	3-1/2	5/16	5/16	2-3/4	1
256U	12-1/2	12-1/2	6-1/4	5	5	4-1/4	17/32	3-3/4	1-3/8	3-1/2	5/16	5/16	2-3/4	1
284U	14	12-1/2	7	5-1/2	4-3/4	4-3/4	17/32	4-7/8	1-5/8	4-5/8	3/8	3/8	3-3/4	1-1/4
286U	14	14	7	5-1/2	5-1/2	4-3/4	17/32	4-7/8	1-5/8	4-5/8	3/8	3/8	3-3/4	1-1/4
324U	16	14	8	6-1/4	5-1/4	5-1/4	21/32	5-5/8	1-7/8	5-3/8	1/2	1/2	4-1/4	1-1/2
324S	16	14	8	6-1/4	5-1/4	5-1/4	21/32	3-1/4	1-5/8	3	3/8	3/8	1-7/8	1-1/2
326U	16	15-1/2	8	6-1/4	6	5-1/4	21/32	5-5/8	1-7/8	5-3/8	1/2	1/2	4-1/4	1-1/2
326S	16	15-1/2	8	6-1/4	6	5-1/4	21/32	3-1/4	1-5/8	3	3/8	3/8	1-7/8	1-1/2
364U	18	15-1/4	9	7	5-5/8	5-7/8	21/32	6-3/8	2-1/8	6-1/8	1/2	1/2	5	2
364US	18	15-1/4	9	7	5-5/8	5-7/8	21/32	3-3/4	1-7/8	3-1/2	1/2	1/2	2	2
365U	18	16-1/4	9	7	6-1/8	5-7/8	21/32	6-3/8	2-1/8	6-1/8	1/2	1/2	5	2
365US	18	16-1/4	9	7	6-1/8	5-7/8	21/32	3-3/4	1-7/8	3-1/2	1/2	1/2	2	2
404U	20	16-1/4	10	8	6-1/8	6-5/8	13/16	7-1/8	2-3/8	6-7/8	5/8	5/8	5-1/2	2
404US	20	16-1/4	10	8	6-1/8	6-5/8	13/16	4-1/4	2-1/8	4	1/2	1/2	2-3/4	2
405U	20	17-3/4	10	8	6-7/8	6-5/8	13/16	7-1/8	2-3/8	6-7/8	5/8	5/8	5-1/2	2
405US	20	17-3/4	10	8	6-7/8	6-5/8	13/16	4-1/4	2-1/8	4	1/2	1/2	2-3/4	2
444U	22	18-1/2	11	9	7-1/4	7-1/2	13/16	8-5/8	2-7/8	8-3/8	3/4	3/4	7	2-1/2
444US	22	18-1/2	11	9	7-1/4	7-1/2	13/16	4-1/4	2-1/8	4	1/2	1/2	2-3/4	2-1/2
445U	22	20-1/2	11	9	8-1/4	7-1/2	13/16	8-5/8	2-7/8	8-3/8	3/4	3/4	7	2-1/2
445US	22	20-1/2	11	9	8-1/4	7-1/2	13/16	4-1/4	2-1/8	4	1/2	1/2	2-3/4	2-1/2
504U	25	21	12-1/2	10	8	8-1/2	15/16	8-5/8	2-7/8	8-3/8	3/4	3/4	7-1/4	2-1/2
504S	25	21	12-1/2	10	8	8-1/2	15/16	4-1/4	2-1/8	4	1/2	1/2	2-3/4	2-1/2
505	25	23	12-1/2	10	8	8-1/2	15/16	8-5/8	2-7/8	8-3/8	3/4	3/4	7-1/4	2-1/2
505S	25	23	12-1/2	10	8	8-1/2	15/16	4-1/4	2-1/8	4	1/2	1/2	2-3/4	2-1/2

table c-2 Adapted from MG 1-11.31.

*Dimension D will never be greater than the values listed, but it may be less so that shims are usually required for coupled or geared machines. When exact dimension is required, shims up to 1/32 in. may be necessary on frame sizes whose dimension D is 8 in. and less; on larger frames, shims up to 1/16 in. may be necessary.

†Effective length of keyway.

standard enclosures for electric motors

illustration c-1

PH Mobile Division Power Unit with Accessories

Open
The open motor is one having ventilating openings which permit passage of external cooling air over and around the windings.

Drip-Proof
The drip-proof motor is an open motor in which ventilating openings are so constructed that drops of liquid or solids falling on the machine at any angle not greater than 15 degrees from the vertical cannot enter the machine.

Guarded
A guarded motor is an open motor in which ventilating openings are limited to specified size and shape to prevent insertion of fingers or rods to avoid accidental contact with rotating or electrical parts.

Splash-Proof
A splash-proof motor is an open motor in which ventilating openings are so constructed that drops of liquid or solid particles falling on the machine or coming toward the machine in a straight line at any angle not greater than 100 degrees from the vertical cannot enter the machine.

Totally-Enclosed
A totally-enclosed motor is a motor so enclosed as to prevent the free exchange of air between the inside and outside of the case, but not airtight.

Totally-Enclosed Nonventilated (TENV)
A totally-enclosed nonventilated (TENV) motor is a totally-enclosed motor which is not equipped for cooling by means external to the enclosing parts.

Totally-Enclosed Fan-Cooled (TEFC)
A totally-enclosed fan-cooled (TEFC) motor is a totally enclosed motor with a fan to blow cooling air across the external frame. It is a popular motor for use in dusty, dirty, and corrosive atmospheres.

Encapsulated
Encapsulated motor is an open motor in which the windings are covered with a heavy coating of material to protect them from moisture, dirt, abrasion, etc. Some encapsulated motors have only the coil noses coated. In others, the encapsulation material impregnates the windings even in the coil slots. With this complete protection, the motors can often be used in applications which formerly demand totally enclosed motors.

Explosion-Proof
An explosion-proof motor is a totally enclosed motor designed and built to withstand an explosion of gas or vapor within it, and to prevent ignition of gas or vapor surrounding the machine by sparks, flashes or explosions which may occur within the machine casing.

three-phase motor design

Design "B" — A Design "B" motor is a 3-phase squirrel-cage motor designed to withstand full-voltage starting and developing lock-rotor and breakdown torques adequate for general application.

Design "C" — A Design "C" motor is a 3-phase squirrel-cage motor designed to withstand full-voltage starting, developing locked-rotor torque for special high torque applications.

Design "D" — A Design "D" motor is a 3-phase squirrel-cage motor designed to withstand full-voltage starting, developing 275 percent locked-rotor torque (generally referred to as a "high slip" motor).

motor starter, conduit and wire size

3 PHASE MOTOR STARTERS

1/2 TO 20 H.P. — MOTOR H.P. 3 ø	1/2		3/4		1		1-1/2		2		3		5		7-1/2		10		15		20	
VOLTAGE	220	440	220	440	220	440	220	440	220	440	220	440	220	440	220	440	220	440	220	440	220	440
Nema Starter Size	00	00	00	00	00	00	00	00	0	00	0	0	1	0	1	1	2	1	2	2	3	2
⊕ Full Load Current	2.0	1.0	2.8	1.4	3.5	1.8	5.0	2.5	6.5	3.3	9.0	4.5	15	7.5	22	11	28	14	40	20	52	26
Fuses – Amps: Std. N.E.C.	15	15	15	15	15	15	15	15	20	15	25	15	40	20	60	30	70	35	100	50	150	70
Fuses – Amps: Dual Element	5	4	5	4	8	4	8	5	12	8	15	10	25	15	30	20	45	25	60	30	80	40
Circuit Breaker Max. Amps.	15	15	15	15	15	15	15	15	15	15	20	15	30	15	50	20	50	30	70	40	100	50
Minimum Wire Sizes: R, RW, T, TW	14	14	14	14	14	14	14	14	14	14	14	14	12	14	10	14	8	12	6	10	4	8
Minimum Wire Sizes: RH	14	14	14	14	14	14	14	14	14	14	14	14	12	14	10	14	8·	12	6	10	6	8

Always specify voltage and frequency.

3 PHASE MOTOR STARTERS

25 TO 200 H.P. — MOTOR H.P. 3 ø	25		30		40		50		60		75		100		125		150		200	
VOLTAGE	220	440	220	440	220	440	220	440	220	440	220	440	220	440	220	440	220	440	220	440
Nema Starter Size	3	2	3	3	4	3	4	3	5	4	5	4	5	4	6	5	6	5	6	5
⊕ Full Load Current	64	32	78	39	104	52	125	63	150	75	185	93	246	123	310	155	360	180	480	240
Fuses – Amps: Std. N.E.C.	175	80	200	100	300	150	350	175	400	200	500	250	600	350	–	400	–	450	–	600
Fuses – Amps: Dual Element	100	50	125	60	175	80	200	100	225	125	300	150	400	200	450	250	600	300	–	400
Circuit Breaker Max. Amps.	125	50	100	70	175	100	200	125	225	125	300	150	400	200	–	250	–	300	–	400
Minimum Wire Sizes: R, RW, T, TW	3	8	1	6	00	4	000	3	0000	2	300	0	500	000	–	0000	–	300	–	500
Minimum Wire Sizes: RH	4	8	3	6	1	6	00	4	000	3	0000	1	350	00	–	000	–	0000	–	350

Always specify voltage and frequency.

SINGLE PHASE MOTOR STARTERS

1/6 TO 5 H.P. — MOTOR H.P. 1 ø	1/6		1/4		1/3		1/2		3/4		1		1-1/2		2		3		5	
VOLTAGE	115	230	115	230	115	230	115	230	115	230	115	230	115	230	115	230	115	230	115	230
⊕ Full Load Current	4.4	2.2	5.8	2.9	7.2	3.6	9.8	4.9	13.8	6.9	16	8	20	10	24	12	34	17	56	28
Fuses – Amps. Std. N.E.C.	15	15	20	15	25	15	30	15	45	25	50	25	60	30	80	40	100	60	–	90
Circuit Breaker Max. Amps.	15	15	15	15	15	15	30	15	40	20	40	20	50	30	70	30	100	50	–	70
Min. Wire Sizes – R, RH, RW, T, TW	14	14	14	14	14	14	14	14	12	14	12	14	10	14	10	14	6	10	–	8

WIRE & CONDUIT SIZES

WIRE SIZE AWG or MCM	14	12	10	8	6	4	3	2	1	0	00	000	0000	250	300	350	400	500	750	1000
MAXIMUM 3 WIRES IN CONDUIT — Wire Capacity: R-RW-T-TW Amps.	15	20	30	40	55	70	80	95	110	125	145	165	195	215	240	260	280	320	400	455
MAXIMUM 3 WIRES IN CONDUIT — Wire Capacity: RH Amps.	15	20	30	45	65	85	100	115	130	150	175	200	230	255	285	310	335	380	475	545
MAXIMUM 3 WIRES IN CONDUIT — CONDUIT SIZE – Inches	1/2	1/2	3/4	3/4	1	1-1/4	1-1/4	1-1/4	1-1/2	2	2	2	2-1/2	2-1/2	2-1/2	3	3	3	3-1/2	4
Volts Drop Per Ampere Per 100 Ft. – 80% P.F.: 1 Phase Volts	.4762	.3125	.1961	.1250	.0833	.0538	.0431	.0370	.0323	.0269	.0222	.0190	.0161	.0147	.0131	.0121	.0115	.0101	.0086	.0081
Volts Drop Per Ampere Per 100 Ft. – 80% P.F.: 3 Phase Volts	.4167	.2632	.1677	.1087	.0714	.0463	.0379	.0323	.0278	.0231	.0196	.0163	.0139	.0128	.0114	.0106	.0091	.0088	.0066	.0061

Capacity of conductors in conduit based on room temperature of 30° C. (86° F.)

⊕ The full load currents shown are average values.

table j (c-3)

solenoid cycle rate

The current flow into an AC solenoid under certain conditions can be shown as follows:

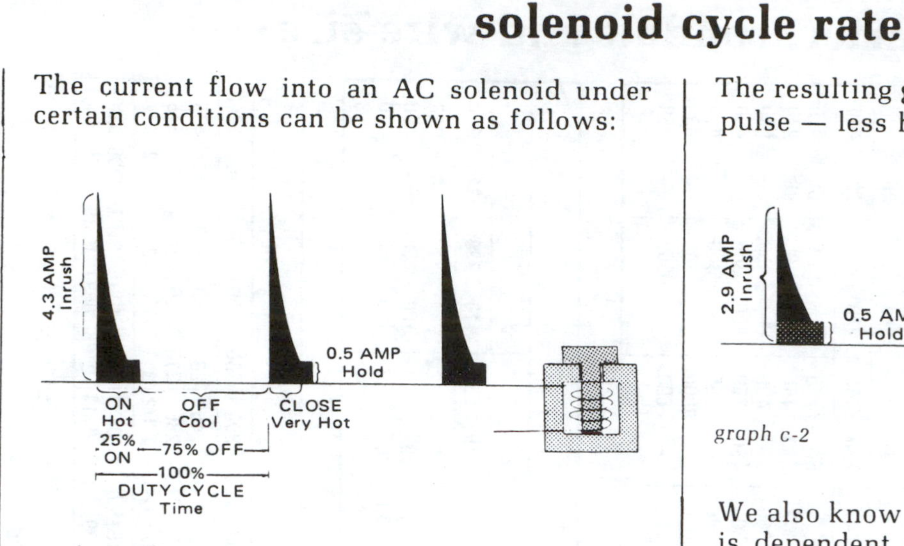

graph c-1

Each time the solenoid is cycled it receives a pulse of high inrush current which gradually declines as the plunger closes. In the full closed position, the solenoid is held energized by a low holding current flowing through the coil until the solenoid is de-engerized.

Current generates heat, and the area under each pulse or cycle represents heat generated in the solenoid. If the total heat generated from the continued cycling of a solenoid exceeds its ability to dissipate that heat, it will soon overheat and fail.

If we increase the cycles per minute of a solenoid, we increase the number of pulses in our illustration and increase the heat generated . . . so you see the cycling rate is limited by the generated heat.

Obviously if we can do anything to reduce the size or duration of these pulses, we can reduce heat and permit faster cycling.

We know that the inrush current decreases with a decreased length of the solenoid's stroke so we can cut down heat by reducing the length of stroke.

The resulting **graph c-2** looks like this. A lower pulse — less heat per each energization.

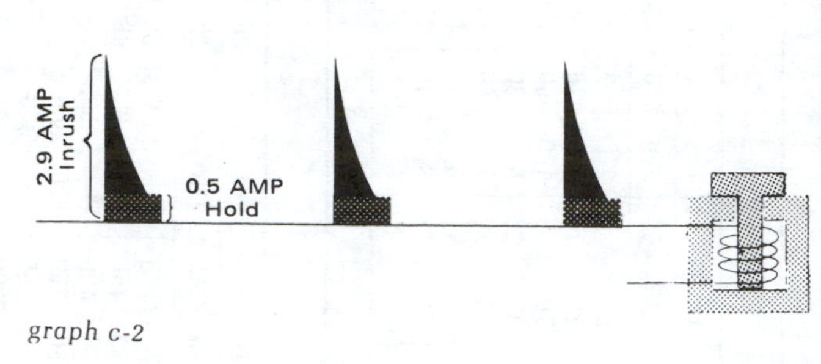

graph c-2

We also know that a solenoid's speed of closing is dependent upon the load it must move. If we reduce that load, we further reduce the heat generating area and our **graph c-3** looks like this:

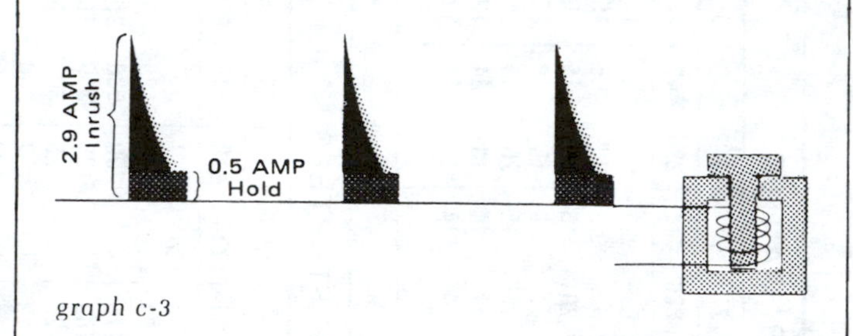

graph c-3

The shaded area represents the reduction in closing time (area of heating) affected by lightening the solenoid's load.

Obviously the percentage of "on time" versus "off time" will also affect the solenoid's temperature. This relationship is known as Duty Cycle. A 40% Duty Cycle is on 40%, off 60%.

Keep in mind that any solenoid can tolerate a higher input heat if it is mounted on a "heat sink" which will substantially increase its heat dissipating ability.

wet armature

Air Gap Valves Almost Always Leak:
In the conventional air gap type valves, dynamic o-ring seals allow a thin film of oil to remain on the push pin as it moves from the oil cavity to the solenoid cavity. Each time the pin returns, fluid is wiped off, accumulating in the solenoid cover (drop by drop). It eventually spills into the valve's electrical cavity and leaks out onto the floor.

Wet Armature Valves Almost Never Leak:
Wet armature valves offer virtually leakproof performance because the hydraulic fluid is effectively contained in an enclosed tube in which the plunger floats freely. There are no dynamic seals since both ends of the solenoid are sealed by static

wet armature continued

o-rings. This eliminates the problem leakage areas.

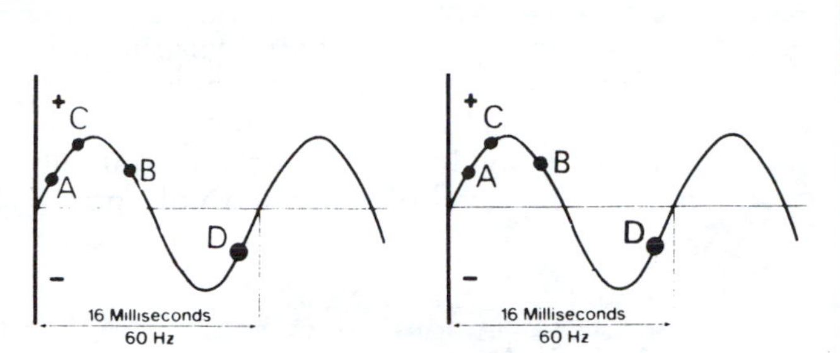

Identical Response Time
Like the air gap, the wet armature solenoid full movement takes from 6 to no more than 16 milliseconds, depending, of course, at which point on the 60-cycle voltage wave the coil is energized. Switch on at point A and full actuation should occur at point B. Energize the solenoid at point C and full actuation should be complete at point D.

Better Heat Dissipation
The wet armature valve employs such an efficient heat dissipation method that the faster it cycles, the cooler it runs! The circulating fluid picks up the heat in the solenoid body and carries it out of the valve to the reservoir. In short, the heat is not just contained in the solenoid, but dissipated throughout the entire valve. Test results are available.

Reliable Override
A small, simple spring is used to the override to prevent cycling with the plunger. This eliminates the potential of leakage through the manual override o-ring seal.

A failed solenoid will not distort the tube; therefore, if you should need the manual override, the plunger is free to move.

The Industrial Standard Of Tomorrow Is Here Today!

Exclusive Dual-Purpose Seal
The wet armature plunger and pin are always totally enclosed in the fluid, thus never exposed to atmospheric contamination.

Also, the wet armature design features exclusive dual-purpose grommet seals that not only seal the common electrical cavity from oil and atmospheric contamination, but also prevent damage to lead wires normally squeezed through drilled holes.

Easy Replacement
Should A Coil Be Damaged Or Burn-Out
If the coil has to be replaced on a wet armature solenoid, simply remove the retainer, and slip the coil from the tube. The new coil easily slips on, and the retainer is replaced.

50 or 60 Hertz Current
The wet armature coil can be used on 50 or 60 Hertz current without rewiring.

50 and 60 cycle solenoids

Coils for solenoids can be wound so that they may be used with either 50 or 60 cycle current. However, for best performance it is recommended that coils be wound for the specific frequency on which they will be used.

So-called "dual frequency" (50-60 cycle) coils are actually wound for 50 cycles and will produce somewhat less force at the higher frequency. Their use on 60 cycles is limited to applications where their reduced force is adequate to operate the mechanism.

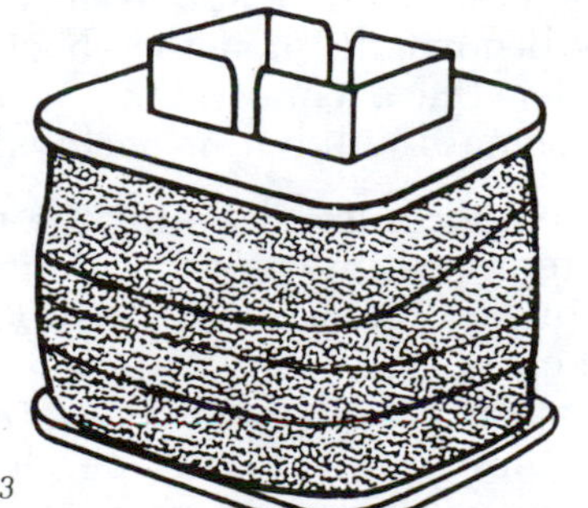

illustration c-3

Sometimes one coil can serve as a "dual frequency" coil if the 50 cycle operating voltage is lower than the 60 cycle nominal voltage. The wet armature has these characteristics.
Example: *120 volts at 60 cycles, 100 volts at 50 cycles.*

In this case, the winding for both specifications is the same so the same coil can be used successfully in both applications.

Looking at the situation in another way, the twenty extra volts available at 60 cycles will provide enough extra power to offset the power loss at 50 cycles mentioned in the previous paragraph.

Tapped or three lead "dual frequency" (50-60 cycle) coils are a more expensive but practical

50 & 60 cycle solenoids continued

solution. These coils have one common lead, with a 60 cycle lead tapped in near the end of the coil and a 50 cycle lead at the coil's end. Such coils are usually wound for 60 cycle voltage at a multiple of 115 or 120 volts, with a 50 cycle tap at a multiple of 110 volts.

Example: *A coil could be wound for 50 cycles at 110 volts with a tap or third lead for 60 cycle, 115 volt operation.*

Ideally a 50 cycle solenoid should be manufactured with more laminations in both plunger and field, but because of the limited demand for 50 cycle solenoids ours as well as other manufacturers use a 60 cycle plunger and field, and alter the coil only.

This compromise results in a 50 cycle solenoid with slightly less than normal force. Force reduction is roughly 10% at 1/2 inch stroke and 5% at 1/4 inch stroke. Holding force is very little affected.

A 50 cycle power supply is not always available for pre-shipment testing of equipment before export. If this happens, 50 cycle power can be simulated on 60 cycles by adjusting your 60 cycle voltage to a level of 6/5 rated voltage.

Specifications for available power supply in most foreign countries are given in a booklet titled, "Electric Current Abroad," published by U. S. Department of Commerce, Business and Defense Services. Available from the Superintendent of Documents, U. S. Government Printing Office, Washington, D.C. 20404 – $.30

solenoid force and voltage

The pull-in force of a solenoid decreases rapidly as the voltage decreases below the coil nominal rating. On the other hand, as the voltage increases over the nominal value, the pull-in force increases, but the solenoid temperature may also rapidly increase.

Low Voltage — From a low voltage standpoint, solenoid size selection should allow for adequate force at some arbitrary low voltage level. This will insure adequate solenoid force even during periods of low voltage and will prevent failure to pull-in and consequent coil burn out.

Design practices vary but low voltage levels are usually set at 90% of rated or nominal levels.

For convenience, our solenoid ratings are listed in our catalogs.

Forces at other reduced voltages can be closely approximated by means of the following formula:

$$F_1 = F \times \left(\frac{E_1}{E}\right)^2$$

Where:

F_1 = solenoid force at a reduced voltage E_1.
F = solenoid force at rated voltage E.

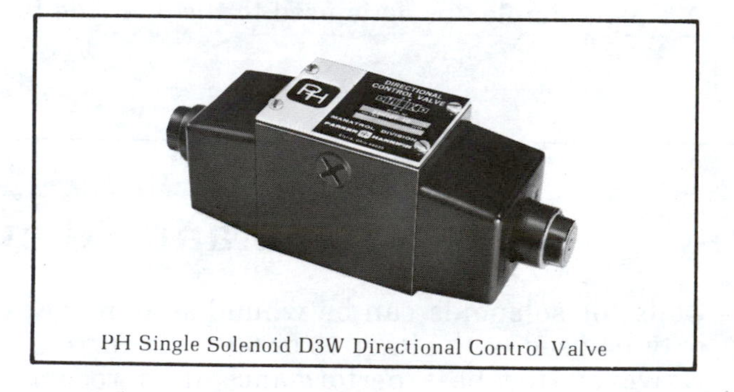

PH Single Solenoid D3W Directional Control Valve

illustration c-4

Forces calculated with this formula will be within 5% of measured forces down to 70% of nominal or rated voltage.

Over Voltage — The extra force resulting from over voltage (or voltage above the coil nominal rating) will normally last for short periods of time. Under these conditions the mechanical life of a solenoid will not be seriously affected.

The coil temperature rise and the consequent ultimate temperature of the solenoid will increase. Unless the ultimate temperature exceeds the class "A", 105°C rating of the insulation, the application will be satisfactory. The solenoid temperature can be lowered by mounting on a surface, such as an aluminum plate, which will conduct the heat away.

trouble shooting solenoid valves

solenoid failures

1. Voltage too low. If voltage will not complete the stroke of alternating current (AC) solenoid it will burn out the coil.

2. Signal to both solenoid of a double solenoid valve simultaneously. One or both of the solenoids will be unable to complete their stroke and will burn out. (Make certain the electrical signal is interlocked so that this condition cannot exist.) (Cut in — Cut out).

3. Covers left off solenoids allowing cast iron dust or other metallic chips to collect within the magnetic field. Plunger cannot complete stroke and solenoid burns out or the contamination could enter the fluid via the push pin. The wet armature solenoid uses a sealed tube assembly and therefore does not contaminate the fluid.

4. Push rod peens to a point where it is not long enough to actuate the valve (replace with new pin).

5. Mechanical damage to leads. (Short circuit, open connections, etc.)

6. Mechanical damage to coils. (Short circuit, open windings, etc.)

7. Tight spool or other mechanical parts of the valve being actuated can prevent the solenoid from completing its stroke and subsequently burning out.

Hydraulic Pressure Control Valve

illustration c-5

8. Replacement springs too heavy in valve. Overloads solenoid and shortens life.

9. Wrong voltage or frequency will either prevent operation because of inadequate capacity to handle the load with the lower voltage or burn out the coil because of improper winding and excessive voltage.

PH D63W Directional Control Valve

illustration c-6

10. Dirty contacts may not supply sufficient current to solenoid to satisfy inrush demands.

11. Low voltage direct current solenoids may be affected by low battery capacity on cold mornings directly after starting cold engine.

12. Long feed lines to low voltage solenoids may cause sufficient voltage drop to cause erratic operation.

13. High ambient temperature or confined area — heat not dissipated properly.

solenoid valve fails to operate

1. Is there an electrical signal to the solenoid or operating device? Is the voltage too low? (Check with voltmeter . . . test light in emergency.)

2. Is the solenoid push rod shifting the pilot spool far enough? (After considerable service the push pin to other parts may have been worn enough to prevent sufficient spool movement.)

3. If the supply to the pilot body is orificed, is the orifice restricted? (Remove orifice and check for foreign matter. Flushing is sometimes necessary because of floating impediment.)

trouble shooting continued

4. Has foreign matter jammed the main spool? (Remove end caps and see that main spool is free in its movements . . . remember that there will be a quantity of fluid escaping when the cap is removed and provide a container to catch it.)

5. Is pilot pressure available? Is the pilot pressure adequate? (Check with gauge on main pressure input port for internally piloted types and in the supply line to the externally piloted type.)

6. Is pilot drain restricted? (Remove pilot drain and let the fluid pour into an open container while the machine is again tried for normal operation. Small lines are often mashed by machine parts banging against them causing a subsequent restriction to fluid flow.)

7. Is pilot tank port connected to main tank port where pressures are high enough to neutralize pilot input pressure? A minimum pressure differential of 65 psi is necessary to shift valve. (Combine pilot drain and pilot tank port and check for operation with the combined flow draining into an open container . . . block line to main tank from pilot valve . . . if this corrects the situation, reroute pilot drain and tank lines.)

8. Are solenoids improperly interlocked so that a signal is provided to both units simultaneously? (Put test light on each solenoid lead in parallel and watch for simultaneous lighting . . . check electrical interlock. This condition probably burns out more solenoids than any other factor.)

9. Has mounting pad been warped from external heating or improper bolt torque? (Loosen mounting bolts slightly and see if valve functions. End caps can also be removed and check for tight spool).

10. Is fluid media excessively hot? (Check for localized heating which may indicate an internal leak . . . check reservoir temperature and see if it is within machine specifications.)

11. Is there foreign matter in the fluid media causing gummy deposits? (Check for contamination . . . make certain seals and plumbing are compatible with the type of fluid being used.)

12. Is an adequate supply of fluid being delivered to actuate the load? (Many times there is sufficient pressure to shift the valve but not enough to actuate the work load. Check pump supply pressure and volume if necessary . . . physical measurement of flow through relief valve with units blocked may be necessary.)

13. Check circuit for possible interlocks on pressure sources to valve or to pilot.

notes

hydraulic filtration

fluids chart d-2
ASTM standard viscosity-temperature chart d-15
explanation of flow vs. pressure differential tables d-16
calculation pressure differential in straight line lengths d-17
calculating pressure differential in fittings and bends d-18
multipliers for fittings and bends ... d-19
Δ P (psi)/ft schedule 40 pipe d-20
Δ P (psi)/ft schedule 80 pipe d-22
Δ P (psi)/ft schedule 160 pipe d-24
Δ P (psi)/ft double extra heavy pipe d-25
Δ P (psi)/ft Tube d-26
Δ P (psi)/ft hose d-32
Δ P (in. hg)/ft schedule 40 pipe d-34
Δ P (in. hg)/ft tube d-36
Δ P (in. hg)/ft hose d-37
filter selection factors d-38
magnets and filtration d-38
filter elements d-39
selection of filter media d-41
suction line filtration d-43
pressure line filtration d-43
return line filtration d-44
hydraulic filtration d-44
sources of dirt d-45
fluid sampling d-45
filter service d-46

FLUIDS CHART[1]

Brand Name		Fluid Type[2]	Specific Gravity	Viscosity (SUS) 100°F	210°F
Amoco (American Oil Co.) Chicago, IL					
American Industrial	15	PB	.893	159	44.2
	21	PB	.887	212	47.7
	51	PB	.891	502	62.5
	75	PB	.892	752	74.6
	95	PB	.892	952	84
	120	PB	.892	1192	95
	150	PB	.894	1504	109
	175	PB	.893	1731	118
	200	PB	.899	2053	132
	240	PB	.895	2446	148
Indoil	15	PB	.872	158	43.6
	21	PB	.875	207	46.6
	31	PB	.881	309	52.5
	51	PB	.886	502	62.5
	75	PB	.889	752	74.5
	95	PB	.892	955	84
Rykon	1	PB	.897	108	40.9
	15	PB	.874	160	45.8
	21	PB	.875	206	50.3
	31	PB	.878	304	59.5
	51	PB	.878	501	74.1
	95	PB	.882	942	99.7
FR Fluid	W0	IE	.925	397	112 @ 175°F
	WG20	WG	1.06	209	102 @ 150°F
	PE15	PE	1.18	160	70 @ 136°F
	PE22	PE	1.17	225	70 @ 149°F
	PE30	PE	1.17	305	70 @ 160°F
	PE55	PE	1.17	550	70 @ 178°F
Arco (Atlantic Richfield Co.) Philadelphia, PA					
Duro AW	S-150	PB	.868	155	70 @ 135°F
	S-215	PB	.868	210	70 @ 140°F
	S-315	PB	.879	310	70 @ 170°F
	S-465	PB	.884	500	70 @ 185°F
Duro AW	1500	PB	.896	1430	103
Duro FR-HD		IE	.930	422	560 @ 77°F
Bel-Ray Co., Inc. Farmingdale, NJ					
Raylene AW	00	PB	.8602	110	
	0	PB	.8633	155	
	1	PB	.8681	210	
	2	PB	.8708	340	
	3	PB	.8762	500	
	LT-75	PB	.8550	56	
	LT-85	PB	.8816	78	
	LT-100	PB	.8681	100	
	LT-175	PB	.8602	175	
	LT-300	PB	.8602	300	
No-Flame	IE	IE	.934	450	
No-Flame G	15	WG	1.06	150	
	20	WG	1.07	200	
	30	WG	1.08	300	

[1] Data in this chart was received from catalog literature or by direct correspondence with the oil refiner. The appearance of a specific fluid in the chart is by no means an indication of quality nor should it be inferred as recommended by Parker-Hannifin Corporation.

[2] IE-Invert Emulsion • PB-Petroleum Base • PE-Phosphate Ester • EB-Phosphate Ester Blend • WG-Water Glycol
E- Emulsion

FLUIDS CHART (CONTINUED)

Brand Name		Fluid Type	Specific Gravity	Viscosity (SUS) 100°F	210°F
BP Canada Montreal, Quebec					
Energol CS	40	PB	.8899	75	36
	50	PB	.8654	108	40.5
	55	PB	.8681	130	42.5
	65	PB	.8708	155	44.6
	80/80S	PB	.8708	205	47.5
	100	PB	.8816	310	52.5
	125	PB	.8816	388	57.5
	150	PB	.8816	470	62.5
	200	PB	.8816	650	71.5
	250	PB	.8816	850	82.5
	300	PB	.8871	1250	105
	425	PB	.8927	1550	117.5
	550	PB	.8927	1950	132.5
	600	PB	.8927	2650	160
Energol EM	65	PB	.9340	155	—
	80	PB	.8984	208	44.5
	85	PB	.9100	250	46
	90	PB	.9100	275	47.5
	100	PB	.9100	310	49
	125	PB	.9100	405	53.5
	150	PB	.9218	510	55.5
	175	PB	.9188	662	64
	200	PB	.9159	775	69
	250	PB	.9159	910	76
	300	PB	.9159	1250	92.5
Energol HL	40/1	PB	.8933	60	—
	40	PB	.8894	72.5	—
	50	PB	.8612	110	—
	65	PB	.8708	150	—
	80	PB	.8751	200	—
	100	PB	.8789	315	—
	125	PB	.8816	380	—
	150	PB	.8833	540	—
	175	PB	.8849	670	71.8
	250	PB	.8860	800	79
	300	PB	.8883	960	86.3
	425	PB	.8939	1530	113.7
	500	PB	.8973	1900	132.1
	600	PB	.9001	2520	158
Energol HLT	55	PB	.8676	128	45.8
Energol TH	38-HB	PB	.8922	63	—
	50-HB	PB	.8633	105	—
	65-HB	PB	.8697	155	—
	80-HB	PB	.8745	210	—
	100-HB	PB	.8805	315	53.8
	125-HB	PB	.8833	405	58.3
	150-HB	PB	.8833	465	61.2
	250-HB	PB	.8883	775	77.7
	300-HB	PB	.8916	1037	90.5
Bray Oil Co. Los Angeles, CA					
Brayco	745	PB	.894	102	38
	756C/D	PB	.868	74	43.5
	760	PB	.854	35.5	30
	762	PB	.855	37.4	30
	776RP	PB	.877	124	41.8
Brooks Oil Co. Cleveland, OH					
FR Fluid	B	IE		375	70
Castrol Oils Inc. Hackensack, NJ					
Hyspin	40	PB	.902	602	34.2
	45	PB	.899	73.5	35.6
	55	PB	.866	112.8	40.9
	70	PB	.873	154	44
	80	PB	.876	203	47
	80SS	PB	.883	247	50.6
	100	PB	.880	286	51.7
	140	PB	.882	434	59.8
	175	PB	.882	543	67.5

FLUIDS CHART (CONTINUED)

Brand Name		Fluid Type	Specific Gravity	Viscosity (SUS) 100°F	210°F
Castrol Oils Inc. (Continued)					
Hyspin AWS	32	PB	.873	160	44.4
	68	PB	.882	286	52.5
	100	PB	.885	427	60.2
	150	PB	.885	697	72.5
Alpha	317	PB	.885	427	60.1
	417	PB	.892	799	77.9
	517	PB	.892	799	77.9
	617	PB	.895	1218	96.4
	717	PB	.900	1821	120.2
	817	PB	.901	2658	149.2
Alpha	LS-68	PB	.892	292	52.4
	LS-150	PB	.899	695	74.5
	LS-220	PB	.903	1093	93.6
	LS-320	PB	.905	1449	108.6
	LS-680	PB	.915	3011	158.1
	LS-1000	PB	.932	4632	178.8
Chemtrend Inc. Howell, MI					
	HF 20	WG	1.07	200	105 @ 140°F
Chevron (Standard Oil Co. of Calif.) San Francisco, CA					
Aviation Fluid	A	PB	.8576	74.2	45.4
	B	PB	.8440	74	45
Tractor Fluid		PB	.8922	216	50.3
OC Turbine Oil	9	PB	.8692	158	43.6
	11	PB	.8740	223	47.7
	13	PB	.8751	290	52
	15	PB	.8767	318	53.2
	19	PB	.8783	418	58.8
	24	PB	.8805	547	65.5
	36	PB	.8844	882	83
EP Hydraulic Oil	9	PB	.8708	152	43.8
	11	PB	.8751	229	48.5
	15	PB	.8783	319	54.5
	19	PB	.8681	418	58.8
	24	PB	.8800	547	65.5
	109	PB	.8905	154	45.5 @ 180°F
EP Industrial Oil	45X	PB	.8789	202	46.8
	55X	PB	.8838	405	—
	68X	PB	.8855	597	68.9
	80X	PB	.8899	940	85.8
Gear Oil	90	PB	.8990	1227	93.5
	140	PB	.9170	5264	203
	250	PB	.9042	7500	315
MS Gear Lub	80	PB	.8783	480	64
	90	PB	.8933	1060	92
	140	PB	.9018	3000	164
Universal Gear Lub	80	PB	.8860	354	55.9
	90	PB	.8973	1101	93.2
	140	PB	.9018	2305	142.8
Gear Compound	60	PB	.8933	301	54.6
	70	PB	.9007	600	71.4
	80	PB	.9047	835	83.9
	90	PB	.9065	1225	102
	120	PB	.9100	1438	113
	140	PB	.9117	2109	141
	160	PB	.9147	2736	164
	240	PB	.9218	3851	206
	400	PB	.9383	11053	407
NL Gear Compound	50	PB	.8762	215	47.8
	60	PB	.8778	315	54
	70	PB	.8783	480	64
	80	PB	.8855	750	77
	90	PB	.8933	1060	92
	120	PB	.8939	1500	110
	140	PB	.8961	2125	136
	160	PB	.9018	3000	164
	250	PB	.9094	4800	258
	290	PB	.9123	7010	279
	400	PB	.9159	10250	410
FR Fluid	8	PE	1.165	151	40
	10	PE	1.135	223	42.5
	13	PE	1.135	327	45
	20	PE	1.135	596	47.5

FLUIDS CHART (CONTINUED)

Brand Name		Fluid Type	Specific Gravity	Viscosity (SUS) 100°F	Viscosity (SUS) 210°F
Conoco (Continental Oil Co.) Houston, TX					
Super Hyd. Oil	15	PB	.860	150	44
	21	PB	.869	210	48
	31	PB	.872	310	54
Polar Start DN	600	PB	.895	270	60.5
Davis-Howland Oil Corp. Rochester, NY					
DSL Oil	41	PB	.847	60	36
	42	PB	.860	100	40
	43	PB	.870	200	51
	44	PB	.865	150	44
	45	PB	.871	200	47
	46	PB	.876	300	54
	47	PB	.882	550	67
	48	PB	.887	750	72
	49	PB	.893	1100	88
DSL Convis OC	150	PB	.865	150	44
	300	PB	.876	300	54
	500	PB	.882	500	62
	1000	PB	.890	1000	85
	1800	PB	.895	1800	120
Drydene (Dryden Oil Co., Inc.) Baltimore, MD					
Blue Hyd. Oil	LT	PB	.8591	162	44.9
	10	PB	.8708	215	48
	20	PB	.8713	303	53
	30	PB	.8762	509	65
EP Gear Lub	1	PB	—	208	47.4
	2	PB	—	302	53
	3	PB	—	508	64.3
	4	PB	—	750	78
	5	PB	—	1074	95
	6	PB	—	1425	112
	7	PB	—	2020	137
	8	PB	—	2967	165
Exxon Co., USA Houston, TX					
Nuto	43	PB	.8702	158	43.8
	47	PB	.8735	215	47.7
	53	PB	.8783	307	53.1
	63	PB	.8822	470	62.4
	76	PB	.8838	722	76.3
	93	PB	.8883	1056	93.3
	113	PB	.8883	1464	113
	146	PB	.8939	2137	143.1
Nuto H	44	PB	.8713	159	44
	48	PB	.8735	212	47.5
	54	PB	.8783	297	52.5
	64	PB	.8833	489	62.8
Terrestic R&O	43	PB	.8591	147	43.6
	47	PB	.8724	220	48.2
	52	PB	.8740	321	54.8
	56	PB	.8762	393	59
	65	PB	.8794	559	69
	85	PB	.8844	864	84.7
Univis	P38	PB	.9291	70	37.5
	J43	PB	.8607	75	43.4
	P48	PB	.8800	165	49.7
	J58	PB	.8602	132	59.5
Imol S	220	PE	1.15	221	42.6
3110 FR Fluid		IE	.928	438	100 @ 175°F
Filmite Oil Corp. Butler, WS					
Flomite Hyd. Oil	65	PB	.880	65	37
	100	PB	.882	103	42
	150	PB	.883	150	45.5
	200	PB	.892	206	50.5
	300	PB	.901	308	59

FLUIDS CHART (CONTINUED)

Brand Name		Fluid Type	Specific Gravity	Viscosity (SUS) 100°F	210°F
Filmite Oil Corp. (Continued)					
Filmite Industrial	50	PB	.890	50	—
	100	PB	.863	100	40
	150	PB	.876	150	44
	200	PB	.876	200	46.5
	300	PB	.884	300	53
	400	PB	.885	400	59
	500	PB	.886	500	65
Fiske Bros. Refining Co. Newark, NJ					
Lubriplate HO	0	PB	.879	160	44
	1	PB	.877	213	48
	2	PB	.881	309	54
	2A	PB	.885	518	66
	3	PB	.889	690	75
	4	PB	.891	840	83
Gulf Oil Co. Houston, TX					
Harmony	41	PB	.864	105	39.7
	44	PB	.869	150	43.3
	47	PB	.873	205	47
	53	PB	.876	305	53.3
	61	PB	.880	460	62
	69	PB	.883	610	69.4
	43AW	PB	.868	150	43.7
	48AW	PB	.875	205	47.6
	54AW	PB	.878	300	53.4
	59EP	PB	.877	409	59.8
	74AW	PB	.886	700	74.6
	75	PB	.887	728	75.8
	76	PB	.893	813	75.3
	77	PB	.886	762	76.7
	88	PB	.890	1019	90.1
	97	PB	.892	1172	97.5
	121	PB	.895	1598	116.6
	151	PB	.900	2492	152.5
	204	PB	.902	3974	205
Endurance	35	PB	.857	57.8	34.6
	37	PB	.871	76.4	37
	39	PB	.886	101.6	38.8
	45	PB	.888	205	46.6
	48	PB	.889	255	49.8
	51	PB	.890	305	52.9
MP Gear Lub	80	PB	.903	372	56.3
	90	PB	.905	997	87.2
	140	PB	.907	2300	143.7
EP Lub	55	PB	.917	362	55.4
	75	PB	.929	894	77.5
	95	PB	.935	1431	97.5
	115	PB	.940	2038	117.5
	125	PB	.941	2501	130.4
	145	PB	.945	3173	150
	250	PB	.954	8314	255
	400	PB	.961	20010	410
	570	PB	.897	701	74.6
	S100	PB	.894	1070	94.5
	S120	PB	.897	1556	116.6
	S150	PB	.901	2302	147.4
Security	39	PB	.863	105.5	39.8
	44	PB	.869	155	43.6
	47	PB	.872	205	47.2
	53	PB	.878	313	53.6
	61	PB	.880	460	61.5
	71	PB	.882	600	68.7
	85	PB	.887	920	86.2
Security	118	PB	.892	1650	118.5
	43AW	PB	.880	154	43.1
	48AW	PB	.884	206	46.5
	54AW	PB	.884	299	52.6
	70AW	PB	.893	550	66.6
	2150RO	PB	.896	2151	141.1

FLUIDS CHART (CONTINUED)

Brand Name		Fluid Type	Specific Gravity	Viscosity (SUS) 100°F	210°F
Gulf Oil Co. (Continued)					
FR Fluid		IE	.948	400	73.1
FR Fluid	G200	WG	1.084	200	69
	P-37	PE	1.211	96	37
	P-40	PE	1.156	151	40
	P-43	PE	1.142	223	42.5
	P-45	PE	1.162	327	45
	P-47	PE	1.143	596	47.7
Houghton (E.F. Houghton Co.) Philadelphia, PA					
Hydro-Drive HP	150	PB	.879	145	43.4
	200	PB	.879	190	47
	300	PB	.879	290	53
Hydro-Drive MIH	LT	PB	.8871	150	43
	10	PB	.8750	200	46
	20	PB	.8711	300	52
	30	PB	.9053	525	62
	40	PB	.9071	750	72.5
	50	PB	.9082	1050	84
Cosmolubric	1702	PB	.882	86	—
	1725	PB	.910	105	44
	1133-A	PB	.913	200	54
	1133-B	PB	.916	650	72
MP Gear Oil	80	PB	.900	398	57
	90	PB	.897	1100	88
	120	PB	.893	1500	110
	140	PB	.897	2300	145
Hi Temp	101	PB	.916	6500	303
	102	PB	.882	1750	125
	103	PB	.919	1100	100
	227	PB	.949	550	56
	228	PB	.947	1250	100
	237	PB	.925	75	—
	303	PB	.928	70	—
	304	PB	.959	118	—
Houghto-Safe	271	WG	1.045	200	80 @ 150°F
	416	WG	1.08	160	110 @ 110°F
	520	WG	1.075	200	90 @ 150°F
	620	WG	1.074	200	89 @ 150°F
	1010	PE	1.20	90	38
	1055	PE	1.145	600	52
	1115	PE	1.165	150	40
	1120	PE	1.15	230	43
	1130	PE	1.145	290	46
	5046	IE	.953	420	210 @ 130°F
	5046-W	E	.96	450	215 @ 130°F
Vital Hyd. Fluid	29	EB	.97	300	50
Hydrotex Dallas, TX					
Deluxe No. 216 H.D. Hyd. Oil - Light		PB	0.866	188	45
217 H.D. Hyd. Oil - Medium		PB	0.873	295	52
218 H.D. Hyd. Oil - Heavy		PB	0.880	483	63
219 H.D. Hyd. Oil - Extra Heavy		PB	0.887	850	80
Hy-Torque Fluid		PB	0.885	278	56
Deluxe Dexron		PB	0.873	190	50
Deluxe No. 750 Type A, Suffix A		PB	0.887	205	51
753 C-2		PB	0.873	190	50
E.H. Kellogg & Co., Inc. Mount Vernon, NY					
Hylube	11	PB	.8485	44	32
	15	PB	.8735	62	35
	23	PB	.8686	130	42
	25	PB	.8729	165	43
	29	PB	.8789	217	48
	33	PB	.8805	326	52
	41	PB	.8822	510	63
	69	PB	.8844	735	72
	72	PB	.8899	1050	85
	75	PB	.8960	1600	116
	76	PB	.9013	1900	130

FLUIDS CHART (CONTINUED)

Brand Name		Fluid Type	Specific Gravity	Viscosity (SUS) 100°F	210°F
Kendall Refining Co. Bradford, PA					
Hyken Super Black	MV	PB	.8876	140.9	48.5
	052	PB	.8877	262.7	55.5
Kenoil R & O	043-EP	PB	.8628	146.6	43.5
	047-EP	PB	.8681	200	47.2
	053-EP	PB	.8762	300	53.1
	065-EP	PB	.8789	500	63.6
	072-EP	PB	.8827	650	71.5
	085-EP	PB	.8860	927	84.5
Gear Lub	80-90	PB	.8816	692	77.8
	90-140	PB	.8927	2481	158.6
	250	PB	.9042	9605	399.5
	800	PB	.9135	24714	785.5
Keystone Precision Lubricants Philadelphia, PA					
KLC	6	PB	.8899	158	42
	5	PB	.8911	200	45
	4A	PB	.8916	300	53
	4	PB	.8956	500	67
	3	PB	.9018	760	80
	2	PB	.9030	1001	89
	1A	PB	.9030	1500	120
	1	PB	.9053	2500	150
Gear Oil WG	5X	PB	.8911	320	54
	4	PB	.8956	462	62
	3	PB	.8973	720	81
	1	PB	.9018	1320	111
	A	PB	.9018	2240	111
	B	PB	.9088	3100	147
WG Special		PB	.9165	5000	220
Gear Lub	1790	PB	—	1000	85
	1791	PB	—	2050	140
Lubrication Engineers Inc. Fort Worth, TX					
Monolec Hyd.	6105	PB	.8602	105	40
	6110	PB	.8708	200	46
	6120	PB	.8762	305	53
	6130	PB	.8844	550	64
Monolec Turb	6401	PB	.8654	150	43
	6402	PB	.8708	200	46
	6403	PB	.8762	305	53
	6404	PB	.8844	550	64
	6405	PB	.8899	750	77
Monolec PF	7500	PB	.8927	220	50
A. Margolis & Sons Corp. Brooklyn, NY					
Silogram Tip	LT	PB	.8665	150	44
	MED	PB	.8708	205	47
	HVY	PB	.8751	325	54.6
Silogram MP AW	157	PB	.8665	150	44
	207	PB	.8708	205	47
	307	PB	.8751	325	54.6
	507	PB	.8816	482	64
	707	PB	.8871	640	74
	907	PB	.8871	926	88
Silogram Hydra Safe	FR150	WG	1.072	150	—
	FR200	WG	1.079	200	—
Metalub (Metal Lubricants Co.) (Continued) Chicago, IL					
Meltran AW	400	PB	.8623	107	40.4
	405	PB	.8693	160	45.2
	410	PB	.8762	215	47.4
	420	PB	.8805	310	54
	425	PB	.8827	410	58.7
	430	PB	.8816	550	67.4
	440	PB	.8838	740	76.9
	450	PB	.8849	970	90.1
	460	PB	.8912	1800	116
	480	PB	.8955	2500	171.1

FLUIDS CHART (CONTINUED)

Brand Name		Fluid Type	Specific Gravity	Viscosity (SUS) 100° F	210° F
Meltac WL	221	PB	.9154	150	45
	222	PB	.9055	350	55
	223	PB	.9042	560	67
	224	PB	.9100	1000	90
Melcolube	100-CP	PB	.8927	200	50
	101-CP	PB	.8927	300	55
	101-1/2-CP	PB	.8933	625	75
	102-CP	PB	.8939	900	90
	103-CP	PB	.8944	1200	100
	104-CP	PB	.8961	1700	120
	105-CP	PB	.8973	2300	160
	106-CP	PB	.8978	2700	180
	107-CP	PB	.8996	5000	210
	108-CP	PB	.9013	7000	250
Melvis	1A	PB	.9117	190	50
	2	PB	.9117	370	60
	3	PB	.9242	700	70
	5	PB	.9206	900	80
	7	PB	.9242	1800	120
	8	PB	.9248	2100	130
Melsyn FR	200	WG	1.07	200	—
Mobil Oil Corp. New York, NY					
Mobil DTE	LT	PB	.8708	150	43.4
	MED	PB	.8762	225	48.5
	HVY MED	PB	.8789	295	52.8
	HVY	PB	.8816	430	60
	EXTRA HVY	PB	.8871	640	70
	BB	PB	.8899	915	85
	AA	PB	.8973	1800	125
	HH	PB	.9001	2400	148
	KK	PB	.9071	6000	250
	11	PB	.864	90	40
	13	PB	.876	150	46
	15	PB	.878	205	52
	16	PB	.881	300	60
	18	PB	.884	485	69
	19	PB	.891	765	89
	24	PB	.871	153	44
	25	PB	.876	225	49
	26	PB	.882	300	53
Mobil Fluid	350	PB	.887	205	52
	423	PB	.8927	267	56
ETNA	24	PB	.8939	150	43
	25	PB	.8950	200	46
	26	PB	.8990	300	52
Mobilube HD	75	PB	.875	161	48
	80	PB	.895	380	58
	80-90	PB	.900	725	76
	90	PB	.894	1200	98
	140	PB	.901	2290	143
Mobilgear	626	PB	.884	315	53
	629	PB	.887	700	74
	630	PB	.892	1000	89
	632	PB	.895	1500	112
	633	PB	.898	1850	127
	634	PB	.902	2150	139
	636	PB	.909	3150	175
Pyrogard	D	IE	.940	550	130 @ 170° F
	53	PE	1.17	215	42
Monsanto (Monsanto Co.) St. Louis, MO					
Pydraul	10-E	PE	1.090	60	35
	29-E-LT	PE	1.100	145	52
	30-E	PE	1.165	150	42
	50-E	PE	1.155	231	46
	65-E	PE	1.152	312	50
	115-E	PE	1.155	565	56
	230-C	EB	1.050	215	49
	312-C	EB	1.040	309	54
	540-C	EB	1.040	491	61
	MC	EB	1.098	320	48
Skydrol		PE	1.098	80	40
Santo Safe WG	15	WG	1.076	150	97 @ 130° F
	20	WG	1.080	200	130 @ 130° F
	30	WG	1.073	275	140 @ 130° F

FLUIDS CHART (CONTINUED)

Brand Name		Fluid Type	Specific Gravity	Viscosity (SUS)	
				100°F	210°F
Nalco (Nalco Chemical Co.)					
Fyre-Safe	211	WG	1.058	200	98 @ 150°F
	225	WG	1.058	225	120 @ 150°F
Ore-Lube (Ore-Lube Corp.) College Point, NY					
Ore-Lube	111W	PB	.8683	150	46.9
	111W-LT	PB	.8951	105	40.0
Phillips 66 (Phillips Petroleum Co.) Bartlesville, OK					
Magnus	150	PB	.870	160	44.5
	215	PB	.875	210	47.5
	315	PB	.881	310	53
	465	PB	.884	475	63
	A150	PB	.871	160	44.5
	A215	PB	.876	210	47.5
	A315	PB	.882	310	53
Sanson & Sons, Inc. Bristol, PA					
No-Gum Hyd. Oil	LT	PB	.8681	155	—
	10	PB	.8708	210	—
	20	PB	.8735	305	—
	30	PB	.8772	515	—
	40	PB	.8805	775	—
	50	PB	.8838	975	—
	60	PB	.8871	1500	—
	70	PB	.8899	2500	—
AW Hyd. Oil	150	PB	.8681	150	—
	215	PB	.8708	215	—
	315	PB	.8735	315	—
	515	PB	.8772	515	—
	775	PB	.8805	775	—
	975	PB	.8835	975	—
	1500	PB	.8871	1500	—
	2500	PB	.8899	2500	—
Hydra-Safe Standard	150	WG	1.07	150	—
	200	WG	1.07	200	—
Hydra-Safe Premium	200	WG	1.08	200	—
Shell Oil Co. Houston, TX					
Tellus	11	PB	.845	40	—
	15	PB	.871	60	35
	23	PB	.867	120	42
	25	PB	.865	150	45
	29	PB	.875	215	47
	33	PB	.875	315	53
	41	PB	.875	465	63
	69	PB	.887	700	73
	71	PB	.889	1000	86
	75	PB	.887	1500	113
	77	PB	.895	2150	140
Hydraulic	21	PB	.865	105	40
	25	PB	.864	150	45
	29	PB	.871	215	47
	33	PB	.876	315	53
	41	PB	.873	465	63
	69	PB	.882	700	73
	71	PB	.882	1000	86
Turbo	25	PB	.872	150	44
	29	PB	.879	215	48
	33	PB	.876	315	52
	37	PB	.882	410	58
	41	PB	.882	465	65
	69	PB	.894	700	71
Tonna	25	PB	.873	150	45
	33	PB	.913	315	56
	71	PB	.916	1000	88
vSI C.O.	27	PB	.871	155	44
	29	PB	.872	215	47
	33	PB	.879	315	53
	41	PB	.878	520	64

FLUIDS CHART (CONTINUED)

Brand Name		Fluid Type	Specific Gravity	Viscosity (SUS) 100°F	210°F
Shell Oil Co. (Continued)					
Delima	69	PB	.887	700	73
	71	PB	.893	1000	87
Thermia	33	PB	.881	275	50
Irus Fluid		IE	.918	410	98 @ 180°F
Vitrea	21	PB	.866	105	39
	25	PB	.870	150	44
	29	PB	.878	215	48
	33	PB	.873	315	53
	41	PB	.876	465	65
	69	PB	.887	700	78
	71	PB	.890	1000	83
	75	PB	.889	1500	118
	77	PB	.896	2150	136
Carnea	15	PB	.910	60	35
	17	PB	.910	75	37
	21	PB	.919	105	38
	25	PB	.915	150	42
	29	PB	.918	215	44
	33	PB	.920	315	48
	41	PB	.925	465	55
	69	PB	.922	700	72
	71	PB	.922	1000	85
	75	PB	.921	1500	95
Omala	33	PB	.884	315	53
	41	PB	.890	500	65
	69	PB	.895	700	78
	71	PB	.893	1000	89
	75	PB	.893	1500	114
	77	PB	.895	2150	145
	81	PB	.900	3150	165
	85	PB	.907	4650	209
	96	PB	.978	44000	550
Sohio (The Standard Oil Co.) Cleveland, OH					
Industron	32	PB	.828	34	—
	34	PB	.861	60	35
	37	PB	.860	80	37
	40	PB	.871	105	40
	44	PB	.876	157	44
	48	PB	.879	212	47
	53	PB	.883	315	53
	66	PB	.890	639	70
	80	PB	.894	950	85
	100	PB	.889	1412	106
	120	PB	.891	2065	132
	156	PB	.898	2900	161
	250	PB	.932	5200	211
Factovis	43	PB	.867	153	43
	47	PB	.874	213	47
	52	PB	.879	315	53
	65	PB	.882	450	61
	80	PB	.879	930	85
Eldoran UTH		PB	.897	278	55.4
Gearep	80W-140	PB	.907	1321	143
	80	PB	.883	350	55
	85	PB	.889	650	72
	90	PB	.893	1025	90
	125	PB	.897	1750	120
	140	PB	.900	2320	143
	170	PB	.902	3110	171
	250	PB	.910	4230	213
	500	PB	.933	—	525
Staysol FR		IE	.9164	440	113 @ 175°F
Stauffer (Stauffer Chemical Co.)					
Fyrguard	150	WG	1.076	150	95 @ 130°F
	200	WG	1.079	200	130 @ 130°F
Fyrquel	90	PE	1.221	96	36.9
	150	PE	1.180	151	40.2
	220	PE	1.170	223	43.2
	300	PE	1.170	327	45.1
	550	PE	1.170	596	47.9
	1000	PE	1.180	969	51.8

filtration

FLUIDS CHART (CONTINUED)

Brand Name		Fluid Type	Specific Gravity	Viscosity (SUS) 100°F	210°F
Sunoco (Sun Oil Co.) Philadelphia, PA					
Solnus	55	PB	.9100	60	—
	70	PB	.9094	70	—
	100	PB	.9076	107	—
	150	PB	.9123	155	—
	200	PB	.9165	207	—
Solnus AC	300	PB	.9248	310	—
	500	PB	.9309	515	—
	650	PB	.9340	675	—
	750	PB	.9371	775	—
	1200	PB	.9402	1250	—
	2200	PB	.9433	2300	—
Suntac	152WR	PB	.9159	210	44
	202WR	PB	.9188	280	46.8
	302WR	PB	.9279	405	51.1
	502WR	PB	.9309	650	57.6
	752WR	PB	.9340	1080	67.5
	1203WR	PB	.9371	2020	83.2
	200HP	PB	.9182	200	43.4
	250HP	PB	.9218	250	45.8
	350HP	PB	.9248	350	49.2
	550HP	PB	.9291	550	55
Sunvis	7	PB	.8473	72.5	36.6
	11	PB	.8571	108	40
	16	PB	.8649	155	43.5
	21	PB	.8692	208	46.8
	31	PB	.8729	308	52.4
	41	PB	.8729	415	59.2
	51	PB	.8751	500	63.3
	75	PB	.8778	750	76.3
	99	PB	.8816	974	87.2
	112	PB	.8849	1465	110
	135	PB	.8888	2079	135
	150	PB	.8911	2572	155
	701	PB	.8602	107	40.2
	706	PB	.8660	155	43.8
Sunvis	747	PB	.8692	208	47.4
	754	PB	.8718	310	53.4
	764	PB	.8778	505	63.8
	775	PB	.8822	725	75
	790	PB	.8849	1050	91
	7100	PB	.8866	1250	100
	7150	PB	.8927	2600	155
	816WR	PB	—	155	44
	821WR	PB	—	210	47.7
	831WR	PB	—	307	53.4
	851WR	PB	—	500	63.4
	865WR	PB	—	650	71.1
Sun R & O	100L	PB	.8591	110	40.3
	150L	PB	.8633	155	43.5
	200L	PB	.8670	208	47.4
	300L	PB	.8718	307	53.2
	500L	PB	.8762	508	64.3
	600L	PB	.8762	620	71.5
	1200L	PB	.8871	1200	98.5
	1800L	PB	.8899	1900	127.6
Lubeway	150	PB	.9248	165	—
	300	PB	.9309	315	—
	1706	PB	.8660	155	—
	1754	PB	.8718	310	—
Hydraulic Oil	2105	PB	.8681	194	52
Sunsafe	F	IE	.921	425	—
Texaco (Texaco, Inc.) Long Island City, NY					
Regal Oil	A R&O	PB	.8681	151	43.8
	B R&O	PB	.8735	208	47.2
	C R&O	PB	.9042	314	47.1
	PC R&O	PB	.8789	310	52.7
	PE R&O	PB	.8822	420	58.4
	ER&O	PB	.9117	500	55.1
	F R&O	PB	.8871	670	70.6
	G R&O	PB	.8905	975	85.0
	H R&O	PB	.8916	1223	95.9

FLUIDS CHART (CONTINUED)

Brand Name		Fluid Type	Specific Gravity	Viscosity (SUS) 100°F	210°F
Texaco (Continued)					
Regal Oil	H	PB	.8922	1230	97.9
	J	PB	.8939	1740	117.5
	K	PB	.8956	2050	137.5
	L	PB	.8961	2480	146.8
Rando Oil	A	PB	.8676	154	43.7
	B	PB	.8729	215	48
	C	PB	.8800	326	53.7
	F	PB	.8894	620	67.8
	G	PB	.8916	920	86.7
Rando Oil HD	150	PB	.8676	146	43.2
	215	PB	.8745	207	47.1
	315	PB	.8800	315	53.2
	700	PB	.8911	720	72.7
	1000	PB	.8944	1004	85.7
	AZ	PB	.8681	154	50.1
Roltex Oil		PB	.9135	103	38
Roll Oil		PB	.8927	57.3	34.2
300 Oil		PB	.8251	39.8	—
Safteytex		PE	1.145	230	—
Hyd. Safety Fluid	200	WG	1.090	208	103 @ 150°F
FR Hydrafluid		IE	.9358	375	150 @ 150°F
Multigear Lub EP	80W	PB	.8990	370	56.8
	90	PB	.9059	991	85.8
	85W-140	PB	.9182	2438	140
Meropa	68	PB	.8899	338	54.4
	150	PB	.8967	725	75.5
	220	PB	.9024	1000	89.4
	320	PB	.9053	1474	112
	460	PB	.9088	2140	138
	680	PB	.9159	3149	175
	1000	PB	.9176	4464	213
	1500	PB	.9248	6605	277
	3200	PB	.9309	17870	450
Geotex HD	30	PB	.8933	552	65
	40	PB	.8973	860	80.2
Ucon (Union Carbide) New York, NY					
Hydrolube	150-CP	WG	1.079	150	78 @ 150°F
	200-CP	WG	1.079	200	95 @ 150°F
	275-CP	WG	1.080	275	124 @ 150°F
	300-CP	WG	1.080	300	150 @ 150°F
	150-LT	WG	1.08	150	78 @ 150°F
	200-LT	WG	1.08	200	95 @ 150°F
	275-LT	WG	1.08	275	124 @ 150°F
	300-LT	WG	1.08	300	150 @ 150°F
Union 76 (Union Oil Co.) Palatine, IL					
Turbine Oil XD	150	PB	.865	150	43
	215	PB	.871	215	47
	315	PB	.876	300	52
	465	PB	.882	435	60.5
	700	PB	.884	700	75
	1000	PB	.887	1000	91
Unax AW	150	PB	.865	150	43
	215	PB	.871	215	47
	315	PB	.876	315	52
	465	PB	.882	465	62
	700	PB	.884	700	75
	1500	PB	.890	1500	114
Unax RX	105	PB	.858	105	39.8
	150	PB	.865	150	43.2
	215	PB	.871	215	47
	315	PB	.876	300	52
	465	PB	.882	435	60.5
	700	PB	.884	700	75
	1000	PB	.887	1000	91
Unax R & O	150	PB	.865	150	43
	215	PB	.871	215	47
	315	PB	.876	300	52
	465	PB	.882	435	60.5
	700	PB	.884	700	75
	1000	PB	.887	1000	91
FR Fluid		IE	.957	423	—

FLUIDS CHART (CONTINUED)

Brand Name		Fluid Type	Specific Gravity	Viscosity (SUS) 100°F	210°F
White & Bagley Co. Worcester, MA					
EP Hyd. Oil	150	PB	.8633	150	44
	225	PB	.8686	225	48
	300	PB	.8708	300	54
Super Hyd. Oil	100	PB	.8612	100	41
	150	PB	.8633	150	44
	225	PB	.8686	225	48
	300	PB	.8708	300	54
	600	PB	.8772	600	69
Penn-Mar Super	150	PB	.8633	150	44
	225	PB	.8686	225	48
	300	PB	.8708	300	54
	450	PB	.8772	450	69
Penn-Mar EP	150	PB	.8633	150	44
	225	PB	.8686	225	48
	300	PB	.8708	300	54
Arthur C. Withrow Co. Los Angeles, CA					
"S" Series	LT	PB	.871	150	43.4
	MED	PB	.871	230	48.6
	HVY-MED	PB	.874	305	53
	HVY	PB	.882	420	59.5
	EXT-HVY	PB	.889	600	68.2
	700	PB	.925	1100	74.5
	1900	PB	.9024	1950	126
Withrolube 624 SE	HYD	PB	.877	270	55
	655 AW HYD-LT	PB	.873	160	44
	656 AW HYD-MED	PB	.876	230	49
	657 AW HYD-HVY-MED	PB	.877	305	53
AP Gear Oil	80	PB	.896	725	76
	90	PB	.896	1100	91
	140	PB	.901	2200	140
Withrolube	632	PB	.925	336	56
	635	PB	.913	650	71
	637	PB	.925	1100	91
	637-1500	PB	.929	1500	112
	637-1800	PB	.930	1800	123
	638	PB	.935	2300	138
Trip-L-Film Custom	10	PB	.880	175	45
	20	PB	.888	325	55
	30	PB	.892	545	66.5
	40	PB	.901	790	78
	50	PB	.902	1375	108
3V3	20	PB	.890	290	54
	30	PB	.894	640	66
Withrolube	628	PB	.898	205	52
Wolverine Oil & Supply Co. Detroit, MI					
A-1 Anti-Wear Hydraulic Oils	150AW	PB	.8686	150-160	44
	200AW	PB	.8708	200-215	47
	300AW	PB	.8724	300-315	53
	500AW	PB	.8762	475-510	64
	600AW	PB	.8762	590-620	69
	1000AW	PB	.8816	950-1000	86

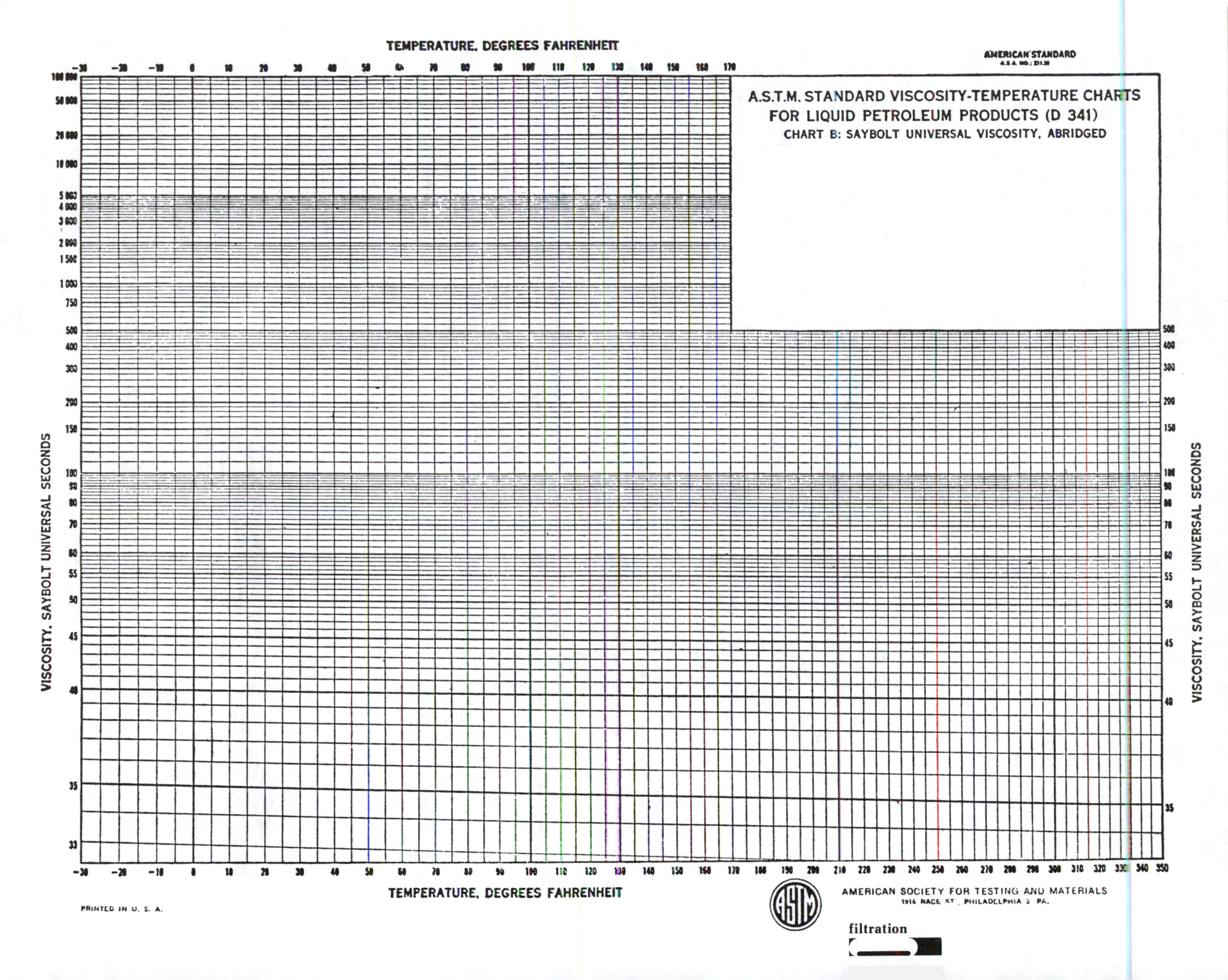
AMERICAN STANDARD
A.S.T.M. STANDARD VISCOSITY-TEMPERATURE CHARTS
FOR LIQUID PETROLEUM PRODUCTS (D 341)
CHART B: SAYBOLT UNIVERSAL VISCOSITY, ABRIDGED
TEMPERATURE, DEGREES FAHRENHEIT
VISCOSITY, SAYBOLT UNIVERSAL SECONDS
AMERICAN SOCIETY FOR TESTING AND MATERIALS
1916 RACE ST., PHILADELPHIA 3 PA.
PRINTED IN U. S. A.
filtration

Explanation of Flow vs. Pressure Differential Tables

Tables 1 through 9 on the following pages indicate estimated pressure differential per linear foot of pipe, tube, and hose for fluid flows up to 200 GPM at ten viscosities. Values were determined by computer and are rounded off to the fourth decimal place.

Tables 1 through 6 are calculated in units of PSI. In tables 7 through 9, pressure units are indicated in terms of inches of mercury (in.Hg); this is used for calculations at the pump suction side.

Fluid flow in a conductor can exist in either a laminar or turbulent form. It can also be in transition between laminar and turbulent. These conditions of flowing fluid are indicated on the tables by a heavy line running through each chart from top to bottom. To the right of the line, flow is laminar. To the left, flow is turbulent. Shaded areas are transitional regions known as critical flow zones.

In the illustration, a reproduction of flow in a circular plastic pipe is shown. A dye is injected at a point mid-stream into the flow. A series of drawings illustrate laminar flow (a), the critical zone (b), and finally turbulent flow (c) as fluid velocity progressively increases.

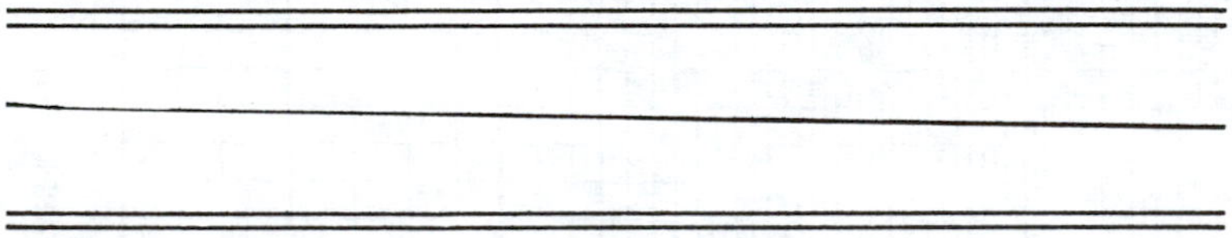

(a) Laminar Flow

(b) Critical Zone

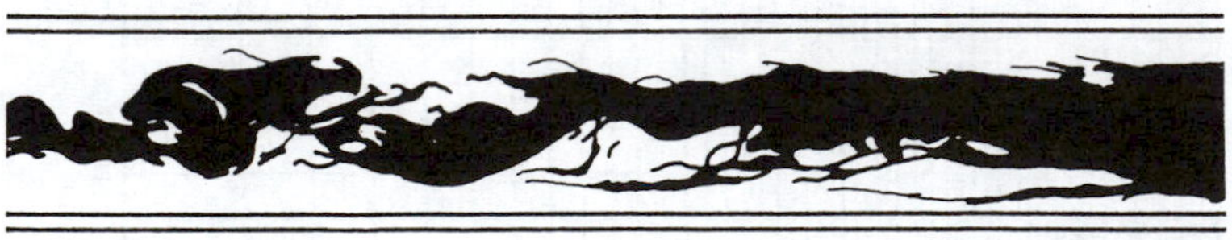

(c) Turbulent Flow

An increase in velocity is not the only cause affecting a change between laminar and turbulent. Factors such as pipe diameter, specify gravity (fluid density), and viscosity also affect laminar and turbulent states.

Reynolds number is a unitless number which takes into account the effects of velocity, viscosity, specific gravity, and pipe diameter on a flowing fluid. With a Reynolds number of less than 2000, flow is laminar. As Reynolds number increases, flow tends to become turbulent, with the critical zone existing between 2000 and 5000. The critical zone is rather complicated and not fully understood; but, under normal conditions, flow will be fully turbulent as Reynolds number equals 5000.

Values given for the critical zone in the tables are for fully developed turbulent flow. This indicates the maximum value of pressure differential which could be expected to occur under the specific flow conditions.

When turbulent flow is obtained, a rapid mixing takes place. Compared with the smooth flowing action of laminar flow, this action causes additional pressure energy to be used.

At times, this results in an apparent contradiction in the tables when going from turbulent flow or the critical zone, to laminar flow (left to right on the chart). For example, with fluid velocity remaining constant, pressure differential at 200 SUS might be more than at 300 SUS for a specific conductor size. This is because flow is turbulent at 200 SUS and laminar at 300 SUS.

Flow GPM	Tube Size	Tube OD	Tube Wall	Tube ID	Velocity FPS	Viscosity — SUS 100	Viscosity — SUS 400
50	1½	1.5	.120	1.260	12.864	.3800	.4708
			.134	1.232	13.455	.4228	.5151
			.156	1.188	14.470	.5026	.5958
			.188	1.124	16.165	.6538	.7435
	1¾	1.75	.065	1.620	7.782	.1152	.1723
			.072	1.606	7.918	.1200	.1784
			.083	1.584	8.140	.1282	.1885
			.095	1.560	8.392	.1378	.2004
			.109	1.532	8.702	.1502	.2154
			.120	1.510	8.957	.1609	.2283
			.134	1.482	9.299	.1758	.2460
			.156	1.438	9.876	.2029	.2775
			.188	1.374	10.818	.2519	.3330

calculating pressure differential in straight line lengths

To calculate fluid flow pressure differential in straight line lengths lying in a horizontal plane, several factors must be known. These are conductor size and length, flow rate, fluid viscosity, and specific gravity.

Assume that 50 GPM is flowing through 45 ft. of 1-3/4 in. tube with a .083 in. wall. Fluid specific gravity is .865. Assume that pressure differential at a viscosity of 100 SUS and 400 SUS must be determined.

Referring to Table 5, we find a flow rate of 50 GPM through 1-3/4 in. tube with a .083 in. wall. The table indicates a pressure differential of .1282 PSI/ft @ 100 SUS and .1885 PSI/ft @ 400 SUS. These values are based on the fluid having a specific gravity of 1.0. Since the fluid in our example has a specific gravity of .865, pressure differential in this instance will be less. To determine how much less, the Specific Gravity Correction Curve is referred to. The curve indicates that the correction factor for a .865 specific gravity is .9. This factor is then multiplied by the pressure differential values for 100 SUS and 400 SUS.

100 SUS ΔP = .1282 PSI/ft x .9 = .1154 PSI/ft

400 SUS ΔP = .1885 PSI/ft x .9 = .1697 PSI/ft

With 45 FT. of tubing, we need only multiply the PSI/FT value by 45 ft. to obtain the total pressure differential.

100 SUS ΔP = .1154 PSI/ft x 45 ft = 5.2 PSI

400 SUS ΔP = .1697 PSI/ft x 45 ft = 7.6 PSI

To determine pressure differential from the tables for horizontal straight line lengths, the following formula can be used:

$$\text{Horizontal Straight Line Length } \Delta P \text{ (PSI)} = \Delta P \text{ (PSI/ft) @ Fluid Viscosity} \times \text{Specific Gravity Correction Factor} \times \text{Line Length (ft)}$$

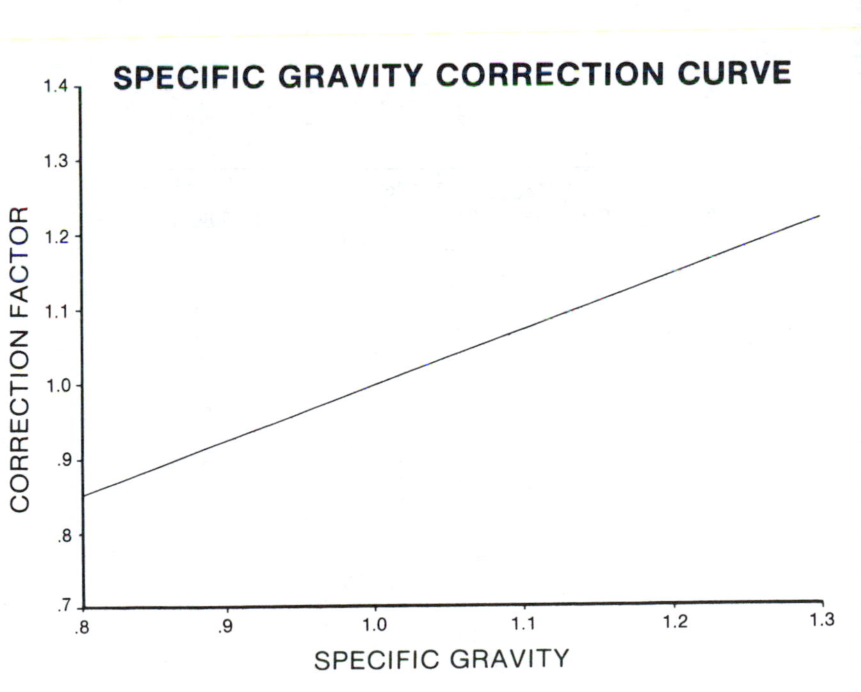

Flow GPM	Pipe — Schedule 80				Velocity FPS	Viscosity — SUS
	Size	OD	Wall	ID		100
25	½	.840	.147	.546	34.253	6.5623
	¾	1.050	.154	.742	18.547	1.4751
	1	1.315	.179	.957	11.150	.4366
	1¼	1.660	.191	1.278	6.252	.1101
30	¾	1.050	.154	.742	22.257	2.0412
	1	1.315	.179	.957	13.380	.5973
	1¼	1.660	.191	1.278	7.503	.1507
	1½	1.900	.200	1.5	5.446	.0704
40	¾	1.050	.154	.742	29.675	3.3851
	1	1.315	.179	.957	17.839	.9890
	1¼	1.660	.191	1.278	10.003	.2473
	1½	1.900	.200	1.5	7.261	.1150
50	¾	1.050	.154	.742	37.094	5.1541
	1	1.315	.179	.957	22.299	1.4820
	1¼	1.660	.191	1.278	12.504	.3643
	1½	1.900	.200	1.5	9.077	.1699
	2	2.375	.218	1.939	5.432	.0500

Flow GPM	Tube				Velocity FPS	Viscosity — SUS
	Size	OD	Wall	ID		100
50	1½	1.5	.120	1.260	12.864	.3800
			.134	1.232	13.455	.4228
			.156	1.188	14.470	.5026
			.188	1.124	16.165	.6538
	1¾	1.75	.065	1.620	7.782	.1152
			.072	1.606	7.918	.1200
			.083	1.584	8.140	.1282
			.095	1.560	8.392	.1378
			.109	1.532	8.702	.1502
			.120	1.510	8.957	.1609
			.134	1.482	9.299	.1758
			.156	1.438	9.876	.2029
			.188	1.374	10.818	.2519

calculating pressure differential in fittings and bends

As indicated, the tables show pressure differential per linear foot of pipe, tube, and hose which are in horozontal plane. They may also be used to determine pressure differential through typical pipe fittings and 90° bends of tube and hose. This is accomplished with the use of a multiplier which transforms a fitting or bend into a length of straight conductor exhibiting an equivalent pressure differential.

Illustrated on the next page is a chart which shows various pipe fittings and 90° bends with their appropriate multipliers. To determine pressure differential, the multiplier is used in the following formula:

$$\text{Fitting or Bend } \Delta P \text{ (PSI)} = \text{Fitting or Bend Multiplier} \times \text{Conductor Inside Dia. (in.)} \times \Delta P \text{ (PSI/ft) @ Fluid Viscosity} \times \text{Specific Gravity Correction Factor}$$

Assume 50 GPM is flowing through 1-1/4 in. schedule 80 pipe. Viscosity is 100 SUS and specific gravity is .865. To calculate pressure differential for one elbow, the chart is referred to for the multiplier which happens to be 2.5 From Table 2, 1-1/4 in. schedule 80 pipe has an inside diameter of 1.278 in. and a pressure differential per foot @ 50 GPM and 100 SUS of .3643 PSI. And, referring to the correction curve for specific gravity, we find that .865 has a correction factor of .9.

To solve for ΔP in one elbow, the formula becomes:

$$\Delta P \text{ 1-1/4 in. Schedule 80 Elbow (PSI)} = 2.5 \times 1.278 \text{ in.} \times .3643 \text{ PSI/ft} \times .9 = 1 \text{ PSI}$$

This indicates that a typical schedule 80 elbow will use approximately 1 PSI as 50 GPM passes through at 100 SUS.

Now, assume that 50 GPM passes through 1-3/4 in. tube with a .083 in. wall at a viscosity of 100 SUS and a specific gravity of .865. Pressure differential for each 90° bend is calculated in the following manner:

$$\Delta P \text{ 90° Tube Bend (PSI)} = 1 \times 1.584 \text{ in.} \times .1282 \text{ PSI/ft} \times .9 = .2 \text{ PSI}$$

Compared to the pipe elbow, this is a substantial pressure saving.

NOTE: If additional information is required regarding the derivation of the Flow vs. Pressure Differential Tables, please contact:

TRAINING DEPARTMENT
Parker Fluidpower
17325 Euclid Ave.
Cleveland, Ohio 44112
1-(216) 531-3000

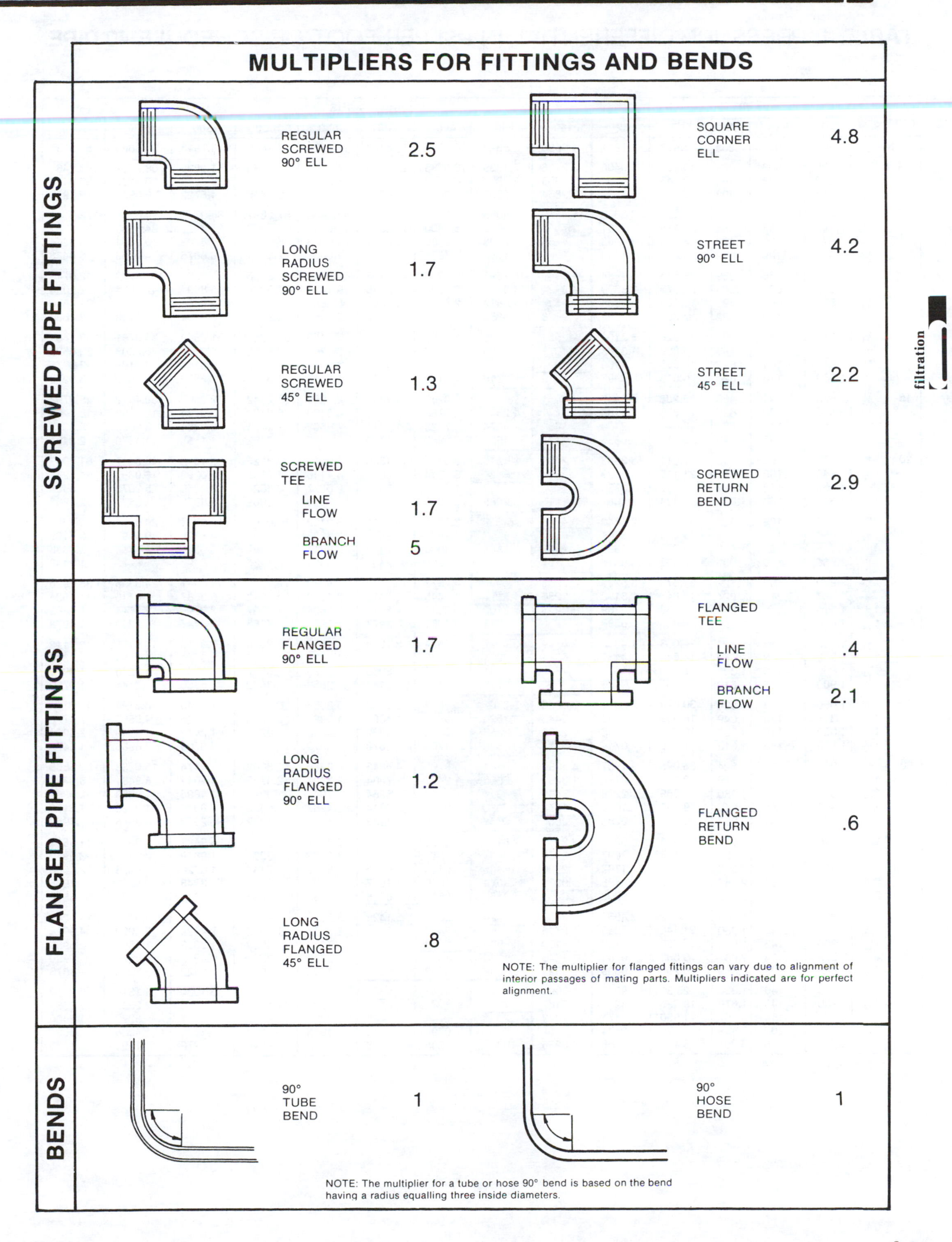
MULTIPLIERS FOR FITTINGS AND BENDS
SCREWED PIPE FITTINGS
REGULAR SCREWED 90° ELL
2.5
SQUARE CORNER ELL
4.8
LONG RADIUS SCREWED 90° ELL
1.7
STREET 90° ELL
4.2
REGULAR SCREWED 45° ELL
1.3
STREET 45° ELL
2.2
SCREWED TEE
LINE FLOW
1.7
BRANCH FLOW
5
SCREWED RETURN BEND
2.9
FLANGED PIPE FITTINGS
REGULAR FLANGED 90° ELL
1.7
FLANGED TEE
LINE FLOW
.4
BRANCH FLOW
2.1
LONG RADIUS FLANGED 90° ELL
1.2
FLANGED RETURN BEND
.6
LONG RADIUS FLANGED 45° ELL
.8
NOTE: The multiplier for flanged fittings can vary due to alignment of interior passages of mating parts. Multipliers indicated are for perfect alignment.
BENDS
90° TUBE BEND
1
90° HOSE BEND
1
NOTE: The multiplier for a tube or hose 90° bend is based on the bend having a radius equalling three inside diameters.

TABLE 1: PRESSURE DIFFERENTIAL IN PSI PER FOOT OF SCHEDULE 40 PIPE

Flow GPM	Pipe — Schedule 40				Velocity FPS	Viscosity — SUS									
	Size	OD	Wall	ID		32	100	200	400	600	800	1000	1500	2000	3000
1.5	1/8	.405	.068	.269	8.467	.7911	1.6098	3.5870	6.7993	10.3107	13.7524	17.1869	25.7913	34.3883	51.873
3	1/8	.405	.068	.269	16.934	3.0406	3.2196	7.1739	13.5985	20.6214	27.5049	34.3738	51.5825	68.7767	103.746
	1/4	.540	.088	.364	9.248	.6451	.9603	2.1397	4.0560	6.1507	8.2038	10.2525	15.3853	20.5137	30.9439
	3/8	.675	.091	.493	5.042	.1372	.2854	.6359	1.2054	1.8278	2.4380	3.0468	4.5722	6.0963	9.1959
5	1/8	.405	.068	.269	28.224	8.3509	11.8789	11.9565	22.6642	34.369	45.8414	57.2897	85.9709	114.628	172.91
	1/4	.540	.088	.364	15.414	1.7231	2.7550	3.5662	6.7600	10.2511	13.6729	17.0876	25.6422	34.1896	51.5731
	3/8	.675	.091	.493	8.403	.3603	.4756	1.0598	2.0089	3.0464	4.0633	5.0781	7.6203	10.1604	15.3264
	1/2	.840	.109	.622	5.279	.1108	.1877	.4183	.7928	1.2023	1.6036	2.0041	3.0075	4.0099	6.0488
7	1/8	.405	.068	.269	39.513	16.1494	21.5145	16.7391	31.7298	48.1166	64.178	80.2056	120.359	160.479	242.074
	1/4	.540	.088	.364	21.579	3.3270	4.9234	4.9927	9.4639	14.3515	19.1421	23.9226	35.899	47.8654	72.2024
	3/8	.675	.091	.493	11.764	.6886	1.1526	1.4837	2.8125	4.2650	5.6886	7.1093	10.6684	14.2246	21.457
	1/2	.840	.109	.622	7.390	.2098	.2628	.5856	1.1100	1.6832	2.2451	2.8058	4.2104	5.6139	8.4683
10	1/4	.540	.088	.364	30.828	6.7104	9.2531	7.1325	13.5199	20.5022	27.3459	34.1751	51.2843	68.3791	103.146
	3/8	.675	.091	.493	16.806	1.3765	2.1278	2.1196	4.0178	6.0928	8.1266	10.1561	15.2406	20.3208	30.6529
	1/2	.840	.109	.622	10.558	.4162	.6991	.8365	1.5857	2.4046	3.2073	4.0082	6.0149	8.0199	12.0975
	3/4	1.050	.113	.824	6.016	.0995	.1219	.2716	.5148	.7807	1.0413	1.3014	1.9529	2.6039	3.9278
15	3/8	.675	.091	.493	25.208	3.0443	4.3451	5.3512	6.0267	9.1392	12.1899	15.2342	22.8609	30.4812	45.9793
	1/2	.840	.109	.622	15.836	.9136	1.4131	1.2548	2.3785	3.6069	4.8109	6.0124	9.0224	12.0298	18.1463
	3/4	1.050	.113	.824	9.024	.2160	.3681	.4074	.7723	1.1711	1.562	1.9521	2.9294	3.9058	5.8917
	1	1.315	.133	1.049	5.568	.0635	.1171	.1551	.2940	.4459	.5947	.7432	1.1153	1.4870	2.2431
20	3/8	.675	.091	.493	33.611	5.3633	7.2292	8.7516	8.0357	12.1856	16.2532	20.3122	30.4812	40.6417	61.3057
	1/2	.840	.109	.622	21.115	1.6029	2.3304	2.8830	3.1714	4.8092	6.4146	8.0165	12.0298	16.0397	24.1951
	3/4	1.050	.113	.824	12.032	.3726	.6057	.5432	1.0297	1.5615	2.0827	2.6028	3.9058	5.2078	7.8556
	1	1.315	.133	1.049	7.424	.1087	.1916	.2068	.3920	.5945	.7929	.9909	1.4870	1.9827	2.9908
25	1/2	.840	.109	.622	26.394	2.4840	3.4718	4.2337	3.9642	6.0115	8.0182	10.0206	15.0373	20.0497	30.2439
	3/4	1.050	.113	.824	15.039	.5744	.8897	1.1112	1.2871	1.9518	2.6033	3.2535	4.8823	6.5097	9.8195
	1	1.315	.133	1.049	9.280	.1667	.2816	.2585	.4900	.7431	.9911	1.2387	1.8588	2.4784	3.7385
	1 1/4	1.660	.140	1.380	5.362	.0416	.0765	.0863	.1636	.2481	.3309	.4136	.6206	.8275	1.2482
30	1/2	.840	.109	.622	31.673	3.5235	4.7953	5.8099	4.7571	7.2138	9.6218	12.0247	18.0447	24.0596	36.2926
	3/4	1.050	.113	.824	18.047	.8193	1.2286	1.5177	1.5445	2.3422	3.1240	3.9042	5.8587	7.8116	11.7834
	1	1.315	.133	1.049	11.136	.2369	.3845	.3102	.5880	.8917	1.1894	1.4864	2.2305	2.9741	4.4862
	1 1/4	1.660	.140	1.380	6.434	.0588	.1046	.1036	.1963	.2977	.3971	.4963	.7447	.9930	1.4978
40	3/4	1.050	.113	.824	24.063	1.4388	2.0349	2.4927	2.0594	3.1229	4.1653	5.2056	7.8116	10.4155	15.7112
	1	1.315	.133	1.049	14.848	.4137	.6384	.7901	.7840	1.1890	1.5858	1.9819	2.9741	3.9654	5.9816
	1 1/4	1.660	.140	1.380	8.579	.1018	.1712	.1381	.2618	.3969	.5295	.6617	.9930	1.3240	1.9971
	1 1/2	1.900	.145	1.610	6.303	.0466	.0826	.0745	.1413	.2143	.2858	.3572	.5360	.7146	1.0780
50	3/4	1.050	.113	.824	30.079	2.2309	3.0956	3.6762	4.3753	3.9036	5.2067	6.5069	9.7645	13.0194	19.639
	1	1.315	.133	1.049	18.559	.6392	.9446	1.1598	.9800	1.4862	1.9823	2.4773	3.7176	4.9568	7.4770
	1 1/4	1.660	.140	1.380	10.724	.1565	.2533	.3158	.3272	.4962	.6618	.8271	1.2412	1.6549	2.4964
	1 1/2	1.900	.145	1.610	7.879	.0714	.1211	.1526	.1766	.2678	.3572	.4465	.6700	.8933	1.3475
60	3/4	1.050	.113	.824	36.095	3.1957	4.3086	5.0364	5.9795	4.6843	6.2480	7.8083	11.7174	15.6233	23.5668
	1	1.315	.133	1.049	22.271	.9132	1.3291	1.5905	1.9043	1.7834	2.3787	2.9728	4.4611	5.9481	8.9724
	1 1/4	1.660	.140	1.380	12.869	.2226	.3468	.4314	.3927	.5954	.7942	.9925	1.4895	1.9859	2.9957
	1 1/2	1.900	.145	1.610	9.455	.1014	.1658	.2080	.2119	.3214	.4287	.5358	.8040	1.0720	1.6170
	2	2.375	.154	2.067	5.736	.0286	.0506	.0641	.0780	.1183	.1578	.1972	.2959	.3946	.5952
80	1	1.315	.133	1.049	29.695	1.6071	2.2152	2.6183	3.1125	2.3779	3.1717	3.9637	5.9481	7.9308	11.9632
	1 1/4	1.660	.140	1.380	17.158	.3897	.5866	.7085	.8505	.7939	1.0589	1.3234	1.9859	2.6479	3.9942
	1 1/2	1.900	.145	1.610	12.606	.1767	.2790	.3405	.2826	.4285	.5716	.7143	1.0720	1.4293	2.1560
	2	2.375	.154	2.067	7.648	.0495	.0833	.1044	.1040	.1577	.2104	.2629	.3946	.5261	.7936
	2 1/2	2.875	.203	2.469	5.360	.0202	.0358	.0452	.0511	.0775	.1033	.1292	.1938	.2584	.3898
100	1	1.315	.133	1.049	37.119	2.4953	3.3292	3.8720	4.5721	5.1429	3.9646	4.9547	7.4351	9.9135	14.954
	1 1/4	1.660	.140	1.380	21.448	.6029	.8718	1.0401	1.2435	.9924	1.3237	1.6542	2.4824	3.3099	4.9928
	1 1/2	1.900	.145	1.610	15.758	.2727	.4130	.4980	.6004	.5357	.7145	.8929	1.3399	1.7866	2.6950
	2	2.375	.154	2.067	9.560	.0761	.1248	.1529	.1300	.1972	.2630	.3287	.4932	.6576	.9920
	2 1/2	2.875	.203	2.469	6.700	.0309	.0525	.0660	.0639	.0969	.1292	.1614	.2423	.3230	.4873
125	1 1/4	1.660	.140	1.380	26.81	.9342	1.3061	1.5348	1.8234	2.0577	1.6546	2.0678	3.1030	4.1374	6.2410
	1 1/2	1.900	.145	1.610	19.697	.4216	.6167	.7357	.8782	.6696	.8931	1.1162	1.6749	2.2332	3.3687
	2	2.375	.154	2.067	11.950	.1171	.1843	.2233	.2701	.2465	.3287	.4108	.6165	.8220	1.2400
	2 1/2	2.875	.203	2.469	8.376	.0474	.0787	.0967	.0798	.1211	.1615	.2018	.3028	.4038	.6091
	3	3.500	.216	3.068	5.424	.0158	.0276	.0347	.0335	.0508	.0677	.0846	.1270	.1694	.2555

TABLE 1 (CONTINUED)

Flow GPM	Pipe — Schedule 40				Velocity FPS	Viscosity — SUS									
	Size	OD	Wall	ID		32	100	200	400	600	800	1000	1500	2000	3000
150	1¼	1.660	.140	1.380	32.172	1.3377	1.8039	2.1393	2.4980	2.8069	1.9855	2.4814	3.7236	4.9648	7.4892
	1½	1.900	.145	1.610	23.637	.6027	.8506	1.0050	1.2007	1.3554	1.0717	1.3394	2.0099	2.6799	4.0425
	2	2.375	.154	2.067	14.340	.1668	.2552	.3069	.3682	.2958	.3945	.4930	.7398	.9864	1.4880
	2½	2.875	.203	2.469	10.051	.0673	.1084	.1317	.1594	.1453	.1938	.2422	.3634	.4845	.7309
	3	3.500	.216	3.068	6.509	.0224	.0384	.0473	.0402	.0609	.0813	.1016	.1524	.2032	.3066
175	1¼	1.660	.140	1.380	37.534	1.8132	2.3892	2.8166	3.2489	3.6549	3.9748	2.8949	4.3442	5.7923	8.7374
	1½	1.900	.145	1.610	27.576	.8159	1.1238	1.3472	1.5570	1.7619	1.2504	1.5626	2.3449	3.1265	4.7162
	2	2.375	.154	2.067	16.730	.2252	.3367	.4002	.4791	.5426	.4602	.5752	.8631	1.1508	1.7359
	2½	2.875	.203	2.469	11.726	.0907	.1427	.1722	.2070	.1695	.2261	.2825	.4240	.5653	.8527
	3	3.500	.216	3.068	7.594	.0301	.0502	.0613	.0745	.0711	.0948	.1185	.1778	.2371	.3577
	3½	4.00	.226	3.548	5.678	.0145	.0251	.0310	.0262	.0397	.0530	.0663	.0994	.1326	.2000
200	1½	1.900	.145	1.610	31.515	1.0614	1.4322	1.6984	1.9647	2.2138	2.4099	1.7858	2.6799	3.5732	5.3900
	2	2.375	.154	2.067	19.120	.2924	.4244	.5045	.6024	.6803	.5260	.6573	.9864	1.3152	1.9839
	2½	2.875	.203	2.469	13.401	.1176	.1812	.2163	.2599	.1937	.2584	.3229	.4845	.6461	.9746
	3	3.500	.216	3.068	8.679	.0389	.0637	.0773	.0934	.0812	.1084	.1354	.2032	.2710	.4088
	3½	4.00	.226	3.548	6.489	.0187	.0317	.0388	.0472	.0454	.0606	.0757	.1136	.1515	.2285

TABLE 2: PRESSURE DIFFERENTIAL IN PSI PER FOOT OF SCHEDULE 80 PIPE

Flow GPM	Pipe — Schedule 80				Velocity FPS	Viscosity — SUS									
	Size	OD	Wall	ID		32	100	200	400	600	800	1000	1500	2000	3000
1.5	1/8	.405	.095	.215	13.254	2.5212	3.9448	8.7899	16.6616	25.2664	33.7004	42.1166	63.2016	84.2687	127.115
	1/4	.540	.119	.302	6.718	.4390	1.0133	2.2579	4.2800	6.4904	8.6569	10.8188	16.2351	21.6467	32.6529
3	1/8	.405	.095	.215	26.509	9.7023	14.3856	17.5797	33.3232	50.5328	67.4008	84.2332	126.403	168.537	254.23
	1/4	.540	.119	.302	13.435	1.6858	2.0266	4.5158	8.5600	12.9807	17.3137	21.6376	32.4701	43.2935	65.3059
	3/8	.675	.126	.423	6.848	.2977	.5266	1.1733	2.2240	3.3726	4.4984	5.6218	8.4363	11.2484	16.9675
5	1/4	.540	.119	.302	22.392	4.5615	6.7730	7.5264	14.2666	21.6346	28.8562	36.0627	54.1169	72.1558	108.843
	3/8	.675	.126	.423	11.414	.7951	.8776	1.9555	3.7067	5.6210	7.4973	9.3697	14.0604	18.7473	28.2792
	1/2	.840	.147	.546	6.851	.2142	.3161	.7044	1.3353	2.0249	2.7008	3.3753	5.0651	6.7535	10.1873
7	1/4	.540	.119	.302	31.349	8.7877	12.2013	10.5369	19.9733	30.2884	40.3987	50.4877	75.7636	101.018	152.38
	3/8	.675	.126	.423	15.980	1.5165	2.3870	2.7377	5.1894	7.8694	10.4962	13.1175	19.6846	26.2462	39.5909
	1/2	.840	.147	.546	9.591	.4079	.4426	.9862	1.8694	2.8349	3.7812	4.7255	7.0912	9.4549	14.2622
	3/4	1.050	.154	.742	5.193	.0859	.1298	.2892	.5481	.8312	1.1086	1.3855	2.0791	2.7721	4.1816
10	3/8	.675	.126	.423	22.828	3.0449	4.4565	3.9110	7.4134	11.242	14.9946	18.7393	28.1209	37.4945	56.5584
	1/2	.840	.147	.546	13.701	.8127	1.3033	1.4089	2.6706	4.0498	5.4017	6.7506	10.1302	13.507	20.3746
	3/4	1.050	.154	.742	7.419	.1692	.3022	.4131	.7830	1.1874	1.5837	1.9793	2.9701	3.9602	5.9737
15	3/8	.675	.126	.423	34.242	6.7598	9.1700	11.0746	11.1201	16.863	22.492	28.109	42.1813	56.2418	84.8377
	1/2	.840	.147	.546	20.552	1.7921	2.6489	2.1133	4.0059	6.0747	8.1025	10.126	15.1954	20.2605	30.5618
	3/4	1.050	.154	.742	11.128	.3690	.6068	.6196	1.1745	1.7811	2.3756	2.9689	4.4552	5.9403	8.9606
	1	1.315	.179	.957	6.690	.1010	.1808	.2239	.4244	.6436	.8585	1.0729	1.6100	2.1467	3.2382
20	1/2	.840	.147	.546	27.403	3.1519	4.3897	5.3667	5.3412	8.0996	10.8033	13.5013	20.2605	27.014	40.7491
	3/4	1.050	.154	.742	14.838	.6386	1.0014	.8262	1.5660	2.3748	3.1675	3.9585	5.9403	7.9203	11.9474
	1	1.315	.179	.957	8.920	.1734	.2965	.2986	.5659	.8582	1.1447	1.4305	2.1467	2.8623	4.3176
	1 1/4	1.660	.191	1.278	5.002	.0401	.0752	.0939	.1779	.2698	.3599	.4498	.6750	.9000	1.3576
25	1/2	.840	.147	.546	34.253	4.8922	6.5623	7.9043	6.6765	10.1245	13.5041	16.8766	25.3256	33.7675	50.9364
	3/4	1.050	.154	.742	18.547	.9864	1.4751	1.8269	1.9575	2.9685	3.9593	4.9481	7.4253	9.9004	14.9342
	1	1.315	.179	.957	11.150	.2664	.4366	.3732	.7074	1.0728	1.4308	1.7882	2.6834	3.5778	5.3970
	1 1/4	1.660	.191	1.278	6.252	.0612	.1101	.1173	.2224	.3373	.4499	.5623	.8473	1.1250	1.6970
30	3/4	1.050	.154	.742	22.257	1.4091	2.0412	2.4994	2.3490	3.5621	4.7512	5.9377	8.9104	11.8805	17.9211
	1	1.315	.179	.957	13.380	.3792	.5973	.7466	.8489	1.2873	1.7170	2.1458	3.2201	4.2934	6.4764
	1 1/4	1.660	.191	1.278	7.503	.0867	.1507	.1408	.2669	.4048	.5399	.6747	1.0125	1.34	2.0364
	1 1/2	1.900	.200	1.5	5.446	.0386	.0704	.0742	.1406	.2133	.2845	.3555	.5335	.7114	1.0730
40	3/4	1.050	.154	.742	29.675	2.4792	3.3851	4.1164	3.1320	4.7495	6.3349	7.9170	11.8805	15.8407	23.8948
	1	1.315	.179	.957	17.839	.6637	.9890	1.2218	1.1319	1.7164	2.2893	2.8611	4.2934	5.7246	8.6352
	1 1/4	1.660	.191	1.278	10.003	.1505	.2473	.3105	.3559	.5397	.7198	.8996	1.3500	1.8000	2.7152
	1 1/2	1.900	.200	1.5	7.261	.0667	.1150	.0989	.1875	.2844	.3793	.4740	.7114	.9485	1.4307
50	3/4	1.050	.154	.742	37.094	3.8489	5.1541	6.0483	7.1967	5.9369	7.9187	9.8962	14.8506	19.8008	29.8685
	1	1.315	.179	.957	22.299	1.0269	1.4820	1.7965	1.4148	2.1455	2.8617	3.5763	5.3668	7.1557	10.794
	1 1/4	1.660	.191	1.278	12.504	.2317	.3643	.4542	.4449	.6746	.8998	1.1245	1.6875	2.2500	3.3940
	1 1/2	1.900	.200	1.5	9.077	.1023	.1699	.2130	.2344	.3555	.4741	.5925	.8892	1.1856	1.7884
	2	2.375	.218	1.939	5.432	.0279	.0500	.0443	.0840	.1273	.1698	.2122	.3185	.4246	.6405
60	1	1.315	.179	.957	26.759	1.4688	2.0675	2.4524	2.9395	2.5746	3.434	4.2916	6.4401	8.5868	12.9528
	1 1/4	1.660	.191	1.278	15.005	.3301	.5025	.6211	.5338	.8095	1.0798	1.3494	2.0250	2.7000	4.0727
	1 1/2	1.900	.200	1.5	10.892	.1454	.2332	.2907	.2813	.4266	.5690	.7111	1.0670	1.4227	2.1461
	2	2.375	.218	1.939	6.518	.0395	.0686	.0866	.1007	.1528	.2038	.2547	.3821	.5095	.7686
80	1	1.315	.179	.957	35.679	2.5886	3.4731	4.0740	4.8148	5.4298	4.5787	5.7221	8.5868	11.4491	17.2703
	1 1/4	1.660	.191	1.278	20.007	.5787	.8523	1.0160	1.2220	1.0794	1.4397	1.7992	2.7000	3.5999	5.4303
	1 1/2	1.900	.200	1.5	14.523	.2539	.3914	.4765	.3751	.5688	.7586	.9481	1.4227	1.8969	2.8614
	2	2.375	.218	1.939	8.691	.0685	.1126	.1412	.1343	.2037	.2717	.3395	.5095	.6794	1.0248
	2 1/2	2.875	.276	2.323	6.055	.0275	.0476	.0602	.0652	.0989	.1319	.1648	.2473	.3298	.4975
100	1 1/4	1.660	.191	1.278	25.008	.8963	1.2668	1.5029	1.7889	1.3492	1.7996	2.2490	3.3750	4.4999	6.7879
	1 1/2	1.900	.200	1.5	18.154	.3923	.5817	.6981	.8385	.7110	.9483	1.1851	1.7784	2.3712	3.5768
	2	2.375	.218	1.939	10.864	.1053	.1691	.2070	.1679	.2546	.3396	.4244	.6369	.8492	1.2810
	2 1/2	2.875	.276	2.323	7.569	.0421	.0703	.0880	.0815	.1236	.1649	.2060	.3092	.4122	.6218
125	1 1/4	1.660	.191	1.278	31.260	1.3904	1.9011	2.2114	2.6264	2.9569	2.2495	2.8113	4.2187	5.6249	8.4849
	1 1/2	1.900	.200	1.5	22.692	.6072	.8699	1.0267	1.2278	.8887	1.1853	1.4814	2.2230	2.9640	4.4710
	2	2.375	.218	1.939	13.580	.1623	.2508	.3028	.3650	.3183	.4245	.5305	.7961	1.0615	1.6012
	2 1/2	2.875	.276	2.323	9.461	.0646	.1056	.1290	.1019	.1545	.2061	.2575	.3865	.5123	.7773
	3	3.5	.300	2.900	6.071	.0210	.0362	.0452	.0419	.0636	.0848	.1060	.1591	.2122	.3200

TABLE 2 (CONTINUED)

Flow GPM	Pipe — Schedule 80 Size	OD	Wall	ID	Velocity FPS	Viscosity SUS 32	100	200	400	600	800	1000	1500	2000	3000
150	1¼	1.660	.191	1.278	37.512	1.9923	2.6281	3.1067	3.5818	4.0376	4.3951	3.3735	5.0624	6.7499	10.1819
	1½	1.900	.200	1.5	27.230	.8687	1.2005	1.4099	1.6802	1.8927	1.4224	1.7776	2.6676	3.5568	5.3652
	2	2.375	.218	1.939	16.296	.2315	.3477	.4164	.4979	.3819	.5094	.6366	.9554	1.2738	1.9215
	2½	2.875	.276	2.323	11.354	.0919	.1453	.1759	.2123	.1854	.2473	.3090	.4638	.6183	.9327
	3	3.5	.300	2.900	7.285	.0298	.0503	.0618	.0503	.0763	.1018	.1272	.1909	.2546	.3840
	3½	4.0	.318	3.364	5.414	.0141	.0249	.0306	.0278	.0422	.0562	.0703	.1055	.1406	.2121
175	1½	1.900	.200	1.5	31.769	1.1768	1.5879	1.8915	2.1817	2.4622	2.6823	2.0739	3.1122	4.1496	6.2594
	2	2.375	.218	1.939	19.012	.3128	.4552	.5405	.6483	.7330	.5943	.7428	1.1146	1.4861	2.2417
	2½	2.875	.276	2.323	13.246	.1239	.1914	.2302	.2760	.2163	.2885	.3605	.5410	.7214	1.0882
	3	3.5	.300	2.900	8.499	.0401	.0658	.0801	.0971	.0891	.1188	.1484	.2228	.2970	.4480
	3½	4.0	.318	3.364	6.316	.0189	.0325	.0399	.0324	.0492	.0656	.0820	.1230	.1640	.2474
200	1½	1.900	.200	1.5	36.307	1.5315	2.0256	2.3941	2.7550	3.0957	3.3652	2.3702	3.5568	4.7424	7.1536
	2	2.375	.218	1.939	21.728	.4063	.5788	.6945	.8156	.9195	.6792	.8489	1.2738	1.6984	2.5620
	2½	2.875	.276	2.323	15.138	.1606	.2413	.2898	.3467	.3927	.3297	.4120	.6183	.8244	1.2436
	3	3.5	.300	2.900	9.714	.0518	.0834	.1010	.1217	.1018	.1357	.1697	.2546	.3394	.5120
	3½	4.0	.318	3.364	7.219	.0244	.0409	.0499	.0606	.0562	.0750	.0937	.1406	.1875	.2828

TABLE 3: PRESSURE DIFFERENTIAL IN PSI PER FOOT OF SCHEDULE 160 PIPE

Flow GPM	Pipe — Schedule 160				Velocity FPS	Viscosity — SUS									
	Size	OD	Wall	ID		32	100	200	400	600	800	1000	1500	2000	3000
3	½	.840	.187	.466	5.643	.1823	.3575	.7966	1.5099	2.2897	3.0540	3.8168	5.7275	7.6367	11.5196
5	½	.840	.187	.466	9.405	.4842	.5958	1.3276	2.5166	3.8162	5.0901	6.3612	9.5459	12.7279	19.1993
	¾	1.050	.218	.614	5.417	.1182	.1977	.4405	.8350	1.2662	1.6889	2.1106	3.1673	4.2230	6.3702
7	½	.840	.187	.466	13.167	.9200	1.5084	1.8587	3.5232	5.3427	7.1261	8.9057	13.3643	17.8190	26.8790
	¾	1.050	.218	.614	7.584	.2241	.2768	.6167	1.1690	1.7727	2.3644	2.9549	4.4342	5.9123	8.9183
10	½	.840	.187	.466	18.809	1.8424	2.7909	2.6552	5.0331	7.6324	10.1801	12.7225	19.0918	25.4557	38.3986
	¾	1.050	.218	.614	10.835	.4447	.7436	.8810	1.6700	2.5324	3.3777	4.2213	6.3346	8.4461	12.7405
	1	1.315	.250	.815	6.149	.1052	.1274	.2838	.5380	.8158	1.0881	1.3598	2.0406	2.7208	4.1042
15	½	.840	.187	.466	28.214	4.0807	5.7149	7.0017	7.5496	11.4486	15.2702	19.0837	28.6377	38.1836	57.5979
	¾	1.050	.218	.614	16.252	.9767	1.5039	1.3215	2.5049	3.7986	5.0666	6.3319	9.5019	12.6691	19.1107
	1	1.315	.250	.815	9.224	.2284	.3878	.4257	.8069	1.2237	1.6321	2.0398	3.0609	4.0812	6.1563
20	½	.840	.187	.466	37.619	7.1951	9.5295	11.4714	10.0662	15.2648	20.3603	25.445	38.1836	50.9114	76.7972
	¾	1.050	.218	.614	21.669	1.7141	2.4813	3.0661	3.3399	5.0648	6.7554	8.4425	12.6691	16.8922	25.4809
	1	1.315	.250	.815	12.299	.3942	.6384	.5676	1.0759	1.6316	2.1762	2.7197	4.0812	5.4416	8.2084
	1¼	1.660	.250	1.160	6.071	.0654	.1189	.1383	.2622	.3976	.5303	.6627	.9945	1.3260	2.0001
25	¾	1.050	.218	.614	27.086	2.6567	3.6978	4.5040	4.1749	6.3310	8.4443	10.5532	15.8364	21.1152	31.8512
	1	1.315	.250	.815	15.373	.6078	.9380	1.1706	1.3449	2.0395	2.7202	3.3996	5.1015	6.8020	10.2605
	1¼	1.660	.250	1.160	7.589	.1000	.1744	.1729	.3277	.4970	.6628	.8294	1.2431	1.6574	2.5002
	1½	1.900	.281	1.338	5.704	.0486	.0886	.0977	.1851	.2808	.3745	.4680	.7023	.9364	1.4125
30	¾	1.050	.218	.614	32.503	3.7694	5.1096	6.1823	5.0099	7.5972	10.1332	12.6638	19.0037	25.3383	38.2214
	1	1.315	.250	.815	18.448	.8671	1.2956	1.5990	1.6139	2.4474	3.2643	4.0795	6.1218	8.1624	12.3126
	1¼	1.660	.250	1.160	9.106	.1418	.2391	.2075	.3932	.5963	.7954	.9940	1.4917	1.9889	3.0002
	1½	1.900	.281	1.338	6.845	.0687	.1212	.1172	.2222	.3369	.4494	.5616	.8427	1.1236	1.6950
40	1	1.315	.250	.815	24.597	1.5231	2.1478	2.6270	2.1518	3.2631	4.3524	5.4393	8.1624	10.8833	16.4168
	1¼	1.660	.250	1.160	12.142	.2469	.3935	.4907	.5243	.7951	1.0605	1.3254	1.9809	2.6519	4.0003
	1½	1.900	.281	1.338	9.126	.1192	.1985	.2500	.2962	.4492	.5991	.7488	1.1236	1.4982	2.2599
	2	2.375	.343	1.689	5.727	.0366	.0658	.0615	.1167	.1769	.2360	.2949	.4425	.5900	.8900
50	1	1.315	.250	.815	30.747	2.3619	3.2680	3.8751	4.6091	4.0789	5.4405	6.7992	10.2030	13.6041	20.5210
	1¼	1.660	.250	1.160	15.178	.3808	.5820	.7190	.6554	.9939	1.3257	1.6567	2.4862	3.3149	5.0003
	1½	1.900	.281	1.338	11.408	.1832	.2920	.3655	.3703	.5615	.7489	.9360	1.4046	1.8727	2.8249
	2	2.375	.343	1.689	7.159	.0560	.0964	.1218	.1458	.2211	.2950	.3686	.5531	.7375	1.1125
60	1	1.315	.250	.815	36.896	3.3837	4.5479	5.3104	6.3001	4.8947	6.5286	8.1590	12.2437	16.3249	24.6252
	1¼	1.660	.250	1.160	18.213	.5434	.8156	.9845	.7865	1.1927	1.5908	1.9881	2.9834	3.9779	6.0004
	1½	1.900	.281	1.338	13.689	.2609	.4034	.4995	.4443	.6738	.8987	1.1232	1.6855	2.2473	3.3899
	2	2.375	.343	1.689	8.591	.0795	.1326	.1659	.1750	.2654	.3539	.4423	.6638	.8850	1.3350
	2½	2.875	.375	2.125	5.427	.0249	.0443	.0368	.0698	.1059	.1413	.1765	.2649	.3532	.5328
80	1¼	1.660	.250	1.160	24.284	.9545	1.3560	1.6154	1.9321	1.5902	2.1201	2.6508	3.9779	5.3038	8.0005
	1½	1.900	.281	1.338	18.253	.4569	.6817	.8160	.9840	.8984	1.1983	1.4976	2.2473	2.9964	4.5199
	2	2.375	.343	1.689	11.455	.1383	.2216	.2713	.2333	.3538	.4719	.5898	.8850	1.1801	1.7800
	2½	2.875	.375	2.125	7.236	.0431	.0729	.0916	.0931	.1412	.1883	.2354	.3532	.4710	.7104
100	1¼	1.660	.250	1.160	30.355	1.4803	2.0330	2.3795	2.8331	3.1972	2.6513	3.3135	4.9723	6.6298	10.0006
	1½	1.900	.281	1.338	22.816	.7071	1.0131	1.2061	1.4394	1.1230	1.4979	1.8719	2.8091	3.7455	5.6498
	2	2.375	.343	1.689	14.318	.2133	.3298	.3963	.4791	.4423	.5899	.7372	1.1063	1.4751	2.2251
	2½	2.875	.375	2.125	9.045	.0661	.1090	.1341	.1164	.1765	.2354	.2942	.4415	.5887	.8880
	3	3.50	.438	2.624	5.932	.0228	.0394	.0495	.0501	.0759	.1013	.1265	.1899	.2562	.3819
125	1¼	1.660	.250	1.160	37.944	2.2987	3.0312	3.5400	4.1664	4.6762	3.3142	4.1418	6.2154	8.2872	12.5008
	1½	1.900	.281	1.338	28.520	1.0963	1.5189	1.7819	2.1117	2.3808	1.8723	2.3399	3.5114	4.6812	7.0623
	2	2.375	.343	1.689	17.898	.3295	.4888	.5851	.7002	.5528	.7374	.9215	1.3829	1.8438	2.7813
	2½	2.875	.375	2.125	11.307	.1017	.1625	.1958	.2371	.2206	.2943	.3678	.5519	.7359	1.1100
	3	3.50	.438	2.624	7.415	.0348	.0590	.0725	.0626	.0949	.1266	.1582	.2374	.3165	.4774
150	1½	1.900	.281	1.338	34.224	1.5702	2.0986	2.4968	2.8770	3.2490	3.5401	2.8079	4.2136	5.6182	8.4747
	2	2.375	.343	1.689	21.477	.4708	.6793	.8030	.9568	1.0815	.8849	1.1058	1.6595	2.2126	3.3376
	2½	2.875	.375	2.125	13.568	.1448	.2233	.2689	.3231	.2648	.3531	.4413	.6623	.8830	1.3320
	3	3.50	.438	2.624	8.898	.0494	.0815	.0986	.1197	.1139	.1519	.1898	.2849	.3798	.5729
	4	4.50	.531	3.438	5.184	.0127	.0223	.0276	.0255	.0386	.0515	.0644	.0967	.1289	.1944
175	1½	1.900	.281	1.338	39.927	2.1288	2.7810	3.2701	3.7666	4.2320	4.5994	3.2759	4.9159	6.5546	9.8872
	2	2.375	.343	1.689	25.057	.6370	.8899	1.0602	1.2475	1.4052	1.0323	1.2901	1.9360	2.5814	3.8938
	2½	2.875	.375	2.125	15.829	.1954	.2945	.3504	.4203	.3089	.4120	.5149	.7767	1.0302	1.5540
	3	3.50	.438	2.624	10.381	.0665	.1064	.1289	.1554	.1389	.1772	.2215	.3323	.4431	.6684
	4	4.50	.531	3.438	6.047	.0170	.0293	.0360	.0297	.0451	.0601	.0752	.1128	.1504	.2268
200	2	2.375	.343	1.689	28.636	.8284	1.1333	1.3467	1.5634	1.7648	1.9229	1.4744	2.2126	2.9501	4.4501
	2½	2.875	.375	2.125	18.091	.2536	.3712	.4431	.5284	.5972	.4709	.5885	.8830	1.1774	1.7760
	3	3.50	.438	2.624	11.864	.0862	.1351	.1616	.1950	.1518	.2025	.2531	.3798	.5064	.7639
	4	4.50	.531	3.438	6.911	.0219	.0369	.0450	.0547	.0515	.0687	.0859	.1289	.1718	.2592

TABLE 4: PRESSURE DIFFERENTIAL IN PSI PER FOOT OF DOUBLE EXTRA HEAVY PIPE

Flow GPM	Pipe — Double Extra Heavy				Velocity FPS	Viscosity — SUS									
	Size	OD	Wall	ID		32	100	200	400	600	800	1000	1500	2000	3000
1.5	½	.840	.294	.252	9.648	1.1121	2.0901	4.6573	8.8281	13.3873	17.8561	22.3154	33.4872	44.6496	67.3516
3	½	.840	.294	.252	19.296	4.2460	4.1803	9.3156	17.6562	26.7747	35.7122	44.6308	66.9745	89.2993	134.703
	¾	1.050	.308	.434	6.506	.2614	.4752	1.0588	2.0070	3.0435	4.0594	5.0732	7.6129	10.1506	15.3116
5	½	.840	.294	.252	32.160	11.7061	16.3358	15.5243	29.4270	44.6245	59.5203	74.3846	111.624	148.832	224.505
	¾	1.050	.308	.434	10.843	.6970	.7919	1.7646	3.3450	5.0724	6.7656	8.4553	12.6882	16.9176	25.5194
	1	1.315	.358	.599	5.692	.1340	.2182	.4863	.9218	1.3979	1.8645	2.3301	3.4967	4.6622	7.0327
7	¾	1.050	.308	.434	15.180	1.3280	2.1103	2.4705	4.6829	7.1014	9.4719	11.8374	17.7635	23.6847	35.7271
	1	1.315	.358	.599	7.969	.2542	.3055	.6808	1.2905	1.9570	2.6103	3.2622	4.8953	6.5271	9.8458
10	¾	1.050	.308	.434	21.685	2.6647	3.9355	3.5293	6.6899	10.1449	13.5313	16.9105	25.3765	33.8353	51.0387
	1	1.315	.358	.599	11.384	.5048	.8368	.9726	1.8436	2.7958	3.7290	4.6602	6.9933	9.3244	14.0654
	1¼	1.660	.382	.896	5.088	.0652	.0872	.1943	.3683	.5584	.7448	.9309	1.3969	1.8625	2.8095
15	¾	1.050	.308	.434	32.528	5.9122	8.0874	9.8412	10.0349	15.2173	20.2969	25.3658	38.0647	50.7529	76.5581
	1	1.315	.358	.599	17.076	1.1097	1.6941	1.4589	2.7654	4.1936	5.5935	6.9904	10.4900	13.9866	21.0981
	1¼	1.660	.382	.896	7.632	.1410	.2472	.2914	.5525	.8377	1.1173	1.3963	2.0953	2.7938	4.2142
	1½	1.900	.400	1.100	5.064	.0500	.0936	.1283	.2432	.3687	.4918	.6147	.9224	1.2298	1.8552
20	1	1.315	.358	5.99	22.768	1.9485	2.7975	3.4493	3.6872	5.5915	7.4580	9.3205	13.9866	18.6488	28.1307
	1¼	1.660	.382	.896	10.176	.2426	.4059	.3885	.7365	1.1169	1.4897	1.8617	2.7938	3.7250	5.6190
	1½	1.900	.400	1.100	6.751	.0855	.1529	.1710	.3242	.4917	.6558	.8195	1.2298	1.6398	2.4735
25	1	1.315	.358	.599	28.460	3.0201	4.1717	5.0696	4.6090	6.9894	9.3225	11.6506	17.4833	23.3110	35.1634
	1¼	1.660	.382	.896	12.720	.3734	.5984	.4857	.9206	1.3961	1.8621	2.3271	3.4922	4.6563	7.0237
	1½	1.900	.400	1.100	8.439	.1309	.2246	.2138	.4053	.6146	.8197	1.0244	1.5373	2.0497	3.0919
30	1	1.315	.358	.599	34.152	4.3271	5.7692	6.9619	5.5309	8.3873	11.1869	13.9807	20.9799	27.9732	42.1961
	1¼	1.660	.382	.896	15.263	.5320	.8201	1.0199	1.1048	1.6753	2.2345	2.7926	4.1906	5.5875	8.4285
	1½	1.900	.400	1.100	10.127	.1859	.3063	.2566	.4863	.7375	.9837	1.2293	1.8448	2.4597	3.7103
	2	2.375	.436	1.503	5.424	.0382	.0698	.0736	.1395	.2116	.2822	.3527	.5293	.7057	1.0645
40	1¼	1.660	.382	.896	20.351	.9326	1.3620	1.6716	1.4730	2.2337	2.9794	3.7234	5.5875	7.4500	11.2379
	1½	1.900	.400	1.100	13.503	.3242	.5079	.6309	.6484	.9833	1.3116	1.6391	2.4597	3.2796	4.9471
	2	2.375	.436	1.503	7.232	.0660	.1139	.0981	.1860	.2821	.3763	.4703	.7057	.9409	1.4193
	2½	2.875	.552	1.771	5.209	.0289	.0526	.0509	.0965	.1463	.1952	.2440	.3661	.4881	.7363
50	1¼	1.660	.382	.896	25.439	1.4444	2.0600	2.4610	2.9428	2.7922	3.7242	4.6543	6.9844	9.3125	14.0474
	1½	1.900	.400	1.100	16.878	.5005	.7486	.9253	.8105	1.2292	1.6394	2.0489	3.0746	4.0995	6.1838
	2	2.375	.436	1.503	9.041	.1013	.1683	.2110	.2325	.3526	.4704	.5878	.8821	1.1762	1.7742
	2½	2.875	.552	1.771	6.511	.0441	.0769	.0636	.1206	.1829	.2440	.3049	.4576	.6101	.9204
60	1¼	1.660	.382	.896	30.527	2.0673	2.8548	3.3648	4.0166	3.3501	4.4691	5.5852	8.3813	11.1750	16.8569
	1½	1.900	.400	1.100	20.254	.7146	1.0526	1.2681	.9727	1.4750	1.9673	2.4587	3.6895	4.9194	7.4206
	2	2.375	4.36	1.503	10.849	.1439	.2310	.2879	.2791	.4232	.5644	.7054	1.0585	1.4114	2.1290
	2½	2.875	.552	1.771	7.814	.0625	.1057	.1327	.1448	.2195	.2928	.3659	.5491	.7322	1.1044
80	1½	1.900	.400	1.100	27.005	1.2565	1.7567	2.0842	2.4849	1.9666	2.6231	3.2782	4.9194	6.5592	9.8941
	2	2.375	.436	1.503	14.465	.2513	.3876	.4720	.3721	.5642	.7526	.9405	1.4114	1.8818	2.8387
	2½	2.875	.552	1.771	10.418	.1086	.1748	.2167	.1930	.2927	.3904	.4879	.7322	.9762	1.4726
	3	3.500	.600	2.300	6.177	.0289	.0499	.0631	.0679	.1029	.1372	.1715	.2574	.3432	.5177
100	1½	1.900	.400	1.100	33.757	1.9500	2.6371	3.0765	3.6470	4.1088	3.2789	4.0978	6.1492	8.1990	12.3677
	2	2.375	.436	1.503	18.081	.3883	.5761	.6914	.8306	.7053	.9407	1.1757	1.7642	2.3523	3.5483
	2½	2.875	.552	1.771	13.023	.1673	.2621	.3163	.3832	.3659	.4880	.6099	.9152	1.2203	1.8407
	3	3.500	.600	2.300	7.721	.0442	.0739	.0923	.0848	.1286	.1715	.2144	.3217	.4290	.6471
	3½	4.000	.636	2.728	5.489	.0187	.0327	.0412	.0429	.0650	.0867	.1083	.1626	.2167	.3270
125	2	2.375	.436	1.503	22.602	.6010	.8615	1.0168	1.2162	.8816	1.1759	1.4670	2.2053	2.9404	4.4354
	2½	2.875	.552	1.771	16.279	.2583	.3886	.4665	.5597	.4573	.6100	.7623	1.1440	1.5253	2.3009
	3	3.500	.600	2.300	9.652	.0679	.1108	.1353	.1060	.1608	.2144	.2680	.4022	.5362	.8088
	3½	4.000	.636	2.728	6.861	.0286	.0489	.0603	.0536	.0812	.1084	.1354	.2032	.2709	.4087
	4	4.500	.674	3.152	5.139	.0138	.0242	.0305	.0301	.0456	.0608	.0759	.1140	.1520	.2293
150	2	2.375	.436	1.503	27.122	.8598	1.1889	1.3960	1.6643	1.8749	1.4111	1.7635	2.6463	3.5285	5.3225
	2½	2.875	.552	1.771	19.534	.3688	.5396	.6392	.7644	.8652	.7320	.9148	1.3728	1.8304	2.7611
	3	3.500	.600	2.300	11.582	.0966	.1525	.1845	.2225	.1929	.2573	.3216	.4826	.6434	.9706
	3½	4.000	.636	2.728	8.233	.0406	.0676	.0820	.0643	.0975	.1300	.1625	.2438	.3251	.4904
	4	4.500	.674	3.152	6.167	.0196	.0339	.0417	.0361	.0547	.0730	.0912	.1368	.1824	.2752
175	2	2.375	.436	1.503	31.642	1.1647	1.5724	1.8731	2.1610	2.4391	2.6572	2.0574	3.0874	4.1165	6.2096
	2½	2.875	.552	1.771	22.790	.4988	.7066	.8365	.9961	1.1235	.8540	1.0673	1.6016	2.1355	3.2212
	3	3.500	.600	2.300	13.512	.1303	.2009	.2414	.2892	.2251	.3002	.3752	.5630	.7568	1.1324
	3½	4.000	.636	2.728	9.605	.0546	.0883	.1071	.1294	.1137	.1517	.1896	.2845	.3793	.5722
	4	4.500	.674	3.152	7.195	.0263	.0444	.0540	.0656	.0638	.0851	.1064	.1596	.2128	.3210
200	2	2.375	.436	1.503	36.162	1.5157	2.0058	2.3710	2.7288	3.0665	3.3336	2.3513	3.5285	4.7046	7.0966
	2½	2.875	.552	1.771	26.046	.6484	.8994	1.0774	1.2473	1.4105	1.5382	1.2198	1.8304	2.4405	3.6814
	3	3.500	.600	2.300	15.443	.1690	.2532	.3040	.3634	.4115	.3431	.4288	.6434	.8579	1.2941
	3½	4.000	.636	2.728	10.977	.0707	.1120	.1351	.1624	.1300	.1734	.2167	.3251	.4335	.6539
	4	4.500	.674	3.152	8.222	.0339	.0559	.0680	.0822	.0729	.0973	.1216	.1824	.2432	.3669

TABLE 5: PRESSURE DIFFERENTIAL IN PSI PER FOOT OF TUBE

Flow GPM	Tube Size	OD	Wall	ID	Velocity FPS	Viscosity — SUS 32	100	200	400	600	800	1000	1500	2000	3000
1.5	1/4	.250	.028	.194	16.257	2.7858	5.9425	13.2413	25.0995	38.0620	50.7672	63.4456	95.2086	126.945	191.489
			.035	.180	18.945	3.9987	8.0442	17.9245	33.9767	51.5238	68.7226	85.8851	128.882	171.843	259.215
	3/8	.375	.028	.319	6.023	.2632	.8142	1.8143	3.4390	5.2151	6.9559	8.6930	13.0450	17.3933	26.2368
			.035	.305	6.583	.3253	.9735	2.1692	4.1119	6.2355	8.3169	10.3939	15.5975	20.7966	31.3706
			.049	.277	7.980	.5138	1.4308	3.1882	6.0434	9.1645	12.2237	15.2764	22.9242	30.5656	46.1066
3	1/4	.250	.028	.194	32.514	9.3704	19.9638	26.4826	50.1989	76.1240	101.534	126.891	190.417	253.889	382.979
			.035	.180	37.890	13.4499	28.6555	35.8489	67.9534	103.048	137.445	171.770	257.764	343.685	518.431
	3/8	.375	.028	.319	12.045	.8853	1.6284	3.6285	8.8780	10.4301	13.9117	17.3859	26.0899	34.7866	52.4737
			.035	.305	13.166	1.0941	1.9471	4.3385	8.2238	12.4709	16.6338	20.7878	31.1949	41.5932	62.7412
			.049	.277	15.960	1.7283	2.8617	6.3764	12.0868	18.3290	24.4473	30.5527	45.8484	61.1312	92.2151
	1/2	.500	.035	.430	6.627	.2142	.4931	1.0987	2.0827	3.1583	4.2126	5.2646	7.9002	10.5336	15.8894
			.049	.402	7.583	.2950	.6455	1.4383	2.7264	4.1345	5.5146	6.8918	10.3421	13.7894	20.8006
			.058	.384	8.310	.3667	.7753	1.7256	3.2743	4.9660	6.6236	8.2777	12.4219	16.5625	24.9836
			.065	.370	8.951	.4374	.8995	2.0043	3.7992	5.7613	7.6844	9.6035	14.4114	19.2151	28.9850
5	3/8	.375	.028	.319	20.075	2.1643	4.6110	6.0475	11.4633	17.3835	23.1862	28.9766	43.4832	57.9776	87.4561
			.035	.305	21.943	2.6747	5.6985	7.2308	13.7063	20.7849	27.7230	34.6464	51.9916	69.3221	104.569
			.049	.277	26.600	4.2252	9.0020	10.6274	20.1447	30.5484	40.7455	50.9212	76.4140	101.885	153.689
	1/2	.500	.035	.430	11.045	.5237	.8218	1.8312	3.4712	5.2638	7.0209	8.7743	13.1670	17.5560	26.4823
			.049	.402	12.638	.7211	1.0758	2.3972	4.5441	6.8908	9.1910	11.4863	17.2368	22.9824	34.6677
			.058	.384	13.850	.8964	1.2922	2.8793	5.4579	8.2766	11.0393	13.7962	20.7031	27.6041	41.6394
			.065	.370	14.918	1.0694	2.2784	3.3405	6.3320	9.6022	12.8074	16.0059	24.0189	32.0252	48.3083
	5/8	.625	.035	.555	6.630	.1559	.2961	.6598	1.2508	1.8967	2.5299	3.1617	4.7445	6.3260	9.5424
			.049	.527	7.353	.1993	.3643	.8117	1.5385	2.3331	3.1119	3.8891	5.8360	7.7814	11.7378
			.058	.509	7.883	.2351	.4186	.9327	1.7680	2.6811	3.5760	4.4690	6.7064	8.9419	13.4883
			.065	.495	8.335	.2684	.4680	1.0428	1.9766	2.9975	3.9980	4.9965	7.4979	9.9972	15.0802
7	3/8	.375	.028	.319	28.105	.38997	8.3085	8.4665	16.0486	24.3369	32.4606	40.5672	60.8765	81.1687	122.439
			.035	.305	30.720	4.8195	10.2680	10.1231	19.1889	29.0989	38.8122	48.5050	72.7882	97.0509	146.396
			.049	.277	37.241	7.6133	16.2204	14.8784	28.2026	42.7678	57.0438	71.2896	106.980	142.640	215.164
	1/2	.500	.035	.430	15.463	.9437	2.0106	2.5637	4.8596	7.3694	9.8293	12.2840	18.4338	24.5784	37.0752
			.049	.402	17.693	1.2994	2.7684	3.3561	6.3617	9.6472	12.8674	16.0809	24.1315	32.1753	48.5347
			.058	.384	19.390	1.6153	3.4414	4.0310	7.6410	11.5872	15.4551	19.3147	28.9843	38.6458	58.2951
			.065	.370	20.885	1.9269	4.1054	4.6767	8.8648	13.4430	17.9303	22.4082	33.6265	44.8353	67.6317
	5/8	.625	.035	.555	9.282	.2808	.4146	.9238	1.7511	2.6554	3.5418	4.4263	6.6423	8.8564	13.3593
			.049	.527	10.295	.3591	.7651	1.1363	2.1539	3.2663	4.3567	5.4447	8.1705	10.8939	16.4329
			.058	.509	11.036	.4236	.9024	1.3058	2.4752	3.7535	5.0064	6.2567	9.3890	12.5186	18.8836
			.065	.495	11.669	.4836	1.0302	1.4599	2.7673	4.1965	5.5973	6.9951	10.4971	13.9961	21.1123
	3/4	.750	.049	.652	6.726	.1307	.2177	.4850	.9194	1.3942	1.8595	2.3239	3.4874	4.6498	7.0140
			.058	.634	7.113	.1492	.2435	.5425	1.0283	1.5594	2.0799	2.5993	3.9006	5.2008	7.8451
			.065	.620	7.438	.1659	.2662	.5932	1.1244	1.7051	2.2742	2.8422	4.2650	5.6867	8.5781
			.072	.606	7.786	.1850	.2917	.6499	1.2319	1.8682	2.4918	3.1140	4.6730	6.2307	9.3987
			.083	.584	8.383	.2205	.3382	.7535	1.4283	2.1660	2.8890	3.6105	5.4180	7.2240	10.8970
10	1/2	.500	.035	.430	22.091	1.7617	3.7533	3.6624	6.9423	10.5277	14.0418	17.5486	26.3340	35.1120	52.9645
			.049	.402	25.275	2.4256	5.1679	4.7945	9.0881	13.7817	18.3820	22.9727	34.4736	45.9648	69.3353
			.058	.384	27.700	3.0153	6.4241	5.7586	10.9158	16.5532	22.0786	27.5925	41.4062	55.2083	83.2787
			.065	.370	29.836	3.5970	7.6636	6.6809	12.6640	19.2043	25.6148	32.0117	48.0378	64.0505	96.6167
	5/8	.625	.035	.555	13.261	.5242	1.1169	1.3197	2.5015	3.7935	5.0597	6.3233	9.4890	12.6519	19.0848
			.049	.527	14.707	.6703	1.4282	1.6233	3.0771	4.6662	6.2238	7.7781	11.6721	15.5628	23.4756
			.058	.509	15.766	.7907	1.6845	1.8654	3.5360	5.3621	7.1520	8.9381	13.4128	17.8837	26.9766
			.065	.495	16.670	.9027	1.9232	2.0856	3.9533	5.9949	7.9961	9.9930	14.9958	19.9944	30.1605
	3/4	.750	.049	.652	9.608	.2439	.5197	.6929	1.3134	1.9917	2.6565	3.3199	4.9820	6.6426	10.0200
			.058	.634	10.162	.2786	.5936	.7750	1.4690	2.2277	2.9713	3.7133	5.5723	7.4297	11.2073
			.065	.620	10.626	.3098	.6600	.8474	1.6063	2.4358	3.2489	4.0602	6.0929	8.1237	12.2544
			.072	.606	11.122	.3453	.7356	.9284	1.7599	2.6688	3.5597	4.4486	6.6758	8.9010	13.4267
			.083	.584	11.976	.4116	.8769	1.0765	2.0405	3.0942	4.1271	5.1578	7.7400	10.3200	15.5671
15	1/2	.500	.035	.430	33.136	3.5817	7.6308	9.3232	10.4135	15.7915	21.0627	26.3229	39.5010	52.6680	79.4468
			.049	.402	37.913	4.9316	10.5069	12.8370	13.6322	20.6725	27.5730	34.4590	51.7103	68.9471	104.003
	5/8	.625	.035	.555	19.891	1.0658	2.2707	1.9795	3.3523	5.6902	7.5896	9.4850	14.2334	18.9779	28.6272
			.049	.527	22.061	1.3629	2.9037	2.4350	4.6156	6.9993	9.3357	11.6671	17.5081	23.3441	35.2134
			.058	.509	23.648	1.6075	3.4248	4.1844	5.3039	8.0412	10.7280	13.4071	20.1192	26.8256	40.4649
			.065	.495	25.005	1.8352	3.9100	4.7771	5.9299	8.9924	11.9941	14.9895	22.4937	29.9916	45.2407
	3/4	.750	.049	.652	14.413	.4959	1.0565	1.0393	1.9701	2.9875	3.9847	4.9799	7.4729	9.9639	15.0300
			.058	.634	15.243	.5664	1.2068	1.1625	2.2035	3.3415	4.4569	5.5699	8.3584	11.1446	16.8110
			.065	.620	15.939	.6298	1.3418	1.2711	2.4094	3.6537	4.8733	6.0903	9.1393	12.1858	18.3816
			.072	.606	16.684	.7020	1.4956	1.3927	2.6399	4.0032	5.3395	6.6729	10.0136	13.3515	20.1400
			.083	.584	17.964	.8368	1.7827	1.6147	3.0607	4.6414	6.1907	7.7367	11.6099	15.4799	23.3506

TABLE 5 (CONTINUED)

Flow GPM	Tube Size	OD	Wall	ID	Velocity FPS	Viscosity — SUS 32	100	200	400	600	800	1000	1500	2000	3000
15	1	1.0	.049	.902	7.530	.1061	.2261	.2837	.5378	.8156	1.0878	1.3595	2.0401	2.7202	4.1032
			.058	.884	7.840	.1168	.2488	.3076	.5830	.8841	1.1792	1.4737	2.2114	2.9486	4.4478
			.065	.870	8.095	.1260	.2684	.3278	.6214	.9424	1.2569	1.5708	2.3572	3.1430	4.7410
			.072	.856	8.362	.1361	.2899	.3498	.6631	1.0056	1.3412	1.6762	2.5153	3.3537	5.0589
			.083	.834	8.809	.1540	.3281	.3882	.7359	1.1159	1.4884	1.8601	2.7914	3.7218	5.6142
			.095	.810	9.338	.1769	.3769	.4363	.8270	1.2542	1.6728	2.0906	3.1372	4.1829	6.3097
			.109	.782	10.019	.2091	.4455	.5022	.9520	1.4437	1.9256	2.4065	3.6112	4.8150	7.2631
			.120	.760	10.607	.2394	.5101	.5630	1.0671	1.6182	2.1584	2.6974	4.0479	5.3972	8.1413
20	⅝	.625	.035	.555	26.521	1.7632	3.7566	4.5898	5.0031	7.5869	10.1194	12.6466	18.9779	25.3039	38.1695
			.049	.527	29.414	2.2548	4.8039	5.8692	6.1541	9.3324	12.4476	15.5562	23.3441	31.1255	46.9512
			.058	.509	31.531	2.6595	5.6661	6.9226	7.0719	10.7242	14.3040	17.8762	26.8256	35.7675	53.9533
			.065	.495	33.340	3.0362	6.4687	7.9033	7.9066	11.9899	15.9921	19.9860	29.9916	39.9888	60.3209
	¾	.750	.049	.652	19.217	.8204	1.7479	2.1355	2.6268	3.9833	5.3130	6.6398	9.9639	13.2852	20.0400
			.058	.634	20.323	.9371	1.9965	2.4393	2.9380	4.4553	5.9425	7.4266	11.1446	14.8594	22.4146
			.065	.620	21.252	1.0419	2.2199	2.7122	3.2125	4.8716	6.4977	8.1204	12.1858	16.2477	24.5088
			.072	.606	22.245	1.1613	2.4743	3.0230	3.5198	5.3376	7.1193	8.8972	13.3515	17.8020	26.8533
			.083	.584	23.952	1.3843	2.9494	3.6035	4.0809	6.1885	8.2542	10.3156	15.4799	20.6399	31.1342
	1	1.0	.049	.902	10.041	.1756	.3741	.3783	.7171	1.0875	1.4504	1.8127	2.7202	3.6269	5.4710
			.058	.884	10.454	.1932	.4117	.4101	.7773	1.1788	1.5722	1.9649	2.9486	3.9314	5.9303
			.065	.870	10.793	.2084	.4441	.4371	.8286	1.2565	1.6759	2.0944	3.1430	4.1907	6.3214
			.072	.856	11.149	.2251	.4797	.4664	.8841	1.3407	1.7883	2.2349	3.3537	4.4716	6.7452
			.083	.834	11.745	.2548	.5428	.5176	.9812	1.4879	1.9846	2.4802	3.7218	4.9624	7.4856
			.095	.810	12.451	.2927	.6236	.5817	1.1027	1.6722	2.2304	2.7874	4.1829	5.5772	8.4130
			.109	.782	13.359	.3459	.7370	.6696	1.2694	1.9249	2.5674	3.2086	4.8150	6.4200	9.6842
			.120	.760	14.143	.3961	.8440	.7506	1.4228	2.1577	2.8779	3.5966	5.3972	7.1962	10.8551
	1¼	1.25	.065	1.120	6.512	.0628	.1338	.1591	.3017	.4575	.6102	.7626	1.1443	1.5258	2.3015
			.072	1.106	6.678	.0667	.1420	.1674	.3172	.4811	.6417	.8019	1.2034	1.6045	2.4203
			.083	1.084	6.952	.0733	.1562	.1814	.3438	.5213	.6954	.8690	1.3041	1.7388	2.6228
			.095	1.060	7.270	.0816	.1738	.1984	.3760	.5702	.7605	.9504	1.4263	1.9017	2.8686
			.109	1.302	7.670	.0926	.1973	.2208	.4185	.6346	.8465	1.0579	1.5875	2.1166	3.1928
			.120	1.010	8.008	.1026	2186	.2407	.4562	.6918	.9227	1.1531	1.7304	2.3071	3.4802
25	⅝	.625	.035	.555	33.151	2.6056	5.5513	6.7824	6.2538	9.4836	12.6493	15.8082	23.7224	31.6299	47.7119
			.049	.527	36.768	3.3319	7.0988	8.6731	7.6927	11.6655	15.5595	19.4452	29.1803	38.9069	58.6890
			.058	.509	39.414	3.9299	8.3729	10.2297	8.8399	13.4052	17.8799	22.3452	33.5320	44.7093	67.4416
	¾	.750	.049	.652	24.021	1.2123	2.5829	3.1557	3.2834	4.9792	6.6412	8.2998	12.4549	16.6065	25.0500
			.058	.634	25.404	1.3848	2.9503	3.6045	3.6725	5.5692	7.4282	9.2832	13.9307	18.5743	28.0183
			.065	.620	26.565	1.5397	3.2804	4.0079	4.0156	6.0895	8.1221	10.1505	15.2322	20.3096	30.6360
			.072	.606	27.806	1.7161	3.6563	4.4671	4.3998	6.6720	8.8991	11.1215	16.6894	22.2525	33.5667
			.083	.584	29.941	2.0457	4.3584	5.3249	5.1011	7.7356	10.3178	12.8945	19.3499	25.7999	38.9177
	1	1.0	.049	.902	12.551	.2595	.5528	.4729	.8964	1.3593	1.8131	2.2658	3.4002	4.5336	6.8387
			.058	.884	13.067	.2855	.6083	.5126	.9717	1.4735	1.9653	2.4561	3.6857	4.9143	7.4129
			.065	.870	13.491	.3080	.6562	.5464	1.0357	1.5706	2.0949	2.6181	3.9287	5.2383	7.9017
			.072	.856	13.936	.3327	.7088	.8660	1.1052	1.6759	2.2353	2.7956	4.1921	5.5895	8.4315
			.083	.834	14.681	.3765	.8021	.9800	1.2265	1.8599	2.4807	3.1002	4.6523	6.2030	9.3570
			.095	.810	15.564	.4325	.9215	1.1258	1.3784	2.0903	2.7880	3.4843	5.2287	6.9716	10.5162
			.109	.782	16.698	.5112	1.0891	1.3306	1.5867	2.4061	3.2093	4.0108	6.0187	8.0250	12.1052
			.120	.760	17.679	.5854	1.2472	1.5237	1.7785	2.6971	3.5974	4.4957	6.7465	8.9953	13.5689
	1¼	1.25	.065	1.120	8.140	.0928	.1977	.1989	.3771	.5718	.7627	.9532	1.4304	1.9072	2.8769
			.072	1.106	8.348	.0985	.2099	.2092	.3966	.6013	.8021	1.0024	1.5042	2.0056	3.0254
			.083	1.084	8.690	.1084	.2309	.2267	.4297	.6517	.8692	1.0863	1.6301	2.1735	3.2786
			.095	1.060	9.088	.1205	.2568	.2479	.4700	.7127	.9506	1.1880	1.7828	2.3771	3.5857
			.109	1.032	9.588	.1369	.2916	.2760	.5231	.7933	1.0581	1.3223	1.9843	2.6458	3.9910
			.120	1.010	10.010	.1516	.3230	.3008	.5702	.8647	1.1533	1.4414	2.1629	2.8839	4.3502
30	⅝	.625	.035	.555	39.781	3.5848	7.6376	9.3314	7.5046	11.3803	15.1791	18.9699	28.4669	37.9558	57.2543
	¾	.750	.049	.652	28.825	1.6680	3.5537	4.3418	3.9401	5.9750	7.9695	9.9597	14.9459	19.9278	30.0601
			.058	.634	30.485	1.9052	4.0591	4.9593	4.4070	6.6830	8.9138	11.1399	16.7169	22.2891	33.6220
			.065	.620	31.877	2.1184	4.5133	5.5142	4.8187	7.3074	9.7466	12.1806	18.2787	24.3716	36.7632
			.072	.606	33.367	2.3611	5.0304	6.1460	5.2797	8.0064	10.6789	13.3459	20.0272	26.7030	40.2800
			.083	.584	35.929	2.8145	5.9964	7.3262	6.1214	9.2827	12.3813	15.4734	23.2199	30.9599	46.7013
	1	1.0	.049	.902	15.061	.3570	.7605	.9292	1.0757	1.6312	2.1757	2.7190	4.0802	5.4403	8.2064
			.058	.884	15.681	.3928	.8370	1.0226	1.1660	1.7682	2.3584	2.9473	4.4229	5.8971	8.8955
			.065	.870	16.189	.4238	.9029	1.1031	1.2429	1.8847	2.5139	3.1417	4.7145	6.2860	9.4821
			.072	.856	16.723	.4577	.9752	1.1915	1.3262	2.0111	2.6824	3.3523	5.0306	6.7074	10.1178
			.083	.834	17.617	.5180	1.1036	1.3484	1.4718	2.2318	2.9768	3.7203	5.5827	7.4437	11.2283
			.095	.810	18.677	.5951	1.2678	1.5490	1.6541	2.5083	3.3456	4.1812	6.2744	8.3659	12.6194

TABLE 5 (CONTINUED)

Flow GPM	Tube Size	Tube OD	Tube Wall	Tube ID	Velocity FPS	Viscosity — SUS 32	100	200	400	600	800	1000	1500	2000	3000
30	1	1.0	.109	.782	20.038	.7033	1.4984	1.8307	1.9040	2.8874	3.8512	4.8129	7.2225	9.6299	14.5262
			.120	.760	21.215	.8054	1.7159	2.0964	2.1343	3.2365	4.3168	5.3949	8.0958	10.7943	16.2827
	1¼	1.25	.065	1.120	9.769	.1277	.2720	.2387	.4525	.6862	.9153	1.1438	1.7165	2.2886	3.4523
			.072	1.106	10.017	.1355	.2887	.2510	.4759	.7216	.9625	1.2029	1.8051	2.4067	3.6304
			.083	1.084	10.428	.1491	.3177	.2721	.5157	.7820	1.0430	1.3035	1.9561	2.6082	3.9343
			.095	1.060	10.906	.1658	.3533	.2975	.5640	.8553	1.1408	1.4256	2.1394	2.8525	4.3029
			.109	1.032	11.506	.1883	.4012	.4902	.6277	.9519	1.2697	1.5868	2.3812	3.1749	4.7892
			.120	1.010	12.012	.2086	.4445	.5430	.6843	1.0376	1.3840	1.7296	2.5955	3.4607	5.2203
	1½	1.5	.065	1.370	6.529	.0490	.1045	.1066	.2021	.3065	.4088	.5109	.7667	1.0223	1.5421
			.072	1.356	6.664	.0515	.1097	.1111	.2106	.3194	.4260	.5324	.7989	1.0652	1.6067
			.083	1.334	6.886	.0556	.1125	.1186	.2248	.3410	.4548	.5683	.8529	1.1372	1.7154
			.095	1.310	7.140	.0606	.1292	.1276	.2418	.3667	.4890	.6112	.9171	1.228	1.8446
			.109	1.282	7.456	.0672	.1432	.1391	.2636	.3997	.5332	.6663	.9999	1.3332	2.0111
			.120	1.260	7.718	.0730	.1555	.1490	.2825	.4294	.5714	.7141	1.0716	1.4288	2.1553
			.134	1.232	8.073	0812	.1730	.1631	.3091	.4687	.6251	.7813	1.1724	1.5632	2.3580
			.156	1.188	8.682	.0965	.2056	.1886	.3575	.5421	.7230	.9036	1.3560	1.8079	2.7212
			.188	1.124	9.699	.1255	.2674	.2353	.4461	.6765	.9023	1.1276	1.6922	2.2562	3.4034
40	¾	.750	.049	.652	38.434	2.7595	5.8792	7.1830	8.4283	7.9666	10.6259	13.2796	19.9278	26.5704	40.0801
	1	1.0	.049	.902	20.081	.5906	1.2582	1.5373	1.4342	2.1749	2.9009	3.6254	5.4403	7.2538	10.9419
			.058	.884	20.907	.6499	1.3847	1.6918	1.5546	2.3575	3.1445	3.9298	5.8971	7.8629	11.8607
			.065	.870	21.586	.7011	1.4938	1.8250	1.6572	2.5130	3.3518	4.1889	6.2860	8.3813	12.6428
			.072	.856	22.298	.7573	1.6134	1.9712	1.7683	2.6815	3.5765	4.4697	6.7074	8.9432	13.4904
			.083	.834	23.489	.8570	1.8258	2.2307	1.9623	2.9758	3.9691	4.9603	7.4437	9.9249	14.9711
			.095	.810	24.902	.9845	2.0974	2.5626	2.2055	3.3444	4.4609	5.5749	8.3659	11.1545	16.8259
			1.09	.782	26.717	1.1635	2.4789	3.0287	2.5387	3.8498	5.1349	6.1473	9.6299	12.8399	19.3683
			.120	.760	28.286	1.3324	2.8388	3.4683	2.8457	4.3153	5.7558	7.1932	10.7943	14.3925	21.7102
	1¼	1.25	.065	1.120	13.025	.2112	.4500	.5498	.6033	.9149	1.2204	1.5251	2.2886	3.0515	4.6031
			.072	1.106	13.357	.2242	.4777	.5836	.6345	.9622	1.2833	1.6038	2.4067	3.2090	4.8406
			.083	1.084	13.904	.2467	.5255	.6421	.6876	1.0427	1.3907	1.7380	2.6082	3.4775	5.2457
			.095	1.060	14.541	.2743	.5845	.7141	.7520	1.1404	1.5210	1.9009	2.8525	3.8033	5.7371
			.109	1.032	15.341	.3115	.6638	.8110	.8370	1.2693	1.6929	2.1157	3.1749	4.2332	6.3856
			.120	1.010	16.016	.3451	.7353	.8984	.9123	1.3835	1.8453	2.3061	3.4607	4.6143	6.9604
	1½	1.5	.065	1.370	8.705	.0811	.1728	.2111	.2695	.4087	.5451	.6812	1.0222	1.3630	2.0561
			.072	1.356	8.886	.0852	.1815	.2217	.2808	.4258	.5680	.7098	1.0652	1.4202	2.1423
			.083	1.334	9.181	.0920	.1961	.2396	.2998	.4546	.6064	.7578	1.1372	1.5162	2.2872
			.095	1.310	9.521	.1003	.2138	.2612	.3224	.4889	.6520	.8149	1.2228	1.6304	2.4594
			.109	1.282	9.941	.1112	.2369	.2894	.3515	.5330	.7109	.8884	1.3332	1.7776	2.6814
			.120	1.260	10.291	.1207	.2572	.3142	.3767	.5712	.7619	.9521	1.4288	1.9051	2.8737
			.134	1.232	10.764	.1343	.2861	.3496	.4121	.6249	.8335	1.0417	1.5632	2.0842	3.1440
			.156	1.188	11.576	.1596	.3401	.4155	.4766	.7228	.9640	1.2048	1.8079	2.4106	3.6362
			.188	1.124	12.932	.2077	.4424	.5406	.5948	.9020	1.2031	1.5035	2.2562	3.0083	4.5379
	1¾	1.75	.065	1.620	6.226	.0366	.0779	.0727	.1378	.2090	.2788	.3484	.5229	.6972	1.0516
			.072	1.606	6.335	.0381	.0812	.0753	.1427	.2164	.2887	.3607	.5413	.7218	1.0888
			.083	1.584	6.512	.0407	.0867	.0796	.1508	.2287	.3050	.3812	.5720	.7627	1.1505
			.095	1.560	6.714	.0438	.0933	.0846	.1603	.2431	.3242	.4052	.6081	.8108	1.2230
			.109	1.532	6.961	.0477	.1016	.9090	.1723	.2614	.3486	.4357	.6538	.8717	1.3149
			.120	1.510	7.166	.0511	.1089	.0963	.1826	.2769	.3694	.4616	.6927	.9236	1.3932
			.134	1.482	7.439	.0558	.1190	.1038	.1968	.2985	.3981	.4975	.7465	.9954	1.5015
			.156	1.438	7.901	.0644	.1373	.1171	.2220	.3367	.4491	.5612	.8422	1.1229	1.6939
			.188	1.374	8.654	.0800	.1704	.2082	.2664	.4039	.5388	.6733	1.0104	1.3472	2.0322
50	1	1.0	.049	.902	25.102	.8727	1.8593	2.2717	2.6655	2.7186	3.6261	4.5317	6.8004	9.0672	13.6774
			.058	.884	26.134	.9604	2.0462	2.4999	2.9333	2.9469	3.9306	4.9122	7.3714	9.8286	14.8259
			.065	.870	26.982	1.0361	2.2073	2.6969	3.1644	3.1412	4.1898	5.2361	7.8575	10.4766	15.8034
			.072	.856	27.872	1.1190	2.3842	2.9129	3.4179	3.3518	4.4707	5.5872	8.3843	11.1790	16.8630
			.083	.834	29.362	1.2664	2.6980	3.2964	3.8679	3.7197	4.9614	6.2004	9.3046	12.4061	18.7139
			.095	.810	31.128	1.4548	3.0994	3.7868	4.4433	4.1806	5.5761	6.9686	10.4573	13.9431	21.0324
			.109	.782	33.397	1.7194	3.6631	4.4755	5.2513	4.8123	6.4186	8.0216	12.0374	16.0499	24.2104
			.120	.760	35.358	1.9689	4.1949	5.1252	6.0137	5.3941	7.1947	8.9915	13.4929	17.9906	27.1378
	1¼	1.25	.065	1.120	16.281	.3121	.6650	.8124	.7542	1.1437	1.5254	1.9064	2.8608	3.8144	5.7538
			.072	1.106	16.696	.3313	.7059	.8625	.7931	1.2027	1.6042	2.0048	3.0084	4.0112	6.0507
			.083	1.084	17.380	.3645	.7766	.9488	.8595	1.3033	1.7384	2.1725	3.2602	4.3469	6.5571
			.095	1.060	18.176	.4054	.8637	1.0553	.9400	1.4255	1.9013	2.3761	3.5656	4.7541	7.1714
			.109	1.032	19.176	.4604	.9809	1.1984	1.0462	1.5866	2.1162	2.6446	3.9686	5.2915	7.9820
			.120	1.010	20.020	.5100	1.0866	1.3276	1.1404	1.7294	2.3067	2.8827	4.3259	5.7679	8.7005
	1½	1.5	.065	1.370	10.881	.1199	.2554	.3120	.3369	.5109	.6814	.8515	1.2778	1.7038	2.5701
			.072	1.356	11.107	.1259	.2681	.3276	.3510	.5323	.7100	.8873	1.3314	1.7753	2.6779
			.083	1.334	11.476	.1360	.2898	.3541	.3747	.5683	.7580	.9472	1.4215	1.8953	2.8589
			.095	1.310	11.901	.1483	.3159	.3860	.4030	.6111	.8150	1.0186	1.5285	2.0381	3.0743
			.109	1.282	12.426	.1643	.3500	.4277	.4393	.6662	.8886	1.1105	1.6665	2.2220	3.3518

TABLE 5 (CONTINUED)

Flow GPM	Tube Size	OD	Wall	ID	Velocity FPS	Viscosity — SUS 32	100	200	400	600	800	1000	1500	2000	3000
50	1½	1.5	.120	1.260	12.864	.1784	.3800	.4643	.4708	.7140	.9523	1.1902	1.7860	2.3813	3.5921
			.134	1.232	13.455	.1985	.4228	.5166	.5151	.7811	1.0419	1.3021	1.9540	2.6053	3.9299
			.156	1.188	14.470	.2359	.5026	.6140	.5958	.9035	1.2050	1.5060	2.2599	3.0132	4.5453
			.188	1.124	16.165	.3069	.6538	.7988	.7435	1.1275	1.5038	1.8794	2.8203	3.7604	5.6724
	1¾	1.75	.065	1.620	7.782	.0541	.1152	.1407	.1723	.2613	.3485	.4355	.6536	.8714	1.3145
			.072	1.606	7.918	.0563	.1200	.1467	.1784	.2705	.3608	.4509	.6767	.9022	1.3610
			.083	1.584	8.140	.0602	.1282	.1566	.1885	.2859	.3813	.4765	.7151	.9534	1.4382
			.095	1.560	8.392	.0647	.1378	.1684	.2004	.3039	.4053	.5065	.7601	1.0135	1.5287
			.109	1.532	8.702	.0705	.1502	.1835	.2154	.3267	.4357	.5446	.8172	1.0896	1.6436
			.120	1.510	8.957	.0755	.1609	.1965	.2283	.3462	.4617	.5770	.8659	1.1545	1.7415
			.134	1.482	9.299	.0825	.1758	.2148	.2460	.3731	.4976	.6219	.9332	1.2443	1.8769
			.156	1.438	9.876	.0952	.2029	.2479	.2775	.4209	.5613	.7015	1.0528	1.4037	2.1174
			.188	1.374	10.818	.1182	.2519	.3077	.3330	.5049	.6735	.8417	1.2630	1.6840	2.5403
	2	2.0	.083	1.834	6.072	.0300	.0639	.0553	.1049	.1590	.2122	.2651	.3979	.5305	.8003
			.095	1.810	6.234	.0319	.0680	.0583	.1106	.1677	.2236	.2795	.4194	.5592	.8436
			.109	1.782	6.431	.0344	.0732	.0621	.1177	.1785	.2380	.2975	.4464	.5952	.8978
			.120	1.760	6.593	.0365	.0777	.0652	.1237	.1876	.2502	.3126	.4691	.6255	.9436
			.134	1.732	6.808	.0394	.0838	.0696	.1319	.2000	.2667	.3333	.5002	.6670	1.0061
			.156	1.688	7.167	.0445	.0947	.1158	.1462	.2217	.2957	.3695	.5545	.7393	1.1152
			.188	1.624	7.744	.0534	.1138	.1391	.1706	.2587	.3451	.4313	.6472	.8629	1.3016
			.220	1.560	8.392	.0647	.1378	.1684	.2004	.3039	.4053	.5065	.7601	1.0134	1.5287
60	1	1.0	.049	.902	30.122	1.2007	2.5581	3.1255	3.6673	3.2624	4.3513	5.4380	8.1605	10.8806	16.4128
			.058	.884	31.361	1.3213	2.8152	3.4395	4.0358	3.5363	4.7167	5.8947	8.8456	11.7943	17.7910
			.065	.870	32.379	1.4254	3.0369	3.7105	4.3537	3.7695	5.0277	6.2833	9.4290	12.5720	18.9641
			.072	.856	33.446	1.5396	3.2802	4.0077	4.7025	4.0222	5.3648	6.7046	10.0611	13.4148	20.2356
			.083	.834	35.234	1.7423	3.7121	4.5353	5.3256	4.4637	5.9537	7.4405	11.1655	14.8873	22.4567
			.095	.810	37.353	2.0015	4.2643	5.2100	6.1133	5.0167	6.6913	8.3623	12.5488	16.7317	25.2389
	1¼	1.25	.065	1.120	19.537	.4294	.9149	1.1178	.9050	1.3724	1.8305	2.2877	3.4330	4.5773	6.9046
			.072	1.106	20.035	.4559	.9712	1.1866	.9517	1.4432	1.9250	2.4057	3.6101	4.8135	7.2609
			.083	1.084	20.856	.5015	1.0685	1.3054	1.5318	1.5640	2.0861	2.6071	3.9122	5.2163	7.8685
			.095	1.060	21.811	.5578	1.1884	1.4519	1.7036	1.7105	2.2815	2.8513	4.2788	5.7050	8.6057
			.109	1.032	23.011	.6334	1.3495	1.6488	1.9346	1.9039	2.5394	3.1736	4.7624	6.3498	9.5784
			.120	1.010	24.024	.7017	1.4950	1.8265	2.1431	2.0753	2.7680	3.4593	5.1911	6.9214	10.4406
	1½	1.5	.065	1.370	13.057	.1649	.3513	.4293	.4042	.6130	.8176	1.0218	1.5334	2.0446	3.0841
			.072	1.356	13.328	.1732	.3689	.4507	.4212	.6387	.8519	1.0647	1.5977	2.1303	3.2135
			.083	1.334	13.772	.1871	.3987	.4871	.4497	.6819	.9095	1.1367	1.7058	2.2744	3.4307
			.095	1.310	14.281	.2040	.4346	.5310	.4835	.7333	.9781	1.2223	1.8342	2.4457	3.6891
			.109	1.282	14.912	.2260	.4816	.5884	.5272	.7995	1.0663	1.3327	1.9998	2.6664	4.0222
			.120	1.260	15.437	.2454	.5229	.6388	.5650	.8568	1.1428	1.4282	2.1432	2.8576	4.3105
			.134	1.232	16.146	.2731	.5818	.7108	.6181	.9374	1.2503	1.5625	2.3448	3.1264	4.7159
			.156	1.188	17.365	.3246	.6915	.8448	.7149	1.0842	1.4461	1.8072	2.7119	3.6159	5.4544
			.188	1.124	19.398	.4222	.8995	1.0990	.8922	1.3530	1.8046	2.2553	3.3844	4.5125	6.8068
	1¾	1.75	.065	1.620	9.338	.0744	.1585	.1936	.2068	.3135	.4182	.5226	.7843	1.0457	1.5774
			.072	1.606	9.502	.0775	.1651	.2018	.2141	.3246	.4330	.5411	.8120	1.0827	1.6332
			.083	1.584	9.768	.0828	.1763	.2154	.2262	.3430	.4575	.5718	.8581	1.1441	1.7258
			.095	1.560	10.070	.0890	.1896	.2316	.2405	.3646	.4864	.6078	.9121	1.2161	1.8345
			.109	1.532	10.442	.0970	.2066	.2524	.2585	.3920	.5229	.6535	.9806	1.3075	1.9723
			.120	1.510	10.748	.1039	.2213	.2704	.2739	.4154	.5540	.6924	1.0390	1.3854	2.0898
			.134	1.482	11.158	.1135	.2419	.2955	.2952	.4477	.5971	.7462	1.1198	1.4931	2.2523
			.156	1.438	11.852	.1310	.2791	.3410	.3330	.5050	.6736	.8418	1.2633	1.6844	2.5408
			.188	1.374	12.981	.1626	.3465	.4234	.3996	.6059	.8082	1.0100	1.5156	2.0209	3.0483
	2	2.0	.083	1.834	7.286	.0413	.0879	.1074	.1259	.1909	.2546	.3182	.4775	.6366	.9603
			.095	1.810	7.481	.0439	.0936	.1143	.1327	.2012	.2684	.3354	.5033	.6711	1.0123
			.109	1.782	7.718	.0473	.1008	.1231	.1412	.2142	.2856	.3570	.5357	.7142	1.0774
			.120	1.760	7.912	.0512	.1069	.1306	.1484	.2251	.3002	.3752	.5630	.7506	1.1323
			.134	1.732	8.170	.0541	.1154	.1409	.1582	.2400	.3201	.4000	.6003	.8004	1.2073
			.156	1.688	8.601	.0612	.1304	.1593	.1754	.2660	.3548	.4434	.6654	.8871	1.3382
			.188	1.624	9.292	.0735	.1566	.1914	.2047	.3105	.4141	.5175	.7766	1.0355	1.5620
			.220	1.560	10.070	.0890	.1896	.2316	.2405	.3646	.4864	.6078	.9121	1.2161	1.8345
80	1¼	1.25	.065	1.120	26.050	.7104	1.5136	1.8493	2.1699	1.8299	2.4407	3.0502	4.5773	6.1031	9.2061
			.072	1.106	26.713	.7542	1.6068	1.9631	2.3035	1.9243	2.5667	3.2076	4.8135	6.4180	9.6812
			.083	1.084	27.808	.8297	1.7677	2.1597	2.5341	2.0854	2.7814	3.4761	5.2163	6.9551	10.4914
			.095	1.060	29.082	.9228	1.9660	2.4021	2.8185	2.2807	3.0420	3.8017	5.7050	7.6067	11.4743
			.109	1.032	30.681	1.0479	2.2326	2.7278	3.2007	2.5385	3.3859	4.2314	6.3498	8.4664	12.7712
			.120	1.010	32.033	1.1609	2.4733	3.0218	3.5456	2.7670	3.6906	4.6123	6.9214	9.2286	13.9208
	1½	1.5	.065	1.370	17.410	.2728	.5813	.7102	.8333	.8174	1.0902	1.3625	2.0446	2.7261	4.1121
			.072	1.356	17.771	.2865	.6103	.7457	.8750	.8516	1.1359	1.4196	2.1303	2.8404	4.2846
			.083	1.334	18.362	.3096	.6596	.8059	.9456	.9092	1.2127	1.5156	2.2744	3.0325	4.5743
			.095	1.310	19.041	.3375	.7190	.8785	1.0308	.9777	1.3041	1.6298	2.4457	3.2609	4.9189
			.109	1.282	19.882	.3740	.7968	.9735	1.1422	1.0660	1.4218	1.7769	2.6664	3.5552	5.3629

filtration

TABLE 5 (CONTINUED)

Flow GPM	Tube Size	OD	Wall	ID	Velocity FPS	Viscosity — SUS 32	100	200	400	600	800	1000	1500	2000	3000
80	1½	1.5	.120	1.260	20.582	.4060	.8650	1.0569	1.2401	1.1424	1.5237	1.9043	2.8576	3.8101	5.7473
			.134	1.232	21.529	.4518	.9625	1.1759	1.3798	1.2498	1.6670	2.0834	3.1264	4.1685	6.2879
			.156	1.188	23.153	.5369	1.1440	1.3977	1.6400	1.4455	1.9281	2.4096	3.6159	4.8212	7.2725
			.188	1.124	25.864	.6985	1.4882	1.8182	2.1334	1.8040	2.4062	3.0071	4.5125	6.0166	9.0758
	1¾	1.75	.065	1.620	12.451	.1231	.2622	.3203	.2757	.4181	.5576	.6969	1.0457	1.3943	2.1032
			.072	1.606	12.669	.1282	.2732	.338	.2854	.4328	.5773	.7215	1.0827	1.4436	2.1775
			.083	1.584	13.023	.1369	.2917	.3564	.3016	.4574	.6101	.7624	1.1441	1.5255	2.3011
			.095	1.560	13.427	.1472	.3137	.3832	.3206	.4862	.6485	.8104	1.2161	1.6215	2.4460
			.109	1.532	13.923	.1604	.3418	.4176	.3447	.5227	.6972	.8713	1.3075	1.7434	2.6298
			.120	1.510	14.331	.1719	.3661	.4473	.3652	.5538	.7387	.9232	1.3854	1.8472	2.7864
			.134	1.482	14.878	.1878	.4002	.4889	.3936	.5969	.7961	.9950	1.4931	1.9908	3.0030
			.156	1.438	15.802	.2167	.4618	.5642	.6620	.6734	.8982	1.1225	1.6844	2.2459	3.3878
			.188	1.374	17.309	.2691	.5733	.7004	.8218	.8079	1.0776	1.3467	2.0209	2.6945	4.0645
	2	2.0	.083	1.834	9.715	.0683	.1454	.1777	.1678	.2545	.3395	.4242	.6366	.8488	1.2804
			.095	1.810	9.974	.0727	.1548	.1892	.1769	.2683	.3578	.4472	.6711	.8948	1.3497
			.109	1.782	10.290	.0783	.1667	.2037	.1883	.2855	.3809	.4760	.7142	.9523	1.4365
			.120	1.760	10.549	.0830	.1769	.2161	.1979	.3001	.4003	.5002	.7506	1.0009	1.5097
			.134	1.732	10.893	.0896	.1909	.2332	.2110	.3200	.4268	.5334	.8004	1.0672	1.6098
			.156	1.688	11.468	.1012	.2157	.2635	.2339	.3547	.4730	.5912	.8871	1.1829	1.7843
			.188	1.624	12.390	.1216	.2591	.3166	.2730	.4140	.5521	.6900	1.0355	1.3806	2.0826
			.220	1.560	13.427	.1472	.3137	.3832	.3206	.4862	.6485	.8104	1.2161	1.6215	2.4460
100	1¼	1.25	.065	1.120	32.562	1.0498	2.2367	2.7327	3.2065	3.5582	3.0509	3.8128	5.7216	7.6288	11.5077
			.072	1.106	33.391	1.1145	2.3744	2.9010	3.4039	3.7773	3.2083	4.0096	6.0169	8.0225	12.1015
			.083	1.084	34.761	1.2261	2.6122	3.1915	3.7448	4.1556	3.4768	4.3451	6.5204	8.6939	13.1142
			.095	1.060	36.352	1.3636	2.9053	3.5496	4.1649	4.6219	3.8025	4.7522	7.1313	9.5084	14.3428
			.109	1.032	33.352	1.5485	3.2992	4.0309	4.7297	5.2486	4.2323	5.2893	7.9373	10.5830	15.9639
	1½	1.5	.065	1.370	21.762	.4032	.8589	1.0494	1.2314	1.0217	1.3628	1.7031	2.5557	3.4076	5.1402
			.072	1.356	22.214	.4233	.9019	1.1019	1.2929	1.0646	1.4199	1.7745	2.6629	3.5505	5.3558
			.083	1.334	22.953	.4575	.9748	1.1909	1.3974	1.1365	1.5159	1.8945	2.8429	3.7906	5.7179
			.095	1.310	23.801	.4987	1.0626	1.2982	1.5233	1.2221	1.6301	2.0372	3.0571	4.0761	6.1486
			.109	1.282	24.852	.5526	1.1774	1.4385	1.6879	1.3325	1.7772	2.2211	3.3330	4.4440	6.7036
			.120	1.260	25.728	.6000	1.2783	1.5618	1.8325	1.4280	1.9047	2.3803	3.5720	4.7626	7.1842
			.134	1.232	26.911	.6676	1.4223	1.7377	2.0390	1.5623	2.0838	2.6042	3.9079	5.2106	7.8599
			.156	1.188	28.941	.7935	1.6905	2.0654	2.4235	2.6893	2.4101	3.0120	4.5199	6.0265	9.0906
			.188	1.124	32.331	1.0322	2.1991	2.6868	3.1526	3.4985	3.0077	3.7588	5.6406	7.5208	11.3447
	1¾	1.75	.065	1.620	15.564	.1818	.3874	.4734	.5554	.5226	.6970	.8711	1.3072	1.7429	2.6291
			.072	1.606	15.836	.1895	.4037	.4933	.5788	.5410	.7216	.9018	1.3533	1.8045	2.7219
			.083	1.584	16.279	.2023	.4311	.5267	.6180	.5717	.7626	.9530	1.4301	1.9068	2.8763
			.095	1.560	16.784	.2175	.4635	.5663	.6645	.6077	.8106	1.0130	1.5202	2.0269	3.0575
			.109	1.532	17.403	.2371	.5051	.6172	.7242	.6534	.8715	1.0891	1.6344	2.1792	3.2872
			.120	1.510	17.914	.2540	.5411	.6611	.7757	.6923	.9234	1.1540	1.7317	2.3090	3.4830
			.134	1.482	18.597	.2776	.5914	.7225	.8478	.7461	.9952	1.2437	1.8664	2.4885	3.7538
			.156	1.438	19.753	.3203	.6824	.8337	.9783	.8417	1.1227	1.4031	2.1055	2.8073	4.2347
			.188	1.374	21.636	.3976	.8471	1.0350	1.2144	1.0099	1.3470	1.6833	2.5261	3.3681	5.0806
	2	2.0	.083	1.834	12.144	.1009	.2149	.2626	.2098	.3181	.4243	.5303	.7958	1.0610	1.6005
			.095	1.810	12.468	.1074	.2288	.2795	.3280	.3353	.4473	.5590	.8388	1.1185	1.6871
			.109	1.782	12.863	.1156	.2464	.3010	.3532	.3569	.4761	.5950	.8928	1.1904	1.7957
			.120	1.760	13.186	.1227	.2613	.3193	.3747	.3751	.5003	.6253	.9383	1.2511	1.8872
			.134	1.732	13.616	.1324	.2820	.3446	.4043	.4000	.5335	.6667	1.0005	1.3339	2.0122
			.156	1.688	14.335	.1496	.3187	.3894	.4569	.4433	.5913	.7390	1.1089	1.4786	2.2303
			.188	1.624	15.487	.1797	.3829	.4678	.5489	.5174	.6902	.8625	1.2943	1.7258	2.6032
			.220	1.560	16.784	.2175	.4635	.5663	.6645	.6077	.8106	1.0130	1.5202	2.0269	3.0575
125	1½	1.5	.065	1.370	27.203	.5958	1.2693	1.5508	1.8196	2.0192	1.7034	2.1288	3.1946	4.2595	6.4252
			.072	1.356	27.767	.6255	1.3327	1.6283	1.9106	2.1202	1.7749	2.2181	3.3286	4.4381	6.6947
			.083	1.334	28.691	.6761	1.4404	1.7599	2.0650	2.2915	1.8949	2.3681	3.5537	4.7382	7.1474
			.095	1.310	29.752	.7370	1.5702	1.9184	2.2510	2.4979	2.0376	2.5465	3.8213	5.0951	7.6857
			.109	1.282	31.066	.8166	1.7399	2.1257	2.4942	2.7679	2.2216	2.7764	4.1663	5.5550	8.3795
			.120	1.260	32.160	.8866	1.8890	2.3079	2.7080	3.0051	2.3808	2.9754	4.4650	5.9533	8.9802
			.134	1.232	33.638	.9865	2.1017	2.5679	3.0130	3.3436	2.6047	3.2552	4.8849	6.5132	9.8249
			.156	1.188	36.176	1.1725	2.4981	3.0521	3.5812	3.9741	3.0126	3.7650	5.6498	7.5331	11.3633
	1¾	1.75	.065	1.620	19.455	.2687	.5725	.6995	.8207	.6532	.8713	1.0889	1.6340	2.1786	3.2863
			.072	1.606	19.795	.2800	.5966	.7289	.8553	.6763	.9020	1.1273	1.6917	2.2556	3.4024
			.083	1.584	20.349	.2990	.6370	.7783	.9132	,7147	.9532	1.1913	1.7876	2.3835	3.5954
			.095	1.560	20.980	.3215	.6849	.8368	.9819	.7597	1.0132	1.2663	1.9002	2.5336	3.8218
			.109	1.532	21.754	.3504	.7465	.9120	1.0701	.8167	1.0894	1.3614	2.0430	2.7240	4.1090
			.120	1.510	22.392	.3753	.7995	.9769	1.1462	.8654	1.1543	1.4425	2.1647	2.8862	4.3537
			.134	1.482	23.247	.4102	.8739	1.0677	1.2528	1.3902	1.2440	1.5547	2.3330	3.1106	4.6922
			.156	1.438	24.691	.4733	1.0084	1.2320	1.4456	1.6042	1.4034	1.7538	2.6319	3.5092	5.2934
			.188	1.374	27.045	.5876	1.2518	1.5295	1.7946	1.9915	1.6837	2.1042	3.1576	4.2101	6.3507

TABLE 5 (CONTINUED)

Flow GPM	Tube Size	OD	Wall	ID	Velocity FPS	Viscosity — SUS 32	100	200	400	600	800	1000	1500	2000	3000
125	2	2.0	.083	1.834	15.180	.1491	.3176	.3880	.4553	.3977	.5304	.6629	.9947	1.3263	2.0007
			.095	1.810	15.585	.1587	.3381	.4130	.4847	.4192	.5591	.6987	1.0485	1.3981	2.1089
			.109	1.782	16.078	.1709	.3641	.4448	.5219	.4462	.5951	.7437	1.1160	1.4880	2.2446
			.120	1.760	16.483	.1813	.3862	.4718	.5536	.4689	.6254	.7816	1.1729	1.5638	2.3590
			.134	1.732	17.020	.1956	.4168	.5092	.5975	.4999	.6668	.8334	1.2506	1.6674	2.5152
			.156	1.688	17.919	.2210	.4709	.5754	.6751	.5541	.7391	.9237	1.3862	1.8482	2.7879
			.188	1.624	19.359	.2656	.5658	.6913	.8112	.6468	.8627	1.0782	1.6179	2.1572	3.2541
			.220	1.560	20.980	.3215	.6849	.8368	.9819	.7597	1.0132	1.2663	1.9002	2.5336	3.8218
150	1½	1.5	.065	1.370	32.643	.8197	1.7463	2.1336	2.5035	2.7782	2.0441	.25546	3.8335	5.1114	7.7102
			.072	1.356	33.321	.8606	1.8336	2.2403	2.6287	2.9171	2.1299	2.6618	3.9943	5.3258	8.0336
			.083	1.334	34.429	.9302	1.9818	2.4213	2.8411	3.1527	3.3881	2.8417	4.2644	5.6859	8.5768
			.095	1.310	35.702	1.1040	2.1603	2.6394	3.0969	3.4367	3.6933	3.0558	4.5856	6.1141	9.2229
			.109	1.282	37.279	1.1236	2.3938	2.9246	3.4317	3.8081	4.0925	3.3316	4.9995	6.6661	10.0554
			.120	1.260	38.592	1.2198	2.5989	3.1753	3.7257	4.1345	4.4432	3.5705	5.3580	7.1440	10.7763
	1¾	1.75	.065	1.620	23.346	.3697	.7877	.9624	1.1292	1.2531	1.0455	1.3066	1.9608	2.6143	3.9436
			.072	1.606	23.755	.3853	.8208	1.0029	1.1767	1.3058	1.0825	1.3528	2.0300	2.7067	4.0829
			.083	1.584	24.419	.4114	.8764	1.0708	1.2564	1.3943	1.1439	1.4295	2.1452	2.8602	4.3145
			.095	1.560	25.176	.4423	.9423	1.1513	1.3509	1.4991	1.2159	1.5195	2.2803	3.0403	4.5862
			.109	1.532	26.105	.4820	1.0270	1.2548	1.4723	1.6338	1.3072	1.6337	2.4516	3.2688	4.9308
			.120	1.510	26.871	.5163	1.1000	1.3440	1.5770	1.7500	1.3851	1.7310	2.5976	3.4635	5.2245
			.134	1.482	27.896	.5643	1.2023	1.4690	1.7236	1.9127	1.4928	1.8656	2.7996	3.7327	5.6306
			.156	1.438	29.629	.6512	1.3874	1.6951	1.9889	2.2071	1.6840	2.1046	3.1583	4.2110	6.3521
			.188	1.374	32.454	.8084	1.7223	2.1043	2.4691	2.7399	2.0204	2.5250	3.7891	5.0521	7.6209
	2	2.0	.083	1.834	18.215	.2051	.4369	.5338	.6264	.4772	.6365	.7954	1.1937	1.5916	2.4008
			.095	1.810	18.702	.2183	.4651	.5683	.6668	.5030	.6709	.8385	1.2583	1.6777	2.5307
			.109	1.782	19.294	.2351	.5009	.6120	.7181	.7968	.7141	.8924	1.3392	1.7856	2.6935
			.120	1.760	19.779	.2494	.5313	.6492	.7617	.8453	.7505	.9379	1.4074	1.8766	2.8307
			.134	1.732	20.424	.2691	.5734	.7005	.8220	.9122	.8002	1.0000	1.5007	2.0009	3.0183
			.156	1.688	21.503	.3041	.6479	.7916	.9289	1.0308	.8869	1.1085	1.6634	2.2178	3.3455
			.188	1.624	23.231	.3654	.7785	.9512	1.1161	1.2385	1.0353	1.2938	1.9415	2.5887	3.9049
			.220	1.560	25.176	.4423	.9423	1.1513	1.3509	1.4991	1.2159	1.5195	2.2803	3.0403	4.5862
175	1½	1.5	.065	1.370	38.084	1.0735	2.2871	2.7943	3.2787	3.6384	3.9101	2.9804	4.4725	5.9633	8.9953
			.072	1.356	38.874	1.1272	2.4014	2.9340	3.4427	3.8203	4.1056	3.1054	4.6600	6.2134	9.3726
	1¾	1.75	.065	1.620	27.237	.4842	1.0316	1.2604	1.4789	1.6411	1.2198	1.5244	2.2875	3.0501	4.6008
			.072	1.606	27.714	.5046	1.0750	1.3134	1.5411	1.7102	1.2629	1.5782	2.3684	3.1578	4.7634
			.083	1.584	28.489	.5387	1.1478	1.4024	1.6455	1.8260	1.3345	1.6678	2.5027	3.3369	5.0336
			.095	1.560	29.372	.5793	1.2341	1.5078	1.7692	1.9633	2.1099	1.7728	2.6603	3.5471	5.3506
			.109	1.532	30.456	.6313	1.3450	1.6433	1.9282	2.1397	2.2995	1.9060	2.8602	3.8136	5.7526
			.120	1.510	31.349	.6762	1.4407	1.7602	2.0653	2.2919	2.4630	2.0195	3.0305	4.0407	6.0952
			.134	1.482	32.545	.7391	1.5746	1.9238	2.2574	2.5050	2.6920	2.1765	3.2662	4.3549	6.5691
			.156	1.438	34.567	.8528	1.8170	2.2199	2.6048	2.8906	3.1064	2.4554	3.6846	4.9128	7.4107
			.188	1.374	37.863	1.0587	2.2556	2.7559	3.2336	3.5884	3.8563	2.9458	4.4206	5.8941	8.8910
	2	2.0	.083	1.834	21.251	.2686	.5722	.6991	.8203	.9103	.7426	.9280	1.3926	1.8568	2.8009
			.095	1.810	21.819	.2859	.6092	.7443	.8733	.9691	.7827	.9782	1.4680	1.9573	2.9525
			.109	1.782	22.510	.3079	.6560	.8015	.9404	1.0436	.8331	1.0412	1.5624	2.0832	3.1424
			.120	1.760	23.076	.3266	.6959	.8502	.9976	1.1070	.8756	1.0942	1.6420	2.1894	3.3025
			.134	1.732	23.828	.3525	.7509	.9175	1.0765	1.1946	.9336	1.1667	1.7508	2.3344	3.5213
			.156	1.688	25.086	.3983	.8486	1.0368	1.2165	1.3500	1.0348	1.2932	1.9406	2.5875	3.9031
			.188	1.624	27.103	.4786	1.0196	1.2457	1.4617	1.6220	1.2078	1.5094	2.2651	3.0201	4.5557
			.220	1.560	29.372	.5793	1.2341	1.5078	1.7692	1.9633	2.1099	1.7728	2.6603	3.5471	5.3506
200	1¾	1.75	.065	1.620	31.128	.6117	1.3032	1.5922	1.8682	2.0731	2.2279	1.7422	2.6143	3.4858	5.2581
			.072	1.606	31.673	.6374	1.3580	1.6592	1.9468	2.1604	2.3217	1.8037	2.7067	3.6089	5.4439
			.083	1.584	32.559	.6806	1.4500	1.7715	2.0786	2.3067	2.4789	1.9060	2.8602	3.8136	5.7527
			.095	1.560	33.568	.7317	1.5590	1.9048	2.2350	2.4802	2.6653	2.0260	3.0403	4.0538	6.1149
			.109	1.532	34.806	.7975	1.6991	2.0759	2.4358	2.7030	2.9048	2.1783	3.2688	4.3584	6.5744
			.120	1.510	35.828	.8542	1.8199	2.2235	2.6090	2.8952	3.1114	2.3080	3.4635	4.6180	6.9660
			.134	1.482	37.195	.9336	1.9891	2.4303	2.8516	3.1644	3.4007	2.4874	3.7327	4.9770	7.5075
			.156	1.438	39.506	1.0773	2.2953	2.8043	3.2905	3.6515	3.9241	4.1490	4.2110	5.6147	8.4694
	2	2.0	.083	1.834	24.287	.3393	.7228	.8831	1.0363	1.1499	.8487	1.0606	1.5916	2.1221	3.2010
			.095	1.810	24.936	.3612	.7695	.9402	1.1032	1.2242	.8946	1.1180	1.6777	2.2369	3.3742
			.109	1.782	25.725	.3889	.8287	1.0124	1.1880	1.3183	1.4167	1.1899	1.7856	2.3808	3.5914
			.120	1.760	26.372	.4126	.8790	1.0740	1.2602	1.3984	1.5028	1.2505	1.8766	2.5021	3.7743
			.134	1.732	27.232	.4452	.9486	1.1590	1.3599	1.5091	1.6218	1.3334	2.0009	2.6679	4.0244
			.156	1.688	28.670	.5031	1.0719	1.3097	1.5367	1.7053	1.8326	1.4779	2.2178	2.9571	4.4607
			.188	1.624	30.974	.6045	1.2880	1.5736	1.8464	2.0490	2.2020	1.7251	2.5887	3.4516	5.2065
			.220	1.560	33.568	.7317	1.5590	1.9048	2.2350	2.4802	2.6653	2.0260	3.0403	4.0538	6.1149

TABLE 6: PRESSURE DIFFERENTIAL IN PSI PER FOOT OF HOSE

Flow GPM	Hose Size	Hose ID	Velocity FPS	Viscosity — SUS 32	100	200	400	600	800	1000	1500	2000	3000
1.5	3/16	.1875	17.427	3.2832	6.8198	15.1960	28.8048	43.6809	58.2617	72.8118	109.264	145.685	219.758
	1/4	.25	9.803	.8372	2.1578	4.8081	9.1140	13.8209	18.4344	23.0381	34.5717	46.0957	69.5328
	5/16	.3125	6.274	.2901	.8838	1.9694	3.7331	5.6611	7.5507	9.4364	14.1606	18.8808	28.4806
3	3/16	.1875	34.855	11.0435	23.5285	30.3921	57.6096	87.3618	116.523	145.624	218.528	291.370	439.516
	1/4	.25	19.606	2.8161	4.3156	9.6162	18.2280	27.6418	36.8687	46.0762	69.1435	92.1913	139.066
	5/16	.3125	12.548	.9757	1.7677	3.9388	7.4662	11.3221	15.1014	18.8728	28.3212	37.7616	56.9613
	13/32	.4063	7.4229	.2804	.6186	1.3784	2.6128	3.9622	5.2848	6.6046	9.9112	13.2149	19.9339
5	1/4	.25	32.676	6.8847	14.6680	16.0271	30.3801	46.0697	61.4479	76.7937	115.239	153.652	231.776
	5/16	.3125	20.913	2.3854	5.0822	6.5647	12.4437	18.8702	25.1691	31.4547	47.2020	62.9359	94.9355
	13/32	.4063	12.372	.6856	1.0310	2.2974	4.3547	6.6037	8.8081	11.0077	16.5186	22.0248	33.2232
	1/2	.5	8.169	.2559	.4495	1.0017	1.8988	2.8794	3.8405	4.7996	7.2025	9.6033	14.4860
	5/8	.625	5.228	.0886	.1841	.4103	.7777	1.1794	1.5731	1.9659	2.9501	3.9335	5.9335
7	5/16	.3125	29.278	4.2982	9.1574	9.1906	17.4211	26.4182	35.2367	44.0366	66.0827	88.1103	132.910
	13/32	.4063	17.320	1.2354	2.6320	3.2163	6.0966	9.2452	12.3313	15.4108	23.1260	30.8347	46.5125
	1/2	.5	11.437	.4610	.9822	1.4024	2.6583	4.0311	5.3767	6.7195	10.0834	13.4446	20.2804
	5/8	.625	7.320	.1597	.2578	.5744	1.0888	1.6511	2.2023	2.7523	4.1302	5.5069	8.3069
	3/4	.75	5.083	.0672	.1243	.2770	.5251	.7963	1.0621	1.3273	1.9918	2.6557	4.0060
10	13/32	.4063	24.743	2.3061	4.9132	4.5947	8.7095	13.2074	17.6161	22.0155	33.0372	44.0496	66.4464
	1/2	.5	16.338	.8606	1.8335	2.0034	3.7975	5.7587	7.6810	9.5992	14.4049	19.2065	28.9720
	5/8	.625	10.457	.2982	.6353	.8206	1.5555	2.3588	3.1461	3.9318	5.9002	7.8670	11.8669
	3/4	.75	7.261	.1254	.2672	.3957	.7501	1.1375	1.5172	1.8961	2.8454	3.7939	5.7229
	7/8	.875	5.335	.0603	.0959	.2136	.4049	.6140	.8190	1.0235	1.5359	2.0478	3.0891
15	13/32	.4063	37.114	4.6885	9.9891	12.2044	13.0642	19.8111	26.4242	33.0232	49.5558	66.0743	99.6696
	1/2	.5	24.507	1.7497	3.7277	4.5544	5.6963	8.6381	11.5215	14.3988	21.6073	28.8098	43.4580
	5/8	.625	15.685	.6062	1.2916	1.2309	2.3332	3.5382	4.7192	5.8978	8.8504	11.8005	17.8004
	3/4	.75	10.892	.2550	.5433	.5936	1.1252	1.7063	2.2759	2.8442	4.2681	5.6908	8.5843
	7/8	.875	8.002	.1226	.2612	.3204	.6073	.9210	1.2285	1.5352	2.3038	3.0718	4.6336
	1	1.0	6.127	.0650	.1385	.1878	.3560	.5399	.7201	.8999	1.3505	1.8006	2.7161
20	1/2	.5	32.676	2.8946	6.1671	7.5348	7.5950	11.5174	15.3620	19.1984	28.8098	38.4131	57.9440
	5/8	.625	20.913	1.0029	2.1368	2.6107	3.1109	4.7175	6.2923	7.8637	11.8005	15.7340	23.7339
	3/4	.75	14.523	.4219	.8988	.7915	1.5003	2.2751	3.0345	3.7923	5.6908	7.5878	11.4457
	7/8	.875	10.670	.2028	.4322	.4272	.8098	1.2280	1.6379	2.0470	3.0718	4.0957	6.1781
	1	1.0	8.169	.1076	.2292	.2504	.4747	.7198	.9601	1.1999	1.8006	2.4008	3.6215
	1 1/8	1.125	6.455	.0615	.1310	.1563	.2963	.4494	.5994	.7491	1.1241	1.4988	2.2609
	1 1/4	1.25	5.228	.0373	.0794	.1026	.1944	.2948	.3933	.4915	.7375	.9834	1.4834
25	5/8	.625	26.141	1.4821	3.1576	3.8578	3.8887	5.8969	7.8653	9.8296	14.7506	19.6675	29.6673
	3/4		18.154	.6234	1.3281	1.6227	1.8753	2.8438	3.7931	4.7404	7.1135	9.4847	14.3072
	7/8	.875	13.337	.2997	.6386	.5340	1.0123	1.5350	2.0474	2.5587	3.8397	5.1196	7.7227
	1	1.0	10.211	.1590	.3387	.3130	.5934	.8998	1.2002	1.4999	2.2508	3.0010	4.5269
	1 1/8	1.125	8.068	.0908	.1936	.1954	.3704	.5617	.7493	.9364	1.4051	1.8735	2.8261
	1 1/4	1.25	6.535	.0551	.1173	.1282	.2430	.3686	.4916	.6143	.9219	1.2292	1.8542
	1 3/8	1.375	5.401	.0350	.0746	.0876	.1660	.2517	.3358	.4196	.6297	.8396	1.2665
30	5/8	.625	31.369	2.0391	4.3443	5.3077	4.6664	7.0763	9.4384	11.7955	17.7007	23.6010	35.6008
	3/4	.75	21.784	.8577	1.8273	2.2325	2.2504	3.4126	4.5517	5.6884	8.5362	11.3816	17.1686
	7/8		16.005	.4124	.8786	1.0735	1.2147	1.8420	2.4569	3.0705	4.6077	6.1435	9.2672
	1	1.0	12.254	.2187	.4660	.5693	.7120	1.0798	1.4402	1.7999	2.7009	3.6012	5.4323
	1 1/8	1.125	9.682	.1250	.2663	.2345	.4445	.6741	.8991	1.1236	1.6862	2.2482	3.3913
	1 1/4	1.25	7.842	.0758	.1614	.1539	.2916	.4423	.5899	.7372	1.1063	1.4751	2.2251
	1 3/8	1.375	6.481	.0482	.1027	.1051	.1992	.3021	.4029	.5035	.7556	1.0075	1.5197
	1 1/2	1.5	5.446	.0319	.0679	.0742	.1406	.2133	.2845	.3555	.5335	.7114	1.0730
40	3/4	.75	29.046	1.4189	3.0231	3.6935	3.0005	4.5501	6.0689	7.5846	11.3816	15.1755	22.8915
	7/8	.875	21.340	.6823	1.4536	1.7760	1.6196	2.4560	3.2759	4.0940	6.1435	8.1914	12.3562
	1	1.0	16.338	.3618	.7709	.9419	.9494	1.4397	1.9203	2.3998	3.6012	4.8016	7.2430
	1 1/8	1.125	12.909	.2068	.4406	.5383	.5927	.8988	1.1988	1.4982	2.2482	2.9976	4.5218
	1 1/4	1.25	10.457	.1254	.2671	.3263	.3889	.5897	.7865	.9830	1.4751	1.9668	2.9667
	1 3/8	1.375	8.642	.0797	.1698	.2075	.2656	.4028	.5372	.6714	1.0075	1.3433	2.0263
	1 1/2	1.5	7.261	.0527	.1123	.0989	.1875	.2844	.3793	.4740	.7114	.9485	1.4307
50	3/4	.75	36.307	2.0968	4.4673	5.4580	6.4043	5.6876	7.5862	9.4807	14.2271	18.9694	28.6143
	7/8	.875	26.675	1.0082	2.1481	2.6244	3.0794	3.0700	4.0948	5.1175	7.6794	10.2392	15.4453
	1	1.0	20.423	.5347	1.1392	1.3918	1.1867	1.7996	2.4003	2.9998	4.5015	6.0020	9.0538
	1 1/8	1.125	16.137	.3056	.6510	.7954	.7409	1.1235	1.4985	1.8727	2.8103	3.7470	5.6522
	1 1/4	1.25	13.071	.1853	.3947	.4822	.4861	.7371	.9832	1.2287	1.8438	2.4584	3.7084
	1 3/8	1.375	10.802	.1178	.2510	.3066	.3320	.5035	.6715	.8392	1.2594	1.6791	2.5329
	1 1/2	1.5	9.077	.0779	.1660	.2028	.2344	.3555	.4741	.5925	.8892	1.1856	1.7884
	1 13/16	1.813	6.213	.0317	.0675	.0579	.1098	.1666	.2222	.2776	.4166	.5555	.8380

TABLE 6 (CONTINUED)

Flow GPM	Hose Size	Hose ID	Velocity FPS	Viscosity — SUS 32	100	200	400	600	800	1000	1500	2000	3000
50	2	2.0	5.106	.0199	.0423	.0391	.0742	.1125	.1500	.1875	.2813	.3751	.5659
60	7/8	.875	32.010	1.3872	2.9554	3.6108	4.2368	3.6840	4.9138	6.1409	9.2153	12.2871	18.5344
	1	1.0	24.507	.7356	1.5673	1.9149	2.2469	2.1595	2.8804	3.5997	5.4018	7.2025	10.8645
	1 1/8	1.125	19.364	.4204	.8957	1.0944	.8890	1.3482	1.7982	2.2473	3.3723	4.4965	6.7827
	1 1/4	1.25	15.685	.2549	.5430	.6635	.5833	.8845	1.1798	1.4744	2.2126	2.9501	4.4501
	1 3/8	1.375	12.963	.1621	.3453	.4219	.3984	.6042	.8058	1.0071	1.5112	2.0150	3.0395
	1 1/2	1.5	10.892	.1072	.2284	.2791	.2813	.4266	.5690	.7111	1.0670	1.4227	2.1461
	1 13/16	1.813	7.456	.0436	.0928	.1134	.1318	.1999	.2666	.3332	.5000	.6666	1.0056
	2	2.0	6.127	.0273	.0582	.0712	.0890	.1350	.1800	.2250	.3376	.4502	.6790
80	1	1.0	32.676	1.2171	2.5930	3.1680	3.7172	2.8794	3.8405	4.7996	7.2025	9.6033	14.4860
	1 1/8	1.125	25.818	.6956	1.4819	1.8106	2.1244	1.7976	2.3976	2.9964	4.4965	5.9953	9.0435
	1 1/4	1.25	20.913	.4217	.8984	1.0977	1.2879	1.1794	1.5731	1.9659	2.9501	3.9335	5.9335
	1 3/8	1.375	17.283	.2681	.5713	.6980	.8190	.8055	1.0744	1.3428	2.0150	2.6866	4.0526
	1 1/2	1.5	14.523	.1774	.3779	.4617	.3751	.5688	.7586	.9481	1.4227	1.8969	2.8614
	1 13/16	1.813	9.941	.0721	.1536	.1877	.1757	.2665	.3555	.4442	.6666	.8888	1.3408
	2	2.0	8.169	.0452	.0964	.1177	.1187	.1800	.2400	.3000	.4502	.6002	.9054
	2 3/8	2.375	5.793	.0200	.0426	.0520	.0597	.0905	.1207	.1509	.2264	.3018	.4553
100	1 1/8	1.125	32.273	1.0278	2.1898	2.6755	3.1393	3.4837	2.9970	3.7455	5.6206	7.4941	11.3044
	1 1/4	1.25	26.141	.6231	1.3276	1.6220	1.9032	1.4742	1.9663	2.4574	3.6877	4.9169	7.4168
	1 3/8	1.375	21.604	.3962	.8442	1.0314	1.2103	1.0069	1.3430	1.6784	2.5187	3.3583	5.0658
	1 1/2	1.5	18.154	.2621	.5584	.6823	.8005	.7110	.9483	1.1851	1.7784	2.3712	3.5768
	1 13/16	1.813	12.427	.1065	.2270	.2773	.3254	.3331	.4443	.5553	.8333	1.1111	1.6760
	2	2.0	10.211	.0668	.1424	.1740	.1483	.2250	.3000	.3750	.5627	.7503	1.1317
	2 3/8	2.375	7.241	.0295	.0629	.0769	.0746	.1131	.1509	.1886	.2830	.3773	.5691
125	1 1/4	1.25	32.676	.9208	1.9618	2.3969	2.8124	3.1210	2.4579	3.0718	4.6096	6.1461	9.2710
	1 3/8	1.375	27.005	.5855	1.2475	1.5242	1.7884	1.9846	1.6788	2.0980	3.1484	4.1979	6.3323
	1 1/2	1.5	22.692	.3873	.8252	1.0082	1.1830	.8887	1.1853	1.4814	2.2230	2.9640	4.4710
	1 13/16	1.813	15.533	.1574	.3354	.4098	.4809	.4164	.5554	.6941	1.0416	1.3888	2.0950
	2	2.0	12.764	.0988	.2104	.2571	.3017	.2812	.3750	.4687	.7034	.9378	1.4147
	2 3/8	2.375	9.052	.0437	.0930	.1137	.0932	.1414	.1886	.2357	.3537	.4716	.7114
150	1 1/4	1.25	39.212	1.2669	2.6991	3.2977	3.8694	4.2939	4.6145	3.6861	5.5315	7.3753	11.1253
	1 3/8	1.375	32.406	.8056	1.7164	2.0970	2.4606	2.7305	2.0146	2.5177	3.7781	5.0374	7.5987
	1 1/2	1.5	27.230	.5329	1.1353	1.3871	1.6276	1.8061	1.4224	1.7776	2.6676	3.5568	5.3652
	1 13/16	1.813	18.640	.2166	.4615	.5638	.6616	.4997	.6665	.8329	1.2499	1.6666	2.5140
	2	2.0	15.317	.1359	.2895	.3537	.4150	.3374	.4501	.5625	.8440	1.1254	1.6976
	2 3/8	2.375	10.862	.0601	.1280	.1564	.1835	.1697	.2263	.2828	.4245	.5659	.8537
175	1 3/8	1.375	37.807	1.0551	2.2478	2.7464	3.2225	3.5760	3.8430	2.9373	4.4078	5.8770	8.8652
	1 1/2	1.5	31.769	.6979	1.4869	1.8166	2.1316	2.3654	2.5420	2.0739	3.1122	4.1496	6.2594
	1 13/16	1.813	21.746	.2837	.6044	.7384	.8664	.9615	.7776	.9718	1.4583	1.9444	2.9330
	2	2.0	17.870	.1780	.3792	.4632	.5435	.6032	.5251	.6562	.9847	1.3130	1.9805
	2 3/8	2.375	12.672	.0787	.1676	.2048	.2403	.1980	.2640	.3300	.4952	.6603	.9960
200	1 1/2	1.5	36.307	.8816	1.8783	2.2948	2.6927	2.9881	3.2112	2.3702	3.5568	4.7424	7.1536
	1 13/16	1.813	24.853	.3584	.7635	.9328	1.0945	1.2146	.8887	1.1106	1.6666	2.2221	3.3520
	2	2.0	20.423	.2248	.4790	.5852	.6866	.7620	.6001	.7499	1.1254	1.5005	2.2634
	2 3/8	2.375	14.483	.0994	.2117	.2587	.3035	.3368	.3018	.3771	.5659	.7546	1.1382

TABLE 7: PRESSURE DIFFERENTIAL IN IN. Hg PER FOOT OF SCHEDULE 40 PIPE

Flow GPM	Pipe — Schedule 40				Velocity FPS	Viscosity — SUS									
	Size	OD	Wall	ID		32	100	200	400	600	800	1000	1500	2000	3000
1.5	¼	.540	.088	.364	4.624	.3457	.9776	2.1783	4.1290	6.2616	8.3514	10.4371	15.6622	20.8830	31.5009
	⅜	.675	.091	.493	2.521	.0759	.2905	.6473	1.2270	1.8608	2.4819	3.1017	4.6545	6.2060	9.3614
	½	.840	.109	.622	1.584	.0243	.1147	.2555	.4843	.7344	.9795	1.2241	1.8370	2.4493	3.6946
3	⅜	.675	.091	.493	5.042	.2793	.5810	1.2947	2.4541	3.7215	4.9637	6.2034	9.3090	12.4120	18.7228
	½	.840	.109	.622	3.167	.0872	.2293	.5110	.9685	1.4687	1.9590	2.4482	3.6739	4.8985	7.3892
	¾	1.050	.113	.824	1.805	.0218	.0745	.1659	.3145	.4769	.6360	.7949	1.1928	1.5905	2.3991
	1	1.315	.133	1.049	1.114	.0067	.0283	.0632	.1197	.1816	.2422	.3026	.4541	.6055	.9134
5	½	.840	.109	.622	5.279	.2255	.3822	.8516	1.6142	2.4479	3.2650	4.0804	6.1232	8.1642	12.3153
	¾	1.050	.113	.824	3.008	.0552	.1241	.2765	.5241	.7948	1.0601	1.3248	1.9881	2.6508	3.9985
	1	1.315	.133	1.049	1.856	.0169	.0472	.1053	.1995	.3026	.4036	.5044	.7569	1.0092	1.5223
	1¼	1.660	.140	1.380	1.072	.0044	.0158	.0351	.0666	.1010	.1348	.1684	.2527	.3369	.5083
7	¾	1.050	.113	.824	4.211	.1033	.1737	.3871	.7337	1.1127	1.4841	1.8547	2.7833	3.7110	5.5979
	1	1.315	.133	1.049	2.598	.0310	.0661	.1474	.2794	.4236	.5650	.7061	1.0600	1.4129	2.1312
	1¼	1.660	.140	1.380	1.501	.0081	.0221	.0492	.0933	.1414	.1887	.2358	.3538	.4717	.7116
	1½	1.900	.145	1.610	1.103	.0038	.0119	.0266	.0503	.0763	.1018	.1273	.1910	.2546	.3841
10	1	1.315	.133	1.049	3.712	.0602	.0945	.2105	.3991	.6052	.8072	1.0088	1.5138	2.0184	3.0446
	1¼	1.660	.140	1.380	2.145	.0155	.0315	.0703	.1332	.2021	.2695	.3368	.5054	.6739	1.0165
	1½	1.900	.145	1.610	1.576	.0073	.0170	.0379	.0719	.1091	.1455	.1818	.2728	.3638	.5487
15	1¼	1.660	.140	1.380	3.217	.0328	.0473	.1054	.1999	.3031	.4043	.5052	.7581	1.0108	1.5248
	1½	1.900	.145	1.610	2.364	.0153	.0255	.0569	.1079	.1636	.2182	.2727	.4092	.5456	.8230
	2	2.375	.154	2.067	1.434	.0045	.0094	.0209	.0397	.0602	.0803	.1004	.1506	.2008	.3029
	2½	2.875	.203	2.469	1.005	.0019	.0046	.0103	.0195	.0296	.0395	.0493	.0740	.0987	.1488
20	1¼	1.660	.140	1.380	4.290	.0556	.1065	.1406	.2665	.4041	.5390	.6736	1.0108	1.3478	2.0331
	1½	1.900	.145	1.610	3.151	.0258	.0341	.0759	.1438	.2181	.2909	.3636	.5456	.7275	1.0974
	2	2.375	.154	2.067	1.912	.0076	.0125	.0279	.0529	.0803	.1071	.1338	.2008	.2678	.4039
	2½	2.875	.203	2.469	1.340	.0032	.0062	.0137	.0260	.0394	.0526	.0657	.0987	.1315	.1984
25	1¼	1.660	.140	1.380	5.362	.0846	.1558	.1757	.3331	.5051	.6738	.8420	1.2636	1.6847	2.5413
	1½	1.900	.145	1.610	3.939	.0391	.0752	.0949	.1798	.2727	.3637	.4545	.6820	.9094	1.3717
	2	2.375	.154	2.067	2.390	.0113	.0157	.0349	.0662	.1004	.1339	.1673	.2510	.3347	.5049
	2½	2.875	.203	2.469	1.675	.0048	.0077	.0172	.0325	.0493	.0658	.0822	.1233	.1644	.2480
	3	3.500	.216	3.068	1.085	.0017	.0032	.0072	.0136	.0207	.0276	.0345	.0517	.0690	.1040
30	1½	1.900	.145	1.610	4.727	.0551	.1026	.1138	.2158	.3272	.4364	.5454	.8184	1.0913	1.6461
	2	2.375	.154	2.067	2.868	.0159	.0315	.0419	.0794	.1204	.1606	.2007	.3013	.4017	.6059
	2½	2.875	.203	2.469	2.010	.0066	.0092	.0206	.0390	.0592	.0789	.0986	.1480	.1973	.2976
	3	3.500	.216	3.068	1.302	.0023	.0039	.0086	.0164	.0248	.0331	.0414	.0621	.0828	.1248
40	2	2.375	.154	2.067	3.824	.0271	.0515	.0559	.1059	.1606	.2142	.2677	.4017	.5356	.8079
	2½	2.875	.203	2.469	2.680	.0112	.0223	.0274	.0520	.0789	.1052	.1315	.1973	.2631	.3968
	3	3.500	.216	3.068	1.736	.0039	.0080	.0115	.0218	.0331	.0441	.0551	.0828	.1103	.1664
	3½	4.000	.226	3.548	1.298	.0019	.0029	.0064	.0122	.0185	.0247	.0308	.0463	.0617	.0931
	4	4.500	.237	4.026	1.008	.0010	.0017	.0039	.0074	.0112	.0149	.0186	.0279	.0372	.0561
50	2	2.375	.154	2.067	4.780	.0413	.0755	.0698	.1324	.2007	.2677	.3346	.5021	.6694	1.0098
	2½	2.875	.203	2.469	3.350	.0170	.0326	.0343	.0650	.0986	.1315	.1643	.2466	.3288	.4960
	3	3.500	.216	3.068	2.170	.0058	.0117	.0144	.0273	.0414	.0552	.0689	.1034	.1379	.2081
	3½	4.000	.226	3.548	1.622	.0028	.0059	.0080	.0152	.0231	.0308	.0385	.0578	.0771	.1163
	4	4.500	.237	4.026	1.260	.0015	.0022	.0049	.0092	.0139	.0186	.0232	.0349	.0465	.0702
60	2½	2.875	.203	2.469	4.020	.0239	.0445	.0412	.0780	.1183	.1578	.1972	.2960	.3946	.5953
	3	3.500	.216	3.068	2.604	.0081	.0159	.0173	.0327	.0496	.0662	.0827	.1241	.1655	.2497
	3½	4.000	.226	3.548	1.947	.0040	.0080	.0097	.0183	.0277	.0370	.0462	.0694	.0925	.1396
	4	4.500	.237	4.026	1.512	.0021	.0044	.0058	.0110	.0167	.0223	.0279	.0419	.0558	.0842
80	2½	2.875	.203	2.469	5.360	.0412	.0729	.0920	.1040	.1578	.2104	.2630	.3946	.5262	.7937
	3	3.500	.216	3.068	3.472	.0139	.0260	.0230	.0436	.0662	.0883	.1103	.1655	.2207	.3329
	3½	4.000	.226	3.548	2.596	.0068	.0131	.0129	.0244	.0370	.0493	.0617	.0925	.1234	.1861
	4	4.500	.237	4.026	2.016	.0036	.0072	.0078	.0147	.0223	.0298	.0372	.0558	.0744	.1123
	5	5.563	.258	5.047	1.283	.0012	.0025	.0031	.0060	.0090	.0121	.0151	.0226	.0301	.0455
100	3	3.500	.216	3.068	4.339	.0212	.0383	.0483	.0545	.0827	.1103	.1379	.2069	.2759	.4161
	3½	4.000	.226	3.548	3.245	.0103	.0192	.0161	.0305	.0462	.0617	.0771	.1157	.1542	.2327
	4	4.500	.237	4.026	2.520	.0055	.0105	.0097	.0184	.0279	.0372	.0465	.0698	.0930	.1403
	5	5.563	.258	5.047	1.604	.0018	.0036	.0039	.0074	.0113	.0151	.0188	.0283	.0377	.0568
	6	6.625	.280	6.065	1.110	.0007	.0015	.0019	.0036	.0054	.0072	.0090	.0135	.0181	.0272

TABLE 7 (CONTINUED)

Flow GPM	Pipe — Schedule 40				Velocity FPS	Viscosity — SUS									
	Size	OD	Wall	ID		32	100	200	400	600	800	1000	1500	2000	3000
125	3	3.500	.216	3.068	5.424	.0323	.0562	.0706	.0682	.1034	.1379	.1723	.2586	.3448	.5201
	3½	4.000	.226	.3548	4.056	.0156	.0281	.0356	.0381	.0578	.0771	.0964	.1146	.1928	.2908
	4	4.500	.237	4.026	3.150	.0083	.0155	.0197	.0230	.0349	.0465	.0581	.0872	.1163	.1754
	5	5.563	.258	5.047	2.004	.0027	.0053	.0049	.0093	.0141	.0188	.0235	.0353	.0471	.0710
	6	6.625	.280	6.065	1.388	.0011	.0022	.0024	.0045	.0068	.0090	.0113	.0169	.0226	.0341
150	3½	4.000	.226	3.548	4.867	.0220	.0388	.0485	.0457	.0694	.0925	.1156	.1735	.2313	.3490
	4	4.500	.237	4.026	3.780	.0117	.0212	.0268	.0276	.0418	.0558	.0697	.1047	.1395	.2105
	5	5.563	.258	5.047	2.405	.0038	.0072	.0093	.0112	.0169	.0226	.0282	.0424	.0565	.0852
	6	6.625	.280	6.065	1.666	.0015	.0030	.0028	.0054	.0081	.0108	.0135	.0203	.0271	.0409
175	4	4.500	.237	4.026	4.410	.0157	.0281	.0348	.0322	.0488	.0651	.0814	.1221	.1628	.2456
	5	5.563	.258	5.047	2.806	.0051	.0095	.0120	.0130	.0198	.0264	.0329	.0494	.0659	.0994
	6	6.625	.280	6.065	1.943	.0021	.0040	.0033	.0062	.0095	.0126	.0158	.0237	.0316	.0477
	8	8.625	.322	7.981	1.122	.0005	.0011	.0011	.0021	.0032	.0042	.0053	.0080	.0105	.0159
200	4	4.500	.237	4.026	5.040	.0202	.0354	.0437	.0368	.0558	.0744	.0930	.1395	.1861	.2861
	5	5.563	.258	5.047	3.207	.0065	.0119	.0151	.0149	.0226	.0301	.0377	.0565	.0753	.1136
	6	6.625	.280	6.065	2.221	.0026	.0050	.0063	.0071	.0108	.0144	.0181	.0271	.0361	.0545
	8	8.625	.322	7.981	1.283	.0001	.0014	.0013	.0024	.0036	.0048	.0060	.0090	.0120	.0182

TABLE 8: PRESSURE DIFFERENTIAL IN IN. Hg PER FOOT OF TUBE

Flow GPM	Tube				Velocity FPS	Viscosity — SUS									
	Size	OD	Wall	ID		32	100	200	400	600	800	1000	1500	2000	3000
1.5	½	.5	.035	.430	3.314	.1297	.5020	1.1185	2.1202	3.2152	4.2884	5.3593	8.0424	9.9081	16.1754
	⅝	.625	.035	.555	1.989	.0386	.1809	.4030	.7640	1.1585	1.5452	1.9311	2.8979	3.5702	5.8285
	¾	.75	.049	.652	1.441	.0180	.0950	.2116	.4011	.6083	.8113	1.1039	1.5215	1.8745	3.0601
3	⅝	.625	.035	.555	3.978	.1298	.3618	.8061	1.5279	2.3170	3.0905	3.8623	5.7959	7.1404	11.6570
	¾	.75	.049	.652	.2883	.0604	.1899	.4232	.8022	1.2165	1.6226	2.0278	3.0430	3.7489	6.1202
	1	1.0	.049	.902	1.506	.0129	.0519	.1155	.2190	.3321	.4430	.5536	.8307	1.0235	1.6708
	1¼	1.25	.065	1.120	.977	.0046	.0218	.0486	.0921	.1397	.1863	.2329	.3495	.4305	.7029
5	¾	.75	.049	.652	4.804	.1476	.3165	.7053	1.3370	2.0275	2.7043	3.3797	5.0716	6.2482	10.2004
	1	1.0	.049	.902	2.510	.0316	.0864	.1926	.3650	.5535	.7383	.9227	1.3846	1.7058	2.7847
	1¼	1.25	.065	1.120	1.628	.0113	.0364	.0810	.1535	.2329	.3106	.3881	.5825	.7176	1.1715
	1½	1.5	.065	1.370	1.088	.0043	.0162	.0362	.0686	.1040	.1387	.1734	.2602	.3205	.5233
7	1	1.0	.049	.902	3.514	.0569	.1210	.2696	.5110	.7749	1.0336	1.2917	1.9384	2.3881	3.8986
	1¼	1.25	.065	1.120	2.279	.0204	.0509	.1134	.2150	.3260	.4348	.5434	.8154	1.0046	1.6401
	1½	1.5	.065	1.370	1.523	.0078	.0227	.0507	.0960	.1456	.1942	.2427	.3642	.4487	.7326
	1¾	1.75	.065	1.620	1.089	.0035	.0116	.0259	.0491	.0745	.0993	.1241	.1863	.2295	.3747
10	1	1.0	.049	.902	5.020	.1063	.1728	.3851	.7300	1.1070	1.4766	1.8453	2.7691	3.4115	5.5694
	1¼	1.25	.065	1.120	3.256	.0380	.0727	.1620	.3071	.4657	.6212	.7763	1.1649	1.4352	2.3430
	1½	1.5	.065	1.370	2.176	.0146	.0325	.0724	.1372	.2080	.2775	.3467	.5203	.6410	1.0465
	1¾	1.75	.065	1.620	1.556	.0066	.0166	.0370	.0702	.1064	.1419	.1774	.2661	.3279	.5353
15	1¼	1.25	.065	1.120	4.884	.0773	.1646	.2430	.4607	.6986	.9317	1.1644	1.7474	2.1527	3.5144
	1½	1.5	.065	1.370	3.264	.0297	.0487	.1086	.2058	.3120	.4162	.5201	.7805	.9616	1.5698
	1¾	1.75	.065	1.620	2.335	.0134	.0249	.0555	.1052	.1596	.2129	.2660	.3992	.4918	.8029
	2	2.0	.083	1.834	1.822	.0074	.0152	.0338	.0641	.0972	.1296	.1620	.2430	.2994	.4888
20	1¼	1.25	.065	1.120	6.512	.1278	.2724	.3240	.6142	.9314	1.2423	1.5526	2.3298	2.8703	4.6859
	1½	1.5	.065	1.370	4.352	.0491	.1046	.1447	.2744	.4160	.5549	.6935	1.0407	1.2821	2.0931
	1¾	1.75	.065	1.620	3.113	.0221	.0332	.0740	.1403	.2128	.2838	.3547	.5323	.6658	1.0706
	2	2.0	.083	1.834	2.429	.0123	.0202	.0451	.0854	.1295	.1728	.2159	.3240	.3992	.6517
25	1¼	1.25	.065	1.120	8.140	.1889	.4025	.4050	.7678	1.1643	1.5529	1.9407	2.9123	3.5879	5.8574
	1½	1.5	.065	1.370	5.441	.0755	.1536	.1809	.3429	.5200	.6936	.8669	1.3009	1.6026	2.6163
	1¾	1.75	.065	1.620	3.891	.0327	.0697	.0925	.1754	.2660	.3548	.4434	.6653	.8197	1.3382
	2	2.0	.083	1.834	3.036	.0182	.0387	.0563	.1068	.1619	.2160	.2699	.4051	.4990	.8147
30	1¼	1.25	.065	1.120	9.769	.2599	.5538	.4860	.9213	1.3971	1.8635	2.3289	3.4948	4.3055	7.0289
	1½	1.5	.065	1.370	6.529	.0998	.2127	.2171	.4115	.6241	.8324	1.0402	1.5610	1.9231	3.1396
	1¾	1.75	.065	1.620	4.669	.0450	.0959	.1110	.2105	.3192	.4257	.5321	.7984	.9836	1.6058
	2	2.0	.083	1.834	3.643	.0250	.0532	.0676	.1281	.1943	.2592	.3239	.4861	.5988	.9776
35	1¼	1.25	.065	1.120	11.397	.3404	.7253	.8861	1.0749	1.6300	2.1741	2.7170	4.0772	5.0231	8.2004
	1½	1.5	.065	1.370	7.617	.1307	.2785	.2533	.4801	.7281	.9711	1.2136	1.8212	2.2437	3.6629
	1¾	1.75	.065	1.620	5.447	.0590	.1256	.1295	.2456	.3724	.4967	.6207	.9315	1.1476	1.8735
	2	2.0	.083	1.834	4.250	.0327	.0697	.0789	.1495	.2267	.3024	.3779	.5671	.6986	1.1405
40	1¼	1.25	.065	1.120	13.025	.4300	.9162	1.1194	1.2284	1.8628	2.4846	3.1051	4.6597	5.7406	9.3718
	1½	1.5	.065	1.370	8.705	.1651	.3518	.4299	.5487	.8321	1.1098	1.3870	2.0814	2.5642	4.1862
	1¾	1.75	.065	1.620	6.226	.0745	.1587	.1481	.2806	.4256	.5676	.7094	1.0646	1.3115	2.1411
	2	2.0	.083	1.834	4.857	.0413	.0880	.0901	.1709	.2591	.3456	.4319	.6481	.7984	1.3035

TABLE 9: PRESSURE DIFFERENTIAL IN IN. Hg PER FOOT OF HOSE

Flow GPM	Hose		Velocity FPS	Viscosity — SUS									
	Size	ID		32	100	200	400	600	800	1000	1500	2000	3000
1.5	13/32	.4063	3.711	.1698	.6297	1.4032	2.6599	4.0336	5.3800	6.7235	10.0896	13.4527	20.2927
	1/2	.5	2.451	.0633	.2746	.6118	1.1598	1.7587	2.3458	2.9316	4.3993	5.8657	8.8481
	5/8	.625	1.568	.0219	.1125	.2506	.4750	.7204	.9608	1.2008	1.8019	2.4026	3.6242
	3/4	.75	1.089	.0092	.0542	.1209	.2291	.3474	.4634	.5791	.8690	1.1587	1.7478
3	1/2	.5	4.901	.2131	.5492	1.2237	2.3195	3.5174	4.6916	5.8632	8.7985	11.7313	17.6961
	5/8	.625	3.137	.0738	.2249	.5012	.9501	1.4407	1.9217	2.4016	3.6039	4.8052	7.2483
	3/4	.75	2.178	.0311	.1085	.2417	.4582	.6948	.9267	1.1582	1.7380	2.3173	3.4955
	7/8	.875	1.600	.0149	.0586	.1305	.2473	.3750	.5002	.6251	.9381	1.2508	1.8868
	1	1.0	1.225	.0079	.0343	.0765	.1450	.2198	.2932	.3665	.5499	.7332	1.1060
5	5/8	.625	5.228	.1805	.3749	.8354	1.5835	2.4012	3.2028	4.0026	6.0065	8.0086	12.0805
	3/4	.75	3.631	.0759	.1808	.4029	.7636	1.1580	1.5445	1.9303	2.8966	3.8622	5.8259
	7/8	.875	2.667	.0365	.0976	.2175	.4122	.6251	.8337	1.0419	1.5635	2.0847	3.1447
	1	1.0	2.042	.0194	.0572	.1275	.2416	.3664	.4887	.6108	.9165	1.2220	1.8433
	1 1/8	1.125	1.614	.0111	.0357	.0796	.1508	.2287	.3051	.3813	.5722	.7629	1.1508
	1 1/4	1.25	1.307	.0067	.0234	.0522	.0990	.1501	.2017	.2502	.3754	.5053	.7550
	1 3/8	1.375	1.080	.0043	.0160	.0357	.0676	.1025	.1367	.1709	.2564	.3419	.5157
7	3/4	.75	5.083	.1368	.2531	.5640	1.0691	1.6212	2.1624	2.7024	4.0553	5.4070	8.1562
	7/8	.875	3.734	.0658	.1366	.3044	.5771	.8751	1.1672	1.4587	2.1889	2.9186	4.4025
	1	1.0	2.859	.0349	.0801	.1785	.3383	.5130	.6842	.8550	1.2831	1.7108	2.5807
	1 1/8	1.125	2.259	.0199	.0500	.1114	.2112	.3202	.4271	.5338	.8010	1.0681	1.6111
	1 1/4	1.25	1.830	.0121	.0328	.0731	.1386	.2102	.2802	.3502	.5256	.7008	1.0571
	1 3/8	1.375	1.512	.0077	.0224	.0499	.0946	.1436	.1914	.2392	.3590	.4786	.7220
	1 1/2	1.5	1.271	.0051	.0158	.0352	.0668	.1013	.1351	.1689	.2535	.3379	.5098
10	7/8	.875	5.335	.1228	.1952	.4349	.8244	1.2501	1.6674	2.0838	3.1271	4.1694	6.2893
	1	1.0	4.085	.0651	.1144	.2549	.4832	.7328	.9774	1.2215	1.8330	2.4440	3.6867
	1 1/8	1.125	3.227	.0372	.0714	.1592	.3017	.4575	.6102	.7626	1.1444	1.5258	2.3016
	1 1/4	1.25	2.614	.0226	.0469	.1044	.1979	.3002	.4003	.5003	.7508	1.0011	1.5101
	1 3/8	1.375	2.160	.0143	.0320	.0713	.1352	.2050	.2734	.3417	.5128	.6837	1.0314
	1 1/2	1.5	1.815	.0095	.0226	.0504	.0955	.1448	.1931	.2413	.3621	.4828	.7282
	1 13/16	1.813	1.243	.0039	.0106	.0236	.0447	.0678	.0905	.1131	.1697	.2262	.3412
	2	2.0	1.021	.0024	.0072	.0159	.0302	.0458	.0611	.0763	.1146	.1528	.2304
15	1 1/8	1.125	4.841	.0757	.1612	.2387	.4525	.6862	.9153	1.1439	1.7165	2.2887	3.4524
	1 1/4	1.25	3.921	.0459	.0703	.1566	.2969	.4502	.6005	.7505	1.1262	1.5016	2.2651
	1 3/8	1.375	3.241	.0292	.0480	.1070	.2028	.3075	.4102	.5126	.7692	1.0256	1.5471
	1 1/2	1.5	2.723	.0193	.0339	.0755	.1432	.2171	.2896	.3619	.5431	.7242	1.0924
	1 13/16	1.813	1.864	.0078	.0159	.0354	.0671	.1017	.1357	.1696	.2545	.3393	.5118
	2	2.0	1.532	.0049	.0107	.0239	.0453	.0687	.0916	.1145	.1718	.2291	.3456
	2 3/8	2.375	1.086	.0022	.0054	.0120	.0228	.0345	.0461	.0576	.0864	.1152	.1738
20	1 1/4	1.25	5.228	.0759	.1617	.2088	.3959	.6003	.8007	1.0007	1.5016	.20022	3.0201
	1 3/8	1.375	4.321	.0483	.1028	.1426	.2704	.4100	.5469	.6835	1.0256	1.3675	2.0628
	1 1/2	1.5	3.631	.0319	.0680	.1007	.1909	.2895	.3861	.4826	.7242	.9655	1.4565
	1 13/16	1.813	2.485	.0130	.0212	.0472	.0895	.1357	.1809	.2261	.3393	.4524	.6825
	2	2.0	2.042	.0081	.0143	.0319	.0604	.0916	.1222	.1527	.2291	.3055	.4608
	2 3/8	2.375	1.448	.0036	.0072	.0160	.0304	.0461	.0614	.0768	.1152	.1536	.2317
25	1 3/8	1.375	5.401	.0713	.1519	.1783	.3380	.5125	.6836	.8543	1.2820	1.7094	2.5785
	1 1/2	1.5	4.538	.0472	.1005	.1259	.2386	.3619	.4827	.6032	.9052	1.2069	1.8206
	1 13/16	1.813	3.107	.0192	.0408	.0590	.1118	.1696	.2262	.2826	.4241	.5655	.8531
	2	2.0	2.553	.0120	.0179	.0398	.0755	.1145	.1527	.1909	.2864	.3819	.5760
	2 3/8	2.375	1.810	.0053	.0090	.0200	.0380	.0576	.0768	.0960	.1440	.1920	.2897
30	1 1/2	1.5	5.446	.0649	.1383	.1511	.2864	.4343	.5792	.7239	1.0862	1.4483	2.1847
	1 13/16	1.813	3.728	.0264	.0562	.0708	.1342	.2035	.2714	.3392	.5090	.6786	1.0237
	2	2.0	3.063	.0165	.0353	.0478	.0906	.1374	.1833	.2290	.3437	.4583	.6913
	2 3/8	2.375	2.172	.0073	.0108	.0240	.0456	.0691	.0922	.1152	.1728	.2304	.3476
40	1 13/16	1.813	4.971	.0436	.0930	.0944	.1789	.2713	.3619	.4522	.6786	.9048	1.3649
	2	2.0	4.085	.0274	.0583	.0637	.1208	.1832	.2444	.3054	.4583	.6110	.9217
	2 3/8	2.375	2.897	.0121	.0258	.0321	.0608	.0921	.1229	.1536	.2304	.3073	.4635
50	2	2.0	5.106	.0405	.0862	.0797	.1510	.2290	.3054	.3817	.5728	.7638	1.1521
	2 3/8	2.375	3.621	.0179	.0381	.0401	.0759	.1152	.1536	.1920	.2881	.3841	.5794
60	2 3/8	2.375	4.345	.0246	.0524	.0481	.0911	.1382	.1843	.2304	.3457	.4609	.6952

hydraulic and lube oil filtration

contamination causes most system failures

The experience of designers and users of hydraulic and lube oil systems have proven the following fact: over 75% of all systems failures are a direct result of contamination! The cost due to contamination is staggering, resulting from:

- Loss of production (downtime)
- Component replacement costs
- More frequent fluid replacement (and disposal)
- Increased overall maintenance costs

The selection and proper use of filtration devices is becoming an important tool in the battle to increase production while reducing manufacturing costs. In one study, a reduction of downtime by 5% resulted in a one year payback of the costs of adding and maintaining properly selected filters. Stated simply, good filtration works to reduce overall operating costs.

how contamination damages systems

Contamination interferes with the four main functions of hydraulic and lubrication fluids:

1. to act as an energy transmission medium
2. to lubricate internal moving parts of components
3. to act as a heat transfer medium
4. to seal small clearances between moving parts

Contamination may be present in many forms. The most common are solid particles (particulate), water and entrained air.

Damage Due to Particles

Solid particles contribute to wear related problems. Particulates greatly interfere with the lubrication properties of fluids. This affects internal moving parts, such as the gear set of a pump or the valve spool in a valve body. As a result, directional, flow or pressure controls may malfunction. Sensitive components, such as servo valves, may jam or stop completely.

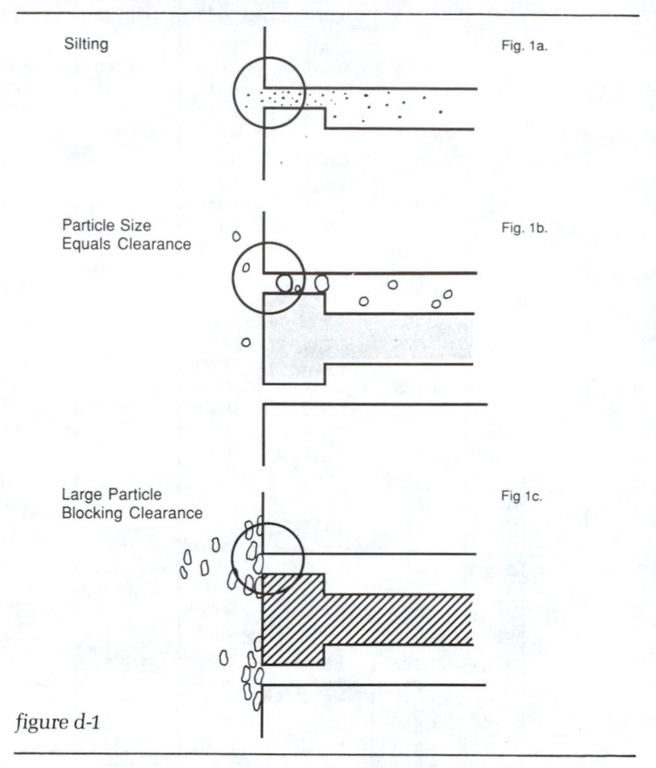

figure d-1

Particles are sometimes categorized by their relative sizes. Extremely fine particles, commonly called "silt," may collect in the space between moving parts in a component. (Figure d 1.a.) This will generally cause a sticking or sluggish action and perhaps even result in a failure to shift. Frequent solenoid burnouts, constant inaccurate positioning and overall wear (leading to a component failure) are the most common results of "silt." Particles approximately equal to the dynamic clearances of a component (see Table d 1.) will interfere with the lubrication process and therefore accelerate abrasive wear. (Figure d 1.b.)

Typical Component Clearances

Component	Micrometers	Component	Micrometers
Slide Bearings (Vane Pump)	.5	Gear Pump (Tooth Tip To Case)	.5-5
Tip of Vane (Control Valve)	.5	Vane Pump (Vane Tip)	.5-1
Roller Element Bearings	.1-1	Piston Pump (Piston To Bore)	5-40
Hydrostatic Bearings	1-25	Servo Valves (Spool Sleeve)	1-4
Gears	.1-1		

table d-1

Particles of this size also contribute to the "wear chain reaction," whereby the abrasive action of particulates helps to make new particles from component surfaces. Left unchecked, this wear particle phenomena will eventually shut a system down.

Larger particles (Figure d 1.c.) restrict or block flow through clearances and orifices. This leads to component malfunctions, higher pressure losses, higher operating temperatures, and often a catastrophic failure causing unscheduled downtime.

Particle sizes are generally measured on the micrometer scale. One micrometer (or "micron") is one-millionth of one meter. The limit of human visibility is approximately 40 micrometers. For comparison of a few familiar items to the micrometer scale, refer to Table d 2. Keep in mind that the damage causing particles in hydraulic or lube oil systems are smaller than 40 micrometers. Therefore, they are microscopic and can not be seen by the unaided eye.

Relative Sizes of Particles and Comparison of Dimensional Units

Sizes of Familiar Objects

Substance	Micron	Inch
	238	.009
	149	.0059
Grain of Table Salt	100	.0039
	74	.0029
Human Hair	70	.0027
Lower Limit of Visibility	40	.00158
White Blood Cells	25	.001
Talcum Powder	10	.00039
Red Blood Cells	8	.0003
	5	.00019
Bacteria (Average)	2	.000078
	1	.000039

table d-2

Fluid Cleanliness Levels

Knowning the cleanliness level of a fluid is the basis for contamination control measures. In order to detect or correct problems, a contamination reference scale is used.

Particle counting is the most common method used to derive cleanliness level standards. Very sensitive optical instruments are used to count the number of particles in various size ranges. These counts are reported as the number of particles greater than a certain size found in a specified volume of fluid.

Several common cleanliness level standards are shown in Table d 3. In this correlation table, the number of particles greater than 5 and 15 micrometers in one milliliter of fluid are shown. The number of 5 micrometer particles is used as a reference point for "silt" particles. Those particles greater than 15 micrometer indicate the quantity of larger particles present, which contribute greatly to wear related and possible catastrophic failures.

The two index numbers shown in the ISO code refer to the number of particles greater than 5 and 15 micrometers. The first index number describes the relative range of those particles greater than 5 micrometer. The second index number indicates the relative range of particles greater than 15 micrometer. This standard has gained wide acceptance, since most particle distributions are easily described by it. For example, a fluid with a very high silt content and a relatively low large particle population may be described by the ISO codes of 25/12 or 23/10. This information may be useful in helping to direct filtration efforts.

Component Cleanliness Level Requirements

Many manufacturers of hydraulic and load bearing equipment specify the optimum cleanliness level required for their components. Subjecting components to fluid with higher contamination levels may result in much shorter component life.

In Table d 4., a few components and their recommended cleanliness levels are shown. It is always best to consult with component manufacturers and obtain their written fluid cleanliness level recommendations. This information is needed in order to select the proper level of filtration. It may also prove useful for any subsequent warranty claims, as it may draw the line between normal use and excessive or abusive operation.

Keep in mind also that new fluid is not necessarily clean fluid. Refineries and mixers of fluids have little control over fluid storage. In many cases, contamination is introduced in the handling and transferring of fluid. For these reasons, most new fluid out of the drum is generally contaminated to ISO 18/15 or SAE class 6 levels — or above. Consequently, all fluid should be filtered before it is put into use.

Cleanliness Level Correlation Table

ISO Code	Particles/Millilitre ≥5 Micrometers	Particles/Millilitre ≥15 Micrometers	ACFTD Gravimetric Level, mg/L	NAS 1638 (1964)	Disavowed "SAE" Level (1963)
26/23	640,000	80,000	1000		
25/23	320,000	80,000			
23/20	80,000	10,000	100		
21/18	20,000	2,500		12	
20/18	10,000	2,500			
20/17	10,000	1,300		11	
20/16	10,000	640	10		
19/16	5,000	640		10	
18/15	2,500	320		9	6
17/14	1,300	160		8	5
16/13	640	80	1	7	4
15/12	320	40		6	3
14/12	160	40			
14/11	160	20		5	2
13/10	80	10	0.1	4	1
12/9	40	5		3	0
11/8	20	2.5		2	
10/8	10	2.5			
10/7	10	1.3		1	
10/6	10	.64	0.01		

table d-3

Fluid Cleanliness Required for Typical Hydraulic Components

Components	Fluid Classification "SAE" Level	Fluid Classification ISO Code
Servo Control Valves	2	14/11
Vane and Piston Pumps/Motors	4	16/13
Directional & Pressure Control Valves	4	16/13
Gear Pumps/Motors	5	17/14
Flow Control Valves, Cylinders	6	18/15

table d-4

Damage Due to Water

There is more to proper fluid maintenance than just removing particulate matter. Water is virtually a universal contaminant, and just like particle contaminants must be removed from operating fluids. Water generated damage includes the following:

- corrosion of metal surfaces
- accelerated abrasive wear
- bearing fatique
- fluid additive breakdown
- viscosity variance

Fluids are constantly exposed to water and water vapor while being handled and stored. For instance, outdoor storage of tanks and drums is common. Water may settle on top of fluid containers and be drawn into the container during temperature changes. Water may also be introduced when opening or filling these containers.

Water can enter a system through worn cylinder or actuator seals or through reservoir openings. Condensation is also a prime water source. As the fluid cools in a reservoir or tank, water vapor will condense on the inside surfaces, causing rust or other corrosion problems.

Microbial growth, slime and the apparent increase of fluid viscosity are also by-products of water contamination. The results are short fluid life, degraded surface finishes and inconsistent fluid performance.

Each fluid has a water saturation point. Above this limit, the fluid can neither dissolve nor hold any more water. The excessive water may be "free" or "emulsified" water. Free water is generally noticeable as a "milkiness," or discoloration, of the fluid. As little as .03% (300 ppm) or water by volume can saturate a hydraulic fluid.

Gross water contamination (over 2% of the system volume) may require settling or centrifuge treatments. Systems contaminated up to 2% by volume of water can be effectively treated with highly absorbent filter media.

Parker has developed Par-Gel™ filter media to help control water related problems. These elements are made in many sizes to fit several standard Parker and competitive housings.

contamination control

Contamination control is best achieved through two activities: exclusion and removal.

Keeping Contamination Out

Exclusion of dirt or water is a very important ingredient in contamination control. Using filters for reservoir air breathers helps exclude airborne dirt. Replacing worn actuator seals will help keep dirt and water out of a system. Capping off hoses and manifolds during handling and maintenance also will prevent contamination from entering the system. Keeping contamination out of a system in the first place is an excellent and effective method in contamination control. (See Table d 5.)

CONTAMINANT EXCLUSIONARY MEASURES *(Table d 5.)*

1. Use reservoirs large enough to disipate heat, promote particle settling and allow entrained air to escape.
2. Specify and use depth-type (cellulose or fiberglass) filters for reservoir breathers.
3. Flush all systems before putting into service and after component failures.
4. Specify rod wipers in cylinders. Replace all worn seals in actuators. Keep the reservoir sealed.
5. Filter all fluid before putting it into service.

Removing Ingressed Contamination

Once contamination is introduced into the system, it must be controlled by the system filters. The goal of the filtration system is to reach and maintain the optimum cleanliness level for the system or a component. The remainder of this discussion will be about solid particulate removal. Media characteristics, selection techniques and filter sizing will be discussed in detail.

filter media

The filter media is that part of the element which removes the contaminant. It usually starts out in sheet form, and is then pleated to expose more surface area to the fluid flow. This reduces pressure drop while increasing dirt holding capacity. In some cases, the filter media may have mulitple layers and mesh backing to achieve certain performance criteria. After being pleated and cut to the proper length, the two ends are fastened together using a special clip, adhesive, or other seaming mechanisms. The most common media include wire meshes, cellulose, microglass composites and other synthetic materials.

The broadest general characteristics of filter media is categorized by the location of the particle retention: surface or depth type media.

Surface Type Media

For a surface type filter media, the fluid stream has a more or less straight through flow path, and the contaminant is caught on the surface of the element which faces the fluid flow. Surface type elements are generally made from woven wire. This filtration process is illustrated in Figure d 2.

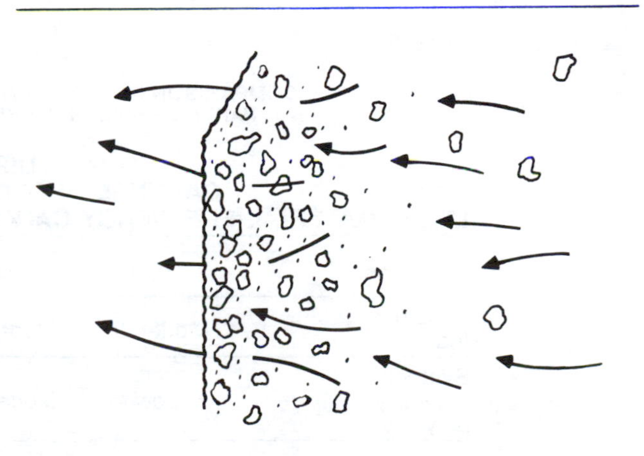

Figure d 2. — Surface Type Media

Particles are trapped on the surface of the element.

filtration

Since the process used in manufacturing the wire cloth can be very accurately controlled, surface type media have a consistent pore size. Because of this, surface type elements are usualy given an absolute pore size rating by the manufacturer. This rating is the diameter of the largest hard spherical particle that will pass through the media under specified test conditions. However, the build-up of contaminant on the element surface will allow the media to capture particles smaller than the pore size rating.

Depth Type Media

For a depth type element, fluid must take a tortuous path through the material which makes up the filter media. Particles are trapped in the maze of openings throughout the media. This process is illustrated in figure d 3.

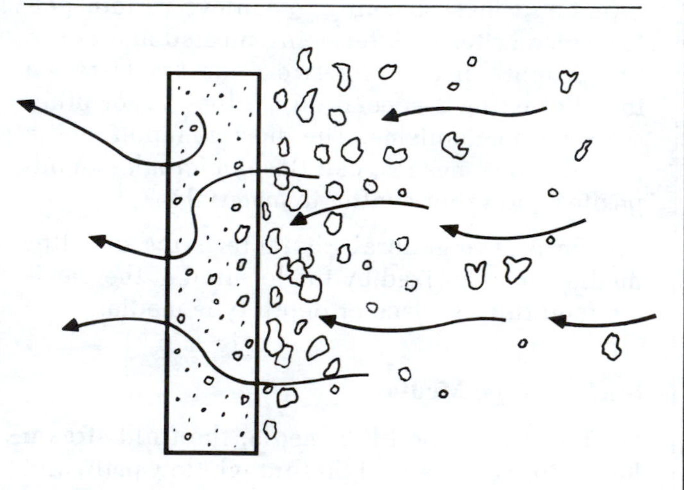

Figure d 3. — Depth Type Media

Particles are trapped throughout the element.

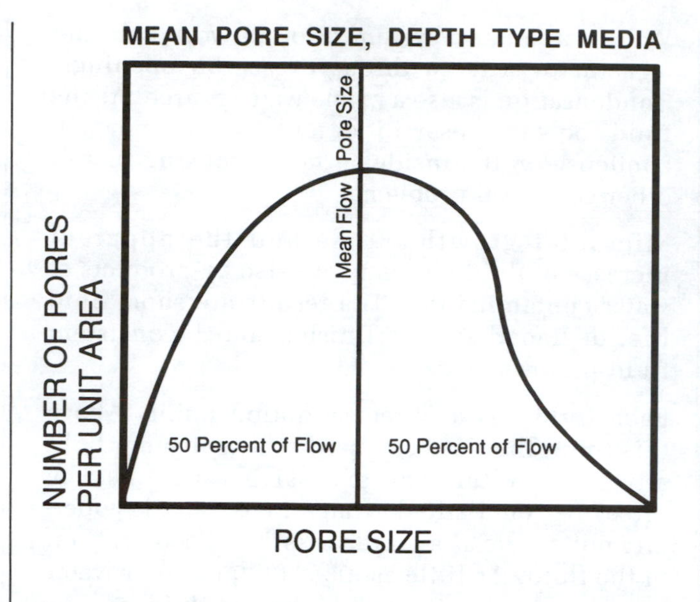

Figure d 4.

Mean Pore Size — Depth Type Media

Because of its construction, a depth type filter media has many pores of various sizes. This is pointed out by the bell curve in Figure d 4. A point on the bell curve is the number of pores per unit area of a given size in a typical depth type element. The shape of the curve in Figure d 4. shows that there are more pores of small sizes than of relatively large sizes. This indicates that a large percentage of flow passes through relatively small holes. However, a mean pore size rating gives no indication of the media efficiency at removing and holding particles of a given size range over the useful life of the element. Table d 6. compares a few characteristics of various depth type media.

table d-6

COMPARISON OF DEPTH TYPE FILTER MEDIA (FOR A GIVEN PARTICLE SIZE, FLOW RATE AND FLUID VISCOSITY)

MEDIA MATERIAL	CAPTURE EFFICIENCY	DIRT HOLDING CAPACITY	PRESSURE DROP	LIFE IN A SYSTEM	OVERALL COST
Fiberglass	High	High	Low	High	Moderate to High
Cellulose (paper)	Moderate	Moderate	Moderate	Moderate	Low
Synthetic (typically polyester)	Low	Moderate	Low	Moderate	Moderate

Nominal and Absolute Ratings

Nominal and absolute micron retention ratings may be arbitrary designations set by the filter manufacturer. For a given media, they may vary widely and thus prove to be relatively useless in describing a filter element's performance. Filters should be rated according to the multi-pass test method (ANSI/B93.31-1973). This test yields the filtration or Beta ratio.

Filtration Ratio

The filtration ratio (also known as the Beta ratio) is a measure of the particle capture efficiency of a filter element. It is therefore a performance rating. The filtration ratio is one of the very important results of the multi-pass test method (ANSI/B93.31-1973).

In this test, a known amount of a special test dust is injected into a filter system. Particle counters measure the number of various sized particles entering and leaving the filter. A ratio is then derived, by dividing the number of upstream particles of a given size by the number of the same sized particles counted downstream or leaving the filter.

For example, if 10,000 particles 10 micrometer and larger were counted upstream of the filter and 5,000 counted downstream, the filtration ratio would be 2 at 10 micrometer and larger.

$$B_{10} = \frac{10{,}000 \text{ (Upstream Count)}}{5{,}000 \text{ (Downstream Count)}} = 2$$

The filtration ratio at 10 micron (B_{10}) would be 2.

The capture efficiency of a filter, expressed as a percent, is easy to calculate. Using the formula

$$\text{Efficiency} = \left(1 - \frac{1}{B_n}\right) \times 100\%$$

for the above filter would show that a ratio of 2 is equal to a 50% capture efficiency. This means that for every two particles greater than 10 micrometers entering the filter, one will pass through the filter.

Table d 7. shows a different particle capture efficiencies for various filtration ratios. Please note that a particle size must be specified in order for the filtration ratio (also called Beta ratio) to make sense.

table d-7

FILTRATION RATIO/EFFICIENCY TABLE *(Table d 7.)*

Filtration Ratio (at a given particle size)	Capture Efficiency (at same particle size)
1.01	1.0%
1.1	9.0%
1.5	33.3%
2.0	50.0%
5.0	80.0%
10.0	90.0%
20.0	95.0%
75.0	98.7%
100	99.0%

Some filter literature may show a rating as: B_x = 2/20/75, x = 6/11/15, which means B_6 = 2, B_{11} = 20, and B_{15} = 75. The first three numbers (2/20/75) are the Beta ratings at the particle size of the second three numbers 6, 11 and 15 micron respectively.

filter housings

The filter housing is a pressure vessel which holds the filter element. It usually consists of two or more subassemblies such as a head and a bowl (or cover) to allow access to the filter element. The housing has inlet and outlet ports allowing it to be installed into a fluid system. Additional housing features may include mounting holes, bypass valves and element condition indicators. Figure d 5. shows a typical filter assembly.

figure d-5

Pressure Ratings

Filter housings are designed for three locations in a circuit: suction, pressure or return lines. One characteristic of these locations is their maximum operating pressures. Suction and return line filters are generaly designed for lower pressures (up to 500 psi). Pressure filter locations may require ratings from 1500 psi to 6000 psi.

The Bypass Valve

The bypass valve is used to prevent the collapse or burst of the filter element when it becomes highly loaded with contaminant. It also prevents pump cavitation in the case of suction line filtration. As contaminant builds up in the element, the differential pressure across the element increases. At a pressure well below the failure point of the filter media and support cylinder, the bypass valve opens, allowing some flow to go around the element. Although the bypass valve is open, a portion of the flow will still flow through the element providing partial filtration.

Some bypass valve designs have a "bypass-to-tank" option. This allows the unfiltered bypass flow to return to tank through a third port, preventing unfiltered bypass flow from entering the system. Other filters may be supplied with a "no bypass" or "blocked" bypass option. This prevents any unfiltered flow from going downstream. In filters with no bypass valves, higher collapse strength elements are required.

Element Condition Indicators

The element condition indicator signals when the element should be cleaned or replaced. The indicator usually has calibration marks which also indicates if the filter bypass valve has opened. The indicator may be directly linked to the bypass valve, or it may be an entirely independent differential pressure sensing device. Indicators may give visual, electrical or both types of signals.

filter locations

The following guidelines may be used to help select the best filter location or combination of locations for a specific system. Filter locations may depend on the placement of the most sensitive component in a system. Sometimes, size and weight restrictions are also involved.

Pressure Filters

Pressure filters are located downstream from the system pump. They are designed to handle the system pressure and sized for the specific flow rate in the pressure line where they are located.

Pressure filters are especially suited for protective sensitive components directly downstream from the filter, such as servo valves. Located just downstream from the system pump, they will protect the entire system from pump generated contamination.

Return Line Filters

When the pump is the most sensitive component in a system, a return line filter may be the best choice. In most systems, the return line filter is the last component through which fluid passes before entering the reservoir. Therefore, it catches wear debris from system working components and particles entering through worn cylinder rod seals before such contaminant can enter the reservoir and be circulated. Since this filter is located immediately upstream from the reservoir, its pressure rating and cost can be relatively low.

In some cases, cylinders with large diameter rods may result in "flow multiplication." The increased return line flow rate may cause the filter bypass valve to open, allowing unfiltered flow to pass downstream. This may be an undesirable condition and care should be taken in sizing the filter.

Duplex Filters

There is another type of filter which is used quite often in hydraulic and lube oil systems: the duplex filter. A duplex filter may be either a pressure or return line design. Its most notable characteristic is continuous filtration. That is, it is made from two or more filter chambers and the necessary valving to allow for continuous, uninterrupted filtration. When one filter element needs servicing, the duplex valve is shifted, diverting flow to the opposite filter chamber. The dirty element can then be changed, while filtered flow continues to pass through the filter assembly. The duplex valve is an open cross-over type, which prevents any flow blockage. Figure d 6. shows a typical duplex filter.

figure d-6

Off-Line or Side Stream Filtering

Off-line or side-stream filtration consists of a pump, filter and the appropriate connections. These components are installed off-line as a small sub-system, separate from the working lines, or included in a fluid cooling loop. Fluid is pumped out of the reservoir, through the filter, and back to the reservoir, in a continuous fashion. By selecting the appropriate pump flow rate and filter efficiency, this type of filtration can keep the fluid as clean as is necessary for most systems.

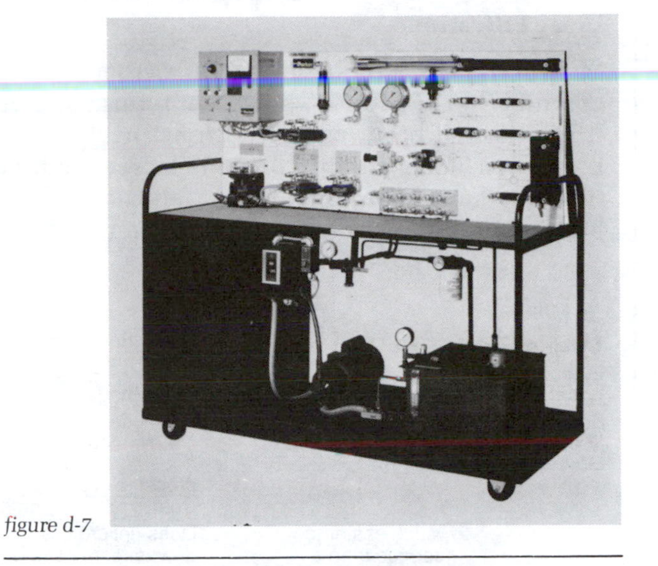

figure d-7

A typical off-line system (in this case, a portable one) is shown in Figure d 7. This is a self contained unit, separate from the main system. It features two filters, an industrial quality gear pump and hose wands for ease of use.

As with a return line filter, this type of system is best suited to maintain overall cleanliness, and does not provide specific component protection. An off-line filtration loop has the added advantage that it is relatively easy to retrofit on an existing system which has inadequate filtration. Also, the filter can be serviced without shutting down the main system.

Suction Filters

Suction filters serve to protect the pump from fluid contamination. They are located before the inlet port of the pump. Some may be inlet "strainers," submersed in the fluid. Others may be externally mounted. In either case, they are relatively coarse filters, due to cavitation characteristics of pumps. For this reason, they are not used as primary protection against contamination. Some pump manufacturers do not allow the use of a suction filter. Always consult the pump manufacturer when in doubt.

Ideal Filtration

Most systems would benefit greatly from having a combination of suction, pressure, return, and off-line filters. If other considerations preclude this idea, then a compromise of some lesser combination may have to be considered. Table d 8. may be helpful in making a filtration location decision.

table d-8

COMPARISON OF FILTER TYPES AND LOCATIONS

Filter Location	Advantages	Disadvantages
Suction (Externally Mounted)	1. Last chance protection for the pump.	1. Must use relatively coarse media, or large housing size to keep pressure drop low. Cost is relatively high.
	2. Much easier to service than a sump strainer.	2. Does not protect downstream components from pump wear debris.
		3. May not be suitable for many variable volume pumps.
Pressure	1. Protects downstream components from pump wear.	1. Housing is relatively expensive because it must handle full system pressure.
	2. Provides last chance protection of specific components.	2. Does not catch wear debris from downstream working components.
	3. Can use high efficiency filter medium with minimum consideration for pressure drop.	3. Does not catch particles entering the system through worn cylinder rod seals, and other working components.
Return	1. Catches wear debris from components, and dirt entering through worn cylinder rod seals before it enters the reservoir.	1. Does not protect specific components.
	2. Lower pressure ratings result in lower costs.	2. No protection from pump generated contamination.
	3. May be in-line or in-tank for easier installation.	3. Return line flow surges may reduce filter performance.
Off-Line	1. Servicing is possible without loss of production.	1. No direct component protection.
	2. Fixed flow rate eliminates surges, allowing for optimal element life and performance.	2. Relative initial cost is high.
	3. Fluid cooling may be easily incorporated.	3. Requries more space than single filter.

the filter selection process

Each system has its own optimum cleanliness level, depending on the number and types of valves, actuators, system pressure and flow rates and other dynamics factors. Properly applied filtration will help achieve this optimum cleanliness level. The primary filtration objective is to reach the desired cleanliness level and maintain it at the lowest possible cost.

Selecting a filter or filters for a system is a two step process. First, the filter element media must be selected. Then, the filter housing, which will contain the element, must be specified. Many factors enter into the decision making processes in both steps. Consulting with filter application specialists will make sure that the filters are properly selected and applied.

Filter Media Selection

To maintain fluid cleanliness levels at an acceptable level, the use of a pressure, return and/or off-line filter is required. Many system designers will make use of a combination of these filters in various branches of the circuit. The nomograph which follows may be used to select media for these filters. This selection process is also available in a slide rule form, or on personal computer diskettes. (Figure d 8. shows the slide rule). The nomograph is shown in Figure d 9.

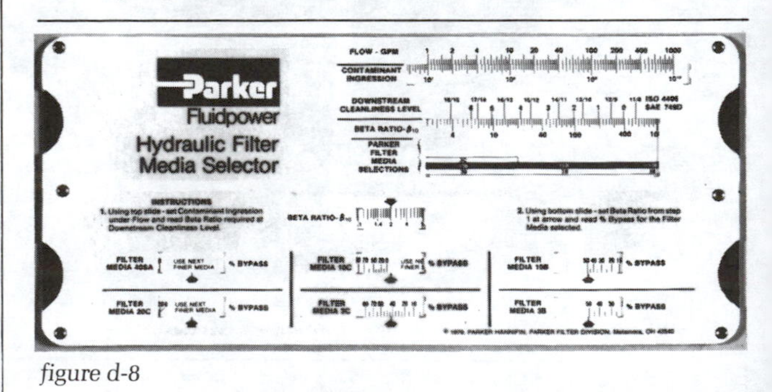

figure d-8

Nomograph Instructions

1. Add all of the flows out of the reservoir.
   a. Be sure to include the flow of any off-line filter system if one will be permanently connected.
   b. If variable volume pumps are used, be sure to use the *average* flow rate, taking into account periods when the pump is compensated.

2. Add all of the flows which return to tank unfiltered (e.g. pump case drains, unfiltered return lines).

3 Calculated percent of bypass (unfiltered flow)

$$\text{Bypass} = \frac{\text{Unfiltered Flow}}{\text{Total Flow}} \times 100\%$$

4. Estimate contaminant ingression rate. Accurate estimates can be obtained by performing a particle count on a fluid sample from an existing system similar to the one being designed or upgraded. (Contact Parker Filter Division for details.) Otherwise, the guidelines in Table d 9. may be used.

5. On the left side of nomograph, find the intersection of the ingression rate line with a vertical line extended from the total average flow value (base of the graph).

6. From this intersection point, draw a horizontal line to intersect with the pivot line.

7. From the pivot point, draw a line through the cleanliness level desired, and intersect the filter media graph at a point on the zero bypass line. (See Table d 4. for cleanliness guidelines).

8. Extend a horizontal line to the right from the point found in Step 7, to intersect with a vertical line extended from bypass percent found in Step 3.

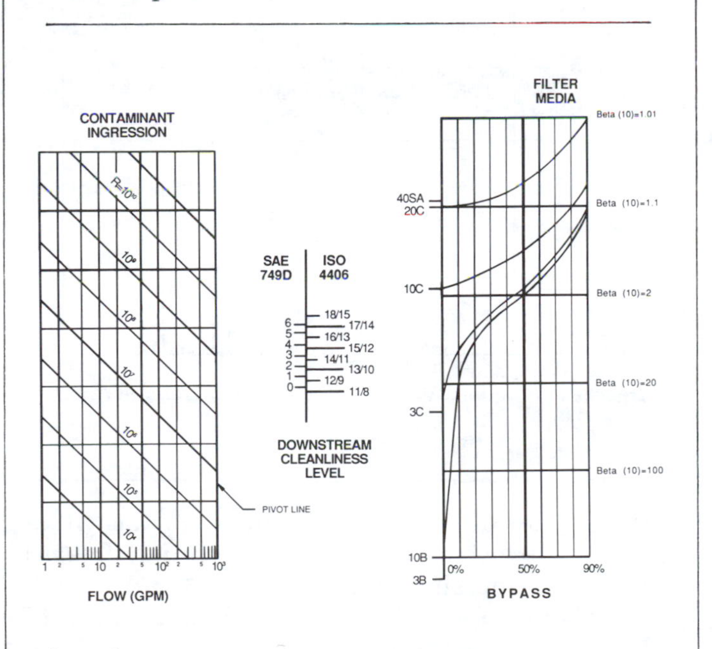

figure d-9

9. Select the media whose curve lies below the intersection point of Step 8.

10. Do not select a media less efficient than one recommended by a component supplier. This might invalidate the warranty.

   a. For example, a pump manufacturer may specify a return filter with a 10 micron "nominal" rated filter.

   b. A servo valve manufacturer may specify a filter with a 3 micron "absolute" rated element.

This approach to media selection ignores the filtration that takes place in the suction filter (usually negligible compared to the contamination control filter). It also assumes that all filters use identical filter media. If a more efficient media is used in one of the filters, fluid cleanliness should be better than the designers' objective.

Typically, contaminant will enter hydraulic fluid at a rate between one million and one billion particles greater than 10 micrometres in size for every minute of operation. The actual rate will depend on the number of components in a system, internal wear rates, seal degradation, and how much airborne contaminant there is around the system. Table d 9. gives ingression rates for typical hydraulic and lubrication systems.

In the absence of any other data, a filter specifier can use the guidelines shown in Table d 9. However, as indicated in Step 4 of the nomograph, a particle count of the fluid in a similar system, operating in a similar environment, provides the basis for a relatively accurate ingression estimate. The specifier also needs to know the Beta Ratios of the filter(s) in that system, and the other variables associated with the nomograph (other than the ingression rate). Using the known variable allows the specifier to "work backwards" on the nomograph to find the contaminant ingression rate.

Parker Filter Division provides particle counting and other fluid analysis services to assist customers in specifying filters.

INGRESSION RATES FOR TYPICAL SYSTEMS *(Table d 9.)*

SYSTEM	INGRESSION RATE[1]
Earthmoving and off-highway (extreme conditions)	$10^9 - 10^{10}$
Farm and other mobile equipment	$10^8 - 10^9$
Typical manufacturing plant	$10^7 - 10^8$
"clean" plant environment (assembly operations)	$10^5 - 10^6$

[1]Number of particles greater than 10 micrometers ingressed into the system per minute from all sources.

Housing Selection

The primary concerns of the housing selection process include mounting methods, porting options, indicator options, and pressure rating. All except the pressure rating depend on the physical system design and the preferences of the designer. Pressure rating of the housing is far less arbitrary. This should be determined before the housing style is selected. Location of the filter in the circuit is the primary determinant of pressure rating. A discussion of the various types of filters may be found in the "Filter Location" section.

It is essential to analyze the circuit for transient and frequent pressure spikes, as well as steady state conditions. Some materials, such as aluminum, have lower fatigue pressure ratings. In circuits with frequent high pressure differences, another material may be required to prevent fatigue related failures.

Selection of Bypass Valve Setting

After the housing style and pressure rating are selected, choose the bypass valve calibration. The bypass valve calibration must be selected before sizing a filter housing. Everything else being equal, the highest calibration available from the manufacturer should be selected. This will provide the longest element life for a given filter size. Occasionally, a lower calibration may be selected to help minimize energy loss in the system or to reduce back-pressure on another component. (In suction filters, usually a 2 or 3 psi bypass valve is used to prevent pump cavitation.)

Certain pressure filters (e.g. those protecting servo valves) may require selection of a blocked bypass or non-bypass option. This prevents unfiltered fluid from reaching the critical components.

Normally, the filter bypass valve is used to limit the differential pressure across the element to a level which prevents collapse of the support cylinder. When no bypass valve is used, the element collects contaminant until the differential pressure across it becomes high enough that the system is no longer functional. In this condition, an operator would be forced to change the element. The support cylinder and end caps must be designed so as to prevent collapsing the element during high pressure differential conditions.

When specifying a non-bypass filter design, make sure that the element has a differential pressure rating close to the maximum operating pressure of the system. When specifying a bypass type filter, it can generally be assumed that the manufacturer has designed the element to withstand the bypass valve differential pressure when the bypass valve opens.

Selection of Housing Size

The filter housing size should be large enough to achieve at least a 2:1 ratio between the bypass valve setting and the pressure drop of the filter with a clean element installed. It is preferable that this ratio be 3:1 or even higher for longer element life.

For example, Figure d 10. illustrates the type of catalog flow/pressure drop curves which are used to size the filter housing. As can be seen, the specifier needs to know the normal operating viscosity of the fluid, and the maximum flow rate through the filter. The maximum flow rate is used, instead of an average, to make sure that the filter does not spend a high portion of time in bypass due to flow surges. This is particularly important in return line filters where flow multiplication from cylinders may increase the return flow rate to twice the pump flow rate.

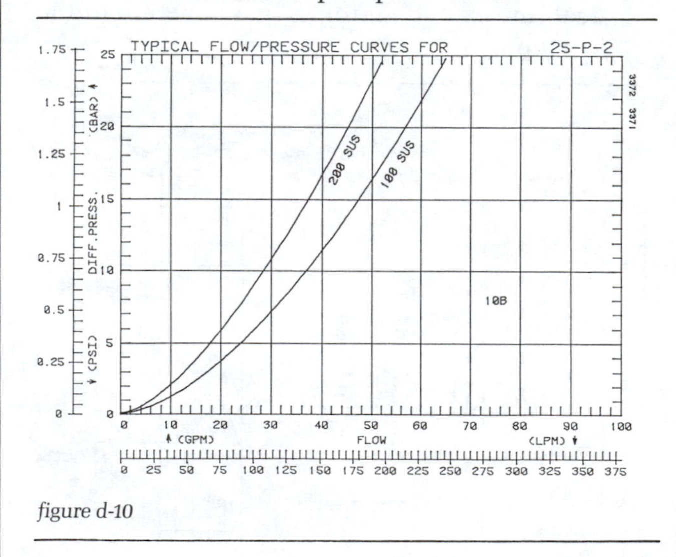

figure d-10

If the filter described in Figure d 10. was fitted with a 50 psi bypass valve, the initial (clean) pressure drop should be between 15 and 25 psi. This is calculated from the 3:1 and 2:1 ratio of bypass valve setting and initial pressure drop.

From the graph and knowing this initial pressure drop limit, the maximum filter flow rate at a given viscosity can be determined. So, for 200 SUS fluid, the flow range would be from 36 to 54 gpm.

In some cases an "oversized" housing may be selected for one or more of the following reasons:

- to minimize pressure drop and limit energy loss
- to use larger elements with higher dirt holding capacity and extend the time between element changes
- to minimize the amount of time spent in bypass due to cold start-up conditions.

fluid sampling and analysis

Fluid analysis is an essential part of any maintenance program. Fluid analysis ensures that the fluid conforms to original specifications, to see if the composition of the fluid has changed, and to determine its overall contamination level.

In taking a fluid sample from a system, care must be taken to make sure that the fluid sample is representative of the system. To accomplish this, the fluid container must be cleaned before taking the sample and the fluid must be correctly extracted from the system. There are two National Fluid Power Association (NFPA) standards for extracting fluid: NFPA T2.9.9-1976 "Method for extracting fluid samples from a reservoir of an operating hydraulic fluid power system" and NFPA T2.9.1-1972, ANSI B93.19-1972 "American National Standard method for extracting fluid samples from the lines of an operating hydraulic fluid power system (for particulate contamination analysis)." Either extracting method may be used.

In either case, a representative fluid sample is the goal. Sampling valves should be opened and flushed for up to fifteen seconds. The clean sample bottle should be kept closed until the fluid and valve is ready for sampling. The system should be at operating temperature for at least 30 minutes before the sample is taken.

Typical Fluid Analysis Tests

The following tests yield useful information about the condition of a fluid:

1. *Viscosity* is the resistance of a liquid to flow. A change in viscosity may mean a decrease in machine performance becuase of increased leakage, higher pressure drops, lack of lubrication, and/or over-heating. It may also mean that an incorrect fluid was used to fill the machine's reservoir.

   A change in viscosity ±10% from original specifications or from a previous sample is considered excessive. In the application of a filter, the viscosity must be known to determine the initial (clean) differential pressure across filter element and housing.

2. *Neutralization Number* is a number which indicates the amount of potassium hydroxide needed to neutralize the acids in a fluid. A check on neutralization number is performed to determine if additives have deteriorated or oxidation is excessive.

3. *Water Content* in hydraulic fluid reduces lubrication and affects additives and promotes oxidation. In filter applications, water content is a consideration in element and housing compatibility. Water content exceeding .5% by volume is considered excessive.

4. *Particle Count* The particles in a 100 ml sample are counted and reported in particle size ranges. Particle counts indicate the overall effectiveness of the system filters. Most reports show both a graph (particle size vs. number of particles). At the very least, a cleanliness level should be assigned to describe the contamination content (e.g. ISO 4406 standard reporting).

5. *Spectrographic Analysis* may be used to determine the presence of required additives, wear generated metals, and other contaminants. These results are reported in "ppm" and may indicate the *sources* of wear problems.

The above tests are available at a nominal charge from the Filter Division, Parker Hannifin Corporation, Metamora, OH. Figure d 11. shows the Par-Test™ fluid analysis kit and report form. The pre-addressed container and pre-cleaned sample bottle make fluid analysis programs easy to start.

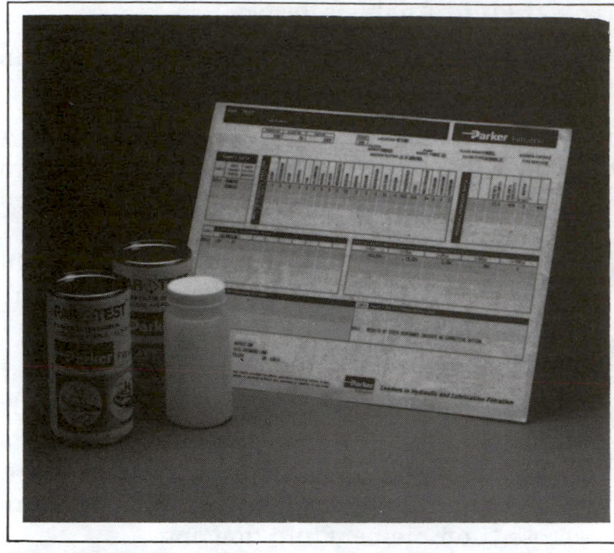

figure d-11

Parker Fluid Analysis Test Kit

Portable Fluid analysis kits are available which contain the basic tools to measure the contamination level of a fluid in plant or in the field. They provide a quick, in-the-field method of checking system fluid on a regular basis. (Figure d 12. shows the Parker portable fluid analysis test kit.) With the use of this kit, a sample "patch" is prepared, showing the relative cleanliness level of the fluid. This patch is compared to "standard" patches of known contamination cleanliness levels. A quick determination of contamination content is possible.

filter and system troubleshooting

Filters are usually very trouble free components. When problems occur, they are generally associated with conditions in the system. Table d 10. below lists some common problems and their solutions.

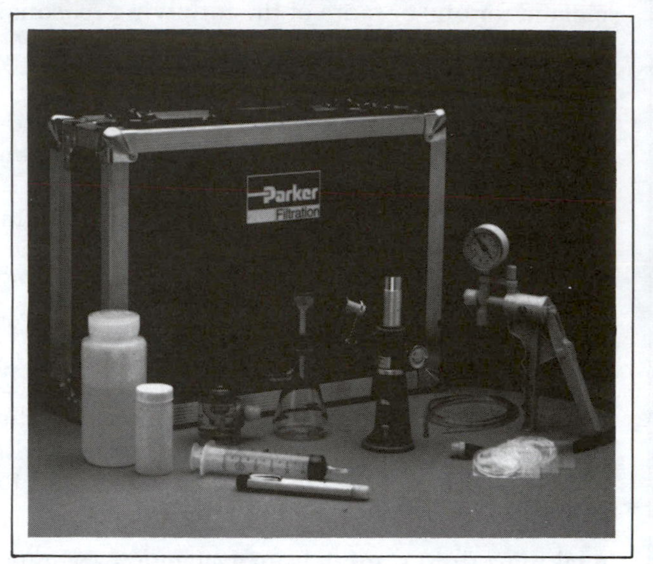

figure d-12

FILTRATION TROUBLESHOOTING *(Table d 10.)*

A. START-UP PROBLEMS

PROBLEM	PROBABLE CAUSES	SOLUTION
1. Filter indicator reads "bypass" immediately upon start-up.	1. Oil viscosity high during cold start-up.	1. Run system until it reaches normal operating temperature.
	2. Highly contaminated system. (if new system has not been flushed, it is common to load a filter element with dirt within minutes of initial start-up.)	2. (a) In new systems, flush with low viscosity fluid at high velocities. (b) Replace filter element several times until fluid is cleaned up. (c) In existing systems, consult specialists and replace filter elements.
	3. Filter bypass setting too low relative to filter pressure drop with a clean element installed (for given conditions).	3. (a) Select a higher bypass setting if available and if circuit conditions will permit. (b) Replace filter with larger one with a lower pressure drop.
	4. Wrong element installed (filter media too fine and/or pressure drop too high.)	4. Replace wrong element with correct one.
	5. Filter indicator out of calibration.	5. Check indicator, Most will read "CLEAN" when system is shut down and there is no flow through filter. Replace or calibrate indicator.
2. Noisy pump, or have to reprime pump repeatedly.	1. Entrained air.	1. (a) Tighten and seal all system fittings and connections. (b) Check and tighten filter bowl or cover. (c) Check filter cover or bowl o-ring for damage; replace if necessary.
	2. Cavitation due to (a) Clogged suction line. (b) Excessive pressure drop in suction line. (Note: Flow velocity in a suction line should not exceed 4 feet per second.) (c) Suction filter bypass setting too high (when added to other suction line losses).	2. (a) Clean sump strainer and remove any obstruction. (b) Check pump inlet with vacuum gauge. If necessary, increase size of suction line and suction filter or sump strainer. (c) Select lower bypass setting and/or larger filter housing.

B. OPERATING PROBLEMS

PROBLEM	PROBABLE CAUSES	SOLUTIONS
1. Filter indicator never reads "BYPASS."	1. Element left out of filter, or not properly installed.	1. Install new element in filter housing. Make sure it is properly positioned so that seals are functional.
	2. Element has been damaged (holes punched in it, or seals removed).	2. Install new element in filter housing with seals properly in place.
	3. Indicator malfunction or out of calibration.	3. Check indicator and replace or recalibrate as necessary.
	4. Oversized filter housing (very low pressure drop for operating conditions).	4. Replace element every six months, even if indicator does not read "BYPASS."
2. Frequent servicing of filter elements required (indicator moves from "CLEAN" to "BYPASS" rather quickly).	1. Same as A.1.1-5 above.	1. Same as A.1.1-5 above.
	2. System has a high contaminant ingression rate.	2. (a) Replace filter with one having an element with larger surface area. (b) Add an off-line, circulating filter loop to the system (portable or permanent). (c) Add filters to pump case drains and elsewhere in system to make sure all fluid returning to reservior is filtered.
3. Fluid samples reveal that fluid is overly contaminated. Frequent component failures occurring.	1. Filter not properly serviced.	1. (a) Check indicators regularly, and replace or clean elements as needed. (b) If filters do not have indicators, replace elements at 250 operating hours intervals. Install gauges or indicators if possible.
	2. Wrong element installed (media too coarse).	2. Replace wrong element with correct one having finer filter medium.
	3. Cylinder rod wiper and other actuator seals worn.	3. Replace cylinder rod seals or other actuator seals.
	4. Pump case drains not filtered, and other lines return flow to reservoir unfiltered.	4. See B.2.b & c above.
4. Directional control valve jammed.	1. Excessive Contamination.	1. See B.3.1-4. above.
	2. Too much back pressure on valve spool.	2. Reduce "BYPASS" setting or return line filter. Check for other line restrictions.

Appendix I

cost per gram analysis

The multipass filter test, described in Standards ANSI B93.31 and NFPA T.3.10.8.8, yields useful data for comparing and selecting filter elements. In this test, elements are challenged with the standard contaminant, Air Cleaner Fine Test Dust. This contaminant is added continuously to the test circuit so that the element becomes increasingly loaded with dirt. Upstream and downstream particle counts are taken to determine the capture efficiency or Beta ratio of the filter element for various particle sizes. The test is over when a predetermined terminal pressure loss through the element is reached.

The data most commonly referred to from the test is the Beta ratio, but other important element performance characteristics are also derived. Two of these are rate dirt holding capacity are apparent capacity and retained capacity.

In the multipass test, the total amount of contamination injected into the test circuit until the element reaches the predetermined terminal pressure drop is called the apparent capacity. However, apparent capacity results may be misleading. For example, a very coarse element may register a high apparent capacity. Or an element may be damaged in such a way that contamination easily passes through it while more contaminant is being injected.

For this reason, it is better to compare filters based on retained capacity, the amount of contamination removed by the filter element. The retained capacity may be thought of as the apparent capacity (the total contaminant injected during the test) less the contaminant in the test system at the end of the test.

Dirt holding capacity, when considered with the cost of an element, yields a valuable filter element selection criterion: economic value. Economic value of an element is defined as the cost per gram of removed contaminant. Beta ratios, dirt holding capacity, and element cost should be considered together with cost per gram to make a proper filter selection. The following examples show how a more complete comparison might be made for filter elements with identical Beta ratios.

	ELEMENT X	ELEMENT Y
Cost	$18.00	$22.25
Beta 10 ratio	75	75
Area - ft²	6	18
Cost per ft²	$ 3.00	$ 1.24

Element Y appears to be a better value; its larger size seems to indicate longer life, with similar Beta ratios, and at less than half the cost per square foot. But a clearer picture forms when retained capacities are reported:

	ELEMENT X	ELEMENT Y
Retained dirt capacity - g	36	20
Cost per gram of dirt	$ .50	$ 1.11

When the cost per gram is computed at $18.00/36 – $.50 per gram for element X, this element's true cost is less than half of Y's! The cost per gram provides a much needed economic value for filter selection. The hidden cost of more frequent service (more downtime) with element Y is also eliminated.

This illustration is based on an actual field example and points out the importance of considering retained dirt capacity as a major factor in filter selection. It also demonstrates that size alone does not determine element life. A large, poorly constructed element could actually provide less effective filtration because contaminant is not retained as well, as in element Y.

To make an intelligent comparison of hydraulic filters, designers should obtain all data relevant to the multipass filter element test, including apparent capacity, retained capacity, Beta ratios, flow rate, and terminal pressure loss. The economic value index, or cost per gram, when combined with other element data, may provide the best method available for selecting filters.

intensifiers

intensifier selection e-2
intensifier clamp circuit e-3
intensifier operating principle e-4
intensifier usage e-5
how to select an air-oil tank e-6

intensifier selector chart

INTENSIFIER RATIOS & VOLUME OUTPUT

BORE SIZE	RAM SIZES									
	1	1-3/8	1-3/4	2	2-1/2	3	3-1/2	4	5	5-1/2
3-1/4	10.57	5.59	3.45	2.64						
4	16.00	8.46	5.23	4.00	2.56					
5	25.00	13.22	8.16	6.25	4.00	2.78	2.04			
6		19.05	11.76	9.00	5.76	4.00	2.94	2.25		
8		33.85	20.90	16.00	10.24	7.11	5.23	4.00	2.56	2.12
10			32.66	25.00	16.00	11.11	8.16	6.25	4.00	3.31
12				36.00	23.04	16.00	11.75	9.00	5.76	4.76
14					31.36	21.78	16.00	12.25	7.84	6.48
Vol. per inch of stroke	.78	1.49	2.40	3.14	4.91	7.07	9.62	12.57	19.63	23.76

table e-1

intensifier selection

Intensifier size is determined by the high pressure requirement of the work cylinder. The ratio between the high pressure required and the available input pressure is termed the intensifier ratio and determines the diameter of the piston and the ram. The intensifier stroke is determined by the volume of high pressure oil required by work cylinder.

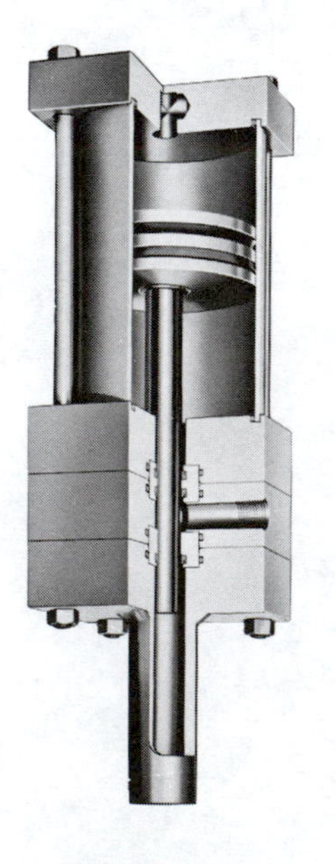

PH Intensifier

Select a single pressure intensifier for applications where the entire stroke of the work cylinder is the load cycle. Use a dual pressure intensifier where low pressure is fed from the air-oil tank through the intensifier to advance the work cylinder to the load; only the remaining stroke of the cylinder is then considered in the intensifier stroke calculations.

intensifier sizing

1. Based on application requirements, select the work cylinder and determine the hydraulic pressure required.

2. Divide the hydraulic pressure required by the input pressure to determine the intensifier ratio.

3. Select intensifier bore and ram size from selector **table e-1.**

4. Calculate the volume of high pressure oil required for full extension of the work cylinder through the high pressure portion of the stroke.

5. To calculate intensifier stroke, divide calculated volume of high pressure oil required by "Volume per Inch of Stroke" shown in **table e-1** for selected ram size.

6. For dual pressure intensifiers only: add 2" to stroke calculated in step 5.

intensifier clamp circuit

An example is the cylinder-powered clamp for bench or machine operations shown in **circuit e-1.** This illustrates a dual pressure intensifier used to supply hydraulic pressure for a simple cyliner-powered clamp. Here the moving portion of the clamp advances rapidly under low pressure to contact work piece. When this contact is made, the intensifier is energized, producing high pressure to hold the work piece firmly. At completion of operations on work piece, the directional control valve in air line is shifted to de-energize the intensifier, and the clamp is retracted for the start of the next cylce.

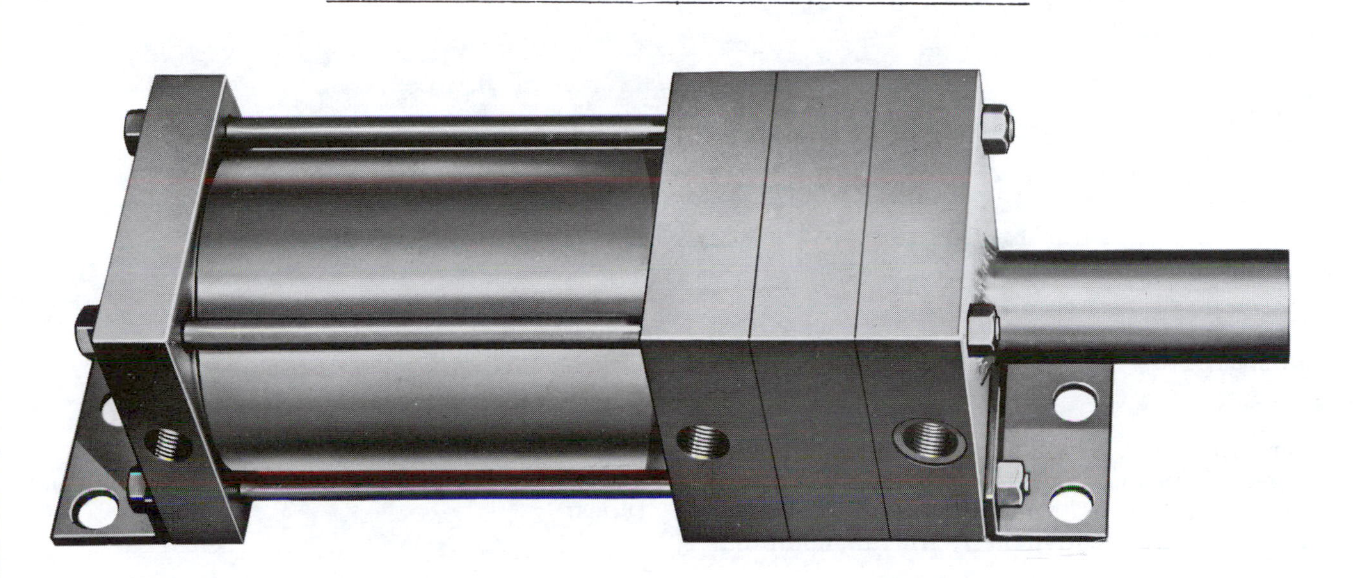

illustration e-1

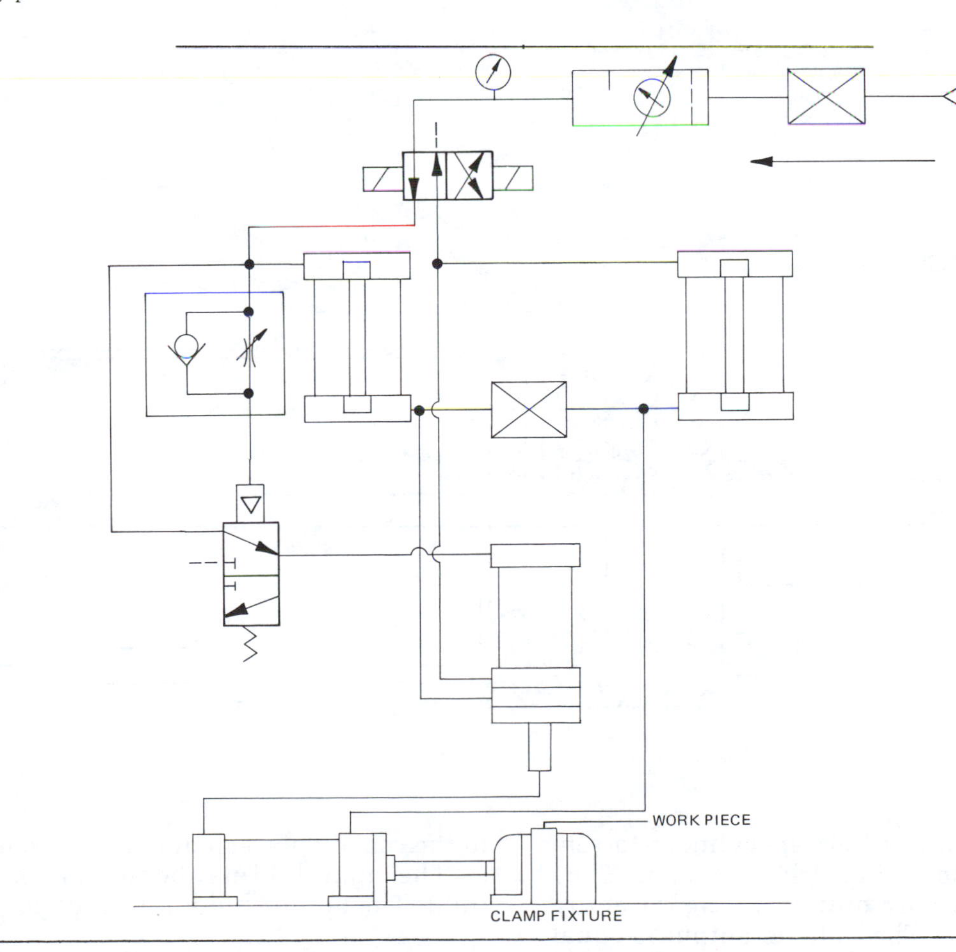

circuit e-1

intensifier operating principle

An intensifier is a fluid power device which will produce an output pressure higher than the source input pressure.

The input fluid may be the same as, or different than, the output fluid. The most common arrangement is to use air pressure from shop lines as the input fluid and a petroleum base hydraulic oil as the output fluid.

Basically, an intensifier is a confined piston subjected to an input pressurized fluid, driving a smaller diameter ram or piston to deliver a volume of fluid at a higher pressure.

The output pressure of an intensifier is proportional to the area of the driving piston divided by the area of the driven ram.

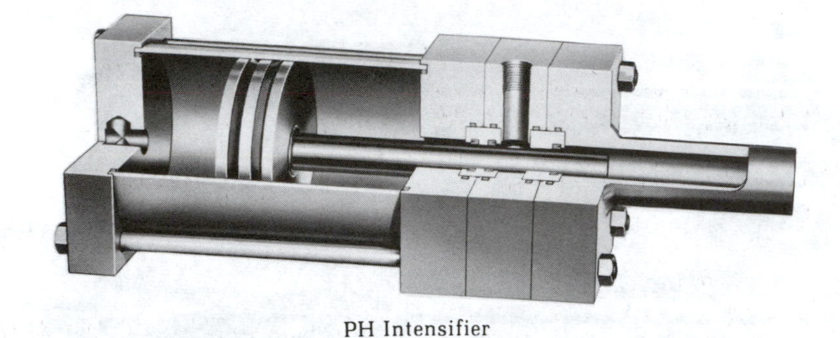

PH Intensifier

illustration e-2

Intensifiers are available in single and dual pressure types consisting of a driving cylinder and a driven ram. Most manufacturers offer bore sizes 3¼" through 14" with ram combinations of 1" through 5½" diameter. Intensifiers are also manufactured in cylinder-to-cylinder types.

single pressure

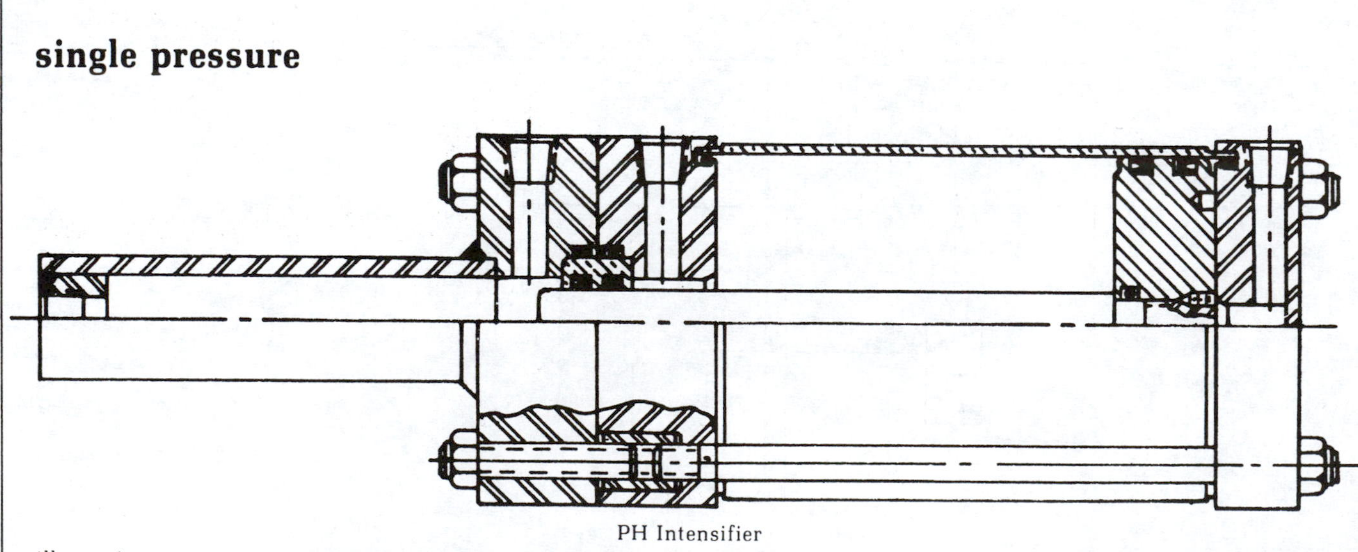

PH Intensifier

illustration e-3

Single pressure intensifiers are cylinder-to-ram type with double-acting driving piston. They provide high pressure output during the entire intensifier stroke. The volume output is equal to the ram displacement in the pressure chamber. The input fluid may be either air or hydraulic fluid. The output fluid is hydraulic.

intensifier operating principle continued

dual pressure

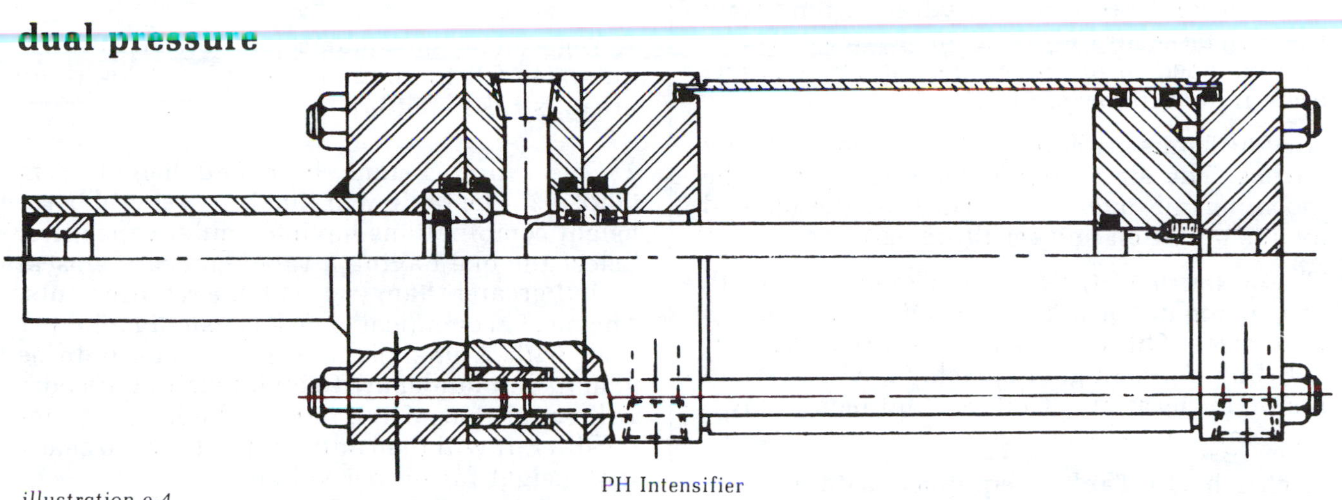

illustration e-4

PH Intensifier

Dual pressure intensifiers are also cylinder-to-ram type with double-acting driving piston. They differ from single pressure units in that they allow low pressure flow to prefill and advance the work cylinder. They then provide high pressure during the work stroke of the cylinder. This high pressure volume output is equal to the ram displacement in the pressure chamber. The input fluid may be either air or hydraulic fluid.

cylinder-to-cylinder intensifiers

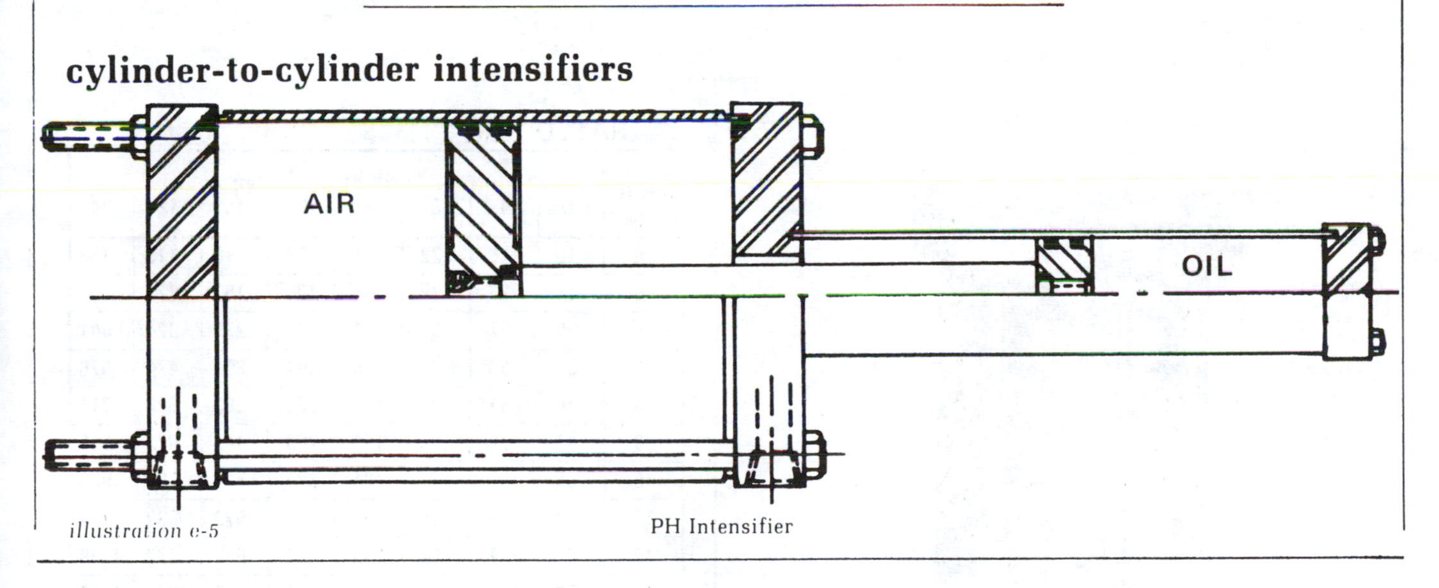

illustration e-5

PH Intensifier

intensifier usage

1. To maintain intensified hydraulic pressure for long periods of time with no added power consumption or energy loss due to heat generation.

2. To provide easily varied high pressure by controlling the input pressure.

3. To allow the use of more compact work cylinders. With intensified pressure, the hydraulic cylinder can do the work of an air cylinder many times larger and heavier. This is particularly true in dual pressure intensifier circuits where low pressure is used for the advance stroke and intensified pressure during the work stroke.

4. To allow the use of lower cost, control components, since in most applications, the controls are in the lower pressure input portion of the circuit.

5. To avoid safety hazards. Since intensifiers can be completely air-operated, they can function safely in any environment.

how to select an air-oil tank

With Air-Oil Tanks, shop line compressed air is converted into smooth hydraulic fluid flow that can be accurately controlled. Air-Oil Tanks may be used in an intensifier circuit or separately in a cylinder circuit.

Air-Oil Tanks are equipped with two fluid flow baffles. The air baffle at the top disperses the incoming air over the surface of the oil and avoids agitation and resulting aeration.

The oil baffle at the bottom provides a smooth proportionate flow pattern. Oil turbulence is minimized. Oil leaves and enters tank without swirling, funneling, or splashing, which would result in aeration of oil, and oil being blown from tank in exhausting air.

Each Air-Oil Tank is equipped with a sight-level gauge marked to show proper oil level.

Air and oil baffles are not identical. It is important to mount the tank in a vertical position with the air port up and the oil port down. The tank will not operate properly in a horizontal position.

PH Air-Oil Tank

illustration e-6

how to select

1. Determine the volume (cu. in.) of fluid required to fill the cylinder at full stroke. (Bore area x stroke length.)

2. Select proper tank bore and height from **table e-2.** The table will illustrate several bore-height combinations having similar capacities. Select the one having a rated capacity closest to but greater than, your volume requirements. The most economical choice is usually a higher tank with a smaller bore. If the tank is to be mounted in tandem with an intensifier, its bore will equal that of the input cylinder of the intensifier. It will then only be necessary to select tank height for proper volume.

3. Describe location of ports and sight gage when ordering.

RATED CAPACITIES – CUBIC INCHES

TANK HEIGHT (Inches)	TANK BORE (Inches)							
	3-1/4	4	5	6	8	10	12	14
4	10	15	22	31	56	78	113	153
5	18	26	39	56	100	157	226	307
6	25	37	56	81	151	225	324	441
7	32	50	73	106	194	294	424	576
8	40	61	93	134	238	363	522	711
9	47	72	110	159	283	441	621	845
10	54	83	127	184	327	510	720	980
11	62	96	144	208	377	579	833	1114
12	69	107	162	233	421	647	932	1249
13	76	118	181	261	465	725	1030	1382
14	84	129	199	286	509	795	1130	1539
15	92	141	216	311	560	864	1229	1671
16	98	152	233	335	604	942	1329	1809
17	106	163	250	360	646	1010	1440	1942
18	114	174	270	388	691	1080	1540	2078
19	120	185	287	414	735	1158	1639	2208
20	128	197	304	438	785	1228	1738	2342
21	136	208	324	466	830	1295	1838	2480
22	143	218	340	490	873	1362	1938	2618
23	151	231	360	519	924	1441	2050	2765
24	158	243	380	547	974	1520	2160	2921

table e-2

hydraulic motors

horsepower - speed - torque table f-2
rated full load torque table f-3
hydraulic motor selection f-3
hydraulic motor calculations f-4
hydraulic motor circuits f-5
bidirectional circuits f-6
closed loop circuits f-9

horsepower — speed — torque table

This table shows the relation between the output horsepower, speed, and torque of a rotary motor. If two of the characteristics are known, the third can be found.

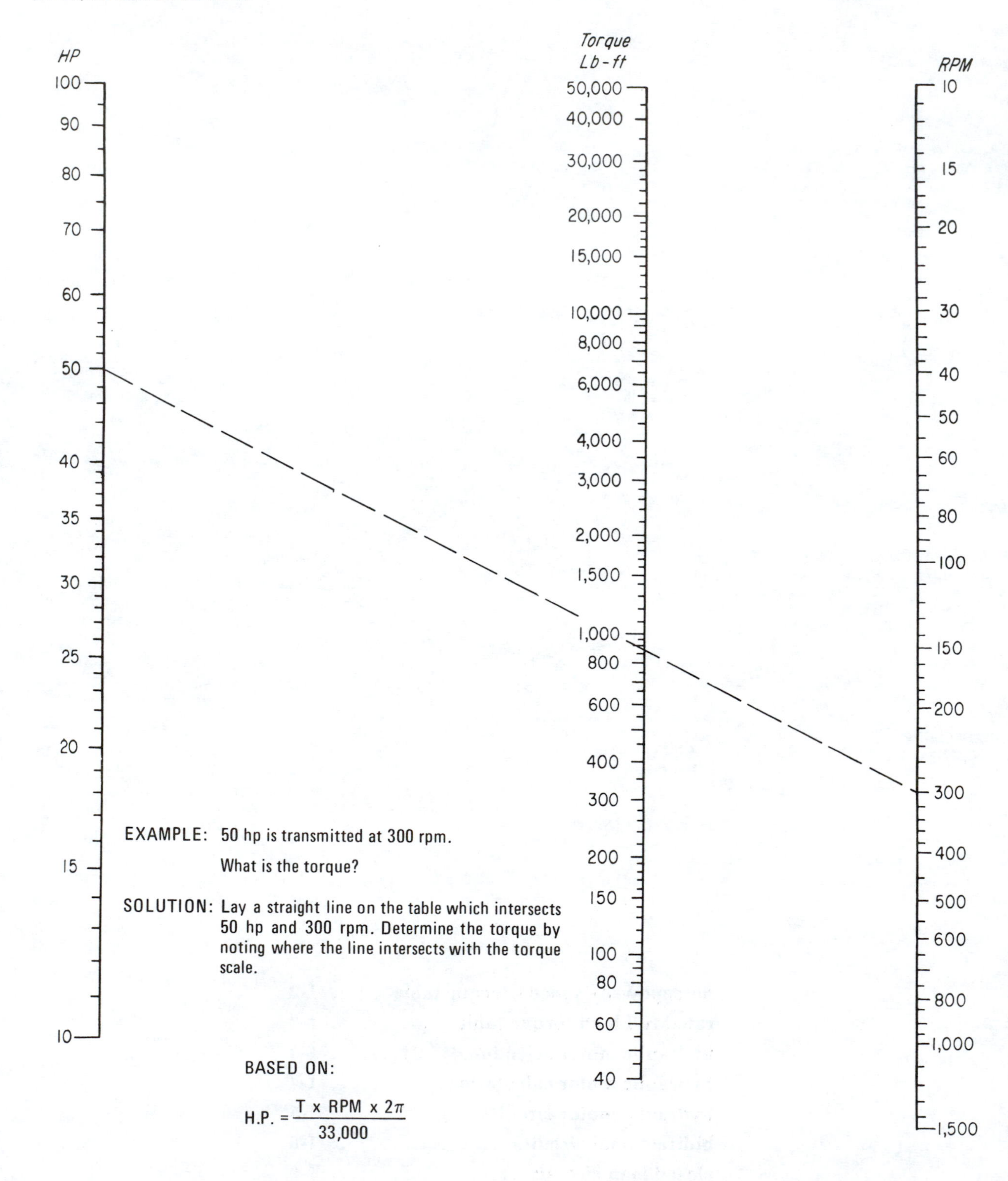

EXAMPLE: 50 hp is transmitted at 300 rpm. What is the torque?

SOLUTION: Lay a straight line on the table which intersects 50 hp and 300 rpm. Determine the torque by noting where the line intersects with the torque scale.

BASED ON:

$$H.P. = \frac{T \times RPM \times 2\pi}{33,000}$$

Friction losses and volumetric losses are not involved because the table covers only the output characteristics.

table f-1

rated full load torque table

HP	100 RPM	500 RPM	750 RPM	1000 RPM	1200 RPM	1500 RPM	1800 RPM	2400 RPM	3000 RPM	3600 RPM
1/4	13.10	2.63	1.76	1.31	1.10	.88	.73	.55	.44	.37
1/3	17.50	3.50	2.34	1.75	1.46	1.17	.97	.73	.58	.49
1/2	26.30	5.25	3.50	2.63	2.20	1.75	1.46	1.10	.88	.73
3/4	39.40	7.87	5.24	3.94	3.28	2.62	2.18	1.64	1.31	1.09
1	52.50	10.50	7.00	5.25	4.38	3.50	2.92	2.19	1.75	1.47
1-1/2	78.80	15.70	10.50	7.88	6.56	5.26	4.38	3.28	2.63	2.19
2	105.00	21.00	14.00	10.50	8.76	7.00	5.84	4.38	3.50	2.92
3	158.00	31.50	21.00	15.80	13.10	10.50	8.76	6.57	5.25	4.38
5	263.00	52.50	35.00	26.30	22.00	17.50	14.60	11.00	8.75	7.30
7-1/2	394.00	78.80	53.20	39.40	32.80	26.60	21.80	16.40	13.10	10.90
10	525.00	105.00	70.00	52.50	43.80	35.00	29.20	21.90	17.50	14.60
15	788.00	158.00	105.00	78.80	65.60	52.60	43.80	32.80	26.50	21.90
20	1050.00	210.00	140.00	105.00	87.60	70.00	58.40	43.80	35.00	29.20
25	1313.00	263.00	175.00	131.00	110.00	87.70	73.00	54.80	43.80	36.50
30	1576.00	315.00	210.00	158.00	131.00	105.00	87.40	65.70	52.60	43.70
40	2100.00	420.00	280.00	210.00	175.00	140.00	116.00	87.50	70.00	58.20
50	2626.00	523.00	350.00	263.00	220.00	175.00	146.00	110.00	87.50	72.80
60	3151.00	630.00	420.00	315.00	262.00	210.00	175.00	131.00	105.00	87.40

Please note that all torque values in this table are in foot pounds. Multiply by 12 to get inch pounds.

For values not in the table, use one of these formulae:

Torque = HP x 5252 ÷ RPM

HP = Torque x RPM ÷ 5252

table f-2

motors

hydraulic motor selection

In selecting a hydrualic motor, first, requires the amount of torque needed to move the work (breakaway torque) to be known. Torque is known as rotary force and is accomplished by developing pressure within a motor housing, against the motor vane, piston or gear face. This then generates a rotary action known as torque. Torque required can be found by multiplying the distance from the center of the hydraulic motor shaft to the end of the rotary arm, times the load or force which the rotary arm is required to move. T = RxF (Torque = Radius x Force). Breakaway torque "is the amount of torque required to initially start a load moving." This torque is generally much higher than "Running torque". "Running torque" is the amount of torque required to keep a load moving after it has been started.

Torque is found in terms of pounds-inch or pounds-foot of torque. Next establish a hydraulic motor system working pressure. With the pressure established, the maximum torque that can be developed by the motor can be found by checking the manufacturers hydraulic motor specifications. This will enable you to determine how much torque can be developed per 100 psi or at the maximum system working pressure. Depending on the system, sometimes it is necessary to oversize the hydraulic motor as much as 40% or more to allow for the "breakaway torque" requirements.

When hydraulic motors are being used at a high rpm or have a rotary mass (such as a fly wheel conveyor system, etc.), a cross-over relief or breaking relief valve should be utilized. These valves will absorb most of the shock created by the centering of the directional control valve and the inertia of the rotating mass.

Hydraulic motors should be operated at the highest practical speed. Generally, 100 rpm is the lowest practical rpm unless the motor is designed for low rpm and high torque applications. Gear reduction is sometimes used to obtain lower output shaft rotation with high torque requirement. With some systems, it's practical to increase the hydraulic motor size considerably. This allows the motor to obtain a lower operating speed and maintain a high torque requirement. Flow controls can be used to control the motor speed. Care should be taken in applying the flow control with each system. In some systems, a "meter out" application may create enough back pressure on the

hydraulic motor selection continued

motor to cancel out some of the torque which the motor otherwise is able to produce. On "Meter in" applications care should be taken on some lower rpm applications, where there may be binding or change in friction; this will affect the constant rpm. With "bleed off" applications, always use a pressure compensated flow control. Some critical requirements may also require the pressure compensated flow control to be temperature compensated. The "bleed off" arrangement creates less heat in the system over the "meter in" or "meter out" arrangements. This is a HP savings.

To calculate the amount of oil required to drive the hydraulic motor at your specified rpm, first check the manufacturers specifications for the motor displacement. This displacement is generally given in cubic inches per revolution. Multiply the number of $in.^3/rev.$ times the number of revolutions you have specified. Take this answer and divide it by 231 in^3 (231 in^3 is the number of cubic inches in one U.S. gallon). This will give you an answer in gallons per minute. The pump is required to deliver this amount of oil (gpm) to drive the motor at your specified rpm. Example: A fluid motor has a displacement of 1.61 $in^3/rev.$ The specifications state that the hydraulic motor should be driven at 1400 rpm. What pump delivery is required to drive this motor at 1400 rpm?

$$1.61\ in.^3/rev. \times 1400\ rpm = 2254$$

$$2254 \div 231 = 9.15\ \text{gpm (delivered from the pump)}$$

A motor efficiency allowance factor should be taken into consideration. This can be found by checking the manufacturers specifications. This can increase the gpm requirement by as much as 10 - 15%.

hydraulic motor calculations

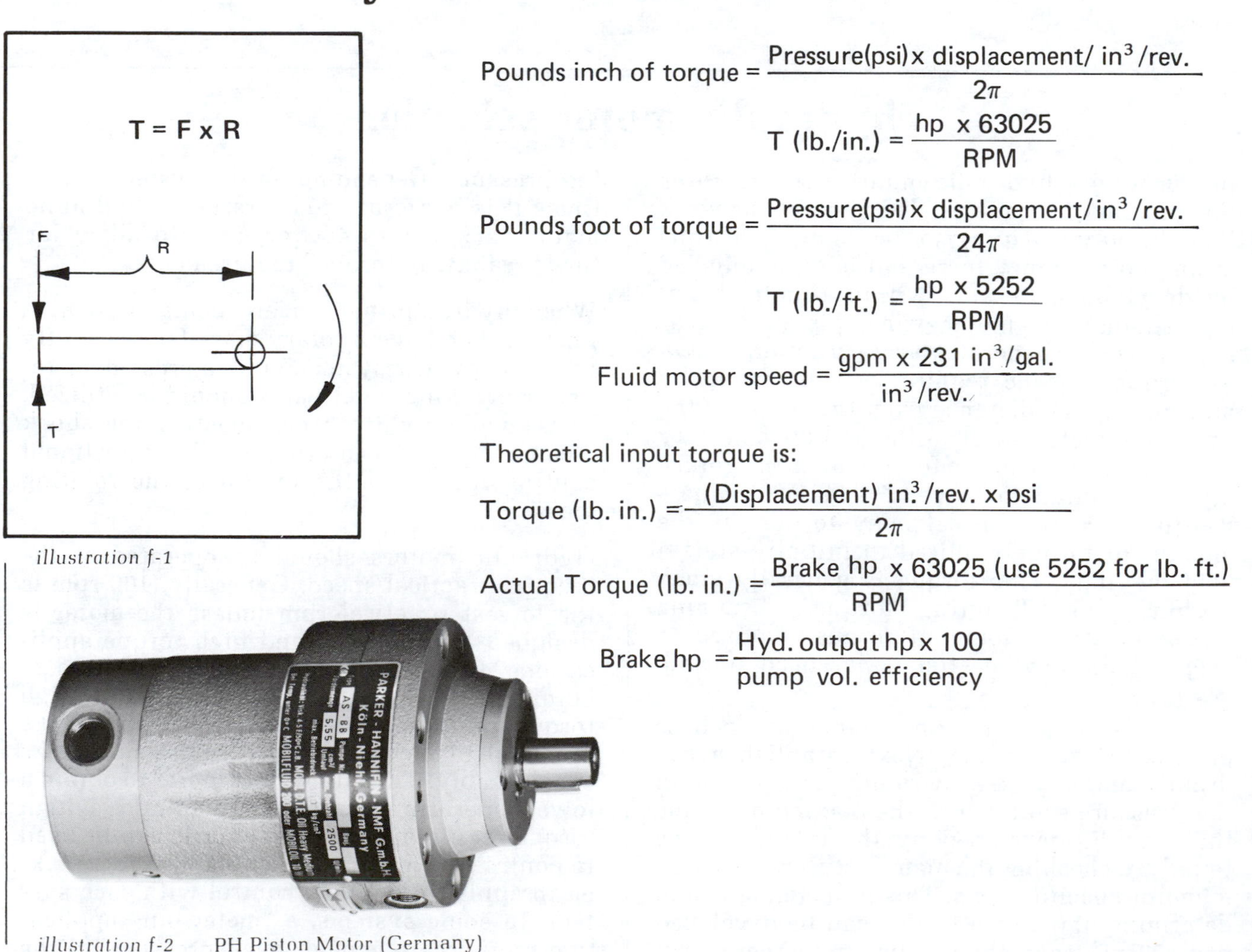

$$\text{Pounds inch of torque} = \frac{\text{Pressure(psi)} \times \text{displacement}/in^3/rev.}{2\pi}$$

$$T\ (lb./in.) = \frac{hp \times 63025}{RPM}$$

$$\text{Pounds foot of torque} = \frac{\text{Pressure(psi)} \times \text{displacement}/in^3/rev.}{24\pi}$$

$$T\ (lb./ft.) = \frac{hp \times 5252}{RPM}$$

$$\text{Fluid motor speed} = \frac{gpm \times 231\ in^3/gal.}{in^3/rev.}$$

Theoretical input torque is:

$$\text{Torque (lb. in.)} = \frac{\text{(Displacement)}\ in.^3/rev. \times psi}{2\pi}$$

$$\text{Actual Torque (lb. in.)} = \frac{\text{Brake hp} \times 63025\ \text{(use 5252 for lb. ft.)}}{RPM}$$

$$\text{Brake hp} = \frac{\text{Hyd. output hp} \times 100}{\text{pump vol. efficiency}}$$

illustration f-1

illustration f-2 PH Piston Motor (Germany)

hydraulic motor circuits

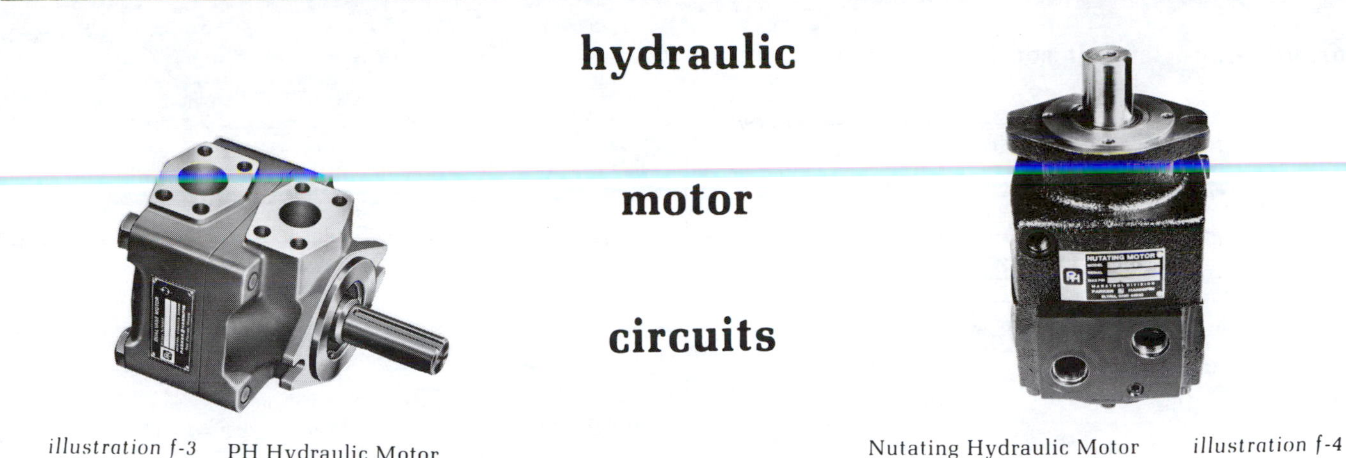

illustration f-3 PH Hydraulic Motor

Nutating Hydraulic Motor *illustration f-4*

unidirectional circuits

The following are partial and complete hydraulic circuits, illustrating some of the ways hydraulic motors can be controlled. Some circuits are condensed to their simplest form, to show one basic idea. These circuits can be combined with more complex hydraulic circuitry to form complete working hydraulic systems.

unidirectional motors

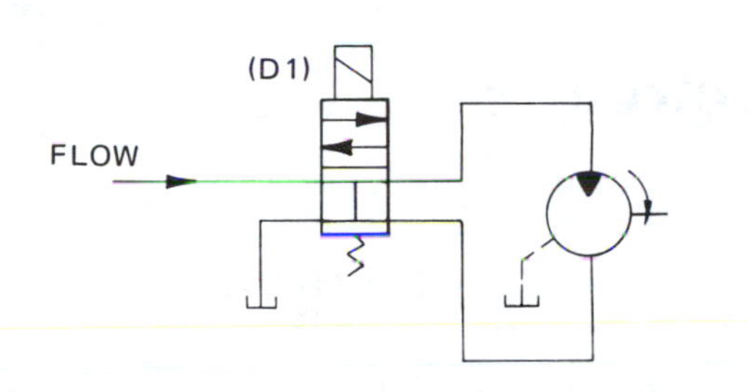

circuit f-1

Circuit f-1 is basically the same as **circuit f-2.** A pressure compensated flow control (PCI) has been added (for variable speed) along with a check valve (C1).

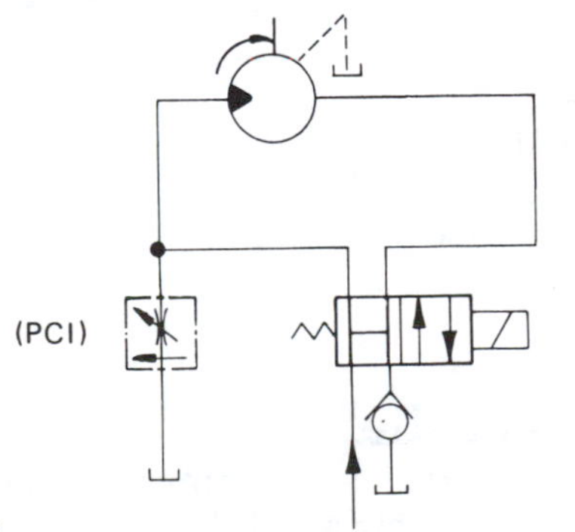

circuit f-2

NOTE: A 65 p.s.i. check. This is required in the return line for PH dual vane motors when in a "free wheeling" position.

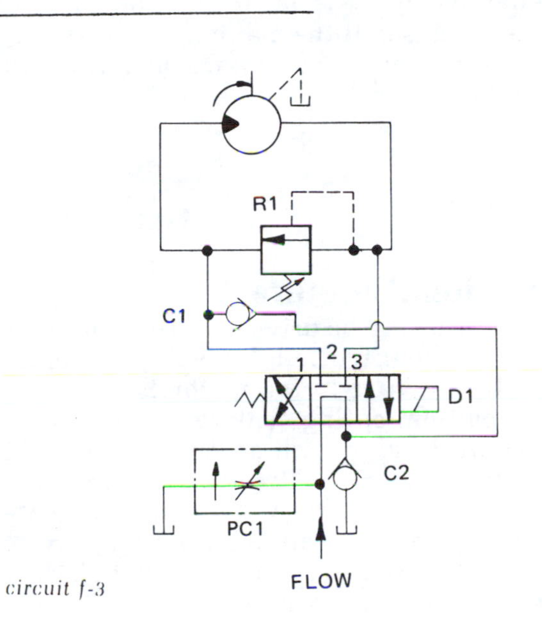

circuit f-3

In **circuit f-3** a pressure compensated flow control valve (PC-1) determines the motor speed and the system relief on the power unit determines the torque. The braking relief (R-1) only performs the braking function.

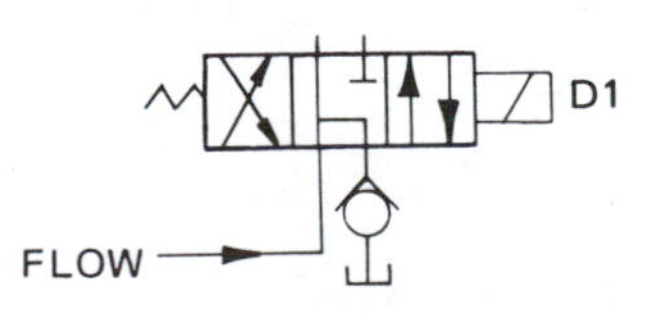

circuit f-4

NOTE: To eliminate the make up check requirement for position (2) in **circuit f-3** a directional valve (D-1) with this center configuration can be used.

motors

hydraulic motor circuit continued

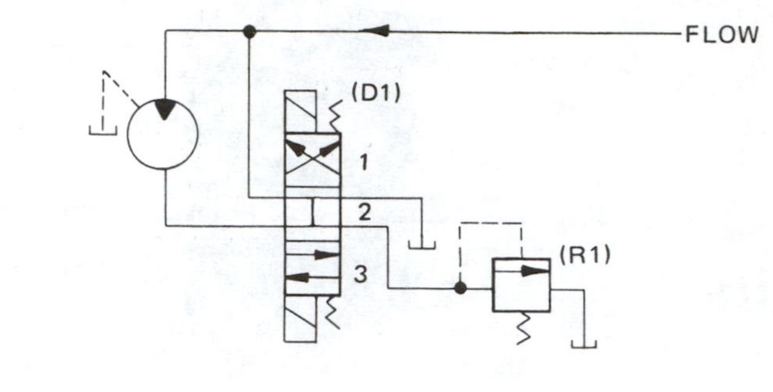

circuit f-5

braking relief unidirectional

Unidirectional rotation with free-wheeling in the directional valve (D1) center position (position 2). The relief valve (R1) acts as a breaking relief when (D-1) is in position (3). (R1) also has to be set at the maximum pressure required to give you the maximum torque you need to do the work.

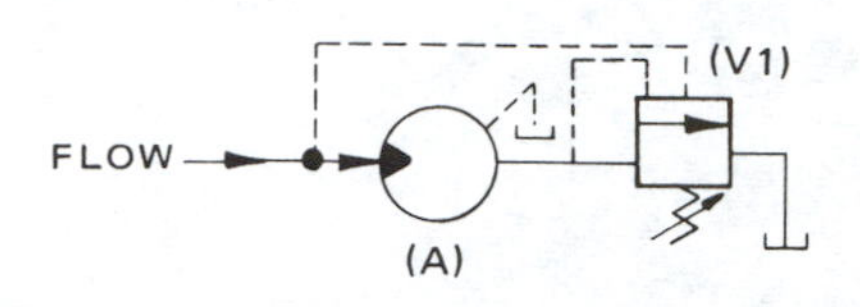

circuit f-6

over-run limiter

Circuit f-6 can be used to slow or stop an over-run condition. The pilot line from the brake valve (V-1) is connected to the inlet side of the motor (A). As the pressure increases the brake valve (V-1) opens and allows the motor (A) to turn. As the pressure drops the counterbalance valve (V-1) closes and brakes the motor (A) to a stop. The motor (A) should be able to take pressure at its outlet port and also be externally drained.

bidirectional circuits

bidirectional motors

These motors may be driven in either direction and may be controlled by various control valves. In **circuit f-7,** the motor is controlled by a 3-position, spring centered, 4-way directional control valve with an open center. This allows the motor to "free wheel" in either direction, when valve (V-1) is in the center position (2) and also allows the pump to unload back to the reservoir. A 65 psi check is required for dual vane motors.

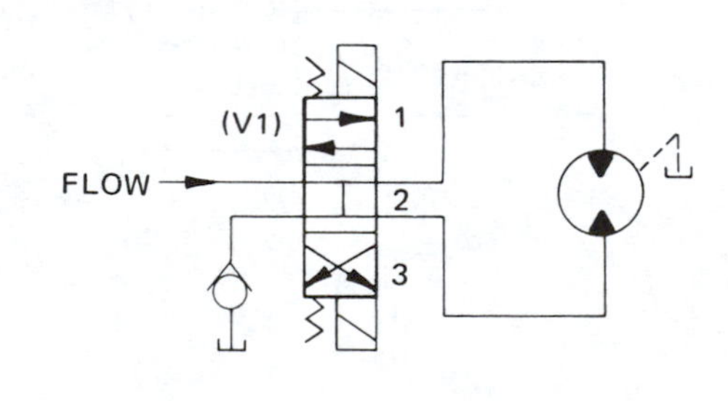

circuit f-7

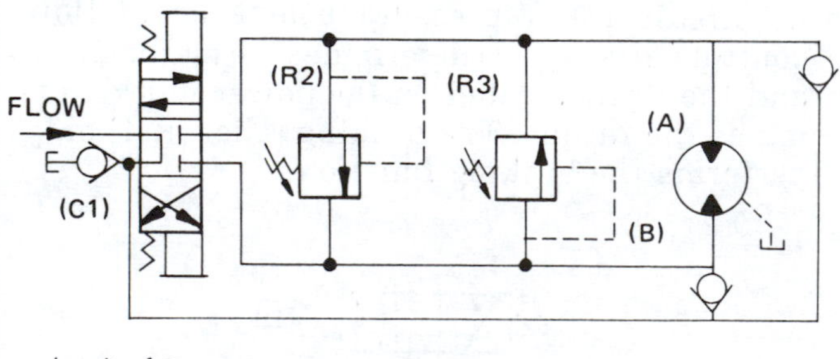

circuit f-8

dual/relief braking

In this circuit **(circuit f-8)** two relief valves are used as braking valves. With this method of braking you are able to obtain one level of braking in (A) direction and another level of braking in (B) direction.

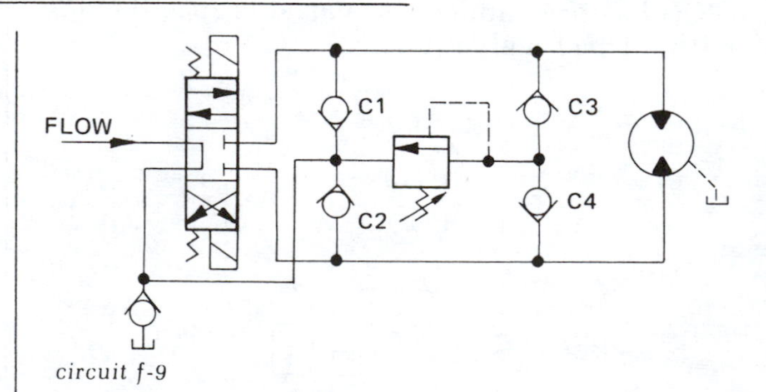

circuit f-9

cross-over relief

This circuit **(circuit f-9)** is similar to **circuit f-8.** The difference is **circuit f-9** has a single relief acting as a braking relief through several check valve (C1, C2, C3 and C4.) This circuit also has a single braking level regardless of rotation.

bidirectional circuits continued

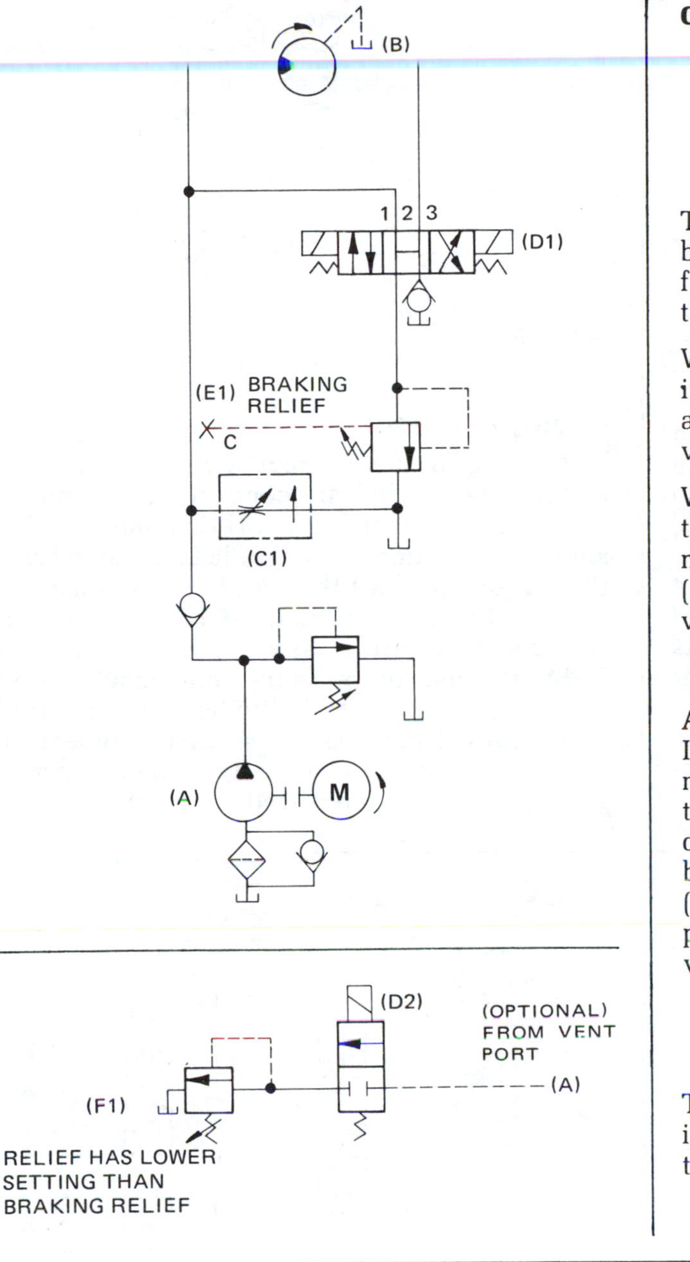

optional - braking or free wheeling

The torque is determined by the setting of the braking relief (E1). This is because the flow from pump (A) flows through the direction control valve (D1) to the braking relief (E1).

With the directional control valve (D1) in position (2), the motor (B) is "freewheeling" or able to coast, since all ports within the control valve (D1) are connected.

With the directional control valve (D1) in position (3), the outlet port of the motor (B) is connected to the braking valve (E1). The pump (A) is unloaded through the directional control valve (D1) back to tank.

Added Feature
If the setting of braking relief (E1) **circuit f-10** must change between controlling pressure to the motor (B) and the braking function - this can be accomplished using the vent port of the braking relief (E1) and connecting relief valve (F1) to this vent port and setting (F1) at a lower pressure setting. Whenever needed energize valve (D2).

This is accomplished by connecting port (A) in **circuit f-11** to port (C) (vent connection) on the braking relief in question in **circuit f-10.**

motor control unidirectional

In **circuit f-12** a motor of any size can be controlled. Relief valve (R-1) is used as the system relief and counterbalance valve (R-2) is used as the braking valve. The directional control valve (D-1) can be 1/4" ported valve because of the small flow of oil flowing from the vent ports of valve (R-1) and (R-2). When (D-1) is in position (2) the motor is free-wheeling. When (D-1) is in position (3) the vent port to (R-1) is blocked and (R-2) is vented allowing the motor to run. When (D-1) is in position (1) (R-1) is vented allowing system oil to flow to the reservoir. Valve (R-2) vent line is then blocked and breaks the motor to a stop.

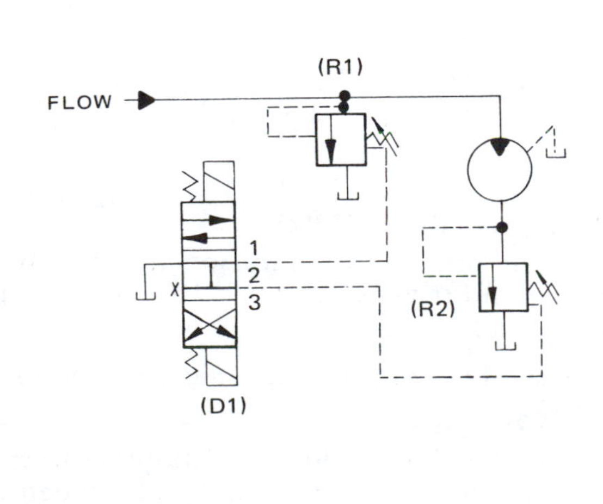

bidirectional circuits continued

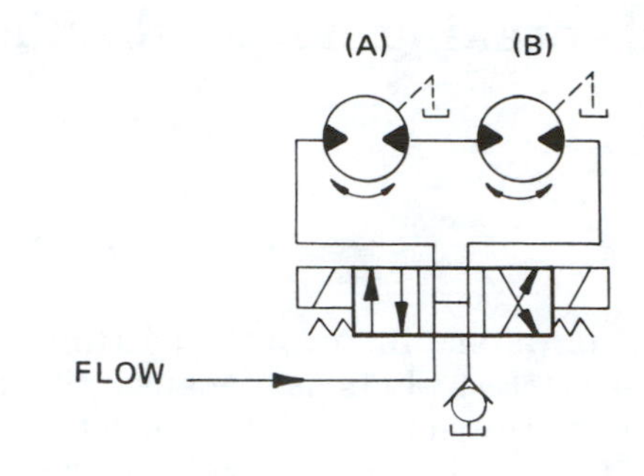

circuit f-13

series motors

In this requirement **circuit f-13,** we have two identical motors in series and will divide the available system pressure in proportion to the work load on each motor. This is due to the pressure that is developed in motor (B) (needed to generate its torque) is also working on outlet port of motor (A). Thus the pressure required to develop the torque in motor (A) equals its required pressure, plus the pressure seen at motor (A) outlet port. NOTE: Motor (A) will have to be able to withstand pressure in its outlet port. Both motors will run at approximately the same speed regardless of their load.

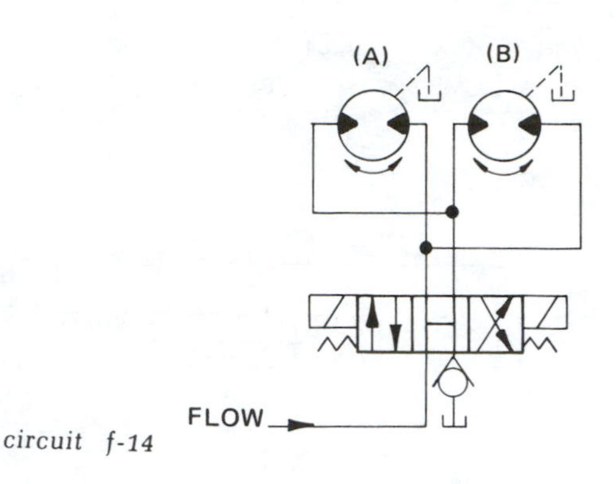

circuit f-14

parallel motors

In this requirement **circuit f-14,** two identical motors are needed, by connecting the motors (A & B) in parallel and driving both with a single fixed pump, we are able to develop twice the torque, but half the speed of a single motor used in the same circuits. If both motors are required to rotate at approximately the same RPM, the use of pressure compensated flow control valves or flow dividers are required. Sometimes it may be necessary to connect the motors together mechanically if the loads vary greatly and the same speed is required.

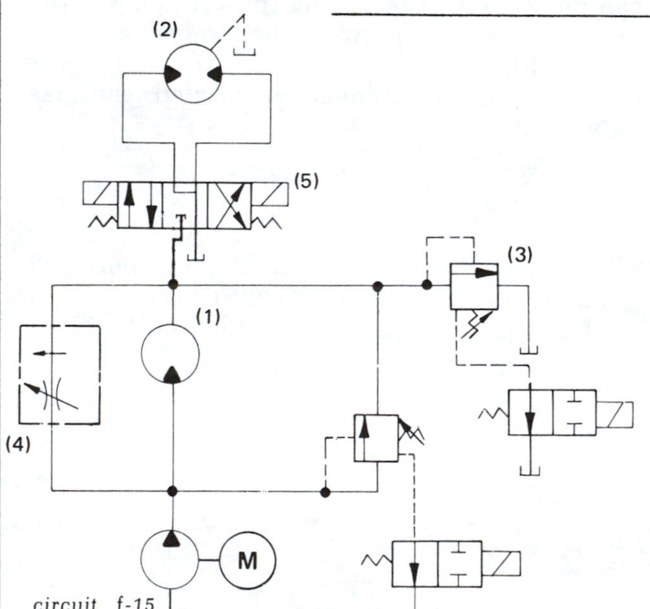

circuit f-15

series motor circuit

Maximum torque is available on motor (1) with relief valve (3) vented. Speed is controlled by flow control valve (4).

Direction of motor (2) is obtained by directional valve (5).

With relief valve (3) loaded, the pressures on motors (1) and (2) can be set independently.

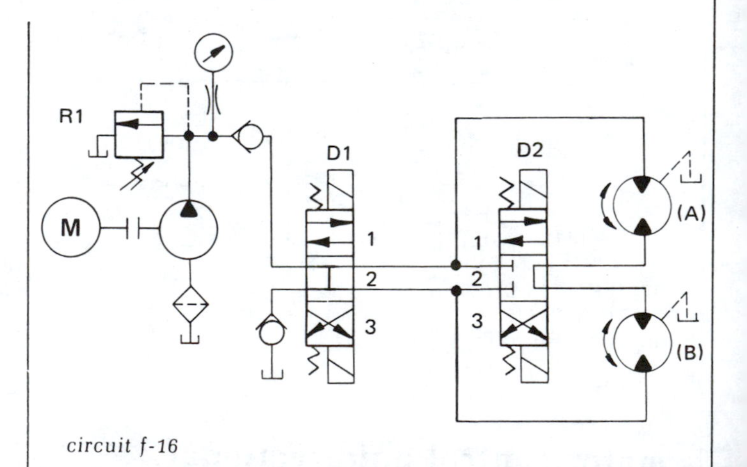

circuit f-16

series/parallel motors bidirectional

In a system that requires two motors giving high-speed, low-torque and low-speed, high-torque, **circuit f-16** may be used. When (D1) is in position (1) and (D-2) is in position (2) the two motors are connected in series and the high speed is available. When (D-1) is in position (1) and (D-2) is in position (3) low speed and high torque is available. With (D-1) in position (2) and (D-2) in position (1) both motors can "free-wheel". Check valve (C-1) gives the required back pressure for the dual vane hydraulic motor.

bidirectional circuits continued

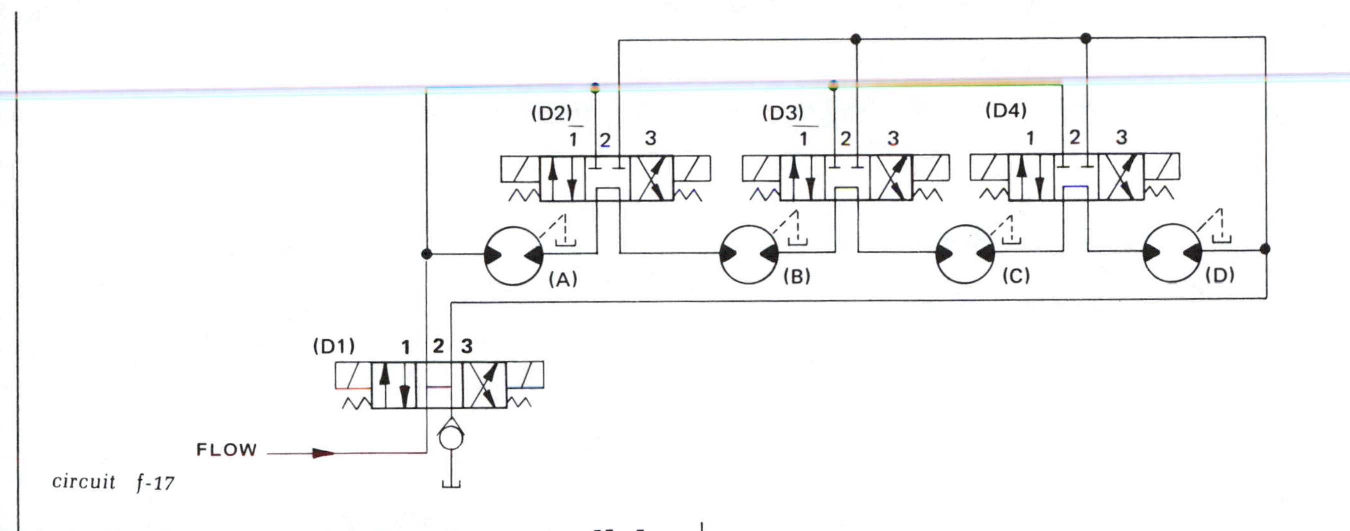

circuit f-17

multiple motors in series/parallel with free wheeling —

This circuit operates the same as **circuit f-16** operates. The difference is this circuit has four motors rather than two. The principal is the same with (D-1) in position (2) and (D-2), (D-3) and (D-4) in position (1) all motors (A), (B), (C), (D) are free-wheeling. With (D-1) in position (1) and (D-2), (D-3) and (D-4) in position (2) all motors (A), (B), (C) and (D) are in series with (D-1) in position (1) and (D-2), (D-3) and (D-4) in position (3) all motors are in parallel by shifting (D-2), (D-3) and (D-4) into various positions we are able to run motors (A), (B), (C), and (D) in series and parallel. (D-1) is used to start, stop or reverse the motors. With this circuit you are able to obtain various speeds and torques.

closed loop circuits

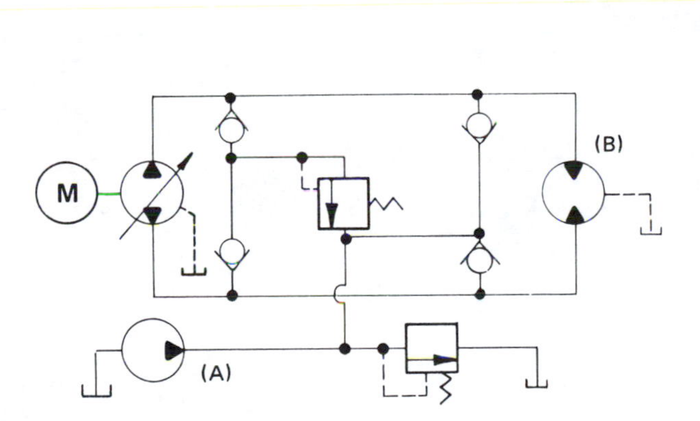

circuit f-18

conventional overcenter rotary drive

Direction and speed of rotation of the motor is determined by the amount of off-center of the variable pump. A single relief valve is used for acceleration and deceleration. A boost pump (A) will be necessary to give positive filling of the circuit.

Regeneration from the hydraulic motor (B) is available for braking.

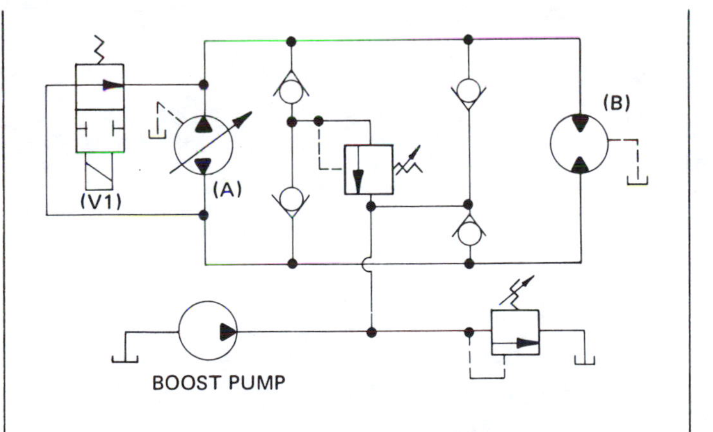

circuit f-19

prevention of creep

Due to the infinate variable speed control it is difficult on some controls and systems to zero the pump (A) to ensure the motor (B) stops positively.

One way to achieve this is to fit a solenoid or mechanical valve (V-1) operated at the point or center which unloads the pump (A) to prevent motor (B) creep.

notes

pneumatics

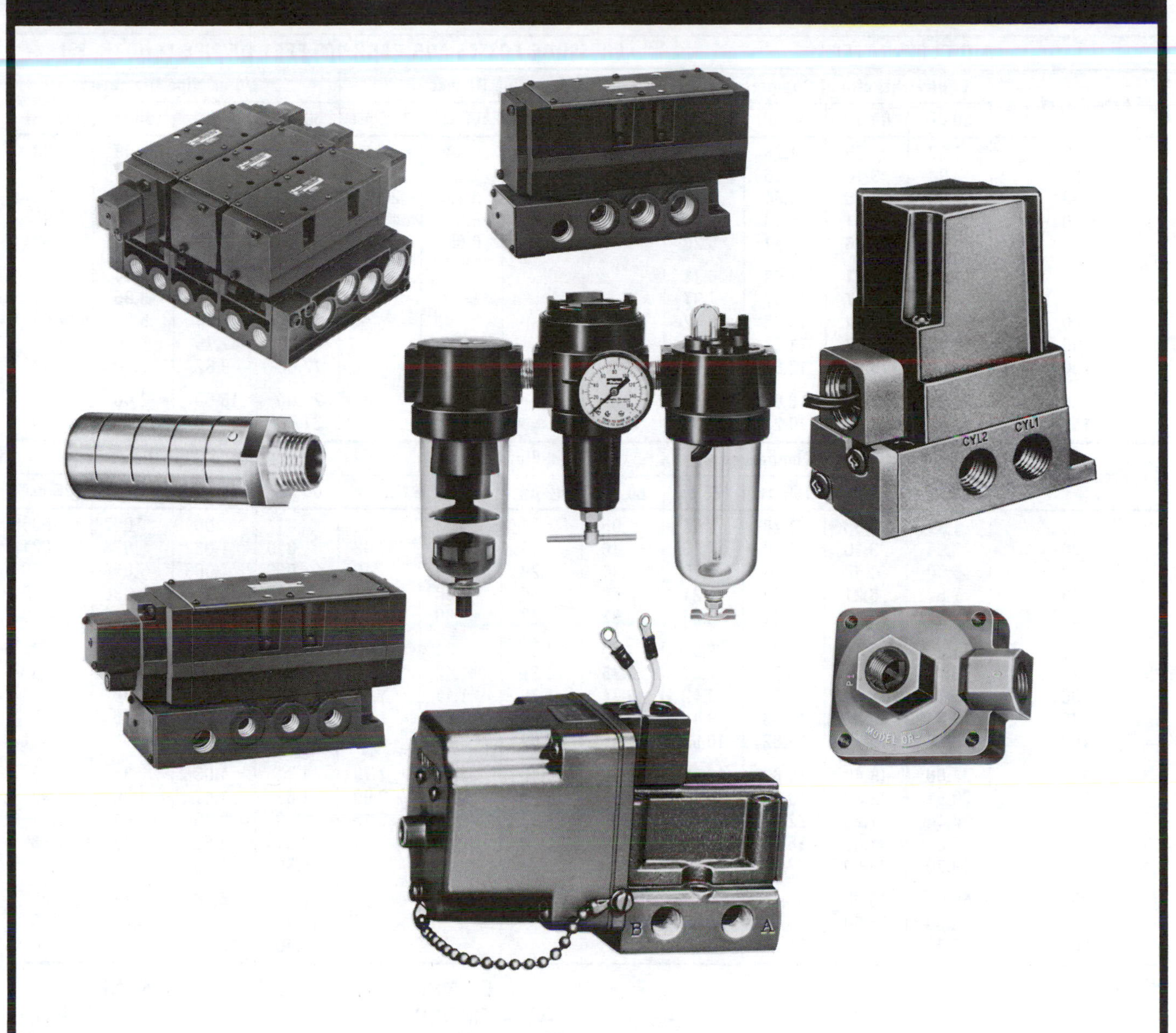

tables & graphs

PRESSURE LOSS DUE TO AIR FRICTION IN PIPES

For Various Flow Rates, for 1/2-in., 3/4-in., 1-in., 1-1/4-in. Nominal Pipe Diameters, at Initial Pressures of 60, 80, 100, 125 psi

AIR FLOW RATES					PRESSURE LOSSES FOR EACH 100 FEET OF PIPE LENGTH, PSI							
Free Air, cfm	Equivalents cfm of Compressed Air				1/2-in. Pipe Diameter				3/4-in. Pipe Diameter			
	60 psi	80 psi	100 psi	125 psi	60 psi	80 psi	100 psi	125 psi	60 psi	80 psi	100 psi	125 psi
10	1.97	1.55	1.28	1.05	.59	.46	.38	.31	.14	.11	.09	.08
20	3.94	3.10	2.56	2.10	2.23	1.74	1.42	1.17	.53	.41	.34	.28
30	5.90	4.66	3.84	3.16	4.94	3.84	3.13	2.54	1.14	.90	.74	.60
40	7.87	6.21	5.13	4.21	8.90	6.93	5.55	4.53	1.99	1.55	1.28	1.05
50	9.84	7.76	6.41	5.26	14.20	10.70	8.65	7.01	3.08	2.42	2.00	1.62
60	11.81	9.31	7.69	6.31					4.45	3.47	2.84	2.33
70	13.78	10.87	8.97	7.37					6.06	4.73	3.85	3.14
80	15.74	12.42	10.25	8.42					7.96	6.14	5.01	4.08
90	17.71	13.97	11.53	9.47					10.00	7.75	6.40	5.17
100	19.68	15.50	12.82	10.52					12.60	9.62	7.80	6.33
125	24.60	19.40	16.02	13.15					21.00	15.50	12.40	9.80
150	29.51	23.30	19.22	15.78					31.50	23.00	18.10	14.40
Free Air, cfm	Equivalents cfm of Compressed Air				1-in. Pipe Diameter				1-1/4-in. Pipe Diameter			
	60 psi	80 psi	100 psi	125 psi	60 psi	80 psi	100 psi	125 psi	60 psi	80 psi	100 psi	125 psi
10	1.97	1.55	1.28	1.05	.05	.04	.03	.02	.011	.0086	.0071	.0058
20	3.94	3.10	2.56	2.10	.16	.13	.10	.08	.040	.032	.026	.021
30	5.90	4.66	3.84	3.16	.34	.28	.23	.19	.086	.068	.056	.046
40	7.87	6.21	5.13	4.21	.59	.46	.38	.31	.146	.116	.096	.079
50	9.84	7.76	6.41	5.26	.92	.73	.60	.49	.22	.18	.146	.120
60	11.81	9.31	7.69	6.31	1.30	1.02	.84	.69	.32	.25	.21	.17
70	13.78	10.87	8.97	7.37	1.75	1.36	1.12	.92	.42	.34	.28	.23
80	15.74	12.42	10.25	8.42	2.24	1.76	1.44	1.18	.55	.44	.36	.30
90	17.71	13.97	11.53	9.47	2.88	2.23	1.85	1.49	.69	.55	.45	.37
100	19.68	15.50	12.82	10.52	3.45	2.69	2.21	1.81	.84	.66	.55	.45
125	24.60	19.40	16.02	13.15	5.38	4.18	3.41	2.79	1.31	1.03	.85	.69
150	29.51	23.30	19.22	15.78	7.81	5.75	4.91	3.99	1.87	1.47	1.20	.99
175	34.44	27.20	22.43	18.41	10.80	8.10	6.80	5.45	2.58	2.00	1.64	1.32
200	39.36	31.00	25.63	21.05	14.50	10.90	8.79	7.11	3.31	2.58	2.12	1.73
250	49.20	38.80	32.04	26.31					5.30	4.05	3.30	2.67
300	59.00	46.60	38.45	31.57					7.51	5.78	4.71	3.83
350	68.90	45.30	44.86	36.83					10.30	7.90	6.45	5.15
400	78.70	62.10	51.26	42.09					13.70	10.30	8.30	6.74

table g-1

W = Air Density (lb/cu. ft.)
P″ = Absolute Pressure (in. Hg)
LFM = Velocity (Linear ft./min.)
CFM = Air Flow (cu. ft./min.)

DENSITY AND VISCOSITY OF AIR AT ATMOSPHERIC PRESSURE (29.92 IN. HG.)

Temperature C.	Temperature F.	Specific Weight lb./cu. ft.	Mass Density slugs/cu.ft.	Viscosity 10^{-6} Poise	Viscosity 10^{-9} lb.-sec./sq. ft.	Kinematic Viscosity Reciprocal sec./sq. ft.
−20	−4	0.0873	0.00272	160.7	336	8100
−10	14	0.0838	0.00261	165.8	347	7520
0	32	0.0807	0.00251	170.9	357	7030
10	50	0.0779	0.00242	175.9	368	6580
15	59	0.0765	0.00238	178.3	373	6380
20	68	0.0753	0.00234	180.8	378	6190
30	86	0.0728	0.00226	185.6	388	5970
40	104	0.0704	0.00219	190.4	398	5500
50	122	0.0682	0.00212	195	408	5200
60	140	0.0661	0.00205	200	418	4910

The effect of pressure and temperature on air density is given by the equation:

$$w = \frac{1.325p''}{460 + {}^\circ F.} \text{ lb./cu. ft.}$$

table g-3

tables & graphs continued

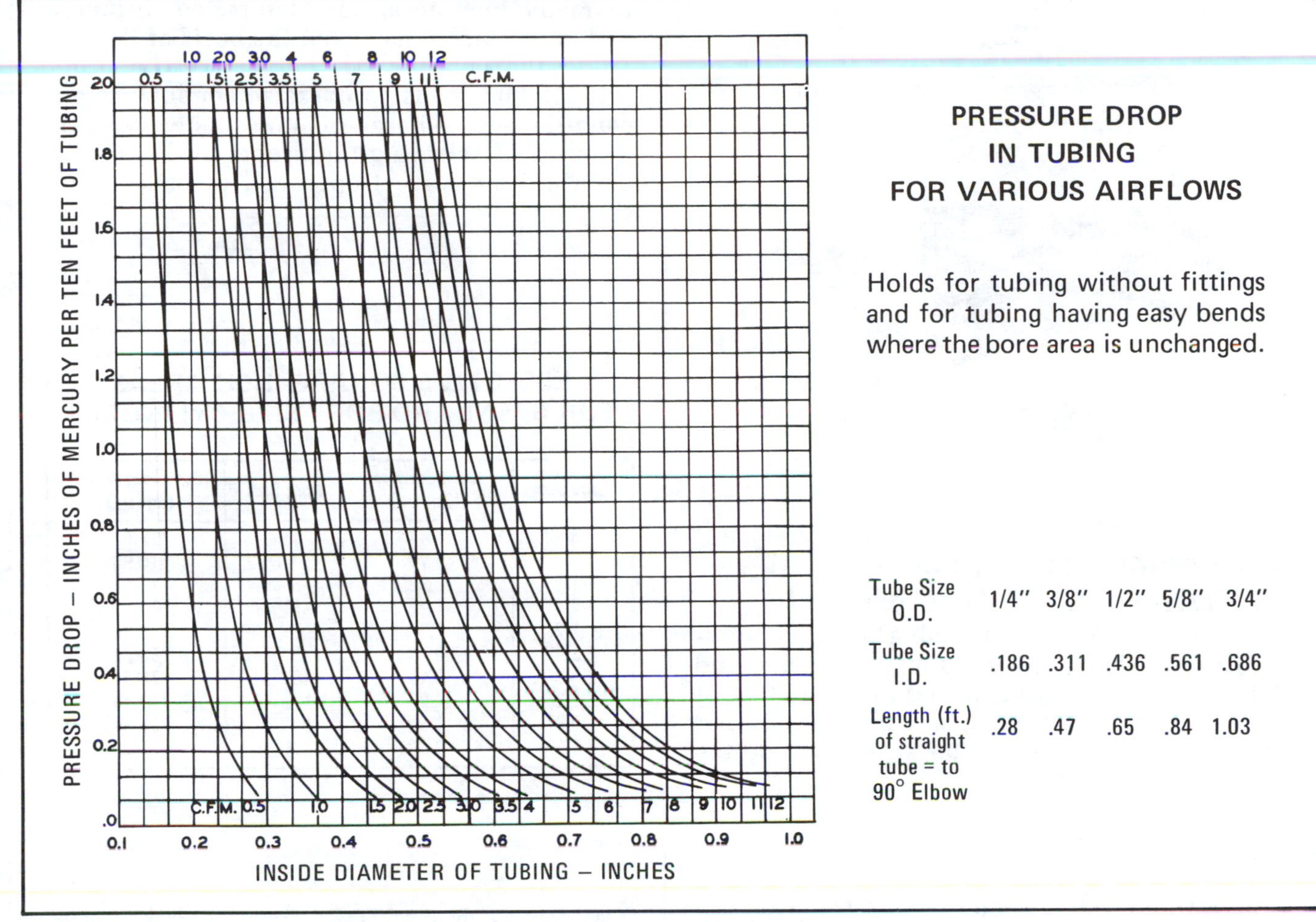

PRESSURE DROP IN TUBING FOR VARIOUS AIRFLOWS

Holds for tubing without fittings and for tubing having easy bends where the bore area is unchanged.

Tube Size O.D.	1/4″	3/8″	1/2″	5/8″	3/4″
Tube Size I.D.	.186	.311	.436	.561	.686
Length (ft.) of straight tube = to 90° Elbow	.28	.47	.65	.84	1.03

graph g-1

CUBIC FEET PER MIN. LINEAR FEET PER MIN. WITH AIR FLOWING THROUGH TUBES

CFM.	Inside Diameter of Tube						
	1/2″	3/4″	1″	1-1/4″	1-1/2″	1-3/4″	2″
5	3660	1630	915	586	403	297	229
10	7320	3260	1830	1172	807	595	458
15		4900	2745	1760	1210	892	686
20		6520	3660	2345	1614	1190	916
25		8150	4575	2930	2017	1487	1144
30		9780	5490	3520	2420	1780	1372
40			7320	4690	3228	2380	1832
50			9150	5860	4034	2974	2288
60				7040	4840	3560	2744
70				8220	5650	4160	3210
80		LFM.		9380	6456	4760	3664
90					7270	5354	4120
100					8080	5958	4576

Formula: $LFM = \frac{576\ CFM}{\pi D^2}$

table g-4

PH Series "SSA" "plug-in" Valves

illustration g-1

air valve selection

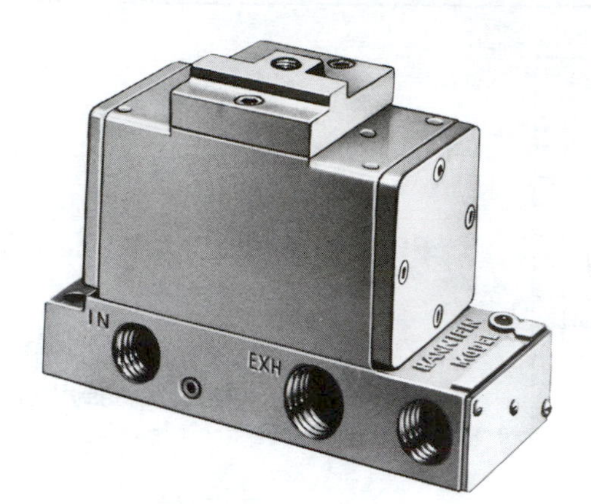

PH Hustler 4 Way Air Valve - Air Operated

illustration g-2

save money and space by sizing your valves properly

For each valve, you can "plug" your requirements into the following simple formula, and determine the Cv needed to do the job. By not oversizing, you'll save space and money, and you'll ensure the valve you select will do the job.

Converting the Job Requirements Into Cv (Capacity Co-efficient).

$$C_v = \frac{\text{Cylinder Area (Sq. In.) (See table g-5)} \times \text{Cylinder Stroke (In.)} \times \text{"A" (See table g-6)} \times \text{Compression Factor (See table g-6)}}{\text{Stroke Time (sec)} \times 29}$$

Let's work through an example—

We want to extend a 3-1/4" bore cylinder which has a 12" stroke in one second, and we have a supply pressure of 80 p.s.i. to do the work. Here's what we know:

Cylinder area for a 3-1/4" bore,
Table g-5 .... 8.30 sq. in.
Cylinder stroke .... 12 in.
Stroke time required in seconds ...... 1 sec.
"A" constant for 80 psi, from
Table g-6 .... .048
Compression factor at 80 psi, from
Table g-6 .... 6.4

In many instances, conductors, conditioners, and valves are selected on the basis of the port size of the actuator to be controlled. Generally, this method leads to oversize conductors and conditioners. There remains, however, the problem of selecting the proper size valves to maintain the correct operating pressure at the actuator.

EFFECTIVE SQUARE-INCH AREAS FOR STANDARD-BORE-SIZE CYLINDERS

Bore Size	Cylinder Area (Sq. In.)	Bore Size	Cylinder Area (Sq. In.)
3/4"	.44	4"	12.57
1"	.79	4-1/2"	15.90
1-1/8"	.99	5"	19.64
1-1/4"	1.23	6"	28.27
1-1/2"	1.77	7"	38.48
1-3/4"	2.41	8"	50.27
2"	3.14	10"	78.54
2-1/2"	4.91	12"	113.10
3-1/4"	8.30	14"	153.94
3-5/8"	10.32		

table g-5

COMPRESSION FACTORS & "A" CONSTANTS

Inlet Pressure (psig)	Compression Factor	"A" Constants For Various Pressure Drop 2 psi ΔP	5 psi ΔP	10 psi ΔP
10	1.6	.155	.102	
20	2.3	.129	.083	.066
30	3.0	.113	.072	.055
40	3.7	.097	.064	.048
50	4.4	.091	.059	.043
60	5.1	.084	.054	.040
70	5.7	.079	.050	.037
80	6.4	.075	.048	.035
90	7.1	.071	.045	.033
100	7.8	.068	.043	.031
110	8.5	.065	.041	.030
120	9.2	.062	.039	.029
130	9.9	.060	.038	.028
140	10.6	.058	.037	.027
150	11.2	.056	.036	.026
160	11.9	.055	.035	.026
170	12.6	.054	.034	.025
180	13.3	.052	.033	.024
190	14.0	.051	.032	.024
200	14.7	.050	.031	.023

NOTE: Use "A" constant at 5 psi ΔP for most applications. On very critical applications, use "A" at 2 psi ΔP. You will find in many cases a 10 psi ΔP is not detrimental, and can save money and mounting space.

table g-6

air valve selection continued

Substituting in the formula, we have:

$$C_v = \frac{8.30 \times 12 \times .048 \times 6.4}{1 \times 29} = 1.05$$

Any valve, therefore, which has a CV of at least 1.05, will extend our cylinder the specified distance in the required time or faster.

SERIES 4510

illustration g-3

4-way, 2-position, 5-ported

choosing the valve design

Your next step is to choose a basic valve design to do the job. For a quick guide to valve designs, see **table g-7**.

CHARACTERISTICS OF THE MAJOR VALVE DESIGNS

A. Piston Poppet 3-Way and 4-Way	1. High flow capacities 2. Minimum lubrication requirements 3. Fast response 4. Self-cleaning poppet seats 5. Pressures of 15 to 150 psig (modifications for vacuum to 250 psig)
B. Spool 3-Way and 4-Way	1. Medium flow capacities 2. Short stroke and fast response 3. Inexpensive 4. Wide range of flow-path configurations, actuators, accessories 5. 4 and 5 position 6. 2 and 3 position
C. Metal-To-Metal ("LAPPED") SPOOL 4-Way	1. Medium flow capacities 2. Susceptible to dirt and varnish build-up causing sticking and malfunction.
D. Rotary Or Reciprocating Disc 4-Way, Manually Operated	1. Inexpensive 2. Versatility in manual actuation
E. Packed Bore 4-Way	1. High flow 2. Wide range of flow-paths 3. 2 and 3 position 4. Pressure to 150 psi and with vacuum.

table g-7

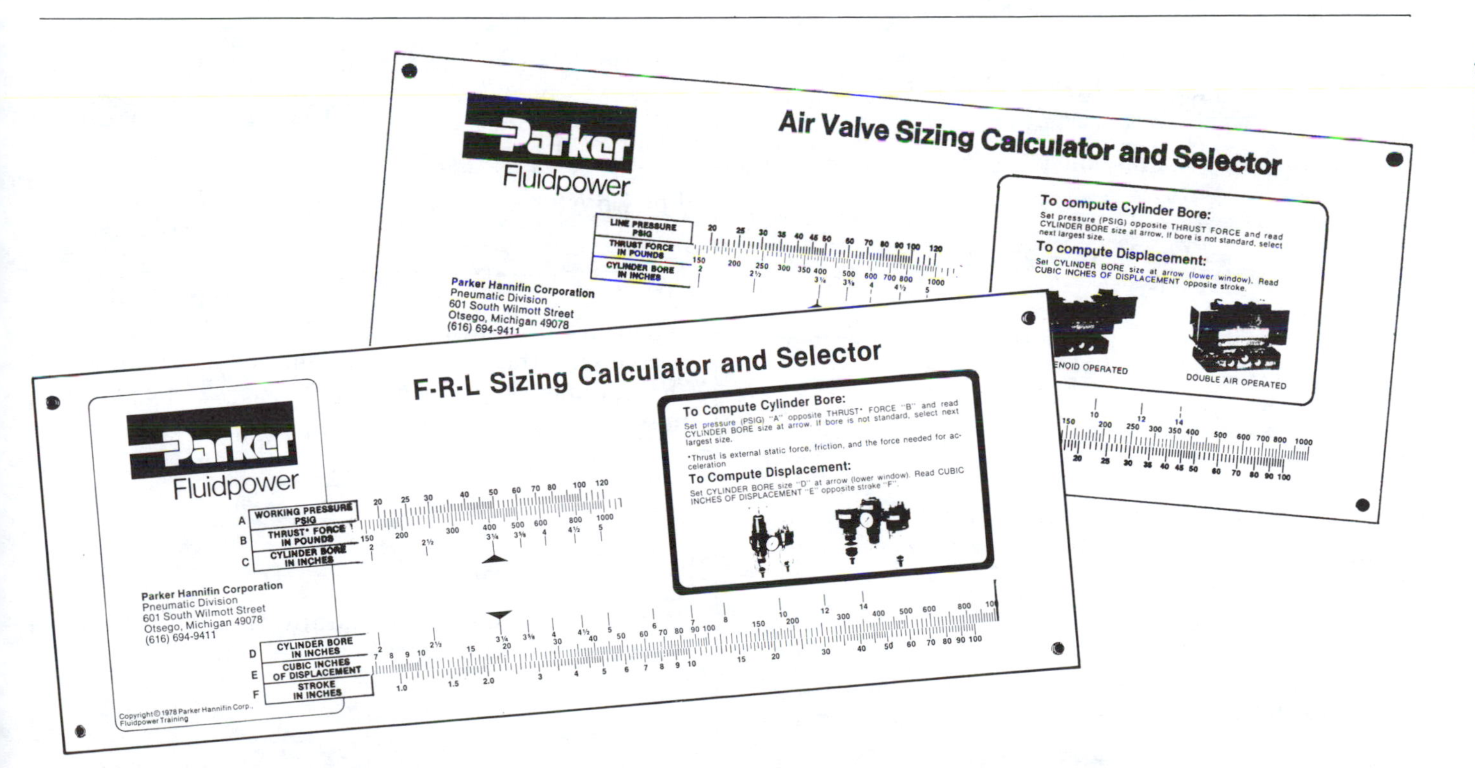

These new Parker Air Valve and F-R-L Sizing Calculators simplify selection.These design aids will select the correct valve or F-R-L to your system requirements. This means a correct and economical initial investment, with known efficiency through the entire life of your air control system. If you are interested in either of these items, contact your Parker Distributor or the Fluid Power Training Department, 17325 Euclid Ave., Cleveland, Ohio 44112.

air formula

V = Volume of free air in cu. ft.
V_c = Volume of compressed air in cu. ft.
V_v = Volume of expanded air in cu. ft.
V_T = Volume of tank in cu. ft.
V_P = Pump capacity in cu. ft. per minute
V_R = Cu. ft. per pump rev. (free air)
P = Gauge pressure in lbs. per sq. in.
P_A = Absolute pressure in lbs. per sq. in.
P_T = Final absolute press. in tank. (inches mercury)
I_M = Inches of mercury, gauge
I_W = Inches of water, gauge
F_w = Feet of water, gauge
F = Linear velocity in ft. per minute
F_S = Linear velocity in ft. per second
a = Area in sq. inches

A = Area in sq. feet
D = Diameter in inches
L = Length in feet
T = Time in minutes
T_A = Absolute Temperature Fahrenheit = 460 + Fahr. Temp.
C = Coefficient of orifice
RPM = Revolutions of pump
π = 3.1416
B = Barometric press. in inches mercury (absolute pressure)
t = Temperature Fahrenheit
W = Weight in pounds
V_{PV} = Pump capacity in cfm. expanded air
CFM = Cubic feet per minute

formula

To find the absolute pressure when the gauge pressure or the inches of vacuum are known, the following formula can be used (at sea level):

$$P_A = 14.7 + P = \frac{14.7\,(29.9 + I_M)}{29.9}$$

To convert inches of mercury to p.s.i. or vice versa:

$$P = \frac{I_M}{2.036} \qquad I_M = 2.036\,P \text{ (at } 32^\circ\text{F)}$$

To convert inches of water to p.s.i. or vice versa:

$$P = \frac{I_W}{27.686} \qquad I_W = 27.686\,P$$

To convert feet of water to p.s.i. or vice versa:

$$P = \frac{F_W}{2.307} \qquad F_W = 2.307\,P$$

To convert feet of water to inches of mercury or vice versa:

$$F_W = \frac{I_M}{.88} \qquad I_M = .88\,F_W$$

To find the volume of a given amount of air when it is under pressure (when atmospheric pressure is 14.7 lbs.):

$$V_c = V\frac{14.7}{P_A} = V\frac{14.7}{14.7 + P}$$

To find the volume of a given amount of air when it is under vacuum (when atmospheric pressure is 14.7 lbs.):

$$V_v = V\frac{14.7}{P_A}$$

To find the velocity in feet per minute of air flowing through a pipe when the volume is known (friction not considered):

$$F = \frac{36CFM}{\pi D^2}$$

For Boyle's law the formula is

$$P_A \times V = \text{constant}$$

For Charles' law the formula is:

$$\frac{P_A \times V}{T_A} = \text{constant}$$

When air flows through an orifice and the diameter is constant, the pressure will vary as the square of the folume: (Upstream conditions define as 1, downstream define as 2).

$$\frac{P_1}{P_2} = \frac{V_1^2}{V_1^2}$$

When the pressure is constant, the volume will vary as the square of the diameter:

$$\frac{V_1}{V_2} = \frac{D_1^2}{D_2^2}$$

When the volume is constant, the pressure will vary as the fourth power of the diameter:

$$\frac{P_1}{P_2} = \frac{D_1^4}{D_2^4}$$

flow of air through orifices

The flow of air through orifices depends upon the air pressure, the orifice diameter and the coefficient of the orifice shape. Many complicated formulae have been devised for calculating the flow and one of them when simplified is approximately as follows:

$$\text{CFM.} = 22 \times c\, d^2 \sqrt{P} = 1096\, c\, A \sqrt{\frac{I_w}{W}}$$

Where: the ambient temperature = 72°F

- c = Coefficient of orifice
- d = Diameter in inches
- P = Pressure in inches water
- A = Area in square feet
- I_w = Inches water gauge
- W = Weight of air in lbs./cu. ft.

Thus it will be seen that the volume varies directly as the coefficient of the orifice, the square of the diameter, the square root of the pressure and the square root of the reciprocal of the weight.

Since air has a weight of .0766 and hydrogen .0053 lbs. per cu. ft. it can be seen from the above formula that the light hydrogen will flow about four times as fast as air.

Table g-8 below gives approximate figures for the flow of air through holes having a coefficient of .60.

By noting the coefficients of the orifices it is evident that the shape of the orifice has a great effect upon the flow of air through the orifice.

Note that the orifice with the rounded entrance and having a coefficient of .98 will pass nearly twice as much air per minute as the orifice with the sharp entrance which has a coefficient of .53.

When dealing with a number of small orifices, the volume of air that will pass through them per minute is the same as will pass through one large orifice having an area equal to the sum of the areas of the small orifices.

FLOW OF AIR THROUGH ORIFICES UNDER PRESSURE

ORIFICE DIA.			PRESSURE DROP-IN (POUNDS PER SQUARE INCH)												
			1/4	1/2	1	2	3	4	5	6	8	10	15	20	25
1/32			.037	.053	.077	.110	.14	.17	.19	.21	.26	.31	.42	.48	.54
	1/16		.150	.210	.300	.430	.55	.65	.74	.84	1.00	1.20	1.70	1.90	2.20
3/32			.330	.470	.670	.980	1.20	1.50	1.70	1.90	2.30	2.70	3.70	4.30	4.80
		1/8	.590	.840	1.200	1.800	2.20	2.50	3.00	3.30	4.00	4.70	6.60	7.70	8.60
5/32			.930	1.300	1.900	2.800	3.40	4.10	4.80	5.30	6.40	7.50	10.00	12.00	13.00
	3/16		1.900	2.400	3.100	4.200	5.10	6.00	7.00	7.50	8.80	11.00	15.00	17.00	19.00
7/32			2.100	2.900	3.700	5.300	6.80	8.00	9.20	10.50	12.00	14.00	20.00	24.00	26.00
		1/4	2.700	3.700	4.500	6.600	8.70	10.00	12.00	13.00	16.00	18.00	27.00	31.00	34.00
9/32			3.500	5.000	7.000	9.600	12.00	14.00	17.00	18.00	20.00	24.00	38.00	44.00	49.00
	5/16		4.000	5.700	7.500	11.000	14.00	17.00	19.00	21.00	25.00	30.00	42.00	47.00	53.00
11/32			4.400	6.300	8.800	13.000	17.00	21.00	24.00	26.00	31.00	36.00	50.00	57.00	64.00
		3/8	5.600	8.000	11.000	17.000	21.00	25.00	29.00	32.00	38.00	47.00	60.00	69.00	77.00
13/32			6.800	9.500	14.000	20.000	25.00	30.00	35.00	37.00	45.00	53.00	70.00	80.00	90.00
	7/16		7.800	11.000	16.000	23.000	29.00	34.00	38.00	42.00	52.00	60.00	82.00	95.00	105.00
15/32			9.500	13.000	18.000	26.000	33.00	40.00	48.00	54.00	66.00	73.00	95.00	109.00	
		1/2	10.000	14.000	20.000	29.000	37.00	44.00	50.00	57.00	71.00	81.00	107.00	123.00	
17/32			11.000	16.000	24.000	34.000	43.00	51.00	58.00	66.00	79.00	92.00	121.00		
	9/16		13.000	18.000	28.000	39.000	49.00	58.00	67.00	75.00	90.00	105.00			
19/32			14.000	21.000	31.000	44.000	55.00	65.00	75.00	84.00	101.00	118.00			
		5/8	16.000	23.000	35.000	49.000	61.00	73.00	83.00	94.00	112.00	132.00			
	11/16		20.000	28.000	41.000	59.000	74.00	89.00	102.00	113.00	138.00	162.00			
		3/4	24.000	34.000	49.000	72.000	90.00	107.00	123.00						
	13/16		29.000	41.000	59.000	86.000	107.00								
		7/8	34.000	48.000	69.000										
		1"	44.000												

CUBIC FEET OF FREE AIR PER MINUTE

Coefficient = .60

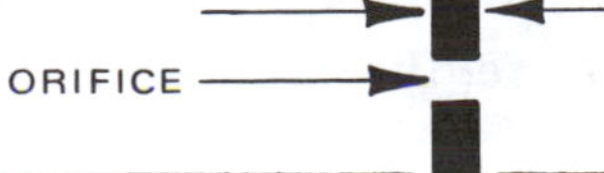

flow of air through pipes

Large Flow or Remote Controlled Regulator

illustration g-4

When air flows through pipes the flow is retarded by the friction of the walls of the pipe. This causes the air flow to slow down near the pipe walls while the velocity is high at the center of the pipe. The result is pressure loss. This loss is proportionately greater when the air pressure is low than when it is high. The loss is much less if the pipe diameter is increased.

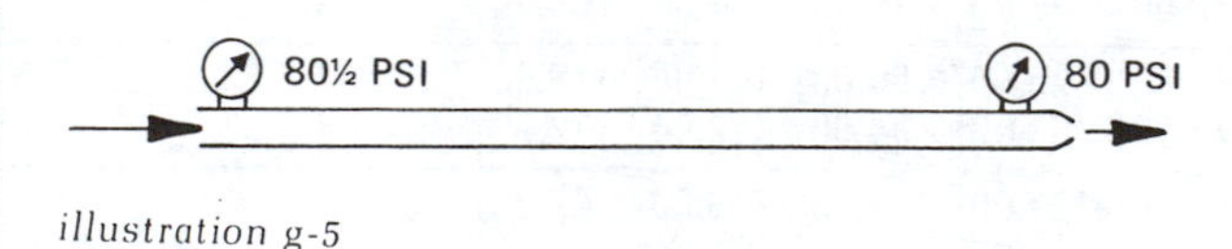

illustration g-5

Referring to the figure above it will be noted that the entrance pressure is 80-1/2 p.s.i. and the exit pressure is 80 p.s.i. In this case the pressure loss due to friction is 1/2 p.s.i., that is, 1/2 p.s.i. is required to overcome the pipe friction.

If an elbow is placed in the line it will cause a loss as much as 5 to 10 feet of pipe would cause. The fractional loss in an elbow depends upon the radius and the following table shows the relative losses:

R/d	COMPARATIVE LOSSES
.5	.83
1.	.28
2.	.21

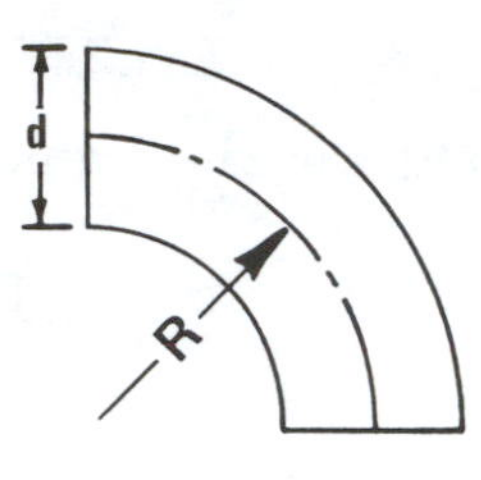

illustration g-6

Thus it will be seen that it is preferable to use large radius elbows to cut down friction loss.

The frictional loss in a tee or a branch is even greater than in an elbow as is seen in the following table:

ANGLE OF CONNECTION	COMPARATIVE LOSSES
15 degrees	.1
45 degrees	1.0
60 degrees	1.7
90 degrees	3.4

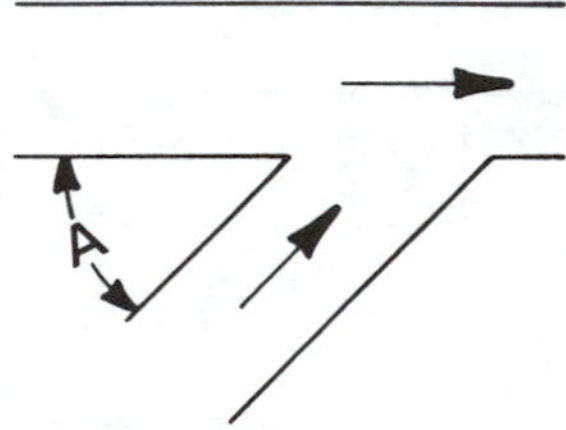

illustration g-7

Thus it will be seen that the friction loss in a 90 degree tee is great and it is therefore preferable to use Y branches whenever possible.

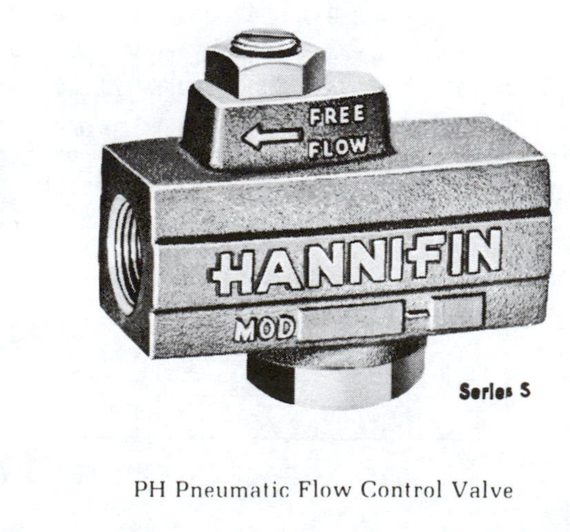

PH Pneumatic Flow Control Valve

illustration g-8

flow of gases through an orifice

The following formula applies to other gases as it takes into consideration the weight of the gas:

$$V = 1096.5\, c\, A\sqrt{\frac{I_w}{W}}$$

For heating air which is lighter, the following formula may be used:

$$W = \frac{B}{346.5 + .7535\, t}$$

Where:

V = Cu. ft. of air per minute under atmospheric pressure
c = Coefficient of orifice (see table g-10)
A = Area of orifice in sq. ft.
I_w = Pressure of air in inches of water
W = Weight of one cu. ft. of air (at temp.)
B = Barometric pressure in inches Hg.
t = Fahrenheit degrees

GAS	WEIGHT PER CU. FT.
Dry air at 72°F	.0746
Dry air at 212°F	.059
Dry air at 500°F	.042
Carbon Dioxide	.113
Hydrogen	.0052
Nitrogen	.072
Oxygen	.0827

table g-9

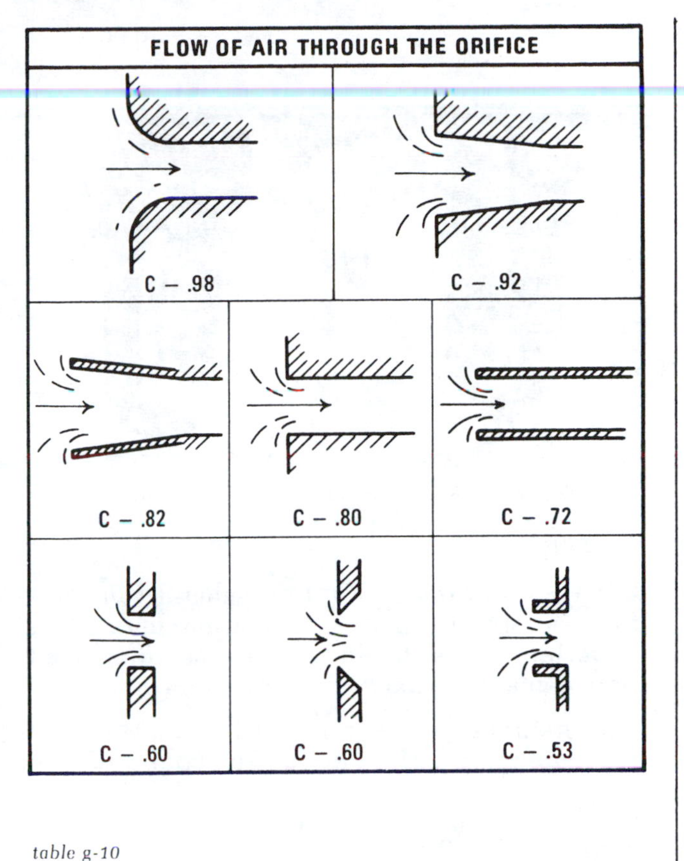

table g-10

flow meter

This instrument, shown below, is used to measure the flow in pipes in which the air is under pressure. It measures the pressure drop across the accurately drilled orifice "H" which has known coefficient. The pressure ahead of the orifice is picked up at "A" and the pressure beyond the orifice is picked up at "B" and the U tube indicates the difference in the pressures or the pressure drop across the orifice.

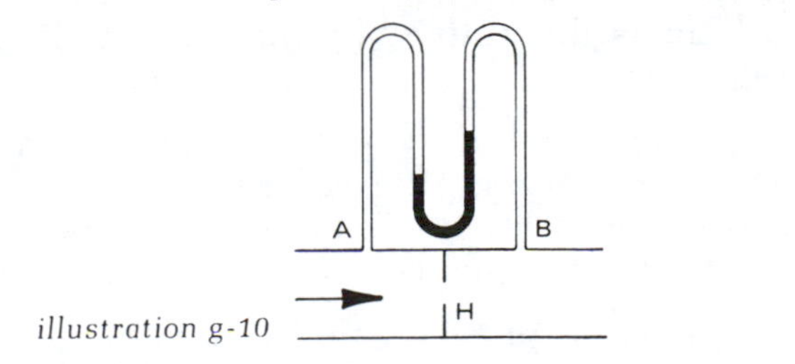

illustration g-10

The cu. ft. per minute is found from the formula:

$$CFM. = \frac{c}{60}\sqrt{I \times P}$$

Where:

c = Coefficient of orifice (table g-10)
I = Gauge reading in inches water
P = Absolute pressure in PSIA (taken upstream)
CFM. = Cubic feet of free air per minute

application

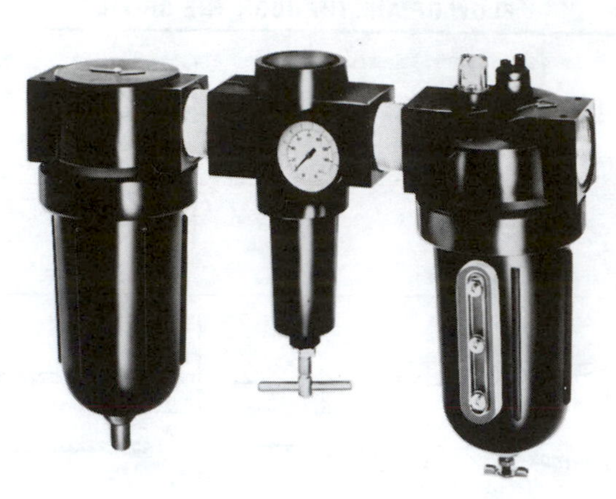

PH Filter-Regulator-Lubricator Combination

illustration g-11

Filter Application — Certain general considerations apply to all air-preparation units. Most important, units should be sized according to airflow demands — *not* according to pipe size.

For instance, an oversized filter may not impart enough swirling action to separate all the droplets. Conversely, an undersized filter creates too much turbulence in the air being processed. This turbulence prevents the liquids from settling into the "quiet zone." Undersized units may also add excess pressure loss to the system. Flow capacities for a given port size also vary among different manufacturers because of variations in internal construction.

Air-drop lines to conditioning units should be attached to the top of the main header to decrease the amount of liquids that might run down to equipment, **illustration g-19.** A separate drain should be provided for those liquids which do accumulate at low points in the main-header line.

Conditioning units should be installed as close as possible to the point of use, consistent with accessibility for inspection and possible servicing, **illustration g-19.** Bends in the air lines should be avoided since they cause additional pressure drop in the system. In addition, momentum forces tend to deposit lubricant at bends. Necessary turns should have as large a radius as possible.

Filters should be installed upstream from, and as close as possible to, the devices to be protected. Filters remove only droplets and solid contaminants; water vapor is not affected. Thus, vapor can condense between the filter and the components being served. To minimize this possibility, filters should be located as close as possible to the point of use.

The filter bowl must hang vertically so that free moisture and oil droplets cling to the inner surface of the bowl as they work their way down to the quiet zone at the bottom of the bowl.

Filters should be located so that manually drained bowls can be observed and drained when necessary. Drain lines should be provided for filters that incorporate automatic drain devices.

Normally, filters have no moving parts to service or adjust. However, the filter element should be replaced when the pressure differential across the filter unit exceeds 10 p.s.i. While a manual drain requires regular attention, it is necessary to drain the collected liquids and solids only when their level approaches the lower baffle.

Regulator Application — A regulator should be located for convenient servicing.

The pressure setting should be adjusted to the value required by typical operating conditions. The "no flow" pressure setting differs from the secondary pressure under flow conditions. A slight decrease in secondary pressure with increasing flow through the unit is normal.

Regulator flow characteristics are affected by the spring rate used to cover the desired pressure range. Generally, a standard unit is rated for 250 p.s.i.g. primary supply pressure; an adjustable secondary pressure is rated from 3 to 125 p.s.i.g. Low-pressure (1-1/2 to 60 p.s.i.g.) and high-pressure (5 to 250 p.s.i.g.) units are also available.

If an application requires 30 p.s.i.g., then the standard unit will probably be satisfactory. If optimum performance is required, the low-pressure unit can be used, since its lower rate spring gives better performance characteristics.

The standard, adjustable secondary pressure range is satisfactory for most industrial pneumatic applications.

All secondary pressure requirements are covered by one pilot-controlled regulator, since the secondary pressure range of a pilot-controlled regulator is determined by the range of its pilot regulator.

If the pressure and flow requirements of a system vary widely, a pilot-controlled regulator with a high-pressure pilot regulator provides the best performance.

If the decrease in pressure under flow conditions seems excessively high for a particular

application continued

application, the supply lines, the regulator unit, or other system components may be undersized for the application. A clogged filter element can also cause an excessive decrease in pressure under flow conditions.

To lower a setting, the regulator should be reset from a pressure below the final secondary pressure desired. For example, to lower the secondary pressure from 80 p.s.i.g. to 60 p.s.i.g., decrease the secondary pressure to 55 p.s.i.g. or less, then adjust it upward to 60 p.s.i.g.

In a nonventing regulator, some secondary air must be exhausted to lower the setting. Many regulators include a constant bleed of secondary air to obtain good pressure-cracking characteristics. A small fixed bleed in the secondary line provides the same effect. This modification is especially important in no-flow conditions, and permits the regulator to provide immediate response.

To determine the required secondary pressure setting, follow the recommendations of the manufacturer of the equipment being served. When recommended pressures are not available, use as low a setting as practical to accomplish the desired performance. Lower pressures minimize wear and conserve air.

PH Filter-Regulator "Piggy Back" and Lubricator Combination

illustration g-12

Pilot-controlled regulators are applied in the same way as standard spring-controlled regulators. The distance between the pilot-controlled regulators and its pilot regulator is not critical.

Regulators do not require routine maintenance in normal service when properly protected by filters. The secondary pressure setting should be checked, however, whenever system requirements change.

lubricators

illustration g-13

The lack of proper lubrication in pneumatic power components creates excessive maintenance costs, production inefficiency and premature failure. Proper lubrication allows components to operate with a minimum of friction and corrosion, and it minimizes wear on the component parts. The most economical way to lubricate pneumatic components is with an air line lubricator. In most cases it is also the most efficient.

The Mist Lubricator is the most commonly used, in-line lubricator. It is the direct flow type. It permits every drop of oil in the sight dome to go downstream into the air line. All the oil observed in the sight dome flows downstream in aerosol form, ranging in size from .01 to 500 micron particles. This lubricator can be filled while the air line is under pressure since it contains a pressurization check valve.

The Micro-Mist Lubricator was developed to fulfill the requirements of the more difficult applications. It is a recirculating flow type lubricator. This means that the oil, once injected into the air, is recirculated into the bowl of the lubricator where the larger oil particles fall out of the air. Only the smaller oil particles that float around in the upper part of the bowl go downstream. The particles going downstream in this type of lubricator range in size from .01 to 2 micron. This lubricator, because of its design, cannot be filled under pressure unless a remote fill button is incorporated into the lubricator.

Oil mist flowing from the two different lubricators will remain suspended over varying distances. Direct flow Mist Lubricators can achieve oil

lubricators continued

particle suspension in the line over a distance of 20 feet, utilizing the greater portion of the oil supply available. By comparison, as shown in Illustration g-14, the recirculating flow Micro-Mist Lubricator can keep the majority of the oil suspended in the piping for approximately one hundred feet. This can be achieved because of the absence of the larger oil particles in the output of the Micro-Mist Lubricator. The larger particles in the Mist Lubricator tend to "eat up" or "coalesce" the smaller particles as that aerosol mixture flows downstream. Because of the absence of these larger particles in the Micro-Mist Lubricator, this "coalescing" or "eating up" affect is not prevalent. Consequently, the oil particles that do come from the lubricator can remain suspended over considerably longer distances of piping.

oil mist in suspension

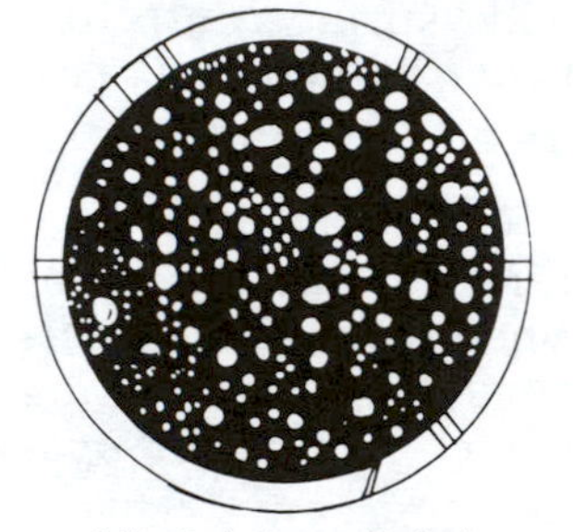

Mist Lubricator Majority Suspended in Piping for 20 Feet

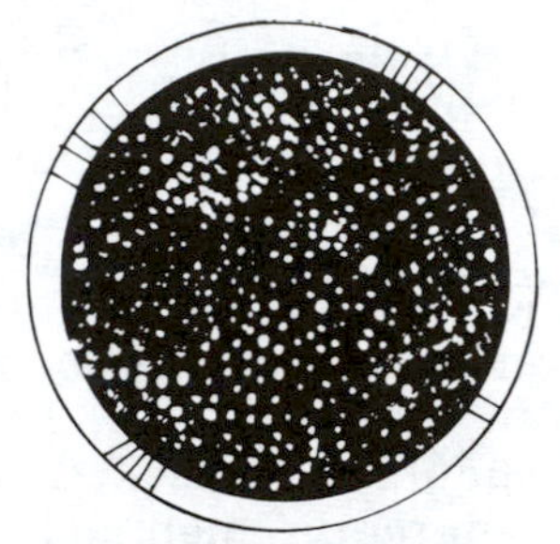

Micro-Mist Lubricator Majority Suspendec in Piping for 100 Feet

illustration g-14

Both lubricators are fitted with an easily adjustable precision needle valve which insures consistent oil delivery. The needle valve and the seal itself are long and tapered to provide a large area through which the oil can flow for a given setting. This design results in a significant reduction in the filtering effect of dirt in the oil between the needle and valve seat to insure that the drip rate will remain constant.

mist lubricator

The Mist Lubricator injects oil into the air. The resulting air-oil mixutre is then carried directly downstream as shown in illustration g-15. Air passes through the vortex generator where the oil drips into the air stream and is atomized. Then the air-oil mixture flows directly into the secondary port cavity and downstream. During very low air flows the air passes through the center of the vortex generator. As the air flow increases, the flexible bypass disc starts to bend, permitting this increasing air flow to bypass the center of the vortex generator. The bypass disc allows the pressure drop to be controlled in the lubricator as the flow changes, thereby providing linear oil delivery. Mist Lubricators start delivering oil when minimum flow of one SCFM is reached. This permits the lubricator to be used for low flow as well as high flow applications. It can be filled while under pressure, simply by hand removing the fill plug. The air in the bowl will automatically exhaust as soon as the "O" ring seal is broken on the fill plug. The lubricator can then be filled with oil and the fill plug replaced and hand tightened. No wrenches or screw drivers are needed. When the pressure in the bowl is exhausted, the pressurization check valve reduces the flow of air to the bowl to a minute induced bleed. It is this bleed that repressurizes the bowl once the fill plug has been replaced.

micro-mist lubricator

The Micro-Mist Lubricator injects oil into the air and then recirculates the resulting air-oil mixture into the bowl of the lubricator, as shown in illustration g-15. It is here that the larger oil particles fall out of the air. Only the smaller oil particles, ranging in size from .01 to 2 micron, go into the air stream.

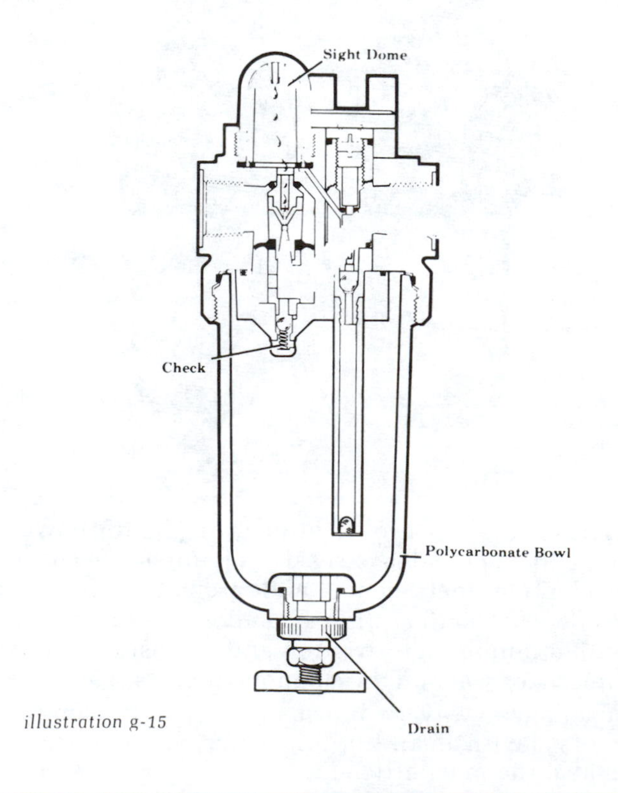

illustration g-15

lubricators continued

This represents approximately 3% of the oil seen dripping in the sight dome. (To get one drop downstream, adjust the lubricator to a drip rate of 33 drops per minute). All of the oil particles above 2 micron in size (which represents approximately 97% of the oil dripping in the sight dome) fall out of the air into the bowl. The vortex generator and bypass disc assembly is the same in both lubricators. Consequently, the Micro-Mist Lubricator will also deliver oil with a minimum of 1 SCFM air flow. The ability to support large reverse flow with no oil delivery in reverse flow conditions is another feature of the Micro-Mist Lubricator. This ability makes the Micro-Mist Lubricator ideal for piping between a valve and a cylinder.

Micro-Mist Lubricators are particularly well suited for the following applications: (1) where instant oil delivery and lubrication are required, whether the air line is wet or dry, (2) in systems where multi-point lubrication is required, such as lines that contain several valves and cylinders, (3) where large oil drops from the air exhaust would contaminate either the environment or the material being processed. The Micro-Mist Lubricator actually permits less oil to be injected into the air line to perform the lubrication job, (4) where the air line includes (after the lubricator) valves, bends and similar baffling points that would tend to separate out heavier particles, (5) where ease of drip rate adjustment is required, (6) for high cycle pneumatic applications, (7) where very low flow rates are necessary, (8) where the oil particles stay in suspension in the air lines for 100 feet or more, (9) where the oil particles must stay in suspension for more than 10 minutes, and (10) where the oil mist becomes a vaporous part of the air and travels as part of the air until it is baffled out by bearings, vanes and turbulence.

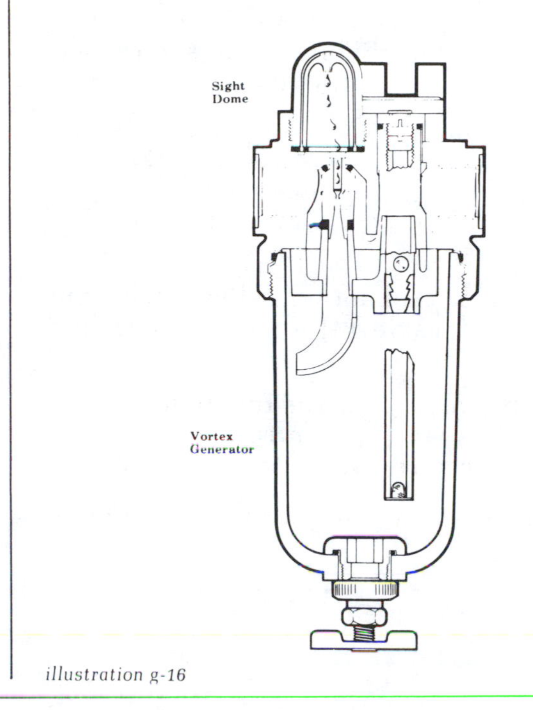

illustration g-16

pneumatic circuitry

application

Provides adjustable decelerating air cushioning for handling high inertia loads at fast stroke speeds.

Valve (1) supplies air to the head end of the Cylinder and simultaneously to the piston operator of valve (5) thru the flow control valve (3). Full exhaust rate from the Cap end of the cylinder permits fast stroke rate until valve (5) is made to operate as pressure builds up in the piston operator. When valve (5) is operated, the cylinder Cap end exhaust is restricted by valve (7). The resulting pressure buildup produces an air cushion that controls the high inertia of the stroke load. On the return stroke, valves (2), (6) and (4) function in a like manner to valves (5), (3) and (7).

* Volumes represent added volume to control air system which should be sufficient to broaden the adjustment range of valves 3 and 6, so that desired settings can be easily accomplished.

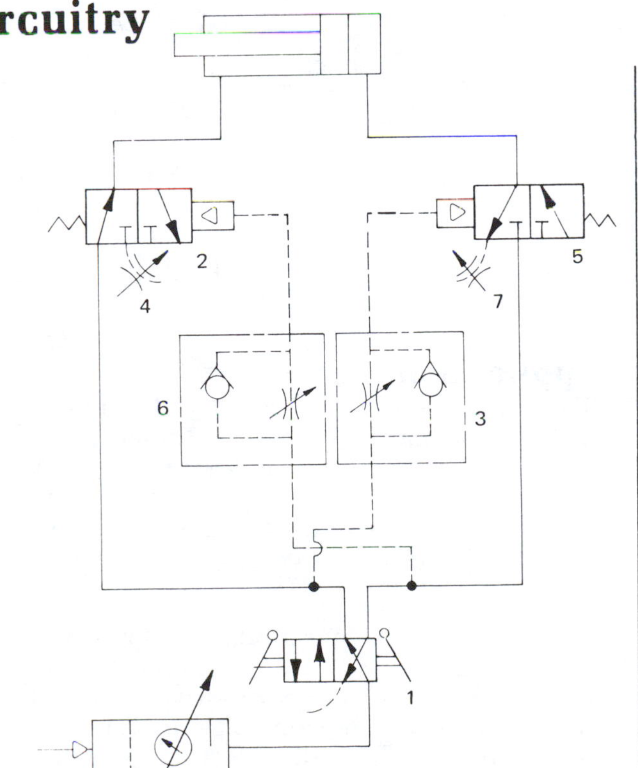

circuit g-1

pneumatic circuitry index continued

application

Remote Control of reversible air motors. Regulated supply air pressure is applied to both air motor ports with valves (3) and (4) in normal position. Balanced pressures does not cause motor rotation to occur.

Operation of pilot valve (1) alone directs pilot pressure to the air operator of valve (3). When operated, valve (3) provides an exhaust path for one port of the air motor and "forward" rotation occurs.

Return of valve (1) to normal relieves the air pilot pressure of valve (3) and air motor rotations stops.

Operation of valve (2) alone functions in a like manner to valve (1) controlling valve (4) which produces "reverse" air motor rotation.

Simultaneous operation of valve (1) and (2) in no rotation.

circuit g-2

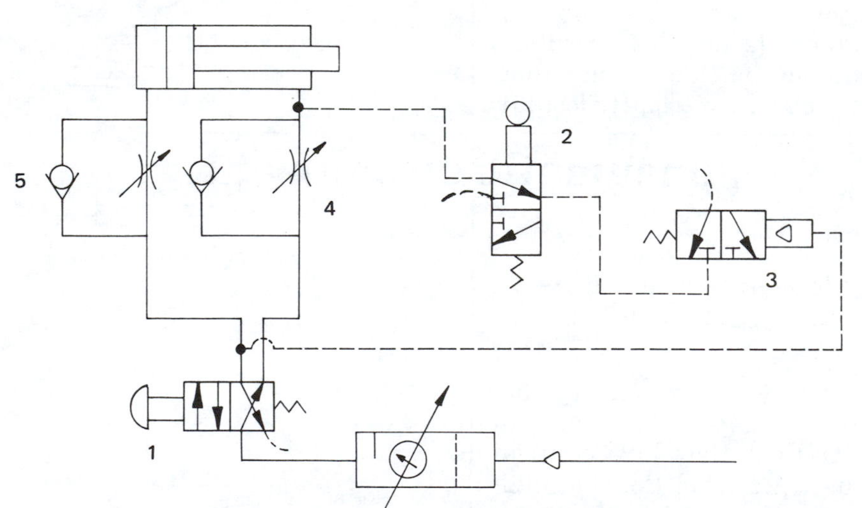

circuit g-3

application

To provide means of rapidly dropping pressure in one end of large air cylinder to produce "no delay" starting when 4-way Control Valve is operated. Problem Occurs most frequently where Flow Control Valves must be set for very low metered flow.

When valve (1) is operated, pressure is directed to the Cap end of the Cylinder and to the piston operator of valve (3). Valve (2) is operated by the piston rod cam. The piston rod cam length or "dwell" will determine the distance of piston rod movement before valve (2) returns to the normally closed position.

The exhaust from the cylinder head end is unrestricted while both valves (2) and (3) are operated and the start of the stroke will be fast. When valve (2) returns to normal position, the speed of the balance of the stroke distance is controlled by valve (4).
Returning valve (1) to normal positon directs pressure to the head end of the cylinder through valve (4). Valve (3) returns to normal position also. The speed of the return stroke of the cylinder is controlled by valve (5).

pneumatic circuitry index continued

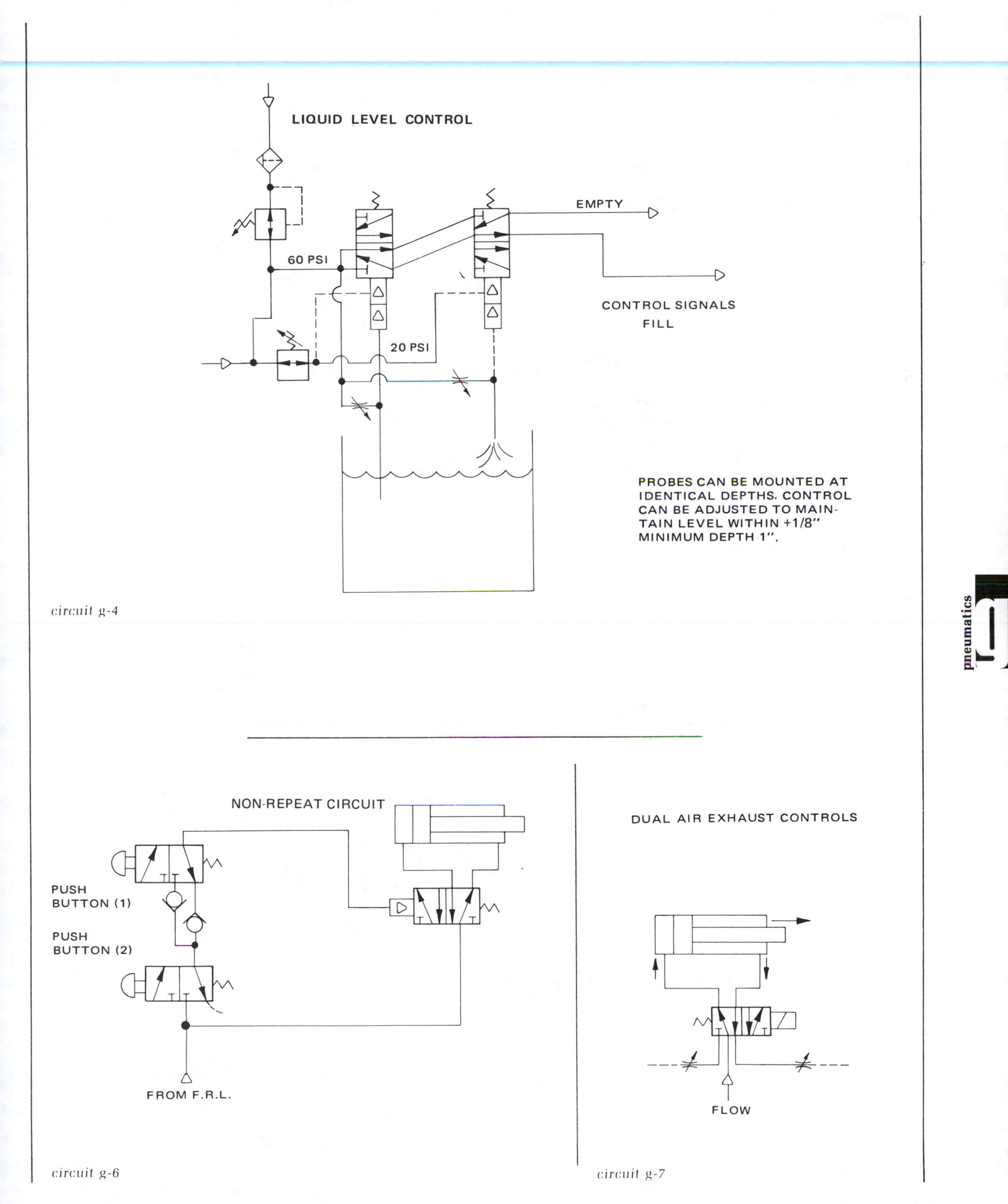

circuit g-4

circuit g-6

circuit g-7

pneumatic circuitry index continued

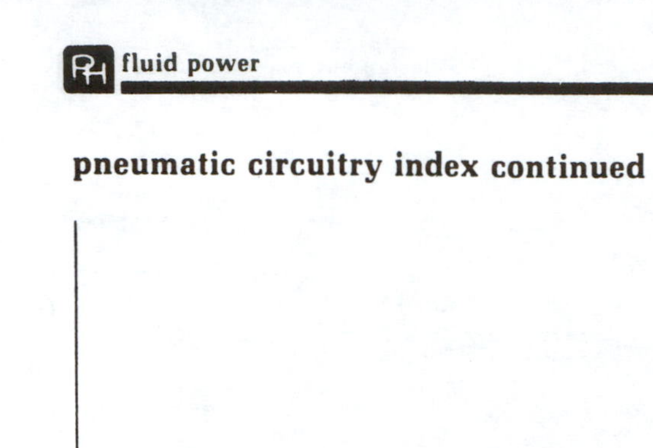

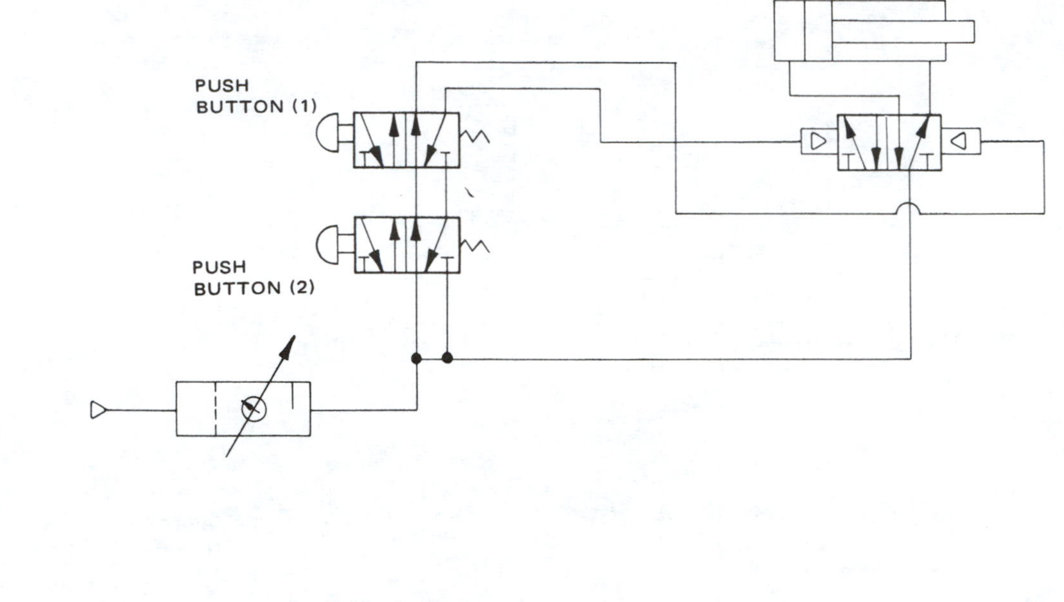

circuit g-8

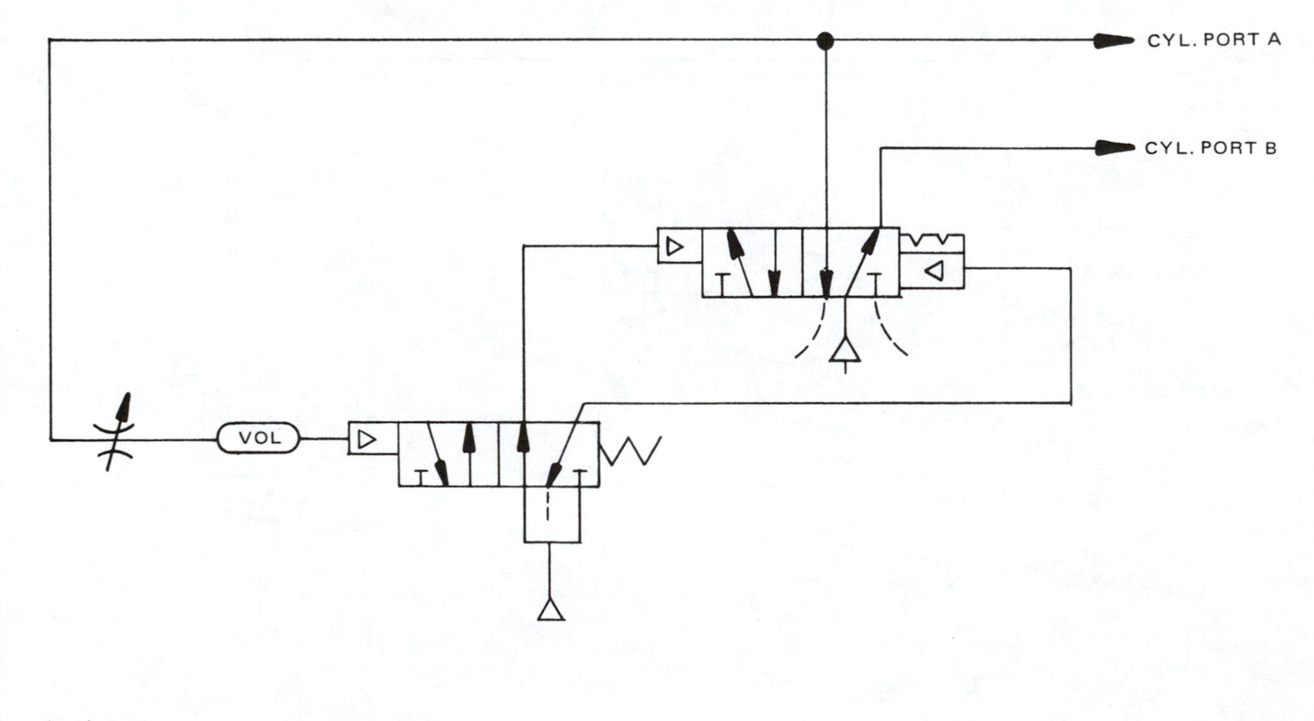

circuit g-9

compressor considerations

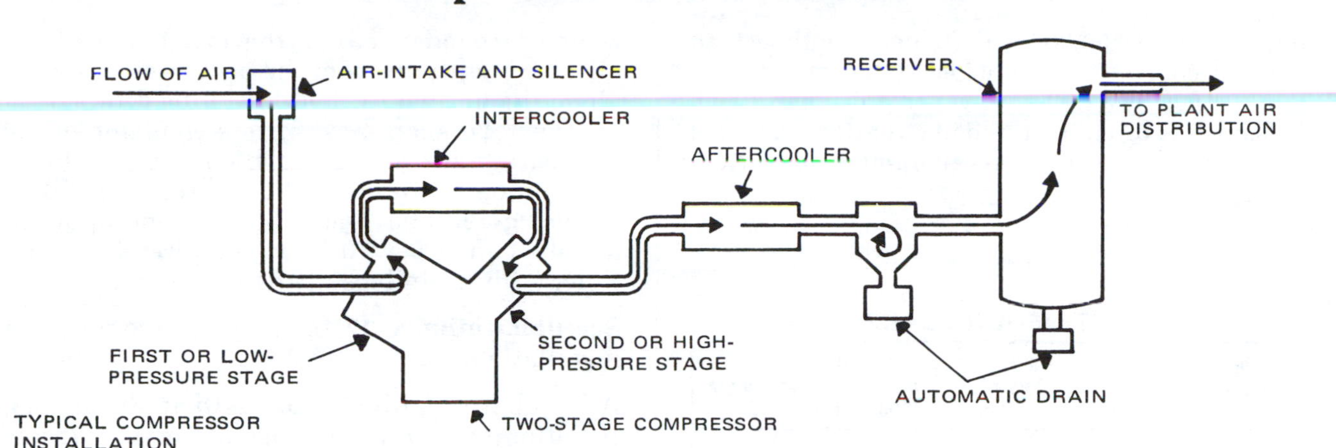

illustration g-17

The first element required for a pneumatic system is a source of compressed air. In selecting a compressor, consider these items:

1. Present needs and possible future requirements in terms of pressure and amount of air.

2. Possibility of splitting the requirement between two compressors, so that when one unit is down for servicing, work stoppage is avoided.

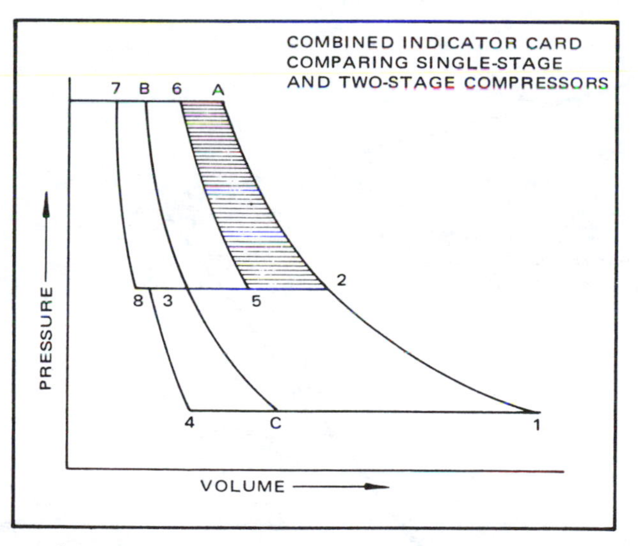

graph g-2

Graph g-2 shows a typical indicator card for a two-stage compressor superimposed over the card for a single-stage compressor. The trace for the single-stage unit is outlined by points 1-A-B-C. The first stage of the two-stage unit is outlined 1-2-3-4 and the second stage by 5-6-7-8. The shaded area represents the work saved by the two-stage over the single-stage compressor. Heat developed during compression in the first stage is transferred through an intercoller to circulating water.

When air demand is 1,000 cfm or greater, a centrifugal compressor should be considered. One of the advantages of this type of unit is its availability as a complete package requiring only a rigid base for installation. Since centrifugal compressors require no lubrication in the air chamber, they are well suited for applications requiring oil-free air.

Regardless of whether the compressor is reciprocating or centrifugal, the air intake should be in a cool, clean location. It should be kept away from steam exhausts, paint sprays, chimneys, and sources of contamination such as sulphur dioxide. This common product of combustion, when inhaled by the compressor, combines with the water in the air line to form sulphurous acid. This acid can damage the entire system from the compressor to the actuator that utilizes the air.

The air should be filtered at the intake. After the air leaves the compressor, it should be passed through an aftercooler which condenses much of the moisture in the air. Water removed by the separator is collécted in a trap and discharged from the system.

The last stop before the air enters the distribution lines is the storage or receiver tank. Size should be large enough to reduce air velocity for further moisture removal and to allow for future requirements. The receiver should be equipped with an automatic drain to keep the tank free of water.

compressed air is not free

Although air is "free," **compressed** air is not. The cost ranges from 25c per 1,000 cu. ft. in large installations down to about 10c per 1,000 cu. ft. in large installations. Air losses through leaks that total the equivalent of a 1/2-inch diameter hole

compressor considerations continued

amount to 11,700,000 cu. ft. per month. At an average cost of 12 1/2c per 1,000 cu. ft., the monthly cost is $1,400. However, since an air leak is clean, it is often tolerated in situations in which a hydraulic leak would receive immediate attention.

COST OF AIR LEAKS

Diameter of Leak (in.)	Air Loss (cfm)	Approximate Power Loss (hp)	Air Loss per Month (1000 cu. ft.)	Cost per Month ($)
1/32"	1	1/5	43	6.5
1/16"	4-1/4	7/8	180	27.0
1/8"	17	3-1/2	730	109.5
1/4"	70	14	3000	450.0
3/8"	150	30	6500	975.0
1/2"	270	54	11700	1755.0

NOTE: Based on cost of 15c per 1000 cu. ft. and orifice coefficient of 0.65.

table g-10

Moisture condensed by the expansion of the air along with oil from the compressor, moves along the bottom of the pipe in the direction of air flow. To keep most of this emulsion out of valves, cylinders and tools, outlets should always to be taken off the top of the lines, **illustration g-19.** The lines should slope approximately 1 inch per 10 ft. in the direction of air flow, with drain legs at the low points.

See **illustration g-17** for a typical compressor installation.

A loop pattern, **illustration g-18,** in the air-distribution system allows use of small pipe and assures more equal distribution throughout the plant. This arrangement provides a parallel path to all work points. At points of heavy momentary demands for air, a receiver can be used to store the energy for peak demands, and thus prevent serious pressure losses.

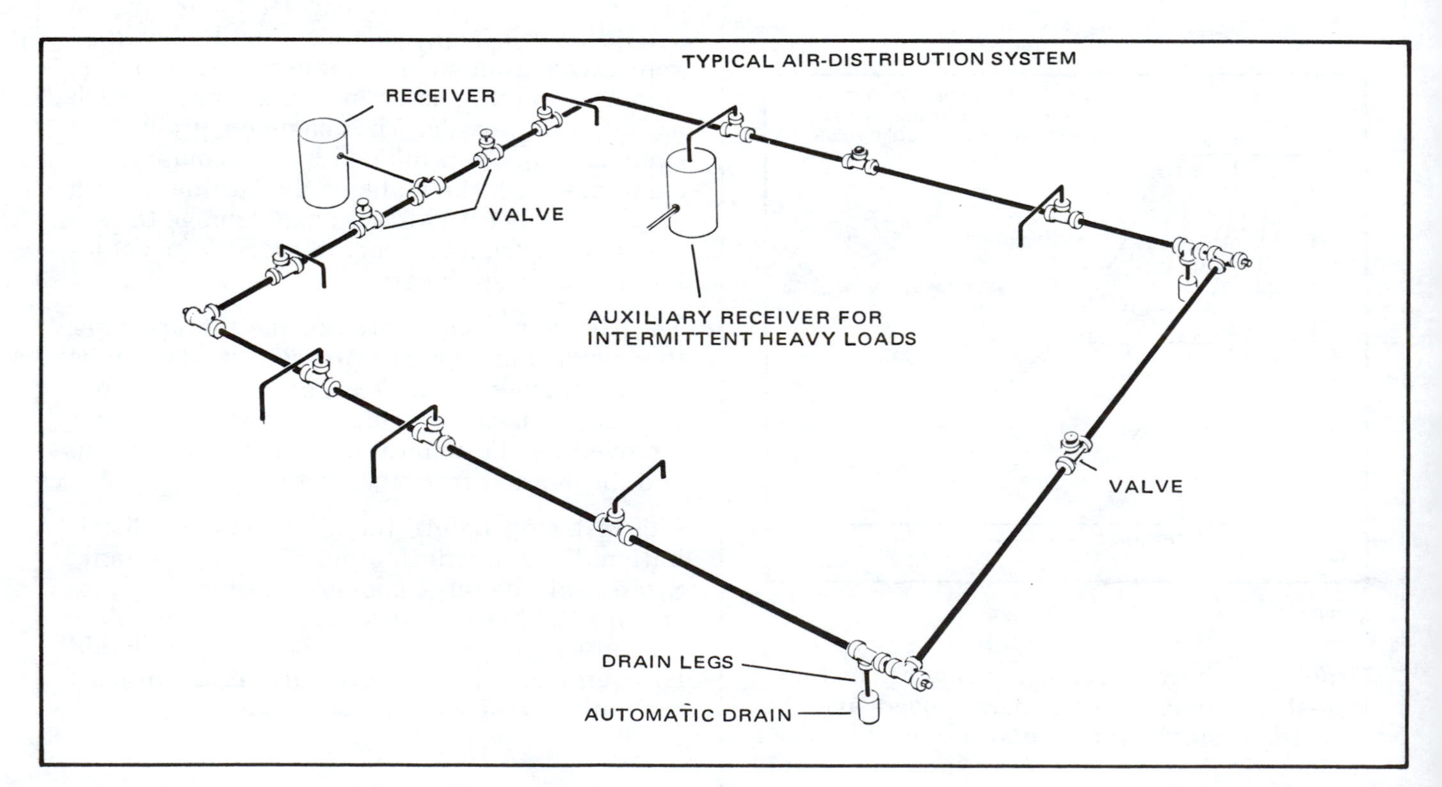

illustration g-18

compressor considerations continued

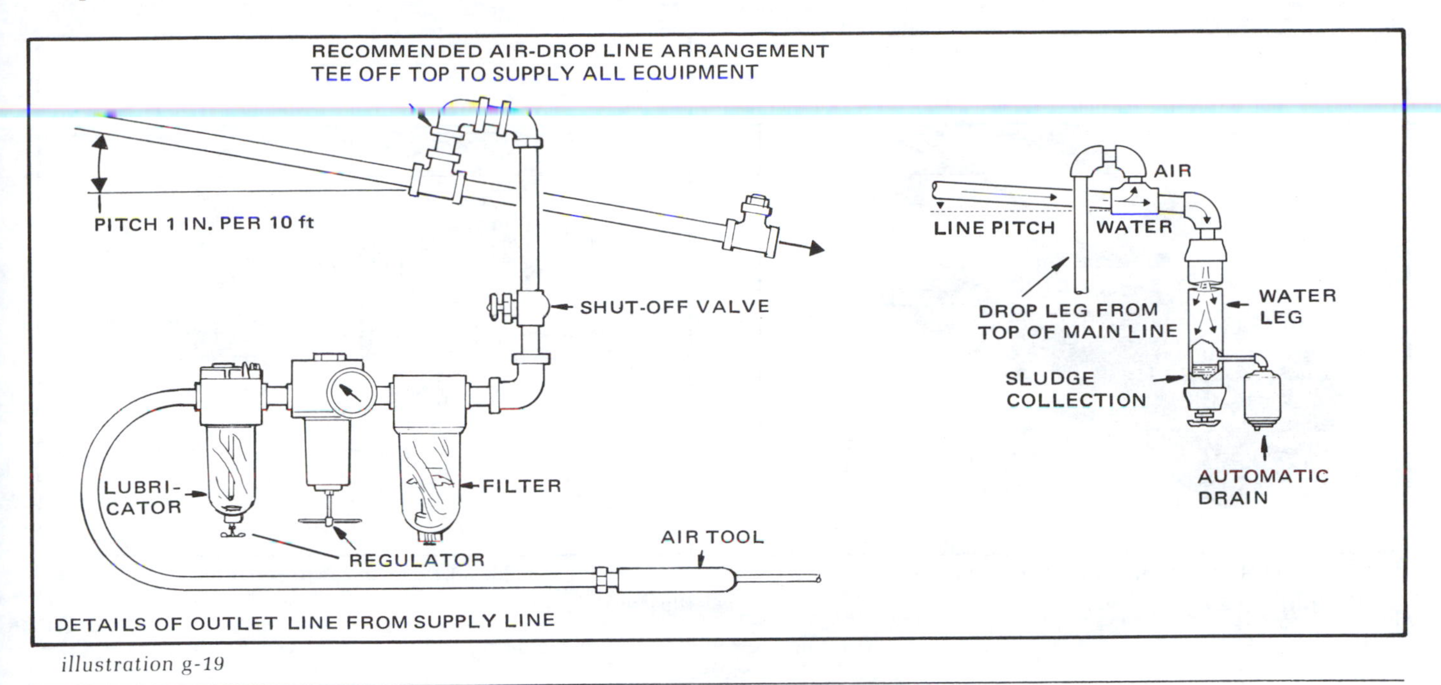

illustration g-19

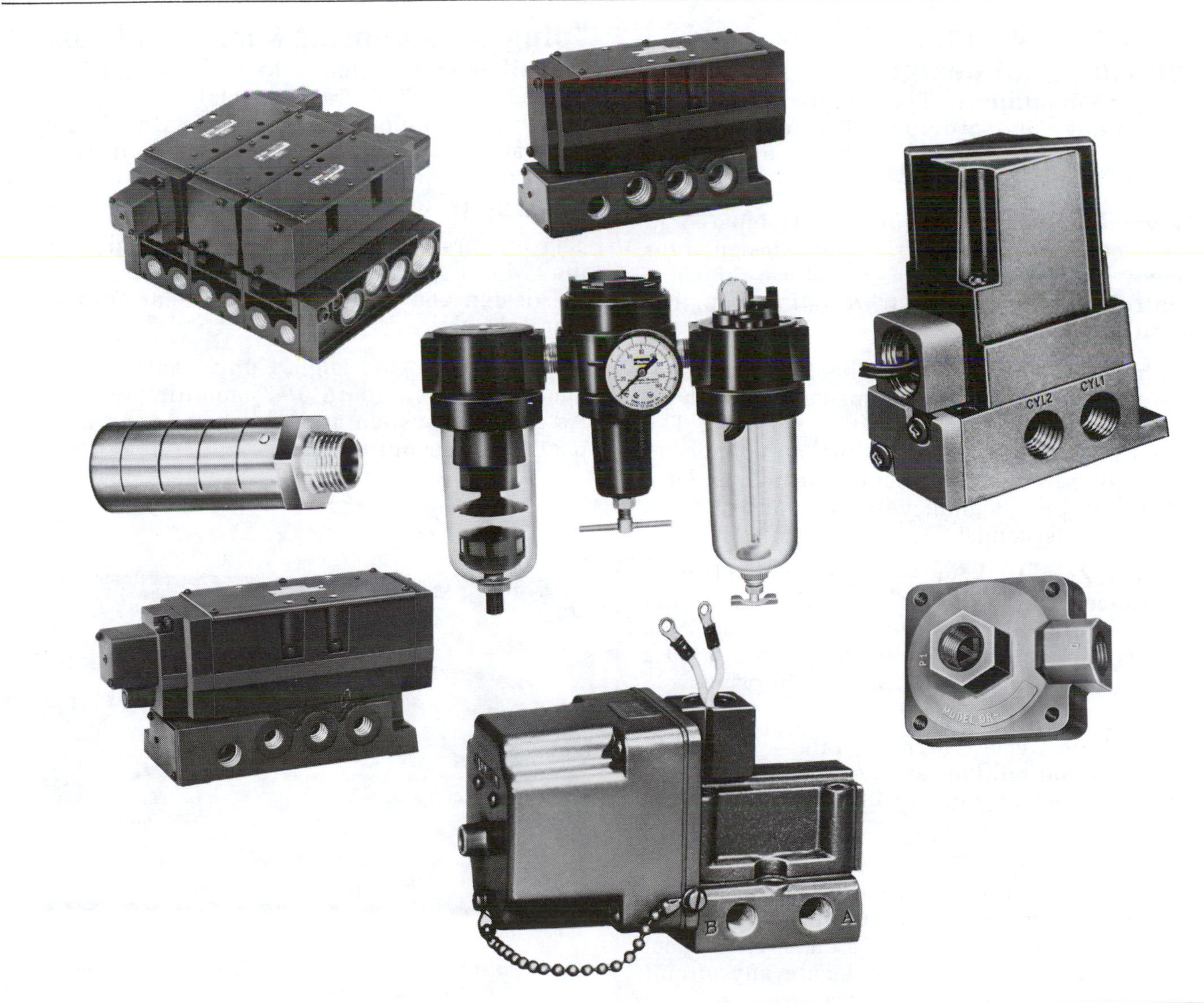

mod-logic

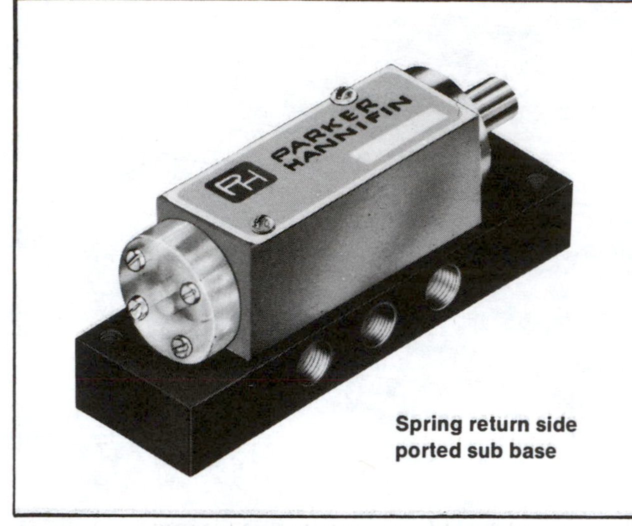

PH Mod-Logic Valves on a Manofold
Illustration g-20

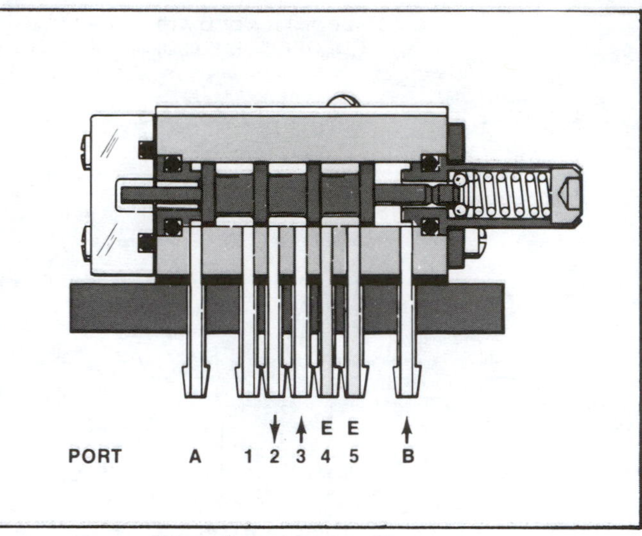

PH Mod-Logic Valve Illustration
illustration g-21

4-way, 5-ported, lapped-spool design

Unique Versatility — The five ports of a Mod-Logic relay valve provide four flow paths that can be used to obtain an almost limitless variety of control functions.

Fewer Components Required — Compared to other controls, particularly those designed for 3-way logic, Mod-Logic can perform the same control functions with up to half the number of circuit components.

High Speed Operation — The Mod-Logic relay valve uses a micro-lapped spool with an extremely short stroke of only 0.1 inch. The spool requires air approaching "no volume" to shift—only .005 cubic inches. Result: under a 50 p.s.i.g. pilot signal, valve response time is only 3 milliseconds!

Almost Zero Leakage — As a result of close tolerance clearance between valve body and spool and miniature unit size, Mod-Logic valves have negligible air leakage. Air loss averages less than 0.002 scfm at 50 p.s.i.g.

100 Million Cycle Operating Life — Valve life averages 100 million test cycles with properly filtered air—either dry or lubricated.

Easy Servicing -- By simply removing two screws, you can take a Mod-Logic component out of a circuit without disconnecting a single line. And Mod-Logic's clear-view spool position indicator lets you locate any circuit malfunction quickly and easily.

"plug-in" mounting & manifold base

Simplified Installation — Mod-Logic's push-on barbed fittings require no special tools, sleeves or sealing compounds. Because of the miniature size of Mod-Logic terminals and 1/16" I.D. tubing, neat, compact harnessing and labeling of circuitry requires a minimum of space.

The universal system of mounting and porting Mod-Logic components also simplifies the design and fabrication of special unitized manifolding.

Modularity — Mod-Logic's universal base dimensions fit standard 3/4" mounting centers. As a result, system assemblies are neat, compact, modular units.

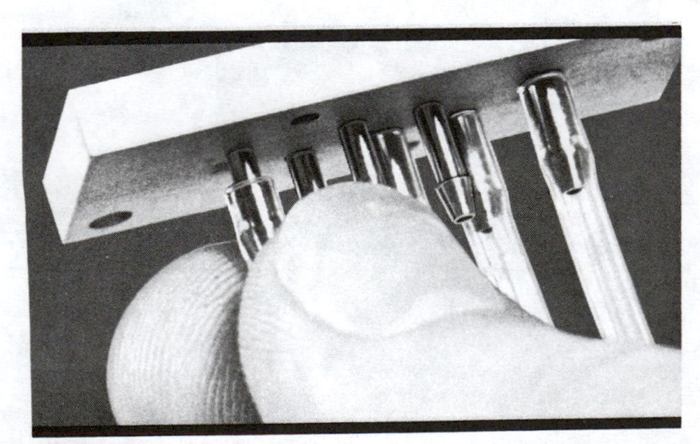

illustration g-22

Unique Mod-Logic barbed terminals retain 1/16" I.D. tubing under a full 100 p.s.i. air pressure, yet tubing slips on easily.

mod-logic continued

mod-logic base markings

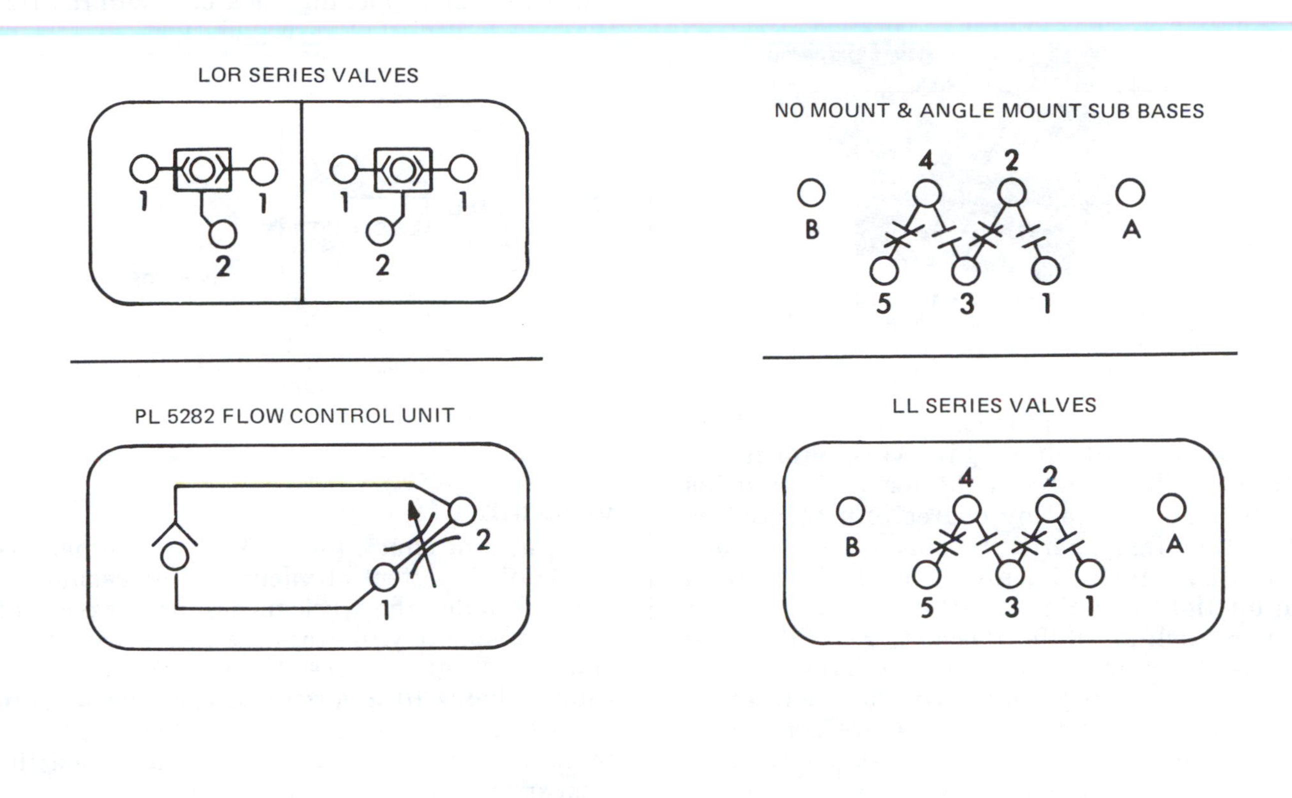

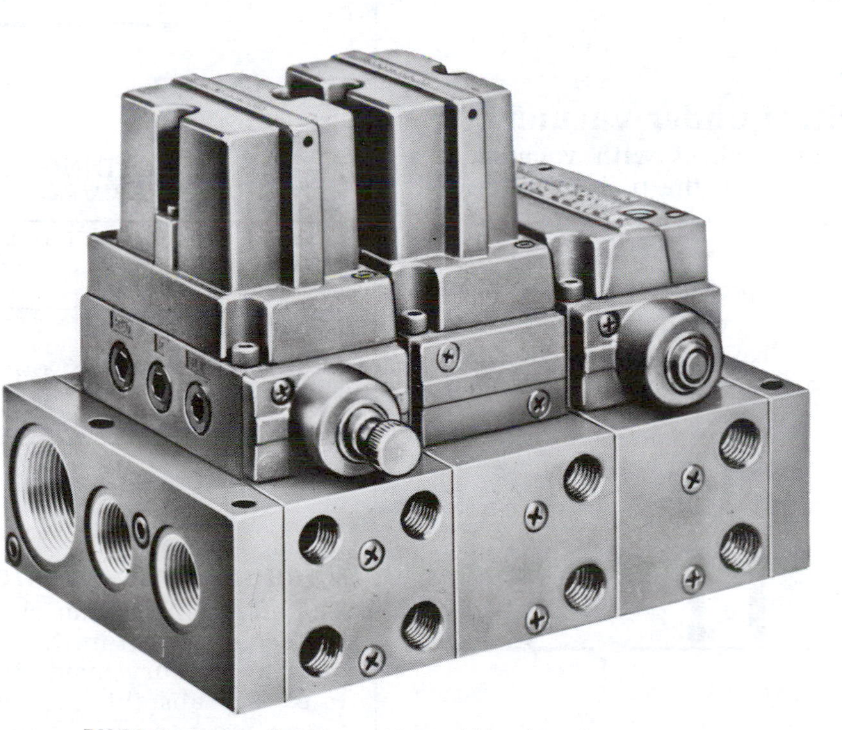

PH SS-Series Manifold Assembly Usable with Mod-Logic.

illustration g-23

vacuum

"4501" Series 1/8 & 1/4 in. NPT
4-Way, 5-Port, 2-Position Double Solenoid

illustration g-24

The term "Vacuum" means a space absolutely devoid of all air or gas. This term is not strictly adhered to and the term "Vacuum" is usually applied to any degree of pressure less than the atmospheric pressure of 14.7 pounds per square inch. If we reduce the pressure in an air tight container until it is, say 9 lbs. per square inch absolute pressure, we will have a partial vacuum. If we reduce the pressure still further until it is zero, we will have a full vacuum or perfect vacuum. When we speak of a very high vacuum we mean a very low absolute pressure. Partial vacuum is usually measured with a mercury manometer gauge and is given in inches of mercury more often than in pounds absolute pressure. Millimeters are sometimes used instead of inches. 25.4 millimeters equal one inch.

flow of air through orifice under vacuum

When picking up a sheet with vacuum, if a small opening is used, the pick-up force will be small because the force is proportional to the hole area.

If a large opening is used, the force will be proportionaltely larger but between pick-ups the vacuum will drop due to the air rushing freely into the large opening.

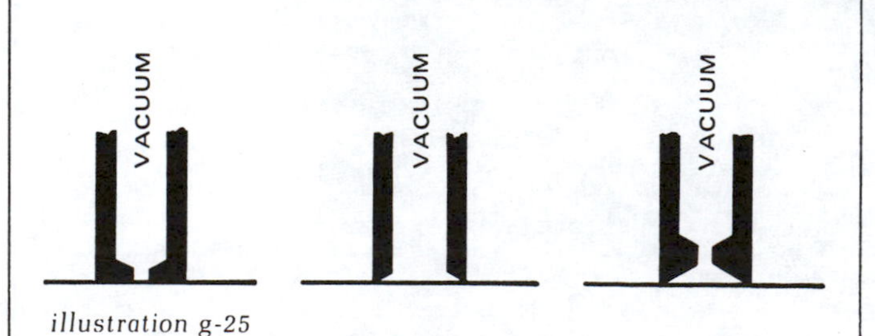

illustration g-25

If a combination of large and small openings is used as shown here, the large area against the sheet will produce a strong pick-up force while the small opening back of it will restrict the inrush of air between pick-ups.

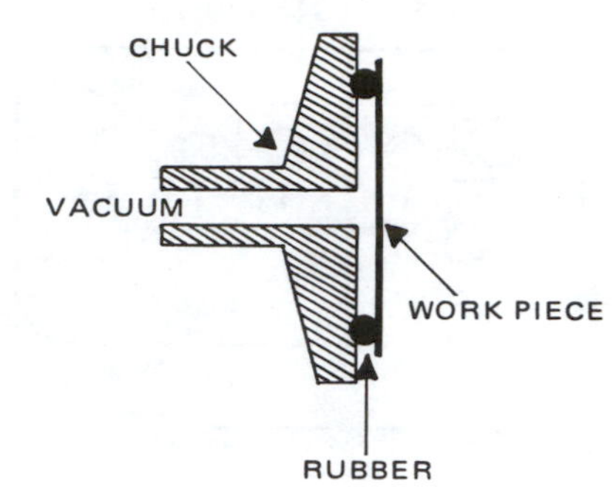

illustration g-26

vacuum lift

The vacuum chuck (or sucker) is another example of the effect of vacuum. If a vacuum is applied under the work piece, the pressure of the atmosphere will press the piece against the chuck with great force. If the area inside the rubber ring is 10 square inches and the vacuum is 10.2 inches (mercury gauge) there will be a force of 5 PSI on each square inch from the atmosphere or a total force of 50 PSI.

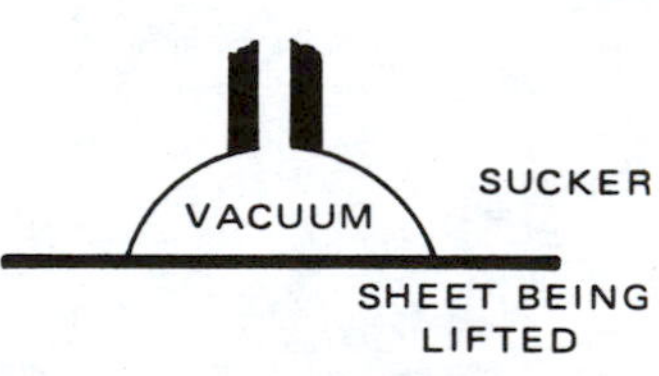

illustration g-27

HOLDING FORCE
CREATED BY VACUUM ON SUCKERS

Vacuum in Inches of Mercury	Diameter of Sucker					
	1"	1-1/2"	2"	3"	4"	5"
5	1.9 psi	4.4	7.8	17.7	31.4	49
10	3.8	8.7	15.4	34.6	62	96
15	5.8	13.1	23.3	52.2	93	145
20	7.6	17.3	30.8	69	123	192
25	9.6	21.8	38.6	87	154	241 lbs.

table g-11

vacuum tank considerations

The effect of vacuum on a tank having a flat surface is shown in the diagram. The vacuum of 10.2" mercury inside the tank is equivalent to 9.7 PSI absolute pressure, while the atmospheric pressure on the outside is 14.7 lbs. per sq. inch, absolute. The difference between the outside pressure and the inside pressure is 5

vacuum continued

PSI therefore is a pressure of 5 PSI on each square inch of surface, tending to push in the tank wall. If the tank wall measured 10″ x 10″ there would be 100 sq. inches and the crushing force would be 500 PSI.

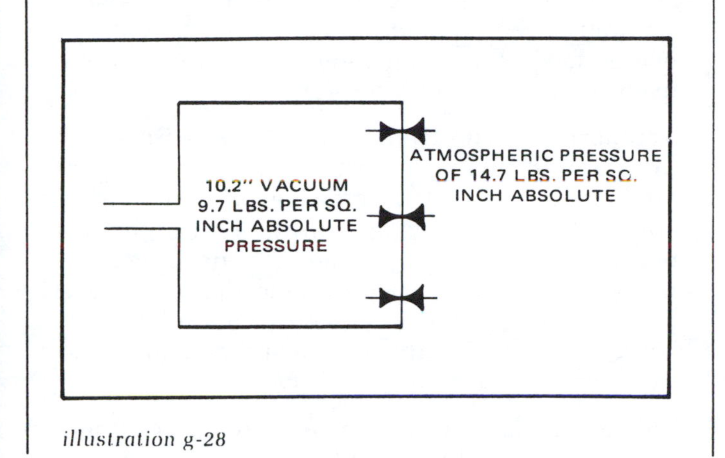

illustration g-28

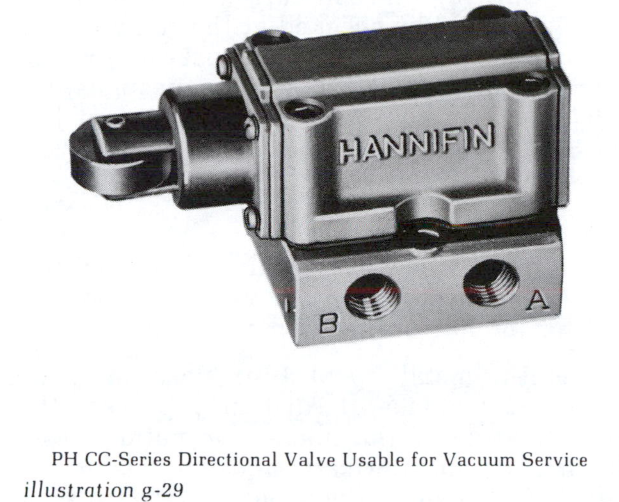

PH CC-Series Directional Valve Usable for Vacuum Service
illustration g-29

notes

air valve troubleshooting

AC solenoids fail

Low Applied Voltage — Not enough magnetic force will be developed to allow the armature of the solenoid to seat. The unit will continuously draw high inrush current and burn out. Voltage should be checked at the coil with the solenoid energized. Possible causes of low voltage include: High resistance connections, too much load on the electrical circuit, and low voltage on the control transformer that powers the solenoid.

Valve Element Stuck — The armature may be held unseated because a valve element won't shift. The solenoid will draw high inrush current for too long a time and burn out.

A metal-to-metal spool type valve may be varnished in place, or dirt may prevent the spool (and hence the solenoid armature) from shifting. Clean, lubricate, and reassemble the valve. If varnish on the spool was the cause of sticking, placing an absorptive type filter in the circuit ahead of the valve may help. It will necessary to periodically replace the filter element.

A packed spool type valve may not shift because swollen seals hold the spool in place, or because dirt prevents the spool from completely shifting. Clean the valve and repack the seals or replace the spool. Where seals are sensitive to the type of air line lubricant used, follow manufacturers' recommendations.

Solenoid Corroded — Use valves with adequate protection against moisture, coolants, etc. which may come in contact with them. Provide sealed electrical connections. Make sure that dirt and moisture covers are securely in place.

Solenoids Energized Simultaneously — On momentary contact, direct-actuated valves, check to make sure that both solenoids are never energized at the same time. This check is easily made by wiring a small indicator lamp temporarily across the coil of each solenoid. If both lamps are lit at the same time, the last solenoid to be energized will burn out. Correct the electrical circuit to prevent this. If the electrical circuit cannot be foolproofed, consider the use of a pilot operated, solenoid controlled valve where the two solenoids do not exert forces against each other.

High Transient Voltages — Solenoid burnout may be caused by high transient voltages that break down coil insulation, causing short circuits to ground. High transient voltages are most common where solenoids are connected to lines operating above 120 volts, which also control motors and other inductive load-type equipment. Switching of such loads can create very high voltage peaks in the circuit. The remedy is to isolate solenoid circuits from main power circuits. Follow industry recommendations — use 120 volt control circuits and observe good grounding practices. Electrical filter networks, which prevent high voltage peaks from reaching solenoids, may also be used.

Temperature Too High or Too Low — Solenoid failures can be expected when a valve is operated above its rated temperature. Insulation may fail because it is not suitable for high temperature use. Specify solenoids designed for the ambient temperature; place the valve in a cooler location; or consider the use of a pilot actuated valve at the hot location, controlled by a remote pilot valve.

Solenoids also can fail when valves are operated at lower than rated temperatues. Metal parts may shrink, and lubricant viscosities may increase, to a point where solenoid motion is retarded or stopped. This can be avoided by moving the valve to a warmer place or by providng a heated enclosure for the valve. If trouble persists with closely fitted valves, change to a type that can work at low temperatures.

dc solenoids fail

Direct current solenoids should never burn out due to low voltage or failure of armatures to seat. This is due to the fact no high inrush current is experienced with a dc solenoid. Maximum current is the holding current, which is the same whether or not the armature is seated. Thus, the coil is never subjected to an inrush current of three to four times its holding current.

Valve failures to shift can, of course, occur with dc as well as ac solenoids. Corrosion and temperature also can effect dc solenoid operation. The general guides for ac solenoids can be applied in these cases. High transient voltage problems can be handled in the same manner as for ac solenoids.

solenoid valve has loud ac hum

If the solenoid noise level is very high and occurs each time the solenoid is energized, check to see that the armature is seating. Most direct solenoid actuated valves are provided with a manual override. If the solenoid noise decreases when the override is operated, incomplete solenoid motion is indicated. Check to determine if rubbing parts can move proper-

air valve troubleshooting continued

ly and that the correct voltage is available at the coil. Extremely loud ac hum can be caused by a broken part within the solenoid. Replace the solenoid.

In marginal noise problems, consider mounting the valve so that the armature works up and down rather than horizontally. Maximum quieting can be obtained by using dc solenoids. This is practical even on large valves of the solenoid controlled, pilot actuated type. Direct current can be obtained by rectifying alternating current at the valve.

internally piloted valve shifts improperly

An internally piloted valve may shift partially, then stall. Air blows steadily through the exhaust port. This is a sign that the pressure at the inlet port of the valve has fallen below the valve's minimum operating pressure. Increase the supply pressure or provide a local air accumulator to maintain pressure at the valve during periods of high flow.

If the valve is provided with an internal pilot exhaust, change it to exhaust externally if possible. Check for restrictions in the supply line. A gage can be used to check the pressure available at the inlet port just before the valve fails to shift. Common causes of restricted supply lines (and remedies) are: Clogged filters (clean periodically); restrictive lubricators (decrease restriction or change to a unit with better flow capabilities); undersized hose or fittings (use conductors and fittings with internal cross-sectional areas as large as the inside diameter of the pipe for which the inlet port is tapped).

On some valve designs, especially momentary contact types, full shifting may be obtained by restricting the exhaust port to allow the valve to maintain a pressure level above the minimum operating pressure. In extreme cases, provide a local accumulator for the pilot circuit or use pilot pressure from a remote source.

valve spews oily air from exhaust port

This is usually a sign of overlubrication of the circuit. Adjust the lubricator oil feed rate to an acceptable level. Where quieting of the exhaust is desirable, an ordinary oil filter can be used to trap most of the exhausting liquids while providing a good degree of muffling action. The drain of the filter should be left open and connected to a sump.

valve occasionally malfunctions-circuit reaction slow

This type of problem can be caused by icing. Rapidly exhausting air often cools entrained moisture below its freezing point, causing ice particles to restrict exhaust flow. Mufflers can also freeze. Check the exhaust ports of the valve and/or muffler for signs of icing. Dry the incoming air or provide means for heating areas which tend to freeze.

notes

power units

general practices in power unit design h-2
reservoir assembly standard jic h-4
drive coupling control drawing h-5
pump risers h-6
estimating heat rise h-7
calculating reservoir heat dissipation h-7
heat generation and control h-9
power unit start-up considerations .. h-10
noise control h-10
troubleshooting areas h-13

general practices in power unit design

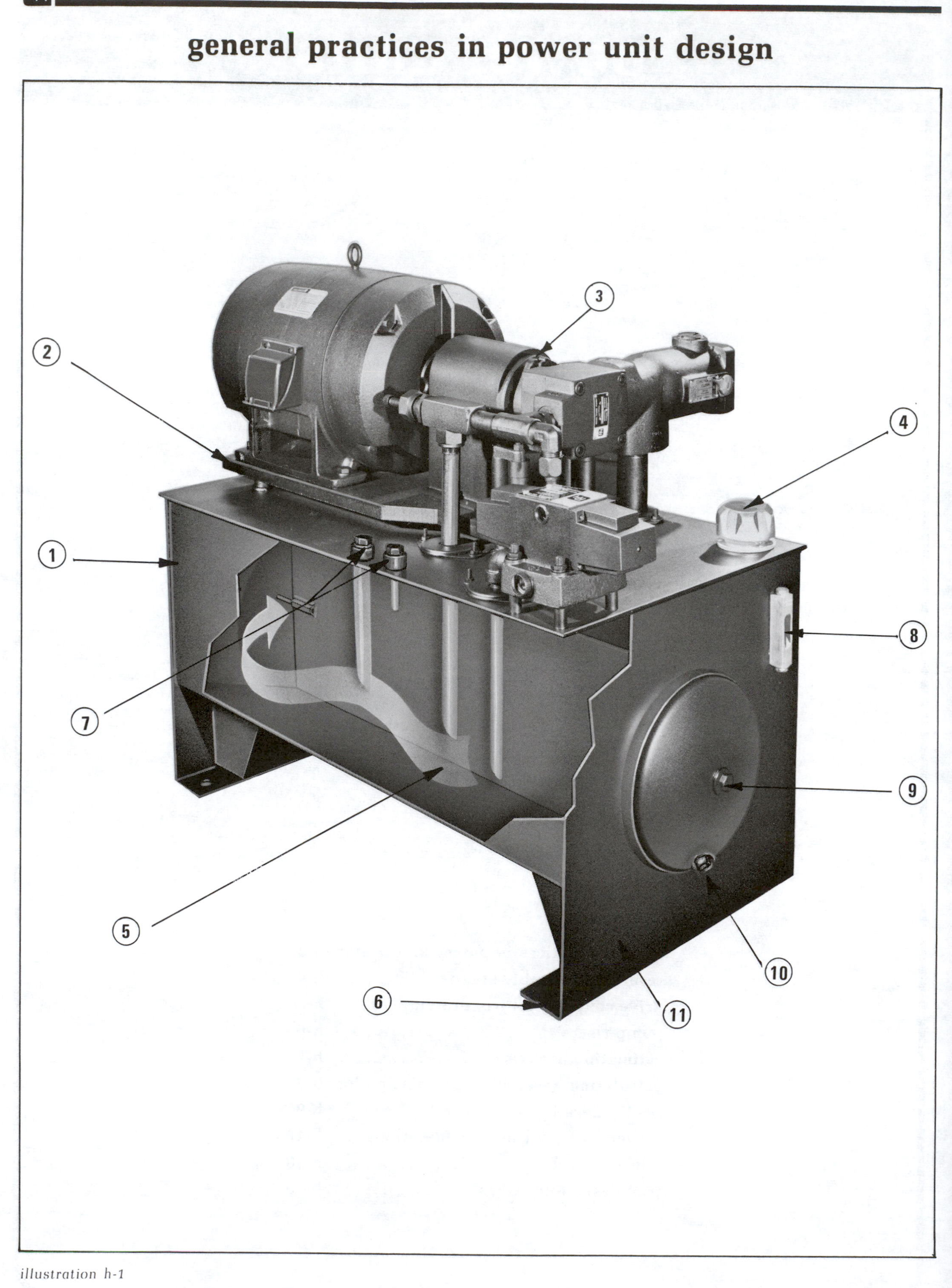

illustration h-1

general practices continued

1. Size the reservoir to approximately three times the capacity of the pump rating. (in gallons per minute). Sometimes when using large bore, large rod and/or long stroke cylinders, a larger reservoir than normal will be required, this is due to the large volume of oil storage required for cylinder extension or retraction. This can also apply if there are a large number of medium size cylinders to be extended or retracted at the same time.

2. Mount the pump, electric motor, and valving to a sturdy plate.

When foot mounted pumps and motors are used, the mounting plate to both the driver and the driven unit should be common and should be rigid enough to withstand the torque loads applied without deflection. The plate also must not be subject to distortion caused by temperature change; i.e., **do not** mount pumps on reservoir tops. Use separate mounting plate, 3/4" minimum for plate.

Drives other than direct, a flexible coupling can be used only with external bearing support.

Failure to adhere to the above may result in leaking pump shaft seals, short bearing life and/or total unit life.

3. All pumps and motors using flexible couplings, should be aligned to coupling manufacturers specifications: the figures listed may be used as guide lines. All Parker Hannifin dual vane pumps and motors should be aligned to .003 T.I.R. The D.H.&M. series Parker Hannifin gear pumps should be aligned to .007 T.I.R. Piston pumps should be aligned to .003 T.I.R. The Vardis Pump should be aligned to .010 T.I.R.

NOTE: *A good, flexible coupling should be used. Couplings should always be sized to handle the total horsepower input . . . not to the size of the electric motor. Many applications overload the electric motor.*

Fit coupling according to coupling manufacturer's recommendations.

The above points are critical to avoid end thrusts. Pumps and motors cannot stand end thrust.

4. Have a filter breather cap and fill port with a strainer. The filter breather allows the oil to rise and lower during system operation.

5. Build baffling into the reservoir. The oil will travel a greater distance (from return to inlet) which gives the dirt time to settle out and allows for greater heat dissipation.

6. Mount the reservoir on legs for better overall air circulation.

7. Have extra return and drain line connections built into the reservoir. (Return lines below fluid level and drain lines [except pump drain lines] above fluid level.)

8. Have an oil level indicator.

9. Easy access to the reservoir interior for cleaning.

10. Have taper to the center of the reservoir with drain plug for ease in cleaning.

11. Use a steel material that is capable of withstanding the oilweight, components mounted on the reservoir and shipment or relocation of the power unit.

reservoir assembly standard j.i.c.

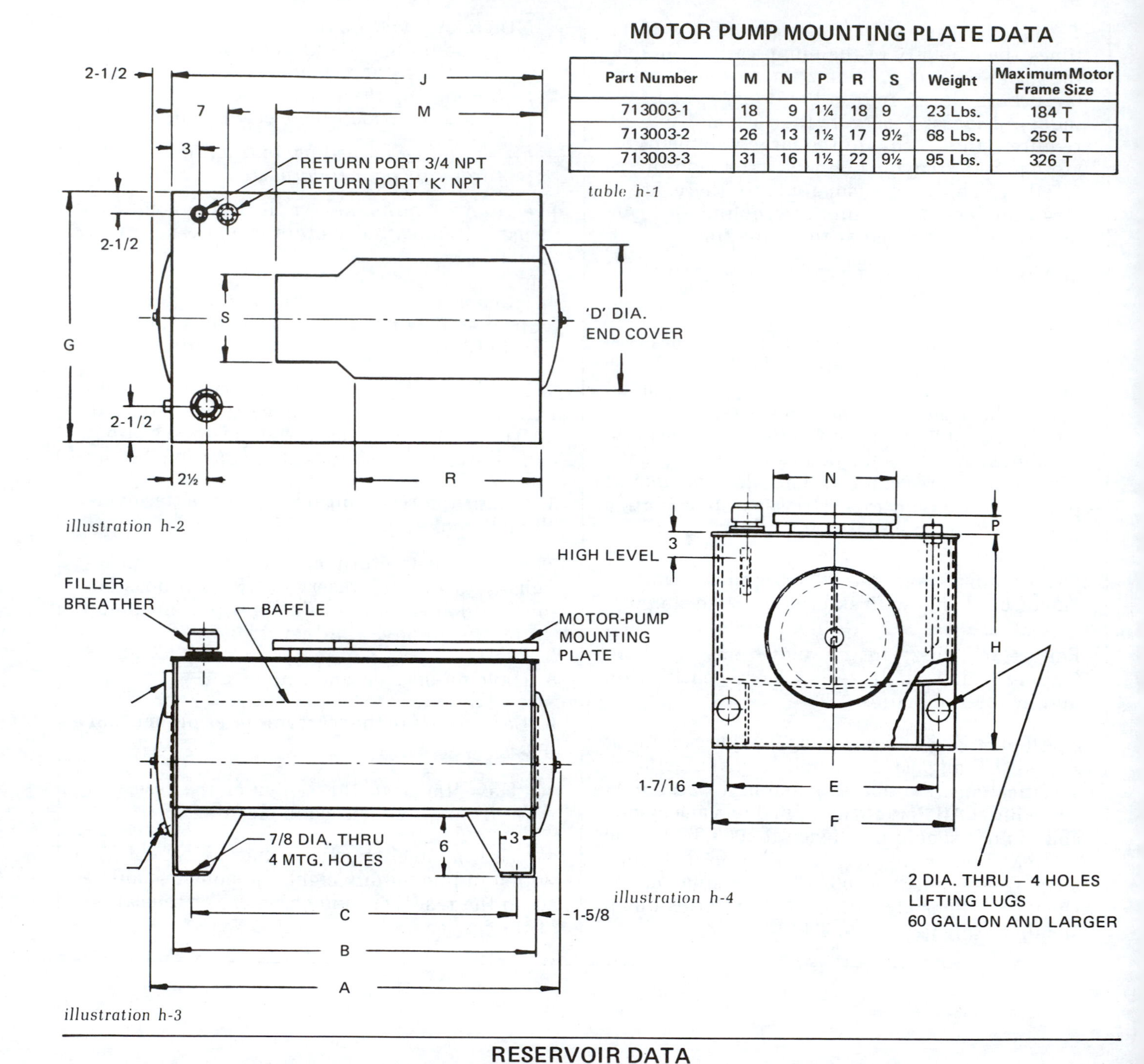

MOTOR PUMP MOUNTING PLATE DATA

Part Number	M	N	P	R	S	Weight	Maximum Motor Frame Size
713003-1	18	9	1¼	18	9	23 Lbs.	184 T
713003-2	26	13	1½	17	9½	68 Lbs.	256 T
713003-3	31	16	1½	22	9½	95 Lbs.	326 T

table h-1

illustration h-2

illustration h-4

illustration h-3

RESERVOIR DATA

<table>
<tr><th rowspan="2">Size Gal.</th><th colspan="2">Tank Vol.-Gal.</th><th rowspan="2">A</th><th rowspan="2">B</th><th rowspan="2">C</th><th rowspan="2">D</th><th rowspan="2">E</th><th rowspan="2">F</th><th rowspan="2">G</th><th rowspan="2">H</th><th rowspan="2">J</th><th rowspan="2">K</th><th rowspan="2">L</th><th colspan="2">Weight Lbs.</th></tr>
<tr><th>High</th><th>Low</th><th>Empty</th><th>Full</th></tr>
<tr><td>10</td><td>13</td><td>10</td><td rowspan="2">29.5</td><td rowspan="2">24</td><td rowspan="2">20.75</td><td rowspan="3">10</td><td>12.63</td><td>15.5</td><td>16</td><td>17</td><td rowspan="2">24.5</td><td rowspan="3">3/4</td><td rowspan="3">3</td><td>90</td><td>190</td></tr>
<tr><td>20</td><td>22</td><td>19</td><td rowspan="2">16.63</td><td rowspan="2">19.5</td><td rowspan="2">20</td><td rowspan="2">20</td><td>115</td><td>297</td></tr>
<tr><td>30</td><td>33</td><td>29</td><td rowspan="2">41.5</td><td rowspan="2">36</td><td rowspan="2">32.75</td><td rowspan="2">36.5</td><td>150</td><td>424</td></tr>
<tr><td>40</td><td>50</td><td>38</td><td rowspan="6">14</td><td rowspan="2">20.63</td><td rowspan="2">23.5</td><td rowspan="2">24</td><td rowspan="2">23</td><td>1</td><td rowspan="4">5</td><td>240</td><td>655</td></tr>
<tr><td>60</td><td>67</td><td>51</td><td rowspan="2">53.5</td><td rowspan="2">48</td><td rowspan="2">44.75</td><td rowspan="2">48.5</td><td>1-1/4</td><td>270</td><td>826</td></tr>
<tr><td>90</td><td>102</td><td>82</td><td rowspan="2">26.63</td><td rowspan="2">29.5</td><td rowspan="2">30</td><td rowspan="2">26</td><td>1-1/2</td><td>425</td><td>1270</td></tr>
<tr><td>120</td><td>128</td><td>103</td><td>65.5</td><td>60</td><td>56.75</td><td>60.5</td><td>2</td><td>480</td><td>1541</td></tr>
<tr><td>180</td><td>206</td><td>149</td><td rowspan="2">77.5</td><td rowspan="2">72</td><td rowspan="2">68.75</td><td>32.63</td><td>35.5</td><td>36</td><td rowspan="2">28</td><td rowspan="2">72.5</td><td>2-1/2</td><td rowspan="2">7</td><td>660</td><td>2370</td></tr>
<tr><td>240</td><td>244</td><td>176</td><td>38.63</td><td>41.5</td><td>42</td><td>3</td><td>745</td><td>2770</td></tr>
</table>

table h-2

drive coupling control drawing

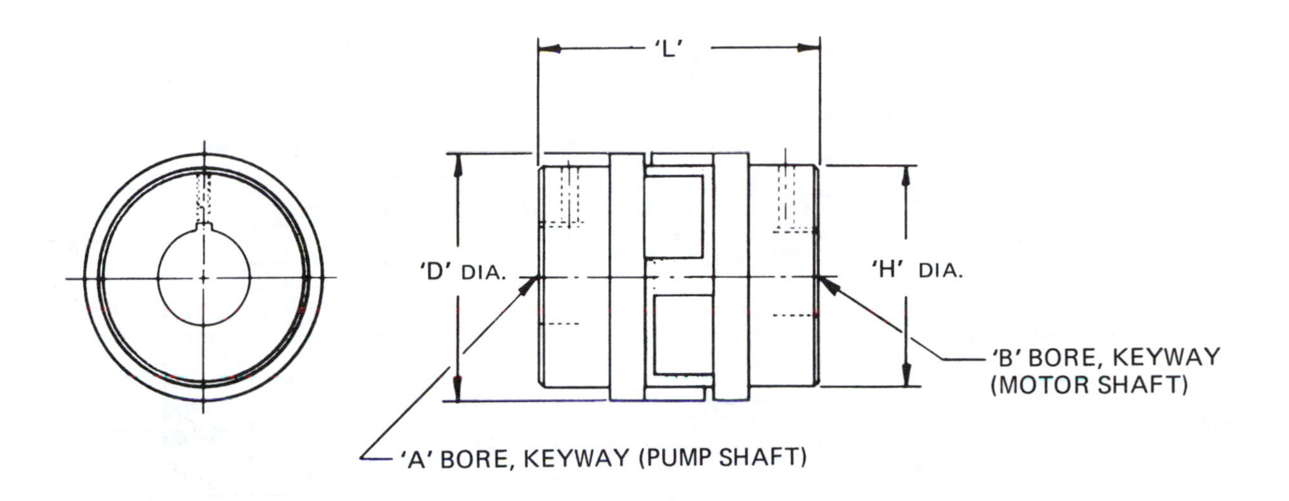

illustration h-5

Model Number	Max.hp@ 1800rpm	'A' Bore	'A' KW	'B' Bore	'B' KW	'D'	'H'	'L'	Love Joy Model	Pumps (Ref.)
C1—1/2—5/8	1	1/2	1/8	5/8	3/16	1-3/8	1-3/8	2	L-070	'D' Series
C2—1/2—5/8	2	1/2	1/8	5/8	3/16	1-3/4	1-9/16	2-1/8	L-075	'D' Series
C2—1/2—7/8	2	1/2	1/8	7/8	3/16				L-075	'D' Series
C3—1/2—7/8	3	1/2	1/8	7/8	3/16	2-1/8	2-1/8	2-1/2	L-095	'D' Series
C3—1/2—1-1/8	3	1/2	1/8	1-1/8	1/4				L-095	'D' Series
C3—3/4—7/8	3	3/4	3/16	7/8	3/16				L-095	'H' Series, 411 & 413-B
C3—3/4—1-1/8	3	3/4	3/16	1-1/8	1/4				L-095	'H' Series, 411 & 413-B
C3—7/8—1-1/8	3	7/8	1/4	1-1/8	1/4				L-095	450 & 'M' Series
C5—1/2—1-1/8	5	1/2	1/8	1-1/8	1/4	2-9/16	2-9/16	2-7/8	L-099	'D' Series
C5—3/4—1-1/8	5	3/4	3/16	1-1/8	1/4				L-099	'H' Series, 411 & 413-B
C5—7/8—1-1/8	5	7/8	1/4	1-1/8	1/4				L-099	450 & 'M' Series
C5—1—1-1/8	5	1	1/4	1-1/8	1/4				L-099	413 'A' Design
C10—3/4—1-3/8	10	3/4	3/16	1-3/8	5/16	2-9/16	2-9/16	3-1/2	L-100	'H' Series, 411 & 413-B
C10—7/8—1-3/8	10	7/8	1/4	1-3/8	5/16				L-100	450 & 'M' Series
C10—1—1-3/8	10	1	1/4	1-3/8	5/16				L-100	413 'A' Design
C10—1-1/4—1-3/8	10	1-1/4	5/16	1-3/8	5/16				L-100	421 & 423
C10—1-1/4—1-1/8	10	1-1/4	5/16	1-1/8	1/4				L-100	421 & 423
C15—3/4—1-5/8	15	3/4	3/16	1-5/8	3/8	3-5/16	3	4-1/4	L-110	'H' Series, 411 & 413-B
C15—7/8—1-5/8	15	7/8	1/4	1-5/8	3/8				L-110	450 & 'M' Series
C15—1—1-5/8	15	1	1/4	1-5/8	3/8				L-110	413 'A' Design
C15—1-1/4—1-5/8	15	1-1/4	5/16	1-5/8	3/8				L-110	421 & 423
C20—3/4—1-5/8	20	3/4	3/16	1-5/8	3/8	3-3/4	3-1/8	4-1/2	L-150	'H' Series, 411 & 413-B
C20—7/8—1-5/8	20	7/8	1/4	1-5/8	3/8				L-150	450 & 'M' Series
C20—1—1-5/8	20	1	1/4	1-5/8	3/8				L-150	413 'A' Design
C20—1-1/4—1-5/8	20	1-1/4	5/16	1-5/8	3/8				L-150	421 & 423
C20—1-3/4—1-5/8	20	1-3/4	7/16	1-5/8	3/8				L-150	431
C20—1-3/4—1-3/8	20	1-3/4	7/16	1-3/8	5/16				L-150	431

table h-3

pump risers

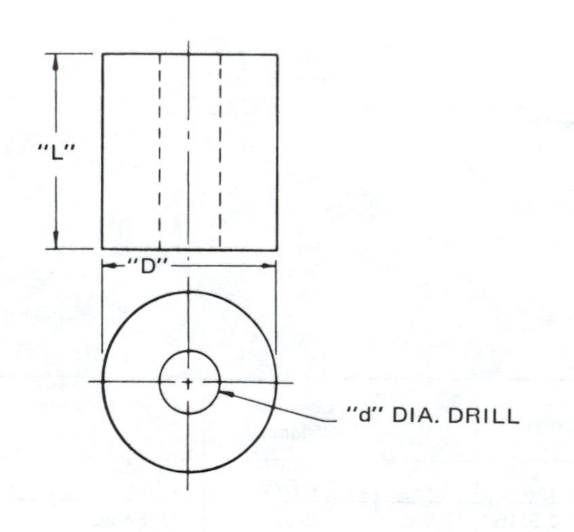

illustration h-6

PUMP MODEL	L +.000 -.010	D	d
411	3/4	1-1/4	13/32
	1-3/4	1-1/4	13/32
	2-1/2	1-1/4	13/32
	3-1/2	1-1/4	13/32
413	3/4	1-3/4	19/32
	1-1/2	1-3/4	19/32
	2-1/2	1-3/4	19/32
	3-1/4	1-3/4	19/32
	4-1/4	1-3/4	19/32
421 & 423	1/2	1-3/4	19/32
	1-1/4	1-3/4	19/32
	2-1/4	1-3/4	19/32
	3	1-3/4	19/32
	4	1-3/4	19/32
	5	1-3/4	19/32
431	1/2	2	27/32
	1-1/2	2	27/32
	2-1/4	2	27/32
	3-1/4	2	27/32
	4-1/4	2	27/32

table h-4

MAT'L – COLD ROLLED STEEL BAR

RISER BLOCK HEIGHTS

PUMP MODEL	MOTOR FRAME SIZE													
	143T	145T	182T	184T	213T	215T	254T	256T	284T	286T	324T	326T	364T	365T
411	3/4	3/4	1-3/4	1-3/4	2-1/2	2-1/2	3-1/2	–	–	–	–	–	–	–
413	–	–	3/4	3/4	1-1/2	1-1/2	2-1/2	2-1/2	3-1/4	3-1/4	4-1/2	4-1/4	–	–
421	–	–	1/2	1/2	1-1/4	1-1/4	2-1/4	2-1/4	3	3	4	4	5	5
423	–	–	1/2	1/2	1-1/4	1-1/4	2-1/4	2-1/4	3	3	4	4	5	5
431	–	–	–	–	1/2	1/2	1-1/2	1-1/2	2-1/4	2-1/4	3-1/4	3-1/4	4-1/4	4-1/4
433	–	–	–	–	1	1	2	2	2-3/4	2-3/4	3-3/4	3-3/4	4-3/4	4-3/4

table h-5

estimating heat rise

When designing a Hydraulic System, a calculation must be made of normal heat generation. Heat generation across an orifice or relief valve may be calculated if the pressure drop across the device and GPM are known or can be measured. The formula is as follows:

Hp = Psi x Gpm ÷ 1714 or
BTU/Hr. = 1-1/2 x Psi x Gpm

Hp = Horsepower
Psi = The total ΔP or pressure drop across that orifice
Gpm = The total gallons per minute flowing through that orifice

This heat may be converted into other units by means of these conversion factors:

1 Hp = 2545 BTU/Hr. = 42.2 BTU/Min.
1 Hp = 746 Watts
1 BTU/Hr. = .016 BTU/Min. = .00039 Hp.

Example: *15 GPM flowing through a relief valve with ΔP, 900 Psi, generates approximately 7.9 Hp. of heat. All of this heat, which is not dissipated through the hose or tubing is carried back to the tank.*

Remember: *Heat is generated only when the hydraulic energy is not going into mechanical energy.*

calculating reservoir heat dissipation

For average calculations, assume heat dissipation from the sides, top and bottom of the reservoir. The surface area of external plumbing may also be used or counted as a dissipating surface. Do not figure the bottom of the reservoir unless it is exposed to free air circulation. The cooling capacity of the reservoir will increase in proportion to the square footage dissipating surface, and also in proportion to the difference between the oil temperature and the ambient air temperature.

For steel reservoirs, this formula will give approximate results as follows:

Hp = (Heat Dissipation) = 0.001 x TD x A

A = Square footage of dissipating surface
TD = Temperature difference between oil and surrounding air
Hp = Cooling capacity expressed in horsepower

There should be a reasonable amount of free air circulation around the reservoir. A forced blast of air directed on the side of the reservoir is a consideration as it can increase the heat dissipating capacity as much as 50%.

Total Gallon Volume	Sq. Ft. Dissipating Surface	Heat Dissipation BTU/hr.	Heat Dissipation HP
6-1/2	4-1/2	575	.23
12	6-3/4	850	.34
17	8-1/4	1,000	.40
33	13-1/4	1,700	.67
80	24	3,000	1.18
120	30	4,000	1.60
250	50	6,000	2.70
500	80	10,200	4.00

table h-6

These figures are approximate because the actual shape of the tank will sometimes influence the dissipation. These values of dissipation are figured very conservatively. These figures are based on an ambient temperature of 100°F. with a maximum oil temperature of 150°F. Sometimes, a considerable amount of heat is dissipated from plumbing, valves, and cylinders. Dissipating capacity will increase at lower ambients.

PH "L" Shaped Power Unit

illustration h-7

heat generation and control

Heat is generated in a hydraulic system whenever oil flows from a higher to a lower pressure, without doing mechanical work. This means, that if a relief valve is allowing the oil to flow back to the tank, and the system pressure is being maintained, the difference in pressure or loss is the difference between the system pressure and the tank line pressure. Pressure losses can occur from flowing through inadequately sized valving or piping, and kinked or sharp bends in hose or tubing. The sun can cause excessive heat. A reservoir should be sized large enough to dissipate normal heat generated in a system. Normal practice is 3 times the pump capacity (rated in GPM). This should allow the oil to maintain an operating temperature of approximately 110°F.-135°F.The reservoir temperature should not be allowed to reach much above 145°F. If a higher temperature is allowed to be reached, the oil can break down and give off an oxidation called varnish. This varnish can clog orifices, produce an acid which can corrode metal parts, produce sludge, and cause metal parts to wear very rapidly. The power unit reservoir temperature should be checked on occasion to make sure the system is operating properly.

High oil temperature can also be caused from metal parts being worn, or from the seals being worn allowing the oil to bypass and cause heat. By checking the reservoir temperature, sometimes a system problem can be corrected before serious damage can be done.

Heat Generating Component

In most systems, the main heat generating component is the relief valve. Generally, the relief valve is used only in a short period of the cycle. It is, therefore, necessary to find the maximum rate of heat generation by using the Hp. formula, then calculate an average for an entire hour.

Example: *In a system, the oil is flowing over a relief valve and generating heat at the rate of 8 Hp. during a portion of the cycle. The oil is flowing over the relief approximately 1/4 the time. This is an average of idle time between cycles, plus operating time and this is over a one hour period. In this example, the average rate of heat generation is 2 Hp.*

Pressure Reducing valves and Pressure Compensated Flow Controls are another source of heat generation. Some Pressure Reducing valves should be checked for heat generation if they are passing high flows or reducing the pressure from an extremely high pressure.

Pressure Compensated Flow Controls, when used to reduce the flow to or from a cylinder or to or from a fluid motor when using a fixed delivery pump, will cause the excess oil not used to move the actuator to flow over the relief valve. Sometimes it is best to use a pressure compensated pump in a system like this or use what is called a bleed off circuit. The bleed off circuit allows the excess oil to flow back to the tank unrestricted. This type of circuit will help reduce heat generation.

Sometimes it is helpful to figure 15%-20% of the electric motor horsepower will go into heat. This will help calculate this hard-to-figure point in a system, such as: miscellaneous valves, fluid friction in pipes, tubing or hose, mechanical friction and slippage of pumps, fluid motor and cylinders. In cases of marginal to high heat generation, it is a good practice to install heat exchanger connections in the main oil return line and relief valve return line. This will allow a heat exchanger to be added at any time.

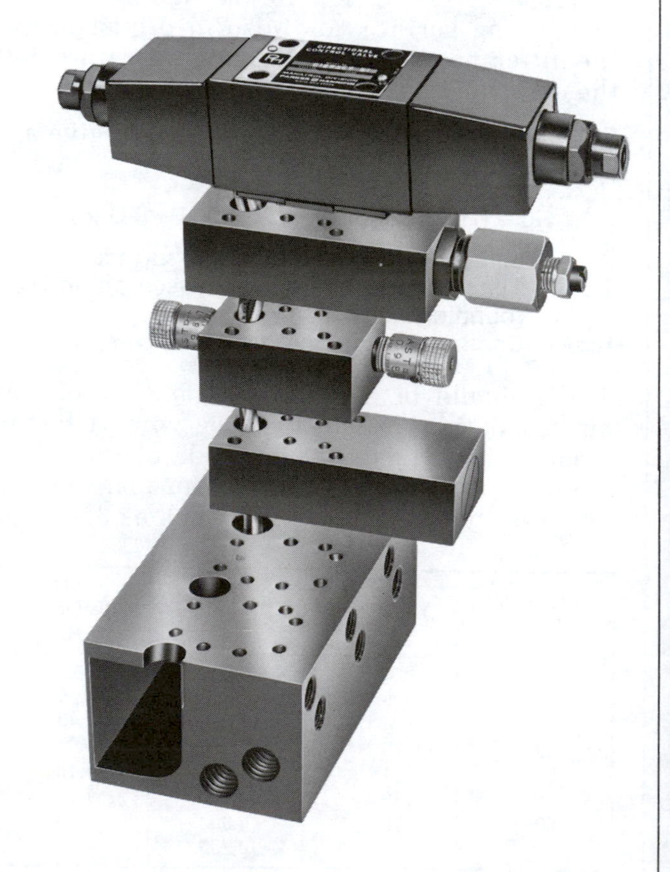

PH Stack Valves

illustration h-9

heat generation & control continued

suggestions for heat reduction

1. Where heat generation may be a problem in some systems, use a generous size reservoir.

2. Unload the pump when system pressure is not required. An unloading valve or a solenoid controlled pilot operated relief may be used to accomplish this.

3. Set the main relief valve for the amount of pressure that is required to do the work. An additional 150 psi should be added to this set point to keep the relief from cracking (opening).

4. If possible, when using pressure compensated flow controls with fixed delivery pumps, connect them in a bleed off arrangement.

5. Reservoirs should be located in the open to insure good air circulation. Reservoirs located outside should be shaded from the rays of the sun.

6. Accumulators may sometimes be used in holding or clamping circuits where high static pressures are needed.

7. On some systems, compressed air to oil intensifiers may be used to maintain system pressure with minimum amount of heat generation.

PH Solenoid Operated Relief Valves
illustration h-9

notes

power unit start-up considerations

initial start-up

Some units can be immediately damaged unless the following initial starting precautions are taken:

Fluid

Clean petroleum oil meeting or exceeding lubricating qualities of SAE 10W with API service classification MS is recommended. Viscosity range 150-250 SSU at 100°F.

Consult Factory for system requiring other fluids.

Case Drain (Variable Volume & Fixed Piston Pump)

The case drain line must not exceed 10 PSI back pressure. The drain line should be at least 3/8 diameter thin wall and piped directly back to reservoir with minimum number of elbows. Suggested maximum line length is 10 feet. This drain line must be extended below fluid level as far from the intake line as possible.

Inlet

Pump vacuum at inlet must not exceed 7 inches HG with petroleum base fluids and 3 inches HG with fire resistant fluids. **Air leak in suction line will cause damage and noise in pump.**

Port Connections

"SAE" or "AN" straight thread fittings should be used on all pump ports. **Tapered pipe thread fittings will distort and damage the pump if tightened too tight.**

Fluid Temperature

Inlet or reservoir fluid temperature must not exceed 145°F or drop below 40°F. If other than noted, consult the power unit manufacturer.

PVV 23 & PVV 33 Pressure Compensated, Variable Volume, Pump

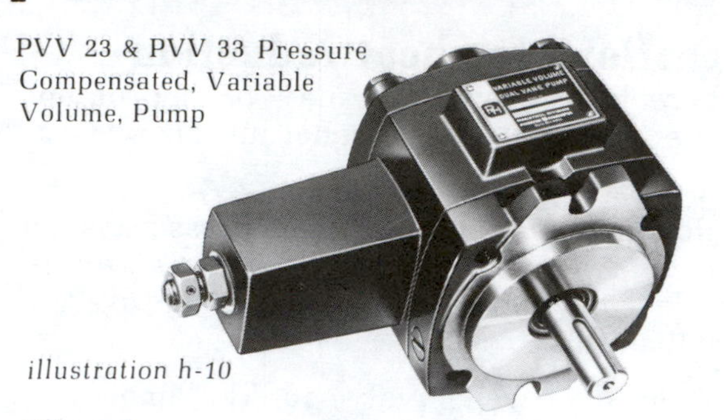

illustration h-10

Filtration

Suction filtration should be $\beta 10$ of at least 5.

Shaft Rotation & Line Up

Check for proper direction of rotation of both motor and pump. Pump and motor shaft alignment must be within specs given on page h-3. Please follow coupling manufacturer's recommended installation instructions to prevent end thrust on pump shaft. Turn pump by hand to assure freedom of rotation. **Pump and motor MUST be on a rigid base.**

If possible, initial start-up should be with an "open" hydraulic circuit. Pressure should be reduced to below 100 PSI, and air bled from the pump outlet line to permit priming. Bleed by loosening an outlet connection until solid stream of fluid appears. Operate pump at reduced pressure level for 2 minutes before increasing pressure adjustment to desired level.

Jog Start

Jog start, on & off, to check circuit connections.

Boost Pump

Where "closed" or boosted circuits are used make sure the boost pump is started first.

noise control

sources of noise

In order to study noise and the power unit, let us first examine the basic components which make up the system. They are as follows: Directional Control Valves, Flow Control Valves, Pressure Control Valves, Relief Valves, Electric Motors, Drive Couplings, Pump, Filters, and the Hydraulic Reservoir.

There are, of course, many combinations of component types and quantities available on a power unit. This simple system will be used for discussion.

Now that we have defined the basic components, we will look at the three sources of noise. They are:

- Noise is generated
- Noise is transmitted
- Noise is amplified

Noise generation basically arises from the vibrations of surfaces in contact with air. Such examples are: fans, bearings, gear teeth, etc.

Noise transmission exists when noise waves generated from one object set up vibrations in other objects. This can be achieved by solid to solid, liquid to liquid and liquid to solid contact.

Noise amplification (magnification) occurs when

noise control continued

sound waves are increased due to the surface or material it comes in contact with. For example, an alarm clock hanging from a string is much quieter than when placed on a table.

The effect of these noise sources vary through a hydraulic system.

The area of greatest potential generation exists at the rotating members of the system, while noise transmission and amplification is most noticed in the hydraulic reservoir. Also, the rotating members, depending on the type of mounting, can be a prime source of noise transmission.

Now that the major sources of noise are defined, let us look at the three ways of reducing noise in a hydraulic power unit. They are:

- Lower generation
- Isolate
- Dampen

To lower generation we should:

1. Convince component manufacturers to design for "quiet" and select components. In other words, select components which in themselves do not generate excessive noise.
2. Use open-dripproof-motors wherever possible to eliminate fan noise encountered in TEFC (Totally Enclosed, Fan Cooled) motors. Some steps have been taken in this area by motor manufacturers. (Eg: plastic fans, oil cooled motors.)
3. Keep motor-pump speed as low as possible, preferably 1200 rpm.
4. Size of hydraulic lines so oil velocity is approximately 15 fps (feet per second) on pressure lines, 8 fps on return lines and 4 fps on suction lines. Avoid sharp bends in piping.
5. Size components properly to avoid high oil velocities (if 5 gpm is required, do not use a 8 gpm pump and dump 3 gpm over the relief valve back to the tank).
6. Use stable control valves to eliminate hunting or chattering. For example: A pilot operated relief versus a direct action relief.
7. Always use a good grade of hydraulic oil with antifoaming additives to prevent air from entering the system.
8. Maintain oil temperature between 100°F and 120°F. This will maintain oil viscosity at a level which has better noise dampening characteristics.
9. Select the proper coupling and maintain pump-motor alignment. Never exceed .003 T.I.R. (Total Indicator Reading).
10. Avoid needle valves or low cracking check valves.
11. Remove all entrapped air prior to starting the system. Be especially careful on the pump suction line.
12. Select a proper reservoir which by size and design will allow air in the oil to escape. Locate suction lines as far from the return lines as possible with adequate baffling.
13. Supercharge pump, if possible, and in no case exceed 7 inches of mercury inlet vacuum.

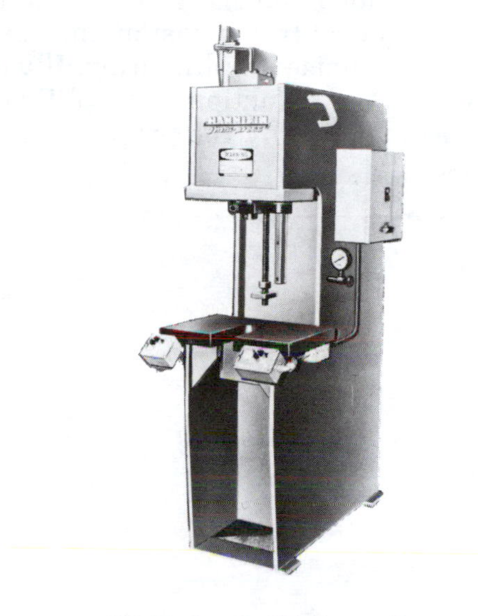

PH Hydraulic "C" Frame Press

illustration h-11

isolating noise

Once the noise generation in a power unit is reduced to a minimum and further noise reduction is desired, attempts to isolate the remaining noise generated must be taken. The following are some of the most common methods of accomplishing this.

1. Select a coupling which will isolate the pump and motor sounds.
2. Install pump-motor mounting plate on isolation mounts.
3. Use flexible hose on pump suction to isolate the pump-motor assembly and prevent noise transmission to the reservoir.
4. Also, for the same reason as listing "3." use a hose assembly on the pump outlet connection.
5. Again, use a flexible hose on the case drain (where applicable). Make sure this drain line terminates below the oil level to prevent siphoning of the pump case.

noise control continued

6. Minimize tubing lengths, use manifolds where possible.

7. When tubing is used, properly secure it, use isolation type mounts, prevent metal to metal contacts.

8. Always use a reservoir which is rigid enough to carry the weight of the components mounted on it.

9. Use acoustic baffling or absorptive materials to reduce noise levels where practical. This will lower noise transmission and reduce the possibility of noise amplification. **Illustration h-12** shows the reduction in amplification which can be achieved by the proper selection of materials.

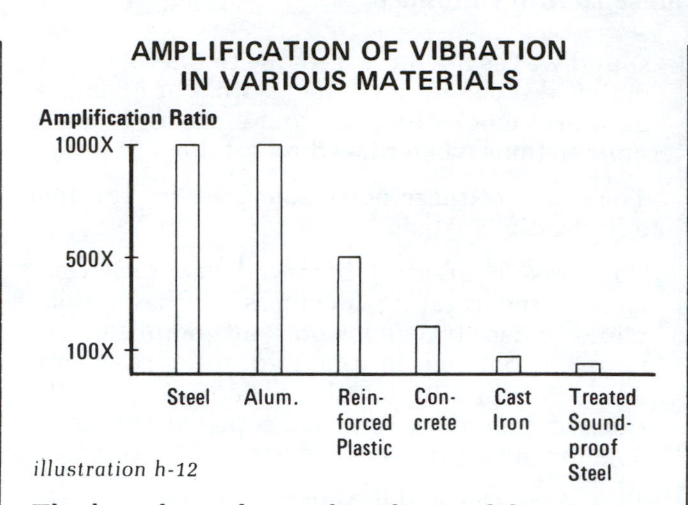

illustration h-12

The bar chart shows that the amplification ratio can be reduced from 1000X for steel and aluminum to 50X with cast iron and 10X with treated soundproof steel.

notes

troubleshooting areas

dirty oil

1. Components not properly cleaned after servicing.
2. Inadequate screening in fill pipe.
3. Air breather left off. (No air breather provided . . . inadequate unit provided . . . insufficient protection of air breather.)
4. Tank not properly gasketed.
5. Pipe lines not properly covered while servicing machine.
6. Improper tank baffles not providing settling basin for heavy materials.
7. Filters elements not replaced at proper intervals, or when indicated.

fire resistant fluids

1. Incorrect seals cause excessive friction which binds spool.
2. Paint, varnish and enamel in contact with fluids can cause sludge deposits on filters and around seal areas.
3. Electrolytic action is possible with some metals. Usually zinke or cadmium.
4. Improper mixtures can cause heavy sludge formations.
5. High temperatures adversely affect some of the fluids, particularly the aqueous base fluids.
6. Adequate identification of tanks containing these fluids should be provided so that they will be refilled with the proper media.
7. As with mineral base oils, nuisance leaks should be remedied at once.
8. Make certain replacement parts are compatible with fluid media.

foaming oil

1. Return of tank line not below fluid level. Broken pipe, line left out between a bulkhead coupling and the bottom of the tank after cleaning tank.
2. Inadequate baffles in reservoir.
3. Fluid contaminated with incompatible foreign matter.
4. Suction leak to pump aerating oil.
5. Lack of anti-foaming additives.

moisture in oils

1. Cooling coils not below fluid level.
2. Cold water lines fastened directly against hot tank causing condensate within tank.
3. Soluble oil solution splashing into poorly gasketed tanks or fill pipes left open.
4. Moisture in cans used to replace fluid in tanks.
5. Extreme temperature differential in certain geographical locations.
6. Drain not provided in lowest point in tank to remove water collected over possibly long operating periods.

overheating of system

1. Water shut off or heat exchanger clogged.
2. Continuous operation at relief setting.
   a. Stalling under load, etc.
   b. Fluid viscosity too high.
3. Excessive slippage or internal leakage.
   a. Check stall leakage past pump, motors and cylinders.
   b. Fluid viscosity too low.
4. Reservoir sized too small.
5. Reservoir assembled without baffling or sufficient baffling.
6. Case drain line from pressure compensated pump returning oil too close to suction line.
   a. Repipe case drain line to opposite side of reservoir baffling.
7. Pipe, tube or hose I.D. to small causing high velocity.
8. Valving too small, causing high velocity.
9. Improper air circulation around reservoir.
10. System relief valve set too high.
11. Power unit operating in direct sun light or ambient temperature is too high.

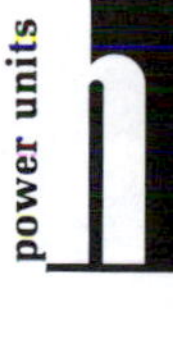

troubleshooting areas continued

foreign matter sources in the circuit

1. Pipe scale not properly removed.

2. Sealing compound (pipe dope teflon tape) allowed to get inside fittings.

3. Improperly screened fill pipes and air breathers.

4. Burrs inside piping components.

5. Tag ends of packing coming loose.

6. Seal extrusions from pressure higher than compatible with the seal or gasket.

7. Human element . . . not protecting components while being repaired and open lines left unprotected.

8. Wipers or boots not provided on cylinders or rams where necessary.

9. Repair parts and replacement components not properly protected while stored in repair depot. (Rust and other contaminants.)

pumps

mounting flanges for pumps and motors i-2
pump selection factors i-5
gallon equivalents in cubic inches ... i-6
measures of volume (compared) i-6
hydraulic pump calculation i-8
double pumps vs. single pumps i-11
pump application i-12
pump circuits i-14
pump operation i-16
basic construction i-18
variable delivery from a gear pump .. i-24
troubleshooting pumps i-25

mounting flanges for pumps and motors

STANDARD KEYWAYS AND SETSCREWS

Diameter of Shaft	Keyway Width x Depth	Diameter of Setscrew	Diameter of Shaft	Keyway Width x Depth	Diameter of Setscrew
5/16" – 7/16"	3/32" x 3/64"	3/16"	2-5/16" – 2-3/4"	5/8" x 5/16"	5/8"
1/2" – 9/16"	1/8" x 1/16"	3/16"	2-13/16" – 3-1/4"	3/4" x 3/8"	3/4"
5/8" – 7/8"	3/16" x 3/32"	1/4"	3-5/16" – 3-3/4"	7/8" x 7/16"	3/4"
15/16" – 1-1/4"	1/4" x 1/8"	1/4"	3-13/16" – 4-1/2"	1" x 1/2"	3/4"
1-5/16" – 1-3/8"	5/16" x 5/32"	5/16"	4-9/16" – 5-1/2"	1-1/4" x 7/16"	3/4"
1-7/16" – 1-3/4"	3/8" x 3/16"	3/8"	5-9/16" – 6-1/2"	1-1/2" x 1/2"	1"
1-13/16" – 2-1/4"	1/2" x 1/4"	1/2"	6-9/16" – 7-1/2"	1-3/4" x 5/8"	1"
			7-9/16" – 8-15/16"	2" x 3/4"	1"

table i-1

TWO-BOLT MOUNTING FLANGES

MOUNTING FLANGE		PILOT DIMENSIONS				FLANGE DIMENSIONS			
USA	SAE	A	W	X Min	Y	K	M	P_1 Rad	B
2F32	A	3.250 / 3.248	0.250 / 0.240	0	0.031	4.192 / 4.182	0.437	0.469	3.750
2F40	B	4.000 / 3.998	0.380 / 0.360	2.000	0.062	5.755 / 5.745	0.562	0.562	4.750
2F50	C	5.000 / 4.998	0.500 / 0.480	2.500	0.062	7.130 / 7.120	0.687	0.625	5.812
2F60	D	6.000 / 5.998	0.500 / 0.480	2.750	0.062	9.005 / 8.995	0.812	0.750	7.875
2F65	E	6.500 / 6.498	0.625 / 0.605	2.750	0.094	12.505 / 12.495	1.062	1.000	10.625
2F70	F	7.000 / 6.998	0.625 / 0.605	2.750	0.094	13.786 / 13.776	1.062	1.000	11.750

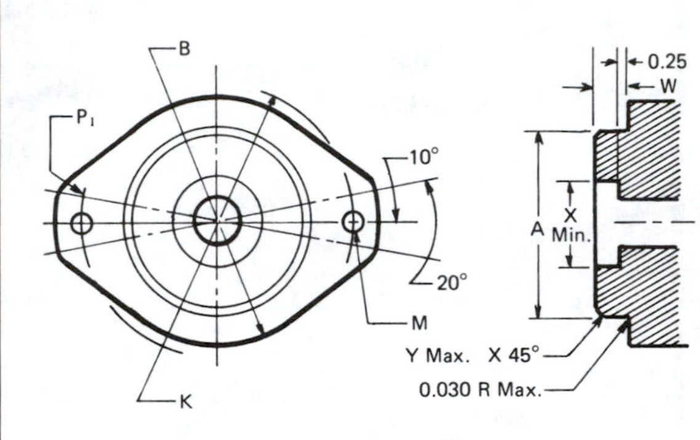

table i-2

FOUR-BOLT MOUNTING FLANGES

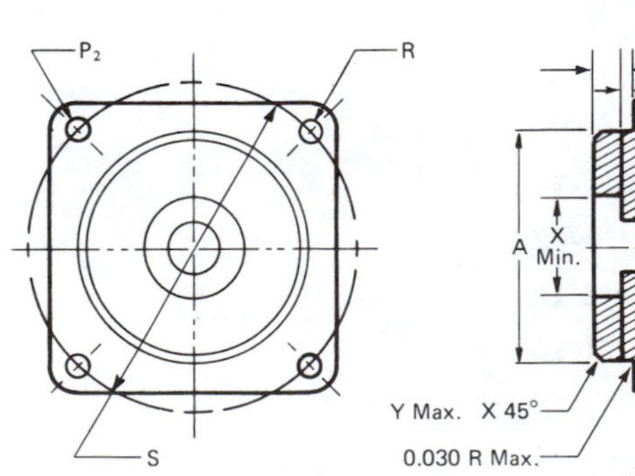

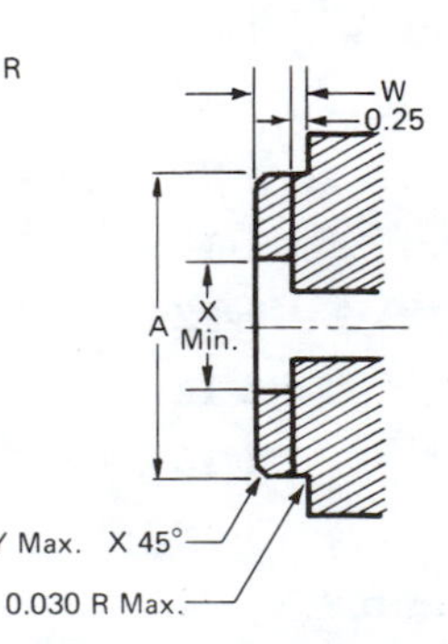

MOUNTING FLANGE		PILOT DIMENSIONS				FLANGE DIMENSIONS		
USA	SAE	A	W	X Min	Y	S	R	P_2 Rad
4F17	–	1.781 / 1.779	0.125 / 0.115	0	0.031	2.843 / 2.833	0.375	0.375
4F32	A	3.250 / 3.248	0.250 / 0.240	0	0.031	4.130 / 4.120	0.438	0.468
4F40	B	4.000 / 3.998	0.380 / 0.360	2.000	0.062	5.005 / 4.995	0.562	0.562
4F50	C	5.000 / 4.998	0.500 / 0.480	2.500	0.062	6.380 / 6.370	0.562	0.625
4F60	D	6.000 / 5.998	0.500 / 0.480	2.750	0.062	9.005 / 8.995	0.812	0.750
4F65	E	6.500 / 6.498	0.625 / 0.605	2.750	0.094	12.505 / 12.495	0.812	0.750
4F70	F	7.000 / 6.998	0.625 / 0.605	2.750	0.094	13.786 / 13.776	1.062	1.000

table i-3

GPM & H.P. FROM CU. IN./REV.

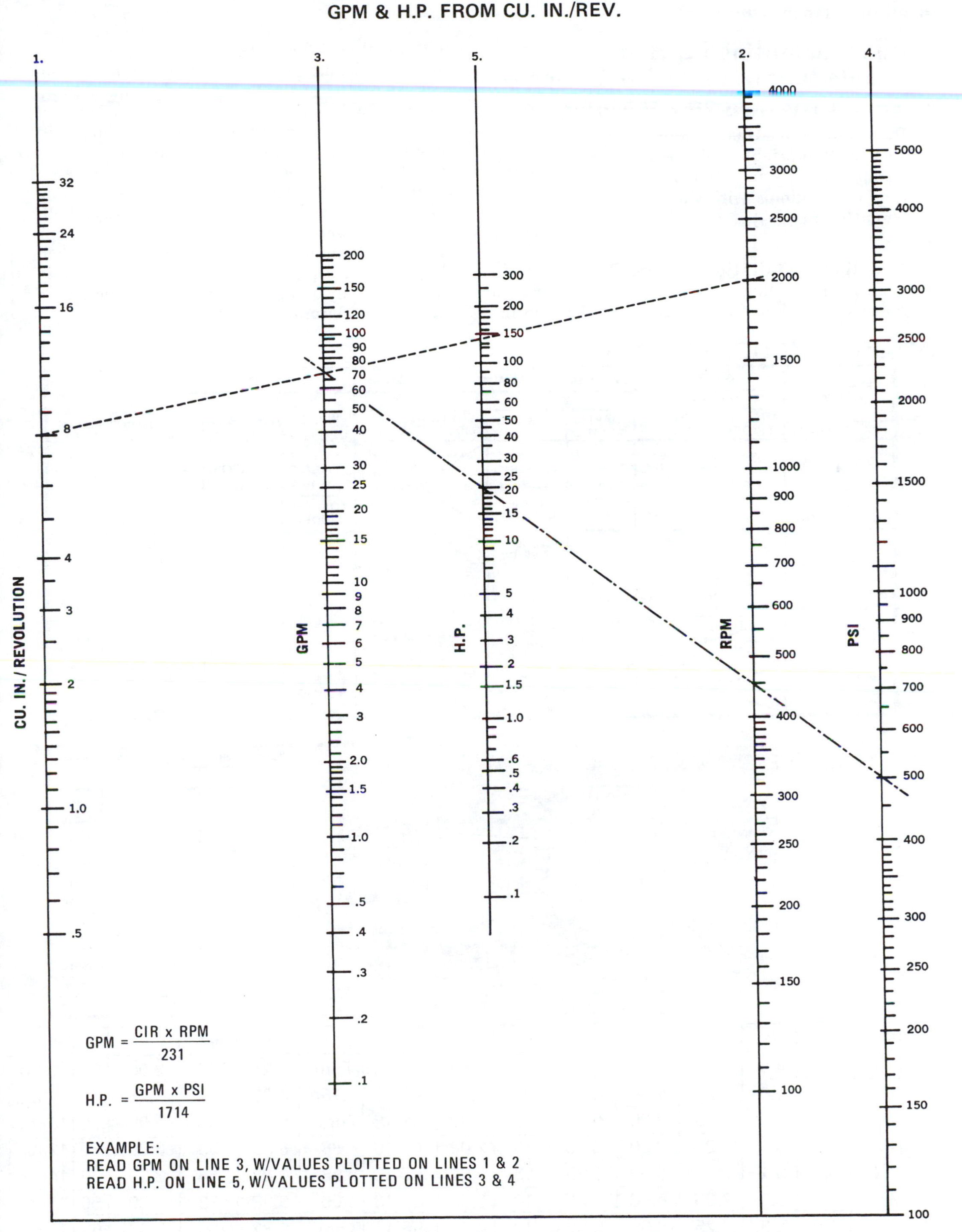

$$GPM = \frac{CIR \times RPM}{231}$$

$$H.P. = \frac{GPM \times PSI}{1714}$$

EXAMPLE:
READ GPM ON LINE 3, W/VALUES PLOTTED ON LINES 1 & 2
READ H.P. ON LINE 5, W/VALUES PLOTTED ON LINES 3 & 4

graph i-1

mounting flanges continued

pump mounting bracket adapts following Parker Pumps — To 1/4 HP thru 20 HP Motors

D, H & M Series Gear Pumps
PVV 23/33 Variable Volume, Pressure Compensated, Vane Pumps
(and other pumps with similar mounting flanges)

Coupling Installation — The Parker totally enclosed pump mounting bracket offers excellent noise absorbtion and safety from the rotating shafts and coupling. The bracket is designed to mount on the motor face with the motor coupling half secure to the shaft. Then the pump, with its coupling half secure on the pump shaft, is mounted and the coupling halves are engaged. This will require proper spacing of the coupling prior to installation and a coupling with an outside diameter less than "P" dimension. If the coupling selected cannot be assembled this way, both coupling halves must be installed on the motor shaft. Next, mount the adapter on the motor. Then the pump can be mounted and the coupling secured to the pump by using the access slot to tighten the pump shaft coupling set screw.

ADAPTER MODEL SELECTION CHART

ELECTRIC MOTOR SIZE		SAE AA MTG		SAE A MTG STD SHAFT		SAE A MTG LONG SHAFT		SAE B MTG		SAE C MTG	
		D GEAR		H GEAR		H GEAR		M GEAR PW23/33 VAR VANE VPO33 VAR PISTON PAF 10/16/20/25 PISTON PAV 16 VAR PISTON		VPO58 VAR PISTON PAF 32/40/50 PISTON PAV 32 VAR PISTON	
FRAME	SHAFT	ADAPTER MODEL	REC CPLG	ADAPTER MODEL	REC CPLG	ADAPTER MODEL	REC CPLG	ADAPTER MODEL	REC CPLG	ADAPTER MODEL	REC COUPLING
56C	5/8	DS	L-075 LOVEJOY	HS	L-075 LOVEJOY	X	X				
143TC	7/8	DS	L-095 LOVEJOY	HS	L-095 LOVEJOY	X	X				
145TC	7/8	DS	L-095 LOVEJOY	HS	L-095 LOVEJOY	X	X				
182TC	1-1/8	DM	AL-100 LOVEJOY	HM	AL-100 LOVEJOY	HML	AL-110 LOVEJOY	MM	AL-100 LOVEJOY		
184TC	1-1/8	DM	AL-100 LOVEJOY	HM	AL-100 LOVEJOY	HML	AL-110 LOVEJOY	MM	AL-100 LOVEJOY		
213TC	1-3/8	DM	300 MAGNALOY	HM	300 MAGNALOY	HML	AL-110 LOVEJOY	MM	AL-100 LOVEJOY	RM	AL-110 LOVEJOY
215TC	1-3/8	DM	300 MAGNALOY	HM	300 MAGNALOY	HML	AL-110U LOVEJOY	MM	AL-100 LOVEJOY	RM	AL-110 LOVEJOY
254TC	1-5/8	X	X	HM	400 MAGNALOY	HML	AL-110U LOVEJOY	MM	400 MAGNALOY	RM	400 MAGNALOY
256TC	1-5/8	X	X	HM	400 MAGNALOY	HML	AL-110U LOVEJOY	MM	400 MAGNALOY	RM	400 MAGNALOY
284TC	1-7/8	X	X	X	X	X	X	MM + RING 753006	AL-150 LOVEJOY	RM + RING 753006	AL-150 LOVEJOY
286TC	1-7/8	X	X	X	X	X	X	MM + RING 753006	AL-150 LOVEJOY	RM + RING 753006	AL-150 LOVEJOY

table i-4

56C FRAME & TC FRAME MOTOR DATA FOR ADAPTER CHART MOTOR SIZE

TYPE	OPDP	OPDP OR TEFC		FRAME SIZE
RPM	3600	1800	1200	
HP	½–1	¼–¾	X	56C
HP	2	1	¾	143TC
HP	3	2	1	145TC
HP	5	3	1½	182TC
HP	7½	5	2	184TC
HP	10	7½	3	213TC
HP	15	10	5	215TC
HP	20	15	7½	254TC
HP	25	20	10	256TC
HP	X	25	15	284TC
HP	X	30	20	286TC

MODEL	A	B	C	D	E	F	G	H	J	L	M	P	T	WEIGHT
DS	6.5	4.5	5.88	4.50	4.04	.21	1.62	3/8-16	Thru	4.25	.44	2.00	.50	8 lbs.
HS	6.5	4.6	5.88	4.50	4.04	.21	2.09	3/8-16	.75	4.25	.44	3.25	.50	7 lbs.
DM	9.0	5.1	7.25	8.50	5.25	.13	1.62	3/8-16	Thru	5.38	.53	2.00	.50	13 lbs.
HM	9.0	5.1	7.25	8.50	5.25	.13	2.09	3/8-16	.88	5.38	.53	3.25	.50	12 lbs.
HML	9.0	5.1	7.25	8.50	6.75	.13	2.09	3/8-16	.88	6.88	.53	3.25	.56	14 lbs.
MM	9.0	6.5	7.25	8.50	6.25	.13	2.88	1/2-13	1.00	6.38	.53	4.00	.50	14 lbs.
RM	9.0	8.38	7.25	8.50	6.69	.13	3.56	5/8-11	Thru	6.82	.56	5.00	.63	17 lbs.

table i-5

pump selection factors

Pressure Capability — One of the most critical factors affecting selection of a pump is its pressure rating. Where there is any doubt concerning the peak pressure to which a pump will be exposed, the pump should be slightly overspecified.

Some personnel often have no idea at what pressure a pump is operating because of the absence of pressure indicators. As a result, pumps often operate at excessive pressure levels. Over-pressure operation results in accelerated wear and ultimate early failure. This problem is common in the industry. To avoid it, systems should be equipped with a remote gage or with a simple means for inserting a gage to perform maintenance checks.

Relief valves are used with fixed-displacement pumps to maintain a fixed pressure setting when the system flow requirement is below the pump output. Plant personnel frequently will adjust this valve to speed up cycle time or to lift a heavier load. This action can result in the pump's rated pressure being exceeded. One solution is to insert "once-adjusted" or sealed relief valves.

Pump Selection — Selection of a pump starts with the selection of a working pressure for the system. Working pressure reflects system cost. A 500-p.s.i. system requires a larger line, larger components, and a larger pump to perform a job than does a 2,000-p.s.i. system. The 2,000-p.s.i. system may have lower initial cost, but if it is not properly engineered, it may require more maintenance. The tendency today, with better maintenance and better components available, is for designers to specify systems with higher pressures to achieve more work at a lower cost.

Once the system pressure has been established, the pump selection can be made. The designer should consider the function (variable or fixed), maximum pressure recommended by the manufacturer, volume, cost, noise, and reliability.

Drive Speed — Excessive operating speeds often result in pump cavitation, which occurs when the suction exceeds a limit the fluid cannot tolerate. Under these conditions bubbles form in the fluid. When these bubbles are exposed to the pressure cavity, they collapse rapidly. This action is called cavitation. It results in noise, erosion, and eventual destruction of the pump.

Essentially, two drive speeds are utilized in industrial pump applications—1,800 and 1,200 rpm. It is recommended that the higher speed be utilized where possible since a greater horsepower-per-dollar value can be achieved at the higher speeds in systems with shaft inputs up to 100 hp. Some pumps are capable of speeds up to 3,600 rpm, but at this speed a noise problem may be encountered, and the danger of cavitation is increased.

To avoid cavitation the pump suction line should (1) be as short as possible, (2) have as large a diameter as practicable, and (3) have a minimum number of elbows.

Environment — The pump is the most critical element in a hydraulic circuit relative to environment. In metal casting applications, for example, the pump will be the first to wear out due to contamination. In general, a system that will provide sufficient protection for the pump also will provide more than adequate protection for all of the other elements.

Filters must be selected carefully to keep contaminants out of the system. A system without a filter is asking for trouble. Therefore a designer should be generous in selecting a filter. However, no matter how good the filter, some foreign materials will find their way into the system. The best that can be achieved is to minimize contaminants.

Reservoir design also is an important factor in cutting down on system contamination. The designer should arrange the breather and fill locations so that they do not collect dirt particles.

Environmental temperatures also affect operation. Some foundries experience extremes in temperature in the molding area. The temperature can go down to 35°F in the winter and up to 110°F in the summer. These variations cause the fluid to undergo viscosity changes that may result in (1) pump cavitation due to high viscosity at low temperatures or (2) excessive internal slippage due to low viscosity at high temperatures. Fluid selection in such

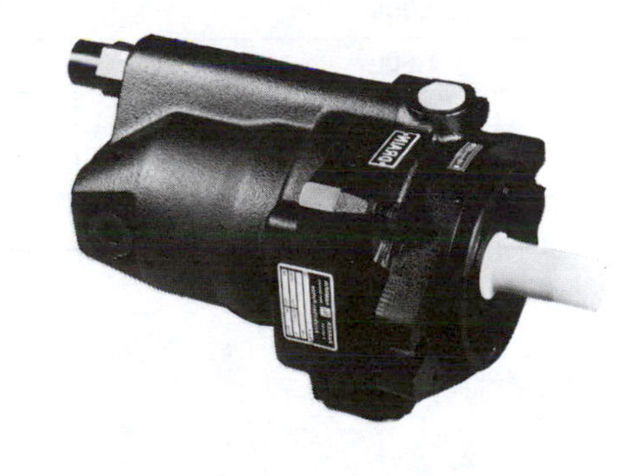

PH Pressure Compensated Pump

illustration i-1

gallon equivalents in cubic inches

GAL.	CU. IN.	GAL.	CU. IN.	GAL.	CU. IN.	GAL.	CU. IN.	GAL.	CU. IN.
1	231	21	4851	41	9471	61	14091	81	18711
2	462	22	5082	42	9702	62	14322	82	18942
3	693	23	5313	43	9933	63	14553	83	19173
4	924	24	5544	44	10164	64	14784	84	19404
5	1155	25	5775	45	10395	65	15015	85	19635
6	1386	26	6006	46	10626	66	15246	86	19866
7	1617	27	6237	47	10857	67	15477	87	20097
8	1848	28	6468	48	11088	68	15708	88	20328
9	2079	29	6699	49	11319	69	15939	89	20559
10	2310	30	6930	50	11550	70	16170	90	20790
11	2541	31	7161	51	11781	71	16401	91	21021
12	2772	32	7392	52	12012	72	16632	92	21252
13	3003	33	7623	53	12243	73	16863	93	21483
14	3234	34	7854	54	12474	74	17094	94	21714
15	3485	35	8085	55	12705	75	17325	95	21945
16	3696	36	8316	56	12936	76	17556	96	22176
17	3927	37	8547	57	13167	77	17787	97	22407
18	4158	38	8778	58	13398	78	18018	98	22638
19	4389	39	9009	59	13629	79	18249	99	22869
20	4620	40	9240	60	13860	80	18480	100	23100

table i-6

measures of volume (compared)

Cu. Yd.	Cu. Ft.	Cu. In.	Gal.	Liquid Qt.	Pint	Gill	Oz.	Drahm	Minim.	METRIC		
										Liter	C. C.	Cu. Meter
1.31	35.310	1728.000	264.00									1.00
1.00	27.000											.76
	1.000	1728.000	7.48	30.00	60.00					28.310		
	.134	231.000	1.00	4.00	8.00	32	128.000			3.790		
		61.000		1.06	2.11					1.000		
		58.000		1.00	2.00	8	32.000				946.00	
		29.000			1.00	4	16.000				473.00	
		1.800					1.000	8	480		29.55	
		1.000					.550				16.40	
							.120	1	60		3.70	
		.061					.034			.001	1.00	

table i-7

pump selection factors continued

situations is critical and may require different types during different seasons.

System Fluids — Although hydraulic fluids do not readily wear out or break down, they should be changed periodically to assure that clean fluid is in the system at all times. Additives in the fluid do break down, and even may be filtered out by a good filtering system. Fluids with anti-wear additives and foam depressants now are common, and they should be specified. Fire resistance of fluids is another consideration. Recommendations of both pump manufacturers and fluid suppliers should be studied carefully before a system is put into operation.

notes

hydraulic pump calculation formula

Hp = Horsepower
RPM = Revolutions per minute
w = Work
d = Displacement — cubic inches
Q = Flow rate — gpm
G.P.M. = Gallons per minute (U.S.)
T = Torque
Sp.Gr. = Specific gravity
F = Force (pounds)
V = Speed (in/min)
π = 3.14
r = Radius
psi. = Pounds per square inch
PSIA = Pounds per square inch-absolute
PSIG = Pounds per square inch-gauge
ΔP = Pressure drop — p.s.i.
231 in.^{3} = 1 U.S. Gallon

PUMP OUTPUT FLOW

$$1 \text{ GAL.} = 231 \; \text{in}^3$$

$$\text{Theoretical output (gpm)} = \frac{\text{Displacement (cu. in.)/rev.} \times \text{rpm}}{231 \text{ cu. in.}}$$

$$\text{G.P.M.} = \frac{\text{RPM} \times d}{231}$$

VOLUMETRIC EFFICIENCY PERCENT

$$\text{volumetric efficiency (pump)} \; \frac{\text{actual displacement (G.P.M.)}}{\text{theoretical displacement (G.P.M.)}} \times 100$$

MECHANICAL EFFICIENCY PERCENT

$$\text{\% Mech. eff.} = \frac{\text{Theoretical torque required to drive the pump (lb. in.)}}{\text{Actual torque supplied to pump (lb. in.)}} \times 100$$

OVERALL EFFICIENCY PERCENT

$$\text{Pump over-all eff. \%} = \frac{\text{Output hyd. Hp}}{\text{Input Hp to pump}} \times 100$$

PUMP OUTPUT HORSEPOWER

$$Hp = \frac{F x v}{396000}$$

PUMP

$$Hp = \frac{\text{G.P.M.} \times \text{PSI}}{1714 \times \text{eff. (pump)}}$$

$$\text{or } Hp = \frac{.000583 \times \text{G.P.M.} \times \text{p.s.i.}}{\text{eff. (pump)}}$$

$$Hp = \frac{T \times \text{RPM}}{5252} \quad \text{Torque in lb. ft.}$$

$$Hp = \frac{T \times \text{RPM}}{63025} \quad \text{Torque in lb. in.}$$

$$Hp = \frac{\text{RPM} \; \Delta \; \text{Pd} \quad (\text{in}^3/\text{rev})}{63025 \quad 2\pi}$$

$$Hp = \frac{T \times \text{RPM} \times 2\pi}{33{,}000 \times 12} \quad \text{Torque in lb.-in.}$$

$$Hp = \frac{\text{volts} \times \text{amperes}}{745.7}$$

POWER

$$\text{power} = \frac{\text{work}}{\text{time}} = \frac{\text{force} \times \text{distance}}{\text{time}}$$

$$\text{power} = \frac{231 \times Q \times \text{p.s.i.}}{12} \quad \frac{\text{ft. - lb.}}{\text{min.}}$$

$$HP = \frac{\text{power}}{33{,}000 \; \frac{\text{ft. - lbs.}}{\text{min.}}}$$

hydraulic pump calculation

hydraulic pump calculation

Pump Pressure Ratings — A 2000 p.s.i. pressure rating does not mean the pump will just develop 2000 p.s.i. in your system. This refers to the maximum pressure which should be developed by the pump allowing a safety factor, without danger of damage to the pumping element over an extended period of time.

By properly setting the system relief valve, you may limit maximum system pressure to any safe value less than the maximum pump rating.

Caution: *A relief valve built into the pump casing should not be used for the main circuit relief valve (other than hand pumps) unless it relieves back into the inlet of the pump, severe local over-heating of the oil can occur in less than one minute. These built-in relief valves are intended only for emergency protection, and should be set higher than the main relief valve setting.*

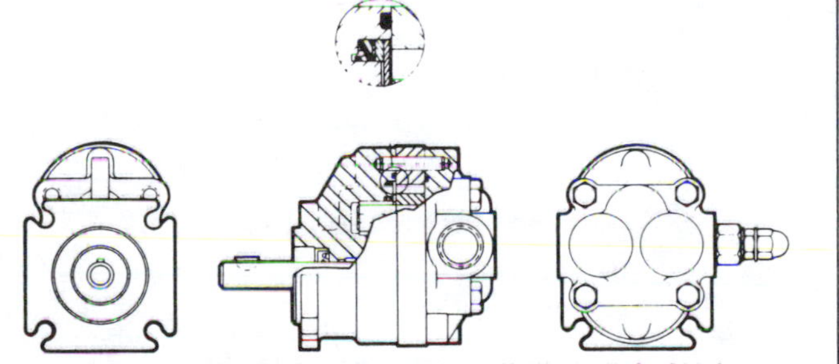

PH 4 Bolt Flange Gear Pump with Built-in Relief Valve
illustration i-2

Select a pump and drive motor to create about 10 to 25 % greater system pressure than needed at the actuator. This is required to take care of pressure losses in valves, piping and friction of the cylinder rod and piston.

sizing pumps

Example of the pumps displacement vs cylinder travel relationship is:

Example: *A cylinder having a 3-1/4" diameter piston is required to stroke 6 inches. By using a hand pump with a displacement of 1.75 cubic inches per pump stroke (pumping oil during the forward pump stroke only, the return stroke in this case is used to refill the pumping chamber) how many pumping strokes are required to move the cylinder piston 6 inches.*

The area of a 3-1/4" diameter piston is 8.29 square inches. The amount of oil required to move the piston 6 inches is 8.29 inches² × 6 inches = 49.6 inches³. The handpump, having a displacement of 1.75 inches³ during our pumping stroke and calculating the number of pumping strokes required is 49.6 inches³ divided by 1.75 inches³ = 28.4 pumping strokes.

Example: *How many GPM is required to move a 5 inch diameter cylinder piston at the rate of 2 inches per second.*

To calculate how much oil (in gallons per minute) is required to move the piston 2 inches every second

Use the formula:

$$\text{VOL (G.P.M.)} = \frac{A \times L \times 60}{231 \times \text{time in sec.}}$$

Where:

A = 5 inch dia. piston = 19.6 in.²
L = Length = 2 inches
231 = number of cubic inches/gallon
time in sec. = 1
60 = 60 sec./min.

The formula is now:

$$\text{VOL (G.P.M.)} = \frac{19.6 \times 2 \times 60}{231 \times 1}$$

$$\frac{19.6 \times 2 \times 60}{231 \times 1} = \frac{2345}{231}$$

$$2345 \div 231 = 10.1 \text{ G.P.M.}$$

Example: *What size pump, rated in GPM, is needed to extend and retract a 5 inch bore, 18 inch stroke cylinder. This cylinder has a 2 inch diameter piston rod and the total cycle time (out and back) is 12 seconds.*

Solution:

Using the formula:

$$\text{VOL (G.P.M.)} = \frac{A \times L \times 60}{231 \times \text{time in sec.}}$$

Find the area of a 5 inch Bore cylinder, the Bore = 19.6 inches². Find the effective area on the return stroke (this is the piston area less the rod area for the return stroke). This is 19.6 inches² - 3.1 inches² = the piston rod area 16.5 inches² which is the effective working area on the return stroke. We are looking for total cycle time extending and retracting add 19.6 inches² + 16.5 inches² = 36.1 inches² total area. Total motion of the cylinder is 18 inches.

hydraulic pump calculation continued

stroke 18 inches extending + 18 inches retracting equals 36 inches (total travel).

$$\text{VOL (G.P.M.)} = \frac{36.1 \text{ inches}^2 \times 18 \text{ inches} \times 60}{231 \text{ inches}^3 \times 12 \text{ seconds}}$$

$$\frac{36.1 \times 18 \times 60}{231 \times 12} = \frac{38988}{2773}$$

$$38988 \div 2773 = 14. \text{ G.P.M.}$$

When a pump is operating in a system, this pump will develop only the amount of pressure required to do the work. For example: when a cylinder is extending and moving a load, the only pressure developed by the pump and noted on the pressure gauge is what is required to overcome the load and friction. Now when the cylinder piston reaches the opposite end of the cylinder and stalls, the pressure in the system can reach an infinite value. When this happens, and if the system relief valve is left out or failed to function properly, the drive motor will stall, a hose, pipe or tube would burst, or possibly the pump or valve casting would burst. For this reason a system relief valve is required in a system, and set at a pressure setting which is equal to or below the lowest pressure rated component in the system. The system relief should be plumbed as close to the pump as possible. A relief or pressure reducing valve can be used in other parts of the system to limit these componenents to a lower pressure.

notes

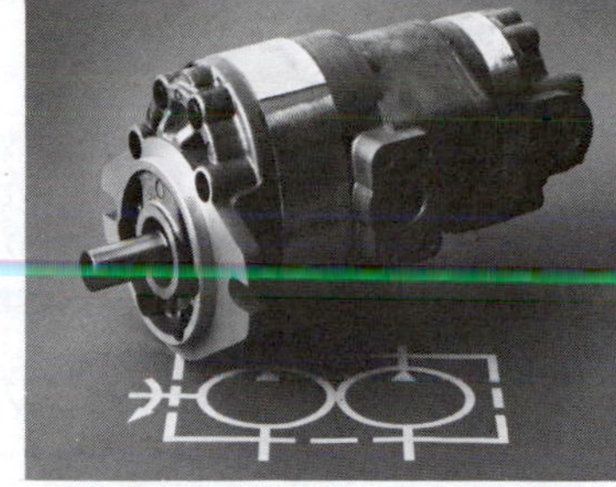

parker double gear pump

illustration i-3

double pumps vs. single pumps

parker single gear pump

illustration i-3a

Some systems require a cylinder to extend at a high rate and then slow down and do their work in the last few inches or less of stroke. These systems generally require a high volume of oil and low pressure to extend the cylinder rapidly. When reaching the last few inches of stroke a low volume of oil and a higher pressure to do the work. (Work such as clamping, pressing, forming, etc.) This can be accomplished by using two single pumps or double pump. When the cylinder is extending, oil is supplied by both pumps at a low pressure.

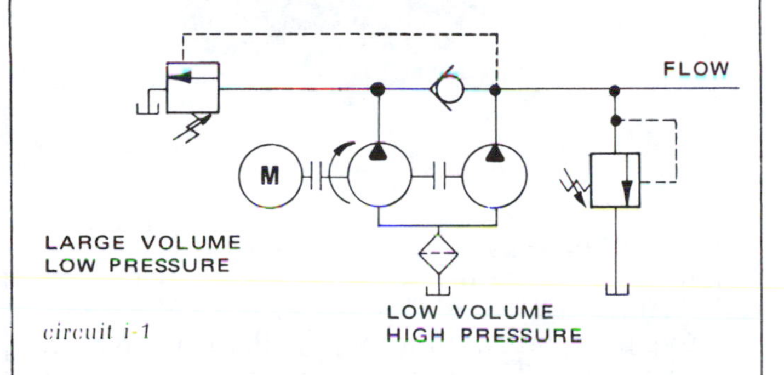

circuit i-1

When the cylinder meets the work resistance the large volume pump is automatically dumped by an unloading valve back to the reservoir, and the work is accomplished with the low volume pump at a higher pressure. The **circuit i-1** shows a typical double pump unloading circuit. Valve (1) is an unloading valve, spool type, with external pilot and internal drain. Valve (2) is a pilot operated relief valve, spool type, capable of handling the combined volume of both pumps. The unloading pressure is usually set where the driving motor has reached its full HP rating, driving both pumps. Then when the high volume pump is unloaded, the driving motor is able to use all its horsepower to drive the small pump to a higher working pressure.

Note: *The driving motor must have enough horsepower to drive both pumps to the unloading pressure or both pumps on the reverse stroke, whichever is the greatest. The motor must also have enough horsepower to drive the low volume, high pressure pump to its required working pressure, while the high volume low pressure pump is being unloaded.*

A hi-lo system cannot be used where high pressure is required during the full working cycle. (Both extending and retracting.)

double pump vs. single pump calculation

Single Pump

Problem — A sheet metal forming machine has a forming cylinder that has a 5 inch bore and a 15 inch stroke. The forming cylinder must extend the full 15 inches in 5 seconds and in the last 1/2 inch of stroke develop 1500 p.s.i.

Problem Solution

What size pump (GPM) and how much HP is required to do this work.

5 inch bore	=	19.6 inches^2
Stroke	=	15 inches
Time	=	5 seconds
Max. Pressure	=	1500 p.s.i.

Using formula

$$\text{VOL (G.P.M.)} = \frac{A \times L \times 60}{231 \times \text{time in sec.}}$$

$$\text{VOL} = \frac{19.6 \times 15 \times 60}{231 \times 5} = 15.3 \text{ G.P.M.}$$

Now using formula

$$\text{Hp} = \frac{\text{G.P.M.} \times \text{p.s.i.}}{1714 \times \text{Pump eff.}} = \text{Hp} = \frac{15.3 \times 1500}{1714 \times 85\%} = 15.7 \text{ Hp}$$

Or using the **"Electric Motor Horsepower" table** in this book **page c-2** the horsepower requirement is 15.7 Hp.

Double Pump Problem & Solution

Using the same requirements as indicated in the single pump system calculation, and making an assumption that 400 p.s.i. is required to extend the forming cylinder 14-1/2", and also return the forming cylinder 15 inches. Let's use the **Horsepower table, page c-2** in this hand book. It takes 4.2 horsepower to pump 15 GPM at 400 p.s.i.

pumps

double pump vs. single pump continued

It also takes 4.2 horsepower to pump 4 GPM at 1500 p.s.i. Using a 11 GPM pump and a 4 GPM pump, pumping a total of 15 GPM at 400 p.s.i. - 14-1/2" unloading the 11 GPM pump automatically through the unloading valve when the pressure starts to rise over 400 p.s.i. This allows the full horsepower to be used to develop the 1500 p.s.i. with the 4 GPM pump.

Since it takes 4.2 horsepower it is suggested if this is a continuous operation a 5 HP motor be used. This will allow a longer motor life and a practical safety factor. Now looking back at the single pump calculation, it takes a 15.7 HP electric motor to do the same job that a 5 HP electric motor can accomplish with a double pump system.

pump application

illustration i-4 PH Gear Pump.

PH Pressure Compensated, Variable Volume, Vane Pump
illustration i-5

Gear Pump Application — Gear pumps may be used singly in constant-volume hydraulic systems, or they may be used in combination with other pumps to satisfy the needs of multiple-function systems. For example, a gear pump may be used as the low-pressure, large-volume pump in a two-pump system. Gear pumps can be used also for auxiliary purposes to supply pressure for control circuits or to supercharge piston pumps.

Combination gear pumps also are available. These include the dual or tandem pump, the piggyback pump, and the flow-divider pump.

The tandem pump contains two pairs of gears with the same pitch diameter, but with different delivery rate. There is a common inlet port, but two separate discharge ports, each tied into a different system. The two deliveries can be tied together to feed a third system with a delivery equal to the sum of the two. Dual pumps often have as many as four pairs of gears, permitting a wide variety of combinations.

The piggyback pump actually is two separate pumps on a common shaft. A smaller pump is mounted onto a larger pump, and their shafts are coupled. The pitch diameters are dissimilar, and there are separate inlet ports. The delivery rates also vary.

Flow-divider pumps have a single inlet, one pair of gears, and two outlet ports. Delivery is controlled by a flow-sensitive valve. The valve modulates, directing the delivery to both outlet ports. The result is that the delivery of the first port does not exceed a desired value, and the second port handles the excess. This type of pump is common in power steering systems.

Advantages — Some advantages of gear pumps are as follow:

1. Low cost relative to power output.
2. Ability to handle foreign matter.
3. High operating speed.
4. Simplicity of design.

Vane Pump Application — Vane pumps generally are used to deliver fluid against system pressures up to 2,500 p.s.i. Variable delivery and fixed delivery vane pumps are being used up to 2,500 p.s.i. Double pumps, which develop two output volumes from a single shaft, can be applied independently to separate systems or combined for high-volume, low-pressure and low-volume, high-pressure applications. The two output cartridges are not usually of the same rating. Another use for double pumps is to fulfill a high-volume requirement. These double pumps go as high as 130 GPM.

pump application continued

Advantages — In general, vane pumps offer these advantages:

1. Low cost relative to power output.
2. Low noise level.
3. Long service life.
4. Ability to handle foreign matter.
5. Can be made with variable displacement.

Piston Pump Application — Piston-type pumps commonly are applied where higher pressures, high volumetric efficiencies, variable displacement, or any combination of these factors is required. Piston-type pumps available are capable of generating pressures as high as 10,000 p.s.i. but the output of most is limited to a maximum of 4,500 p.s.i.

Advantages — In general, piston pumps provide these advantages:

1. Higher working pressures
2. High volumetric efficiency.
3. Can be made to have variable displacement.

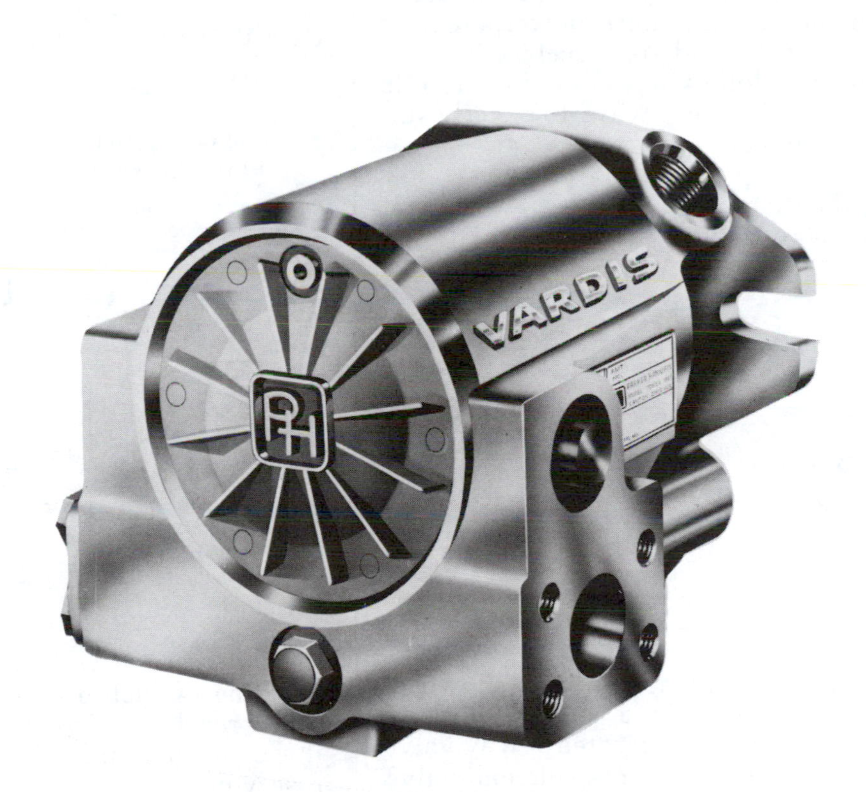

PH Pressure Compensated, Variable Volume, Piston Pump

illustration i-6

pump circuit

Double Gear Pump

illustration i-7

When Hydraulic Pumps are not needed to develop pressure, such as idle time when placing or removing parts from the machine, the oil should be allowed to flow freely, unrestricted, with little or no back pressure, back to the reservoir.

The pump, when spilling fluid across the relief valve, is consuming high horsepower, creating unnecessary heat, and is accelerating the pump wear. The following unloading circuits may be modified in many ways to exactly suit a particular application.

Unloading the Pump With a Two Way Valve

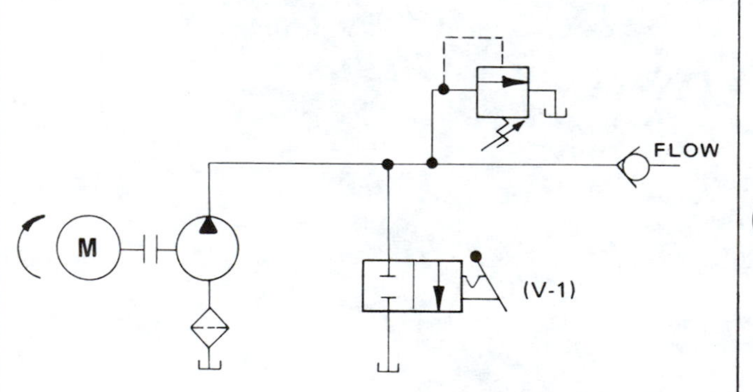

circuit i-2

The easiest way to unload a pump, shown in above circuit, is to unload the oil directly to tank with a manually operated 2-way valve (V-1). A better way is to use a solenoid valve energized automatically by a limit switch. The switch is actuated by the machine at the end of the cycle, and de-energizes the solenoid valve.

Low Pressure Unloading

This **circuit (i-3)** can be used for counterbalancing a cylinder against downward drift due to gravity. The system relief (R-1) is set for the systems working pressure. The remote relief (R-2) (direct operated) vents (R-1) to a lower pressure when valve (V-3) is in a de-

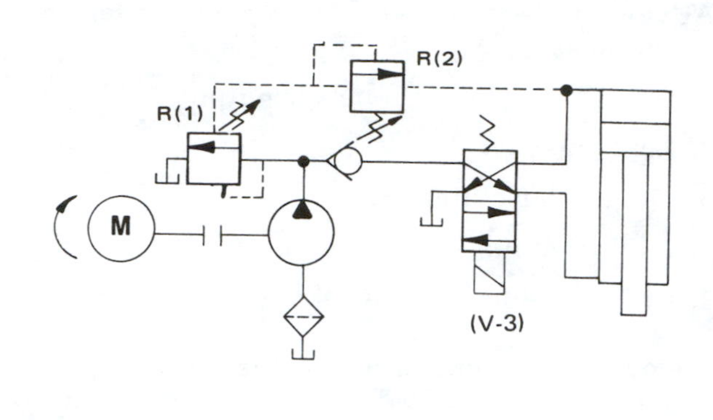

circuit i-3

energized position. When (V-3) is energized a pressure is created in the discharge port of (R-2) allowing (R-1) to operate at whatever setting (R-1) is set. The remote relief (R-2) can be 1/4" ported and should be set at only whatever pressure is needed to return the cylinder.

Cam Valve Unloading

PH Cam Operated Directional Valve

illustration i-8

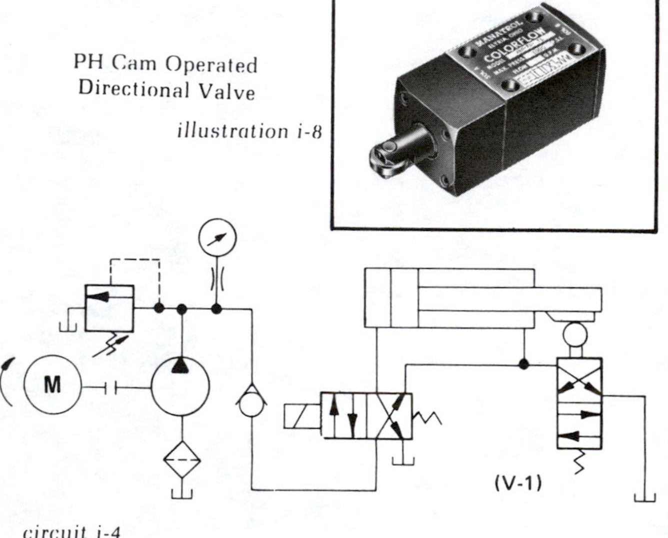

circuit i-4

A cam operated valve (V-1) is mounted on the machine in such a position that when the cylinder reaches its retracted position it makes the cam, and unloads the pump back to the reservoir.

Venting the Relief Valve to Unload

Venting the relief valve (R-1) is a very common practice. This can be accomplished by using a manual or solenoid operated valve connected to the vent part of the system relief (V-1). The vent valve may be 1/4" ported. This size vent valve will vent the largest of pilot operated reliefs, which will unload the corresponding pumps.

pump circuit continued

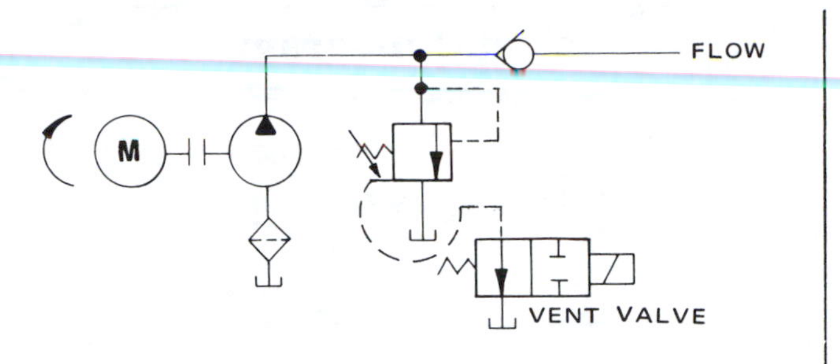

circuit i-5

Note: *The vent system will not work with a direct acting relief, only a pilot operated relief. The pilot operated relief must be able to handle the full pump (or pumps) flow. The only back pressure seen in the system is what developed by the relief valve vent spring.*

Tandem Center Directional Control Valve Unloading

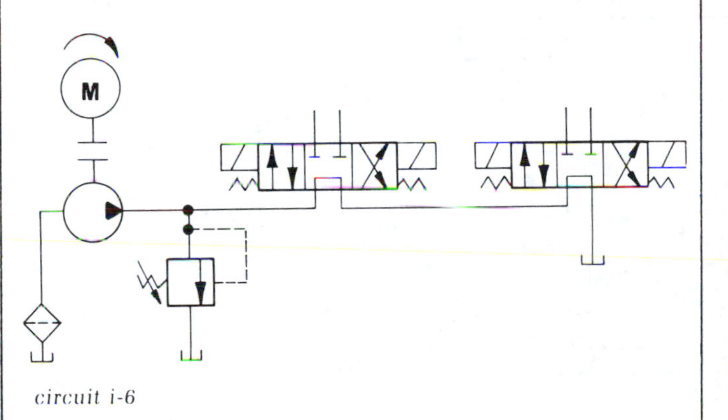

circuit i-6

This circuit should not be used with more than 3 directional valves. Using over 3 valves the pressure drop becomes very high. Some of this pressure drop can be overcome by greatly oversizing the directional valves if three or more are needed.

High Pressure Forward, Low Pressure Return

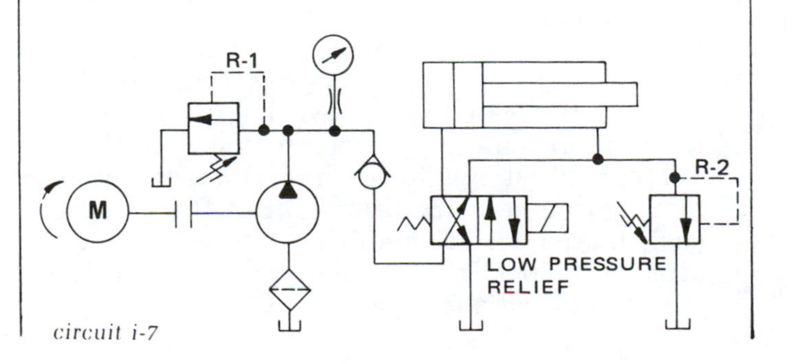

circuit i-7

R-1 is set for high pressure forward, and R-2 for low pressure return.

Hi-Lo System With Set-Up Control

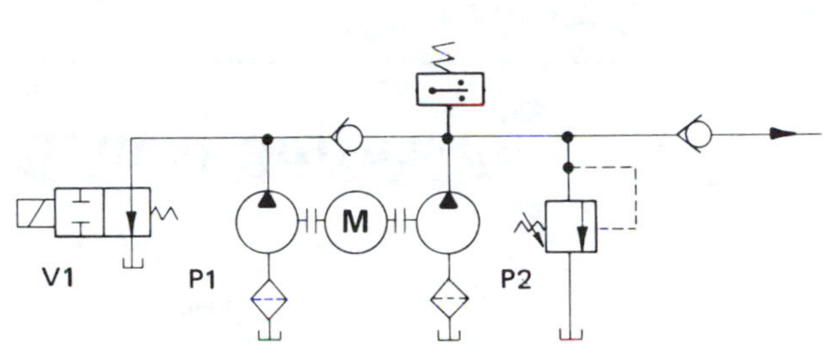

circuit i-8

This is a Hi-Lo system with an optional set-up control. During set-up operations the high volume pump (P-1) is unloaded by energizing (V-1) with the manual switch. Set up operation can be made with a slow operating speed using only the low volume pump (P-2). After set-up and returning the manual switch to it's normal position, the pressure switch will energize the unloading valve at a set lower pressure to allow the high volume low pressure pump to be unloaded. The high pressure low volume pump develops higher working pressure at a slower speed. Both pumps will be used to return the cylinder.

Note: *(V-1) should be a normally open valve. This gives an added safety feature in case of solenoid or electrical problems.*

pump circuit continued

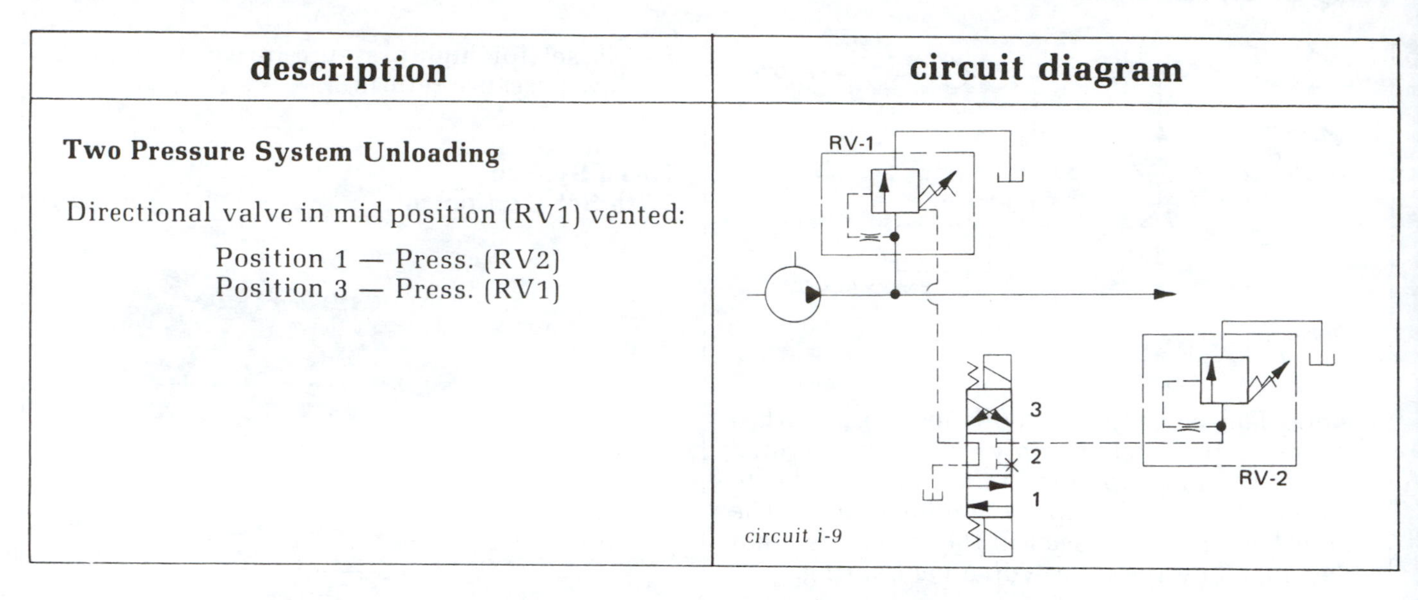

description	circuit diagram
Two Pressure System Unloading Directional valve in mid position (RV1) vented: Position 1 — Press. (RV2) Position 3 — Press. (RV1)	RV-1 3 2 1 RV-2 *circuit i-9*

operating principles and construction

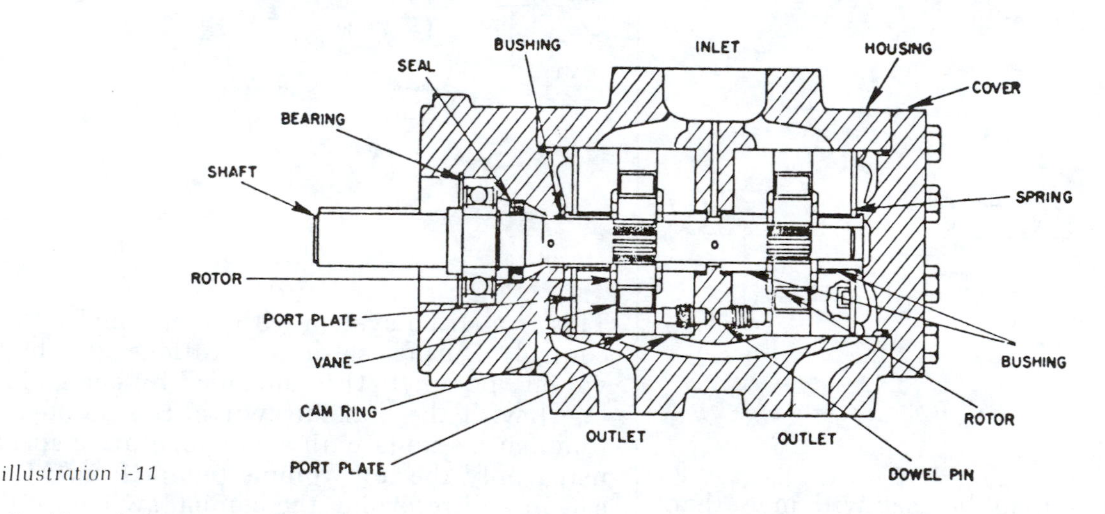

illustration i-11

Double outlet vane pump. The two pumps, on a single shaft, can develop two output volumes to separate systems, or the two volumes can be combined for high-volume, low-pressure and low-volume, high-pressure use.

pump operation

The ever widening application of hydraulic systems to the operation and automatic control of machinery and material handling equipment can be attributed to their many advantages. These include smoothness of operation, relative simplicity, substantial force available, and ease of adaptability. Foundries have put hydraulic systems to good use in meeting the modern trend toward upgrading operations.

In the typical application, power generally is transmitted to a hydraulic system from a motor or engine drive. This mechanical energy is converted to fluid energy through a pump, the heart of any hydraulic system.

Contrary to a common misconception, the pump does not pump pressure. Its job is to set fluids into motion. During operation the pump takes fluid from its inlet port and pushes it out the outlet port. With each increment of revolution, it moves a fixed amount of fluid. As the fluid flows into the system, it encounters a resistance to flow that creates pressure. Resistance is created by the workload and by the internal resistance of components such as valves and elbows in the line. Since the pump tends to move a fixed amount of fluid during each increment of revolution, its operation is affected only slightly when resistance to flow is encountered at the outlet port.

pump operation continued

Pumps are designed with a positive internal seal to prevent fluid flowing backward from the outlet port to the lower pressure inlet port as the system pressure builds up. Backward flow, called slippage, however, cannot be prevented completely by any design. Thus, as pressure increases, the volumetric output of the pump decreases due to slippage.

Pump Ratings — Pumps are rated according to their volumetric output at a given pressure and stated revolutions per minute. They also are rated by displacement per revolution. The volumetric output of a pump, referred to as its delivery, is the amount of liquid the pump can deliver at its outlet port per unit of time, at a given drive speed. Delivery usually is expressed in terms of gallons per minute. Since changes in pump drive speed affect delivery, the SAE Test Code defines delivery for mobile pumps as the flow of fluid developed by the pump at 1,000 RPM and 100 p.s.i., with the fluid at 120°F. Industrial pumps abide by similar rating standards.

Fixed vs Variable Delivery — Depending on the application, hydraulic systems require a pump with a constant delivery or one with output that can be varied. Flow variations may be achieved in two ways: 1. By varying the input speed to the shaft. 2. By maintaining a fixed speed and mechanically varying the pump displacement.

Fixed-displacement pump flow can be varied only by varying the speed. These pumps are slightly less complicated than variable displacement pumps and most often are used when system flow demands are relatively contant. In systems with varying flow demands, the variable-displacement pump begins to show some of its advantages. Demands placed on the pump by the different system elements result in excessive heat loss problems. This problem can be circumvented by using more than one fixed-displacement pump unit in a circuit. A typical application requiring more than one fixed-displacement pump is a ram that must first be moved quickly and then held in a fixed position at the end of the stroke.

Variable-displacement pumps usually are run at constant speed. They are so designed that the delivery can be varied easily from zero to maximum through built-in mechanical means that control the working arrangement of the pump's internal operating mechanism. Since variable-displacement pumps are more complex than fixed-volume units, their use is limited to applications that demand the variable feature.

A variable-displacement pump with pressure compensation has the advantage that it pumps only that amount of fluid that the circuit requires. If a 20-GPM pump is required to pump only 3 GPM under certain conditions, it will deliver only 3 GPM, and at the same time consume only a reduced amount of power. Applying the principle of variable volume permits a wider range of uses.

The ability to vary the volume of fluid in accordance with circuit demands permits the user to reduce the horsepower consumption, eliminate some valves, and, by the same token, generally reduce heat problems. Although variable-displacement pumps have a higher initial cost and require more maintenance, these advantages have contributed to their growing popularity.

Pump Types — Practically all hydraulic pumps fall within three classifications of design: gear, vane, and piston. All three types may be used when a constant displacement is required, but only vane and piston pumps are designed to produce a variable displacement.

basic construction

gear pumps

Gear pumps are the simplest type of fixed-displacement hydraulic pump available. This type consists of two external gears, generally spur gears, within a close-fitting housing. One of the gears is driven directly by the pump drive shaft. It, in turn, then drives the second gear. Some designs utilize helical gears, but the spur gear design predominates.

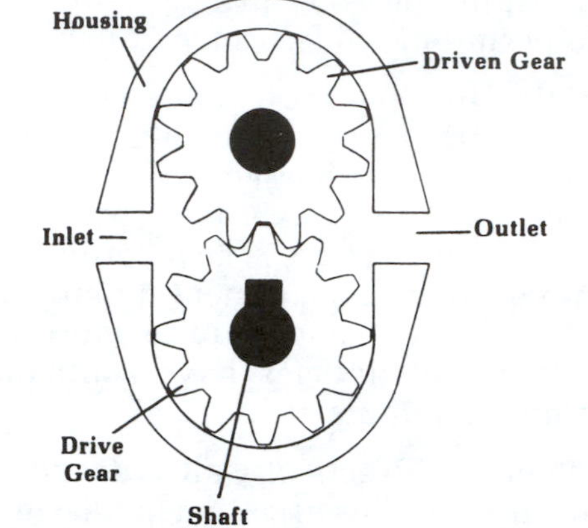

illustration i-12

Gear pump operation. Fluid flow is around the outside of the gears, between the gear teeth and the housing.

Gear pumps operate on a very simple principle, **illustration i-12.** As the gear teeth unmesh, the volume at the inlet port expands, causing fluid to flow from the inlet to the discharge side of the pump. Fluid flow is around the outside of the gears between the gears and the housing. The continuous action of the fluid being carried from the inlet to the discharge side of the pump forces the fluid into the system. Fluid from the discharge side is prevented from returning to the inlet side by the close meshing of the two gears, and the small clearance between the gears and housing.

Volumetric Efficiency — This depends primarily on closeness of fit between the gear faces and the wear plates, between the gears and housing, and between the meshing hears. Normally, the clearance between the wear plates and the gear faces is maintained by pressure loading the wear plates toward the gears. Many schemes have been developed to accomplish this. Closeness of fit also affects the maximum pressure that can be developed.

Factors that limit the amount of pressure the pump can withstand include the ability of the bearing to withstand the unbalanced loads created by pressure within the pump, and strength of the shaft.

Operating range of gear pumps generally is between 1,500-2,500 p.s.i., although pressures up to 4,000 p.s.i. can be achieved. Delivery can range from 2 to 100 GPM. At the lower delivery rate, pump speed can be as high as 6,000 RPM. At the higher delivery rate, pump speed is limited to 2,500 RPM.

Vane Pumps

The vane pump, **illustration i-13,** consists of a housing that contains a slotted rotor mounted on a drive shaft. The housing has an internal surface that is eccentric or offset with respect to the drive shaft axis. In some models this inside surface consists of a cam ring that can be rotated to shift the relationship between rotor and inside surface. The slots in the rotor are rectangular and extend radially from a center radius to the outside diameter of the rotor and from end to end. A rectangular vane that is essentially the same size as the slot is inserted in the slot and is free to slide in and out.

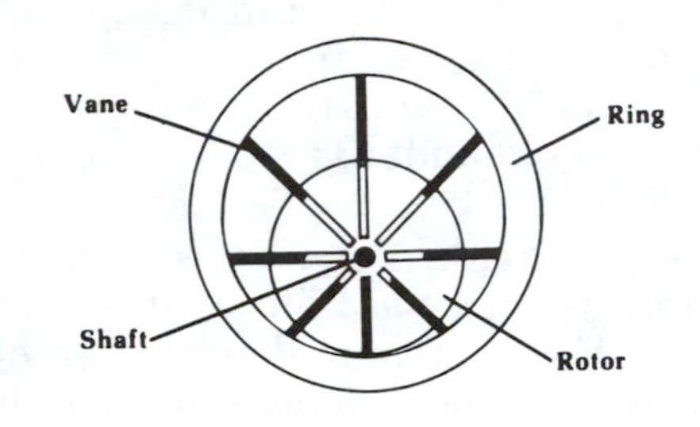

illustration i-13

Vane pump. Rotor has slots in which vanes can slide outward under centrifugal force.

As the rotor turns, the vanes thrust outward, and the vane tips track the inner surface of the housing, riding on a thin film of fluid. Two port or end plates that engage the end faces of the ring provide axial retention.

Centrifugal force generally contributes to the outward thrust of the vanes. As they ride along the eccentric housing (cam ring) surface, the vanes move in and out of the rotor slots. The vanes divide the area between the rotor and casing into a series of chambers. The sides of each chamber are formed by the two adjacent vanes, the port or end plates, the pump casing, and the rotor. These chambers change in volume depending on their respective position about the shaft.

As each chamber approaches the inlet port, its vanes move outward and its volume expands, causing fluid to flow into the expanded chamber. Fluid is then carried within the chamber around to the discharge port. As the chamber approaches the discharge port, its vanes are pushed inward, the volume is reduced, and the fluid is forced out the discharge port.

basic construction continued

The variable-volume vane pump can be adjusted to discharge a different volume of fluid while running at constant speed, simply by shifting the cam ring with respect to the rotor. When the pump components are in a position such that the individual chambers achieve their maximum volume as they reach the inlet port, the maximum volume of fluid will be moved. If the relationship between housing and rotor is changed such that the chambers achieve their maximum or zero volume as they reach the inlet port, the pump delivery will be reduced to zero.

Volumetric Efficiency — Since the vane pump housing or cam ring must be shifted to change the eccentricity and vary the output, variable-displacement vane pumps cannot have the close end fit common to fixed-displacement pumps. Volumetric efficiency is in the range of 90 to 95%. These pumps retain their efficiency for a considerable length of time since compensation for wear between the vane ends and the housing is automatic. As these surfaces wear, the vanes move farther outward from their slots to maintain contact with the housing.

Vane pump speed is limited by vane peripheral speed. High peripheral speed will cause cavitation in the suction cavity, which results in pump damage and reduced flow.

An imbalance of the vanes can cause the oil film between the vane tips and the cam ring to break down, resulting in metal-to-metal contact and subsequent increased wear and slippage. One method applied to eliminate high vane thrust loading is a dual-vane construction.

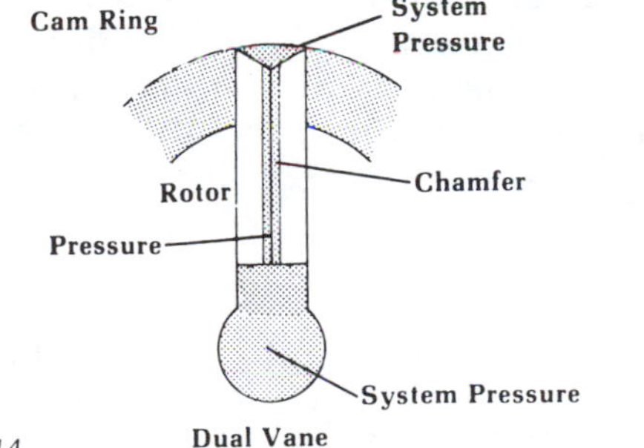

illustration i-14

Dual-vane construction. Two independent vanes are mounted in each slot. Channel between them balances the hydraulic pressure on each individual vane.

In the dual-vane construction, two independent vanes are located in each rotor slot, **illustration i-14.** Chambered edges along the sides and top of each vane form a channel that essentially balances the hydraulic pressure on the top and bottom of each pair of vanes. Centrifugal force causes the vanes to follow the contour of the cam-shaped ring. There is just sufficient seal between the vanes and ring without destroying the thin oil film.

Piston-Type Pumps

Two basic types of piston or reciprocating pumps are the radial piston and the axial types. In most piston pumps, a number of pistons reciprocate in a rotating cylinder barrel. This type of pump is so constructed that the pistons retract while passing the inlet port, expanding the volume and permitting the fluid to flow in through the inlet port and into the pumping chambers. The pistons then extend as they approach the outlet, decreasing the volume and causing the fluid to flow out the outlet port.

Axial piston pumps are available in two basic designs: 1. In-line design. 2. Angle or bent-axis design.

In-Line Piston Pump — This type of pump, **illustration i-15,** includes a housing, a drive shaft, a "rotating group," a shaft seal, and the valve plate. The valve plate contains the inlet and outlet ports and functions as the back cover. The rotating group, which is splined to the drive shaft, is on the same centerline as the drive shaft. It includes the pistons, angled swash plate, and shoe plate. Other terms for swash plate are tilt box, tilting blocks, cam plate, and wobble plate.

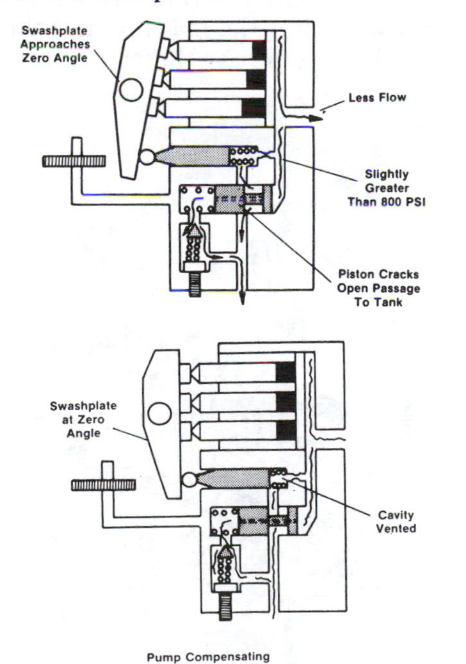

illustration i-15

In-line axis piston pump. A. Maximum swash plate angle for maximum displacement. B. Decreased swash plate angle for partial displacement. C. Zero swash plate angle for zero displacement.

basic construction continued

When the cylinder block rotates, the pistons reciprocate as they ride over the angle plate. The pistons are held against it by the shoe plate. The angle of the swash plate controls the delivery. Where the swash plate is fixed, the pump is of the constant-displacement type.

In the variable-displacement, inline piston pump, the swash plate is mounted on a pivoted yoke. As the swash plate angle is increased, the cylinder stroke is increased, resulting in a greater flow. A pressure compensator control can position the yoke automatically to maintain a constant output pressure.

There is a variation of the swash plate in-line pump. It is a design where the swash plate turns, but the cylinder barrel remains stationary. The plate is canted so that it wobbles as it turns. This action pushes the pistons in and out of the stationary cylinder barrel.

This type of in-line pump contains a separate inlet and outlet check valve for each piston since the pistons do not move past the inlet and outlet port.

Bent-Axis Piston Pump—Illustration i-16 shows a bent-axis piston pump, which contains a cylinder block assembly in which the pistons are equally spaced around the cylinder block axis. Cylinder bores are parallel to the axis. The cylinder block is linked to the driving mechanism in such a way that the driving mechanism and cylinder block rotate at the same speed.

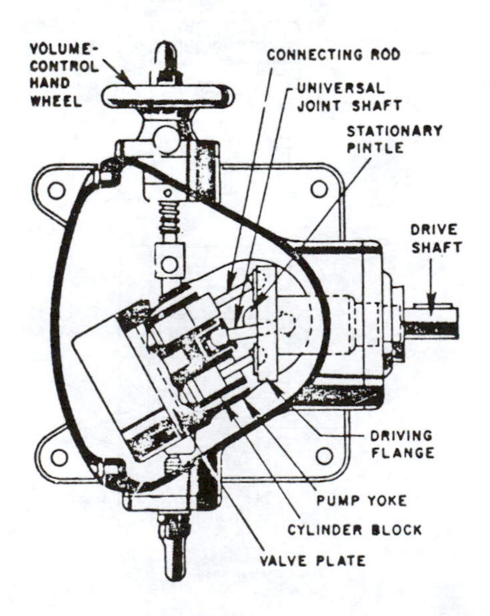

illustration i-16

Bent-axis piston pump. This can provide variable volume by varying the piston strokes. The cylinder block is carried in a yoke that can be set at the desired angle with the drive shaft.

As the shaft rotates, distance between any one piston and the valving surface changes continually. Each piston moves away from the valving surface during one half of the revolution and toward the valving surface during the other half. The inlet chamber is in line as the piston moves away, and the outlet chamber is in line as the piston moves closer, thus drawing liquid in during one half of the inlet chamber as the pistons are moving away from the pintle and open to the pintle outlet chamber as the pistons are moving toward the pintle.

Therefore, during rotation, pistons draw liquid into the cylinder bores as they pass the inlet side of the pintle and force that liquid out of the bores as they pass the outlet side of the pintle. Through a ring-shifting mechanism, output can be varied between zero and full volume. It also is possible to reverse the direction of fluid flow without stopping the pump or changing its direction.

Radial Piston Pumps — Radial piston pumps, are arranged so that the line of piston motion is perpendicular to the centerline of the shaft. This type consists of a rotating cylinder block that contains radial cylinders and is connected to a drive shaft, and an outer casing (rotary ring), the centerline of which is offset from the cylinder block centerline. The outer ends of the pistons are held in contact with the rotor ring by centrifugal force and fluid pressure. A stationary pintle (spindle) ports the inlet and outlet flow.

Since the rotor centerline is offset with respect to the cylinder block centerline, and the pistons must follow the rotor orbit, the pistons are forced toward the pintle during half of a cylinder block revolution and away during the other half of the revolution. The pintle is ported so that cylinder block ports are open to the pintle revolution and forcing it out during the other half.

Pistons are driven through connecting rods attached to the driving plate by ball and socket joints, or by a wobbler plate. The angle between the centerline of the cylinder block and that of the drive shaft determines the volume output. If the angle is fixed, the pump is of the variable-displacement type. With the latter type, flow of fluid can be reversed by changing the angle from one side to the other.

Various controls such as screw adjustment, electric servo, and compensator pistons can be applied to change the angle and control a variable displacement.

trouble shooting pumps

pump makes excessive noise

1. Check for vacuum leaks in the suction line. (Such as leak in fitting or damaged suction line.)

2. Check for vacuum leaks in the pump shaft seal if the pump is internally drained. Flooding connections with the fluid being pumped may cause the noise to stop or abate momentarily. This will locate the point of air entry.

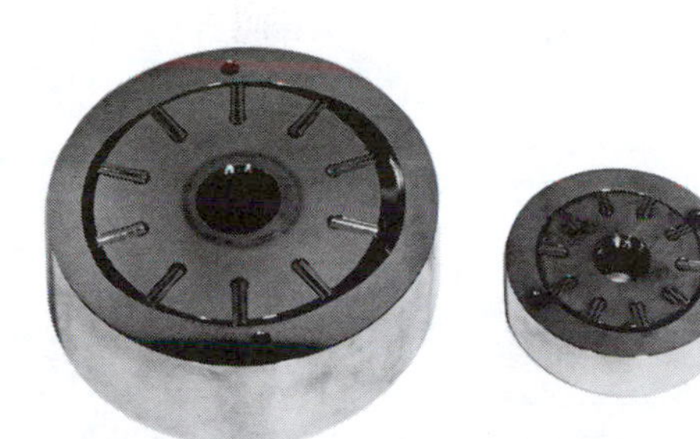

illustration i-21

3. Check alignment with drive mechanism. Misalignment will cause wear and subsequent high noise level in operation.

4. Check manufacturers specifications relative to wear possibilities and identification of indications of wear as high operating noise level, etc.

5. Check compatibility of fluid being pumped with manufacturers recommendations.

6. Relief or unloading valve set too high. Use reliable gauge to check operating pressure. Relief valve may have been set too high with a damaged pressure gauge. Check various unloading devices to see that they are properly controlling the pump delivery.

7. Aeration of fluid in reservoir (return lines above fluid level).

8. Worn or sticking vanes (vane type pump).

9. Worn cam ring.

10. Worn or damaged gears and housing.

11. Worn or faulty bearing.

12. Reversed rotation.

13. Cartridge installed backwards or improperly.

14. Plugged or restricted suction line or suction strainer.

15. Plugged reservoir filter breather.

16. Oil viscosity too high or operating temperature to low.

17. Oil pour point too high.

18. Air leak in suction line or fittings also causing irregular movement of control circuit.

19. Loose or worn pump parts.

20. Pump being driven in excess of rated speed.

21. Air leak at pump shaft seal.

22. Oil level to low and drawing air in through inlet pipe opening.

23. Air bubbles in intake oil.

24. Suction filter too small.

25. Suction line too small.

26. Pump housing bolts loose or not properly torqued.

27. Drain port improperly positioned allowing air to be trapped in pump housing.

pump failure to deliver fluid

1. Low fluid level in reservoir.

2. Oil intake pipe suction strainer plugged.

3. Air leak in suction line and preventing priming.

4. Pump shaft turning too slowly.

5. Oil viscosity too high.

6. Oil lift too high.

7. Wrong shaft rotation.

8. Pump shaft or parts broken.

9. Dirt in pump.

10. Variable delivery pumps. (Improper stroke.)

oil leakage around pump

1. Shaft seal worn.

2. Head of oil on suction pipe connection - connection leaking.

3. Pump housing bolts loose or improperly torqued.

4. Case drain line too small or restricted. (Shaft seal leaking.)

5. Cracked housing, over pressurized.

excessive pump wear

1. Abrasive dirt in the hydraulic oil being circulated through the system.

2. Oil viscosity too low.

3. System pressure exceeding pump rating.

troubleshooting pumps continued

4. Pump misalignment or belt drive too tight.
5. Air being drawn in through inlet of pump.

pump parts inside housing broken

1. Seizure due to lack of oil.
2. Excessive system pressure above maximum pump rating.
3. Excessive torquing of housing bolts.
4. Solid matter being drawn in from reservoir and wedged in pump.
5. Lack of clearance between rotor and ring.

notes

connectors

brass

inverted flared fittings .... j (a-2)
compression: threaded sleeve and standard fittings ...... j (a-2)
sae 45° flared fittings .... j (a-3)
nt fittings .... j (a-4)
dubl-barb fittings .... j (a-4)
poly-tite fittings .... j (a-4)
pipe fittings .... j (a-5)
air brake fittings .... j (a-5)
drain cocks and needle valves .... j (a-6)

inverted flared fittings

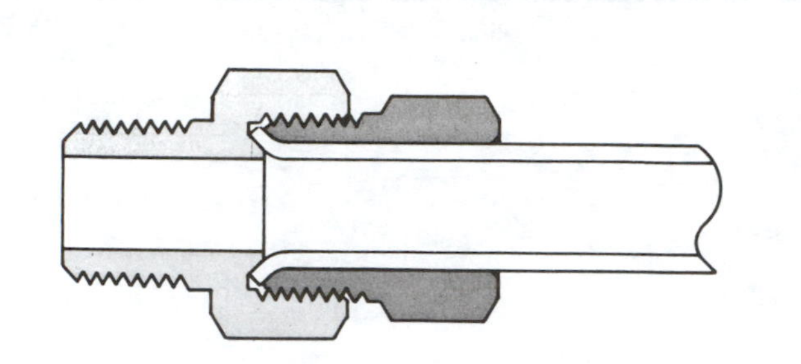

illustration j (a-1)

Advantages — Built to resist mechanical pull-out. This economical fitting can be assembled and disassembled repeatedly. Readily available in a broad selection of styles to fill your specific needs. Extruded from CA 360 or CA 345 brass rod.

Approvals — Approved by Underwriters' Laboratories (No. AU 1432) for hazardous liquid, fuel equipment. Meets specifications of the Society of Automotive Engineers (S.A.E.) automotive tube fitting standards and specifications.

Applications — Use with copper, brass, aluminum and welded steel hydraulic tubing that can be flared. Manufactured especially for hard-to-hold liquids and gases.

Working Pressure Ranges — Temperature and type of tubing used are important factors. However, the following table is a good guide for proper selection. Temperature 73°F with copper tubing.

PSI	Tube O.D. (in.)	Tube Wall (in.)
2800	1/8	.030
1900	3/16	.030
1400	1/4	.030
1200	5/16	.032
1000	3/8	.032
750	1/2	.032
650	5/8	.035
550	3/4	.035

table j (a-1)

Temperature Ranges — From minus 65° to plus 250°F.

Vibration — Will withstand minimal vibration movements.

Assembly Instructions

1. Cut tubing squarely.
2. Place nut onto tubing. (Threaded end of nut toward tubing end.)
3. Flare tube end with flaring tool.
4. Clamp tube flare between nut and flare seat of body by screwing nut on finger tight. Tighten slightly with wrench for leakproof, metal-to-metal joint.

compression: threaded sleeve and standard fittings

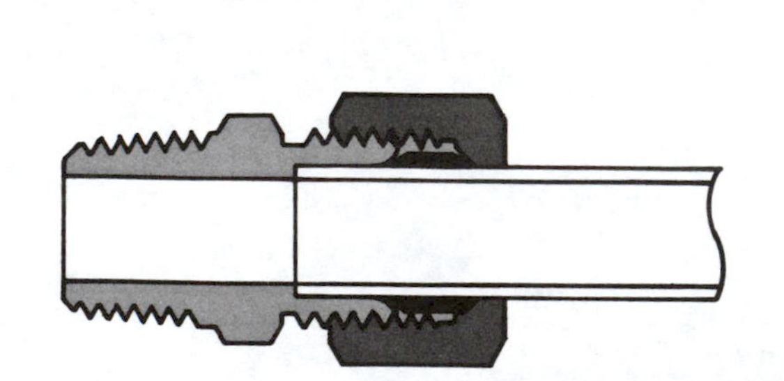

illustration j (a-2)

Advantages — No flaring, soldering or other preparation of tubing necessary to assemble. This economical fitting is readily available in a broad selection of styles. Manufactured from CA 360 or CA 345 brass rod.

Approvals — Compression fittings approved by Underwriters' Laboratories (No. AU1432) for hazardous liquid, and fuel equipment. Both compression and threaded sleeve fittings meet specifications of the Society of Automotive Engineers (S.A.E.) automotive tube fitting standards.

Applications — Use with copper, brass, and aluminum. Not recommended for steel tubing. Manufactured for low- and medium-pressure tubing connection work where excessive vibration or tube movement is not involved.

Working Pressure Ranges — Temperature and type of tubing used are important factors; however, the following table is a good guide for proper selection. Temperature 73°F.

psi	Tube O.D. (in.)	Tube Wall (in.)
400	1/8	.030
400	3/16	.030
300	1/4	.030
300	5/16	.032
200	3/8	.032
200	1/2	.032
150	5/8	.035
100	3/4	.035
75	7/8	.035

table j (a-2)

Temperature Ranges — From minus 65° to plus 250°F.

compression continued

Vibration — Fair resistance to vibration or tube movement, depending on applications involved.

Assembly Instructions — Slide nut, then sleeve onto tubing. The thread end of the nut must face out. (On the threaded sleeve fitting, the lead end of the nut incorporates the sleeve.)

Insert the tube into the fitting. Be sure the tube is bottomed on the fitting shoulder.

Assemble nut to body, and tighten "hand-tight." Then follow the number of wrench turns as indicated in the table below.

Fitting Size	Tube Size	Turns required to seal from hand-tight: Compression Fittings	Turns required to seal from hand-tight: Threaded Sleeve Fittings
2	1/8	1-1/4	1-1/2
3	3/16	1-1/4	1-1/2
4	1/4	1-1/4	1-1/2
5	5/16	1-3/4	1-1/2
6	3/8	2-1/4	1-1/2
8	1/2	2-1/4	1-1/2
10	5/8	2-1/4	
12	3/4	2-1/4	
14	7/8	2-1/4	

table j (a-3)

sae 45° flared fittings

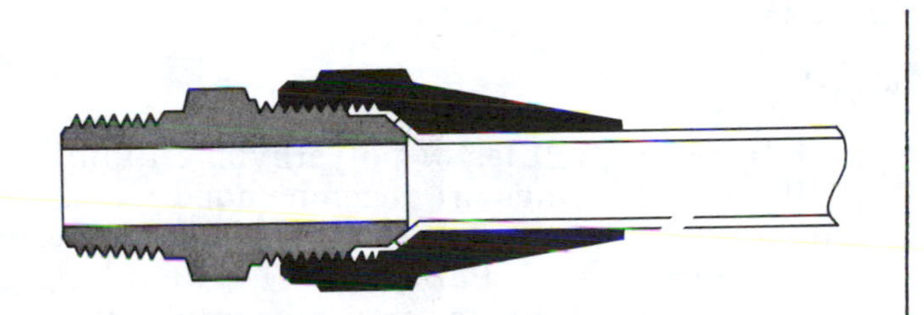

illustration j (a-3)

Advantages — This economical fitting resists mechanical pullout. Can be assembled and disassembled repeatedly. Manufactured from CA 360, CA 345 or CA 377 brass rod.

Approvals — Approved by Underwriters' Laboratories (No. AU1432) for hazardous liquid, fuel equipment. Meets specifications of the Society of Automotive Engineers (S.A.E.) automotive tube fitting standards and specifications.

Applications — Use with copper, brass, aluminum and welded steel hydraulic tubing that can be flared. Manufactured especially for hard-to-hold liquids and gases.

Working Pressure Ranges — Temperature and type of tubing used are important factors. However, the following table is a good guide for proper selection. Temperature 73°F with copper tubing:

PSI	Tube O.D. (in.)	Tube Wall (in.)
2800	1/8	.030
1900	3/16	.030
1400	1/4	.030
1200	5/16	.032
1000	3/8	.032
750	1/2	.032
650	5/8	.035
550	3/4	.035
450	7/8	.035

table j (a-4)

Temperature Ranges — From minus 65° to plus 250°F.

Vibration — Short nut may be used when vibration is minimal. Otherwise the long nut offers additional vibration capacity.

Assembly Instructions

1. Cut tubing squarely.

2. Place nut onto tube. Open threaded end of nut toward end of tube.

3. Flare tube end with flaring tool to provide 45° flare.

4. Clamp tube flare between nut and nose of fitting body by screwing nut on finger tight. Tighten slightly with wrench for leakproof, metal-to-metal joint.

Flare fittings are easy to disassemble and may be reassembled repeatedly, always for a leakproof connection.

nt fittings

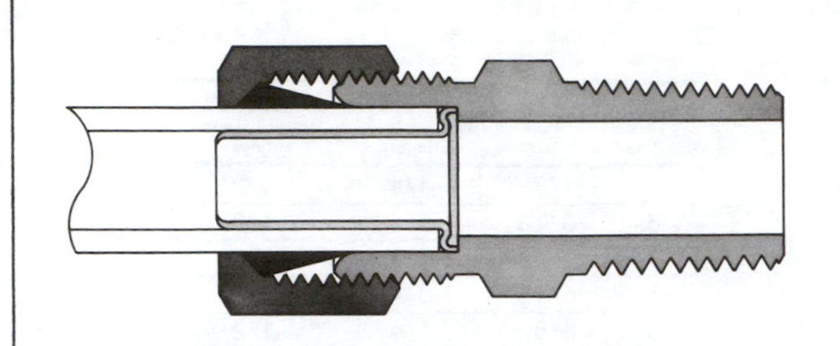

illustration j (a-4)

Advantages — The Parker NT is the only fitting designed especially for nylon tubing with a built-in tube support to prevent tube collapse and a special acetal resin sleeve to prevent tube notching. When the nut is tightened, the sleeve (a very high-impact material with fine memory qualities) and the nylon tubing cold flow to mold to the contour of the fitting. This feature couples the nylon tubing beyond the yield strength of either single- or double-ply tubing and beyond the tubing's burst pressure.

Parker exclusive nylon tube fittings provide more flow clearance than conventional fittings. Readily available in sizes and styles to fill your specific needs. Manufactured from CA 360, CA 345 or CA 377 brass.

Approvals — Fitting bodies and nuts meet S.A.E. specification J-512.

Applications — Use with Fast & Tite™ or other high-quality nylon tubing.

Assembly Instructions — Insert tube to bottom in fitting. Then tighten with a wrench. The built-in tube support will withstand the necessary torque required on the nut to make a secure connection. (The 1/8 and 3/16 tube sizes do not require the built-in tube support, but care should be taken not to over-torque these two sizes when tightening nuts.)

dubl-barb fittings

illustration j (a-5)

Advantages — The double barbs hold the tubing like a Chinese finger trap — actually tighten the grip when the tubing tries to pull loose. In addition to the styles shown, custom DUBL-BARB fittings to meet your exact requirements are available. Extruded from CA 360 or CA 345 brass rod.

Applications — Because of the many available variations in qualities of polyethylene tubing, DUBL-BARB fittings are recommended for use with Parker Fast & Tite polyethylene tubing (or an equal grade). Parker Fast & Tite tubing is highly resistant to environmental stress cracking which is necessary for long life when coupled with expansion fittings.

Working Pressure and Temperature Ranges — In tube sizes -4 to -6, working pressures up to 150 psi are practical at temperatures ranging from minus 65° to plus 90°F. On tube size -8 working pressures to 100 psi at temperatures ranging from minus 65° to plus 75°F.

Assembly Instructions — Simply push tube over the two barbs — be sure tube has a flat, "square" end.

poly-tite fittings

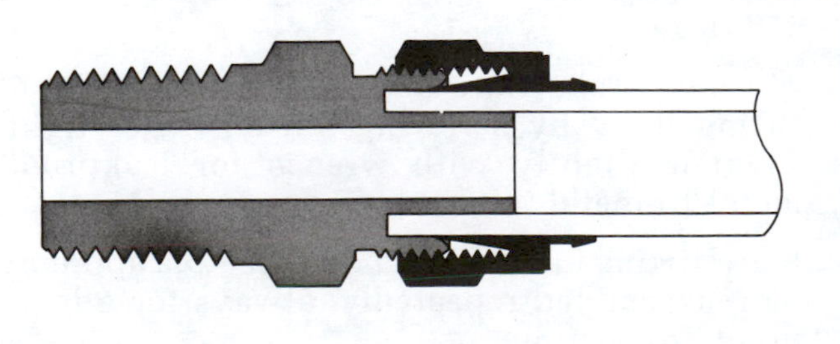

illustration j (a-6)

Advantages — A compact brass compression fitting designed to speed any installation. Body, nut and sleeve are preassembled, ready to use. Poly-Tite, with an exclusive acetal resin sleeve, holds plastic tubing where it belongs, even when pressure exceeds tubing burst point. 'Poly-Tite sleeves have high strength, excellent dimensional stability and resilience to resist creeping; sleeves recover from compression

poly-tite fittings continued

and remain exceptionally strong. Poly-Tite fittings can be assembled and disassembled repeatedly. Manufactured from CA 360, CA 345 or CA 377 brass rod.

Applications — Use with Parker Fast & Tite® or other high-quality polyethylene tubing for instrument, pneumatic, lubricant, fuel and coolant lines. For use with soft metal tubing, use a brass sleeve.

Working Pressure and Temperature Ranges — Up to 160 psi at temperatures ranging from 0° to plus 175°F.

Assembly Instructions — Insert tube end until it bottoms in the Poly-Tite fitting and tighten the knurl/hex nut finger-tight. No tools needed.

pipe fittings

illustration j (a-7)

Advantages — All fitting threads are made to Dryseal standards. Connectors, unions, nuts and extruded elbows and tees are machined from CA 360 and CA 345 brass rod; forged elbows and tees machined from CA 377 brass.

Approvals — Meets specifications of the Society of Automotive Engineers (S.A.E.) automotive pipe fitting standards.

Applications — Use with brass, copper, or iron pipe. Manufactured for low- or medium-pressure line connection work.

Temperature and Working Pressure Ranges — From minus 65° to plus 250°F at 1000 psi.

Vibration — Fair resistance to vibration and pipe movement depending upon conditions.

air brake fittings

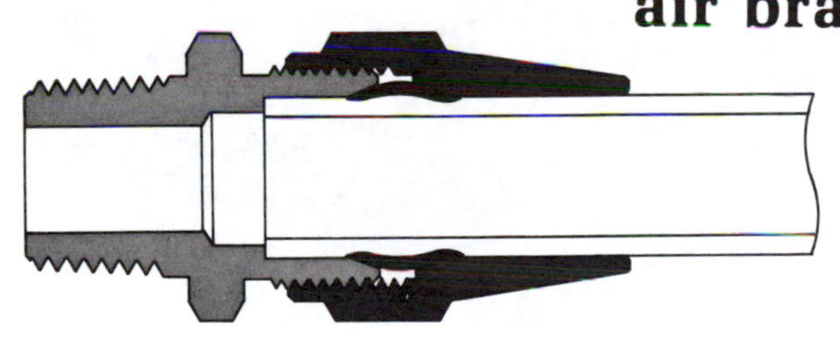

illustration j (a-8)

Advantages — You can assemble tubing to Parker Air Brake fittings without having to flare or solder the tubing. These economical fittings are extruded from CA 360 or CA 345 brass, and are readily available in a broad selection of styles.

Approvals — Meet specifications of the Society of Automotive Engineers (S.A.E.) automotive tube fitting standards for extruded fittings.

Applications — Use with copper tubing in air brake systems.

Working Pressure Ranges — Up to 150 psi.

Temperature Ranges — Parker Air Brake fittings will withstand temperature variations from minus 65° to plus 250°F.

Assembly Instructions

1. Cut tubing squarely.
2. Slide nut and and sleeve onto tubing.
3. Insert tubing into fitting until bottomed on seat. The nut should be screwed down finger-tight, then wrench-tightened as indicated below (this will also allow for the maximum number of remakes):

Tube Size	Additional Number of Turns from Hand-Tight
1/4, 3/8	1-3/4
1/2, 5/8, 3/4	3-1/4

table j (a-5)

drain cocks and needle valves

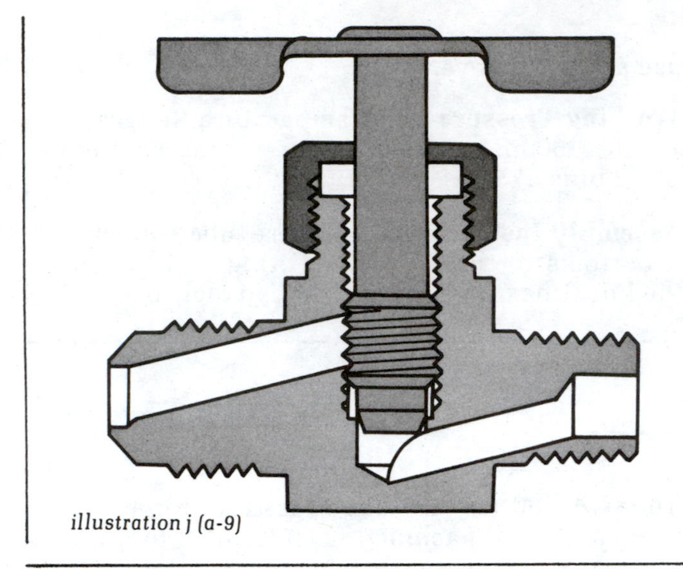

illustration j (a-9)

Drain Cocks Advantages — Drain cocks have both external-seats and internal-seats. Hand-tightening insures a positive seal.

Needle Valves Advantages — Have metal-to-metal seats, with fine-thread screwdown. This enables positive sealing as well as precise adjustments to hold any amount of flow up to the valve capacity.

Working Pressure Ranges — Both drain cocks and needle valves are designed to withstnad up to 150 psi working pressure.

Needle Valves Installation Instructions — Series NV valves should always be installed with the pressure against the seat.

notes

connectors

quick couplings

coupling seal material selection ..... j (b-2)
quick couplings .... j (b-3)
single shut-off .... j (b-3)
double shut-off .... j (b-3)
straight-thru .... j (b-3)

coupling seal material selection

fluorocarbon rubber (FPM)

Working temperature range -20 to +400 degrees F.

FPM is recommended for:
- Petroleum oils
- Di-ester base lubricants
- Silicate ester base lubricants
- Silicone fluids and greases

nitrile or buna n (nbr)

Working temperature range -65 to +250 degrees F.

NBR is recommended for:
- General purpose sealing
- Petroleum oils and fluids
- Water
- Silicone greases and oils
- Di-ester base lubricants
- Ethylene glycol base fluids

ethylene propylene rubber (EPM)

Working temperature range -65 to +300 degrees F.

EPM is recommended for:
- Phosphate ester base hydraulic fluids (Skydrol, Cellulube, Phydraul)
- Steam (to 400 degrees F.)
- Water
- Silicone oils and greases
- Diluted acids
- Diluted alkalies
- Ketones
- Alcohols
- Automotive brake fluids

chloroprene rubber (CR)

Working temperature range -65 to +300 degrees F.

CR is recommended for:
- Refrigerants (Freon, NH^3)
- High aniline point petroleum oils
- Mild acid resistance
- Silicate ester lubricants

steel, carbon alloy

Applications: General purpose use and high pressures for maximum wear and strength.

brass

Applications: Non-magnetic, non-sparking, fresh water marine service, high temperature and water systems.

stainless steel

Applications: Corrosive material handling in chemical, pharmaceutical, beverage and food processing. Excellent for salt water marine applications.

coupling selection

System Pressure and Pressure Surges

Pressure rating is invariably the most important factor in selecting a quick disconnect coupling, because if the system pressure is higher than the rating of the coupler, it cannot be used. Quick disconnect couplings are pressure-rated according to operating, proof, and burst pressures.

Flow Requirements

Pressure drop is usually very important as this factor can drop the overall efficiency of the system.

System media and temperature

The compatibility of materials and seals with the system media and temperature governs the choice of metallic components and sealing materials.

Environmental and Functional Considerations

Environmental considerations such as corrosion, maintainability and cleanliness, along with functional characteristics such as forces required for connection and disconnection, fluid loss at disconnect, air inclusion during connection, end terminations, weight and size, vacuum rating, coupling means and type, are all factors affecting the selection of a quick disconnect coupling.

Performance data to assist in the selection of a quick disconnect coupling is available from the manufacturer's specifications.

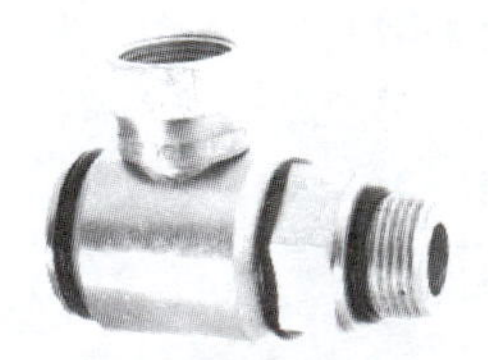

illustration j (b-1)

illustration j (b-2)

coupling seal material selection continued

quick couplings

Quick disconnect couplings are used for ease of rapidly and repetitively connecting and disconnecting fluid lines without the use of tools. One half is normally attached to a flexible line.

Quick disconnect couplings consist of two mating parts: The *female* half (sometimes referred to as the "socket" or "coupler body") and the *male* half (often referred to as the "plug" or "nipple body").

Couplings are locked with detented balls, pins, dogs, pawls, cams or screw threads. The ball type lock is the most widely used.

The three most popular types of quick disconnect couplings are the single shut-off, double shut-off, and straight-thru. Standard materials are brass, aluminum, steel and stainless steel.

single shut-off

Single shut-off couplings are designed basically for use in *pneumatic* systems using standard shop air at 300 psig or less. The female half includes a shut-off valve, but the male half does not, hence the term "single shut-off."

The female half should be installed on the upstream (supply) end of the line to shut off the air supply when the coupling is disconnected.

There are many types and styles of popular interchangeable designs available, as outlined in Parker catalog 3803.

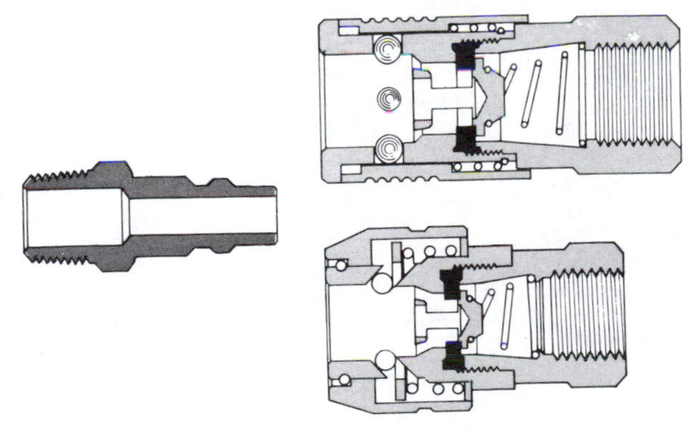

illustration j (b-3)

double shut-off

Often referred to as *hydraulic* couplers, the double shut-off designs have shut-off valves in both male and female halves.

Since it includes positive shut-off in both halves, the double shut-off type virtually eliminates fluid loss upon disconnecting thus this type is ideal for applications on lines carrying liquids.

Quick disconnects using precision steel balls as a valving means are popular on agricultural and mobile equipment for remote hydraulic lines. The steel ball offers long and reliable service for these rugged applications.

Several designs are available for both "open" and "closed" center hydraulic systems, which allow for connecting under pressuized conditions.

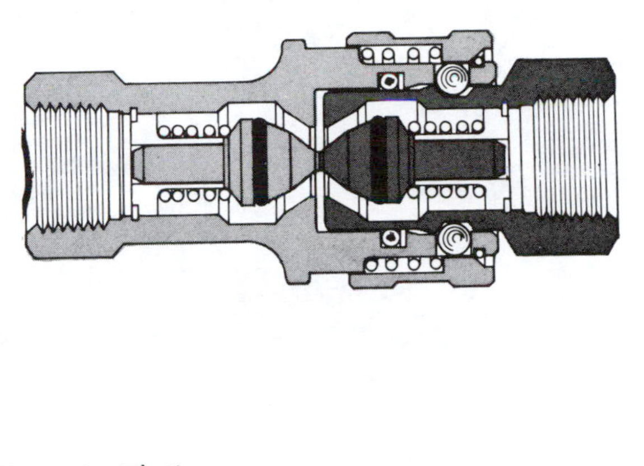

illustration j (b-4)

straight-thru

Since there is no shut-off valve in either coupling half, the straight-thru coupling provides free, unrestricted flow with very little pressure drop.

As the straight-thru coupling lacks a means for shutting off flow automatically upon disconnect, a shut-off valve should be installed in the line.

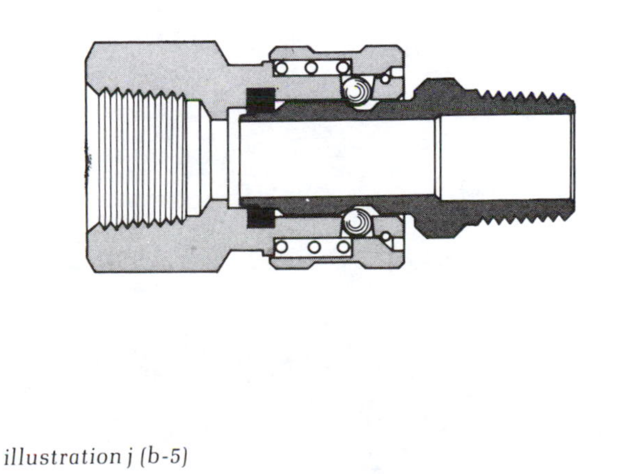

illustration j (b-5)

notes

connectors

hose

hose selection by pressure . . . . j (c-2)
flow capacity of parker hose assemblies j (c-3)
hose pressure drop . . . . j (c-4)
hose selection considerations . . . . j (c-7)
hose application considerations . . . . . . j (c-9)
basic hose construction . . . . j (c-11)
hose fittings types . . . . j (c-13)

hose selection by pressure

CONSTRUCTION		HOSE DIAMETERS (I.D. IN INCHES)															
		3/16	1/4	5/16	3/8	13/32	1/2	5/8	3/4	7/8	1	1 1/8	1 1/4	1 3/8	1 1/2	1 13/16	2
Low Pressure — Hi-Temp 300°F Rating, One Braid Strainless Steel Wire, Polyester Braid Cover SAE J1402 Type D, Class II	SS25	1500	500	500		500	450	450		250		250					
Medium Pressure — SAE—100R5, J1402 Type D, Class I Cotton Wire — Cotton Braided Hose — With Rubber Cover	201 221*	3000	3000	2250		2000	1750	1500		800		625		500	350		
Low Pressure Hydraulic and Freon 12 Refrigerant Service — Multiple Cotton Braid, Perforated Rubber Cover, 250°F	231	500	500	500		500	500	500		400		400					
Freon 12 and 22 Refrigerant Service & Low Pressure Hydraulic Service Seamless Nylon Inner Tube, Fiber Braid Reinforcement, Synthetic Rubber Cover, 212°F	235	875	875	750		625	625	625		440		375		375			
Freon 12 Refrigerant Service — One Wire Braid, Red Impregnated Fabric Cover	241	500	500	500		500	500	500		500		500		300		300	
Low, Medium Pressure, J1402 Type D, Class I Hi-Temp 300°F Rating, Cotton, Wire, Cotton Construction	261	2500	2250	2000		1850	1400	1200		600		500		250		250	
High Pressure LPG Service (250-350 PSI) −40° to +250°F. One S.S. Wire Braid and Two Textile Braid Reinforcement. Fabric cover U.L. approved and labeled.	SS25UL	350	350	350		350	350	350									
High Pressure LPG Service (250-350 PSI) −40°F to +212°F. Nylon innertube, one ply or more Textile Braid Reinforcement. Synthetic Perforated Rubber Cover UL Approved and Labeled.	2LPG	350	350	350		350	350	350									
High Pressure Service — SAE 100R2 — Type AT — Two Wire Braid, Rubber Cover (No Skive)	301	5000	5000		4000		3500		2500		2000						
*With Inner Tube for Phosphate Ester Fluids (No Skive)	304*	4500	4500		3500		3000		2500		1875		1625		1375		1125
Nylon Inner Tube for Greater Fluid Compatibility	305		5000		4000		3500		2250		2000		1625		1250		1000
Super No Skive	381		5800		5000		4500		3200		2550		2300		1875		1650
Very High Pressure Service — Four Wire Spiral Wound Hose, Rubber Cover (No Skive) SAE 100 R9AT	341				4500		4000		3000		2500						
Medium High Pressure Service — SAE 100R1 Type T — One Wire Braid, Rubber Cover (No Skive)	421	3000	2750	2500	2250	2250	2000	1500	1250		1000						
Extreme High Pressure Service — Four Heavy Wire Spiral Wound, Rubber Cover (No Skive)	741								5000		4000		3000		2500		
Extreme High Pressure Service — Four Heavy Wire Spiral Wound, Rubber Cover SAE100R10 — No Skive	751										4000		3000		2500		2500
Low Pressure Service — Push-Lok One Fiber Braid Rubber Cover — Multiple Color (Non-Hydraulic Service)	801		200		200		200	200	200								
Low Pressure Service — Push-Lok One Fiber Braid — NHAMA-STD-11 Cotton Cover (Non-Hydraulic Service) Rubber Cover NHAMA-STD-10.	821 831*		350		300		300	250	250								
Low Pressure — Pump Inlet Line Fiber Braid or Weave With One Wire Spiral To Prevent Collapse SAE 100R4	811									100		70		50		50	100
	881									300		250		200		150	
Extreme High Pressure Service — 4 heavy spiral wound rubber cover SAE 100 R12	771 77C				4000		4000		4000		4000		3000		2500		2500

table j (c-1)

flow capacity of Parker hose assemblies at various flow velocities

(Ordinary Hydraulic Oil, 200°F and less)

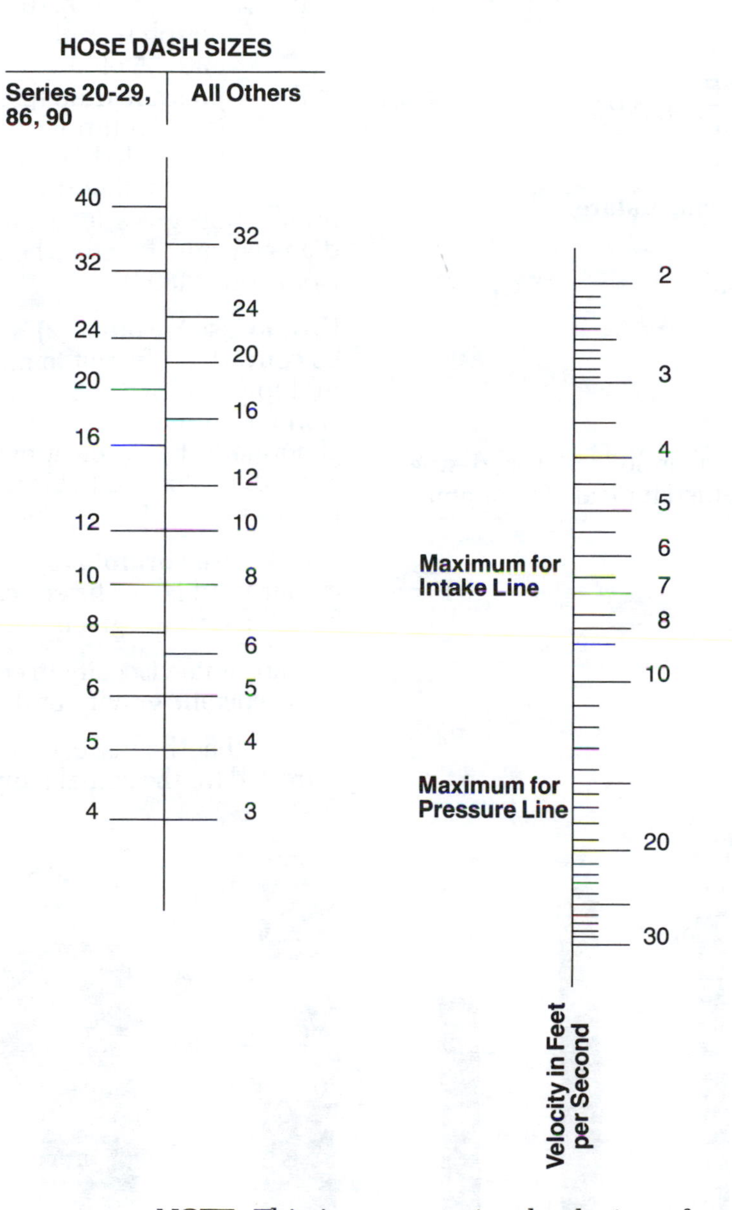

NOTE: This is a conventional substitute for calculation of energy losses. For pressure drop calculation, see the following pages.

table j (c-2)

hose pressure drop

Pressure drop in fluid lines is one of the easier forms of energy loss to calculate with reasonable accuracy, provided the necessary parameters are at hand. The method and values presented below come from over one thousand actual laboratory test points, although they also have been found to agree closely with the D'Arcy and Fanning formulas.

For Number 10 Oil at 180°F.

1) $\triangle P_{10} = (L + .55\, nB)\dfrac{\triangle P}{\triangle L} + \triangle P_{2F}$

For Other Fluids and / or Temperature.

2) $\triangle P = K \times (\triangle P_{10})$ where,

3) $K = .7\mu^{.25}\, SG^{.75}$

Note: The acutual "Number 10" oil tested was Texaco A (R and O), assumed typical of the lighter hydraulic oils.

How to Use Formula (1)

L = length of hose between hose fittings, in ft.
nB = number of bends between 45° and 180°. Ignore bends appreciably less than 45°, but determine the **number** of bends rather than their exact measurement. Note that nB is multiplied by .55 to convert to equivalent length in feet.

$\dfrac{\triangle P}{\triangle L}$ = pressure drop (psi) per foot of hose (graph I), multiply by the sum L and .55 of nB.

$\triangle P_{2F}$ = pressure drop (psi) of hose fittings (regular fittings), or (Mandrel-assembled fittings with oversized bore.) (Graph II or III).

Summation gives total pressure drop from hose diameter and lengths, bends, and hose fittings, No. 10 oil at 180°F.

How to Use Formula (2)

To convert a different temperature, fluid, or both, multiply $\triangle P$ by "K". This is an accurate conversion for all hydraulic lines except when they are abnormally large, creating laminar flow. (In this case, the calculated $\triangle P$ will be **higher** than the true value.)

How to Use Formula (3)

If your fluid is not listed, calculate your own "K" from the formula given.

μ = absolute viscosity in centipoises
SG = specific gravity, or density in gm per cc

Be sure that the viscosity and specific gravity are corrected for the actual temperature.

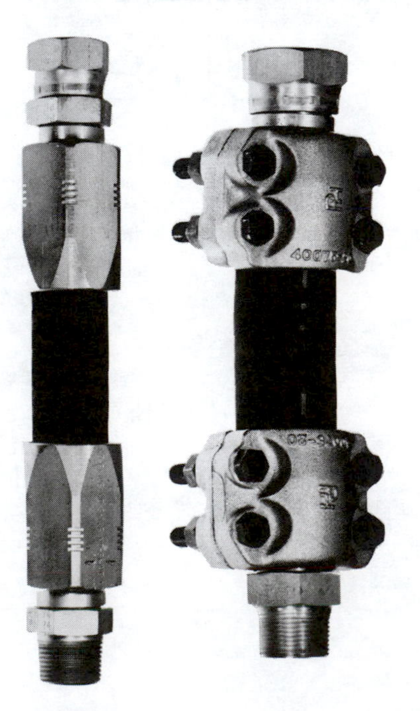

illustration j (c-1)

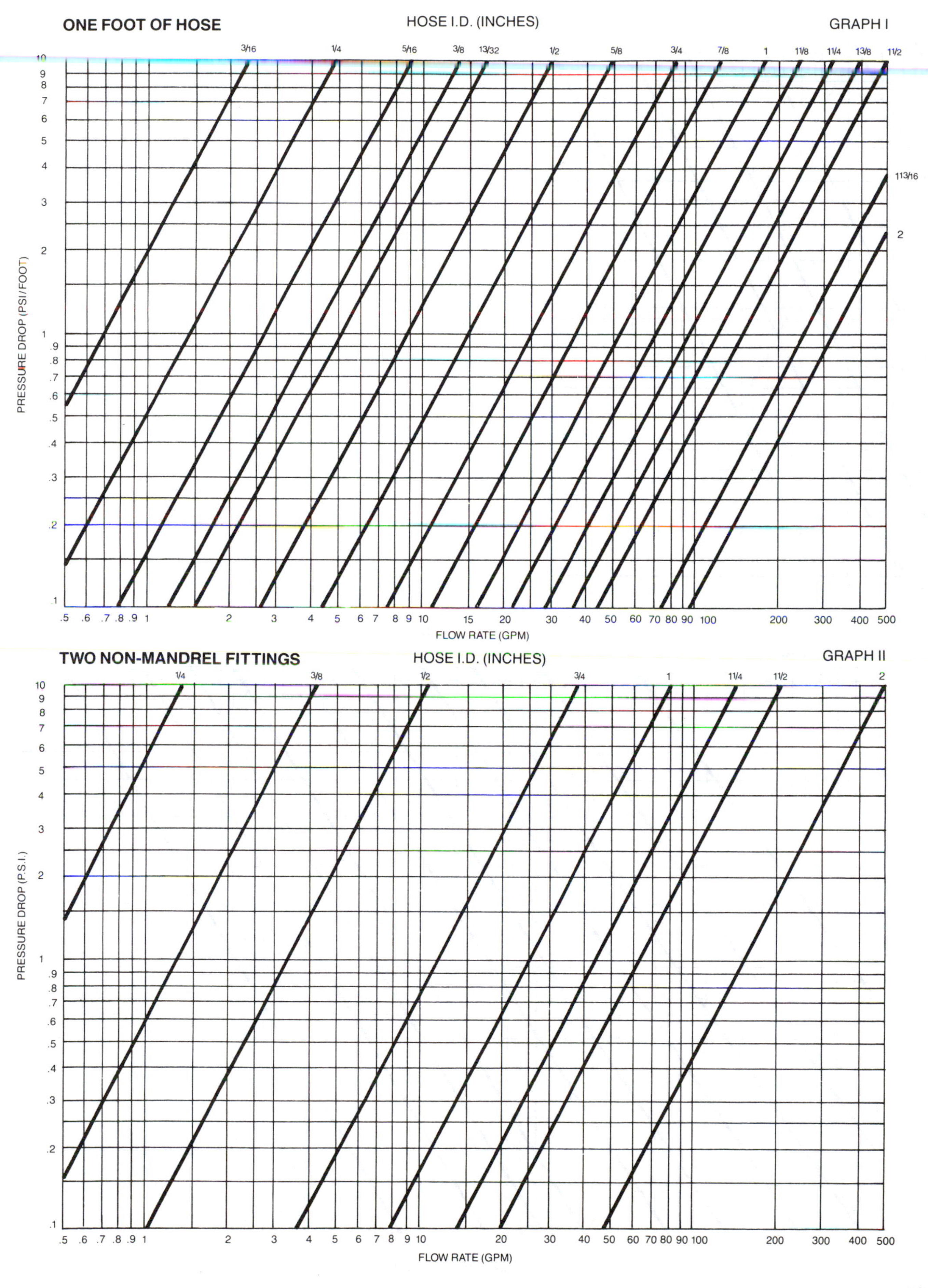
ONE FOOT OF HOSE
HOSE I.D. (INCHES)
GRAPH I
3/16 1/4 5/16 3/8 13/32 1/2 5/8 3/4 7/8 1 1 1/8 1 1/4 1 3/8 1 1/2 1 13/16 2
PRESSURE DROP (PSI / FOOT)
FLOW RATE (GPM)
TWO NON-MANDREL FITTINGS
HOSE I.D. (INCHES)
GRAPH II
1/4 3/8 1/2 3/4 1 1 1/4 1 1/2 2
PRESSURE DROP (P.S.I.)
FLOW RATE (GPM)

connectors

hose selection considerations

- Line Size
- System Pressure
- Surge Pressure Peaks
- Temperature Range — Internal & Ambient
- Fluid Effect on Hose Material
- Service Life
- Cost
- Availability
- Installation

  Stationary - Moving (flexing) - Bend radius - Routing - Amount of Slack - Use of Angle Adaptors and End Fittings - Use of Straight Thread Fittings - Use of Hose Supports - Control of Twisting - Control of Abrasion
- Ease of Maintenance

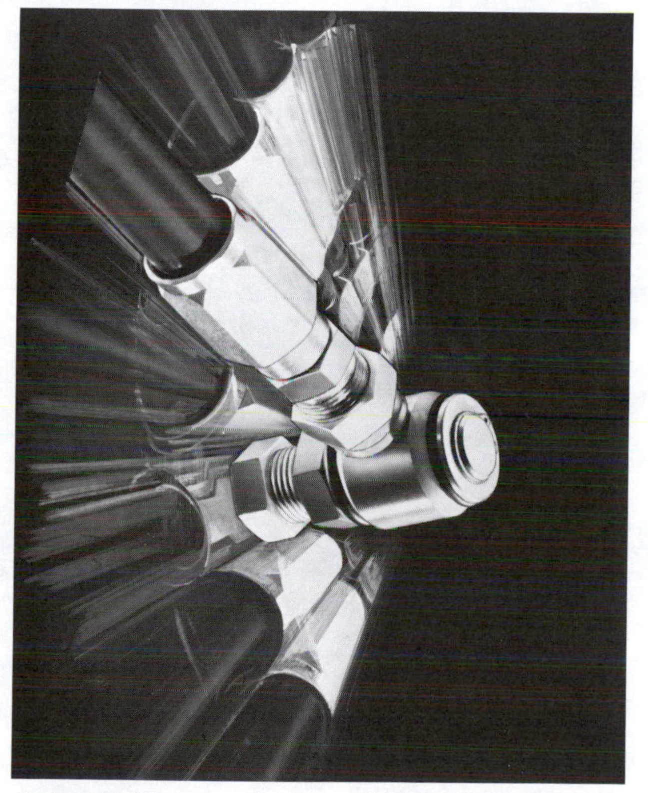

illustration j (c-2)

Ambient Temperatures

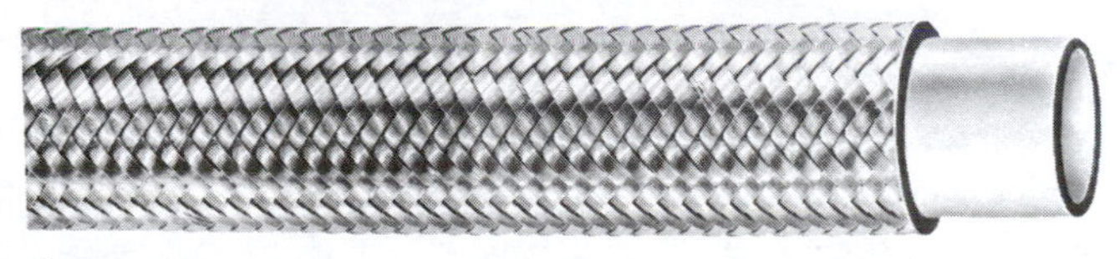

illustration j (c-3)

Very high or low ambient (outside of hose) temperatures will affect cover and reinforcement materials, thus influencing the life of the hose.

Bend Radius
Recommended minimum bend radius is based on maximum operating pressures with moderate flexing of the hose. Safe operating pressure decreases when bend radius is reduced below the recommended minimum.

Burst Pressure
These are test values only and apply to hose assemblies that have not been used and have been assembled for less than one month.

Chemical Resistance
Consider chemical resistance of hose cover and tube stock.

Electrical Conductivity
Hose is available for electrically conductive applications. Hose is also available for non-conductive applications. Each must be specified.

Line Size (See pages on line sizes and pressure drop.)

Undersized pressure lines produce excessive pressure drop which results in energy loss and heating.

Undersize suction lines can cause cavitation at the pump inlet.

Flow rates desired and operating pressures establish required hose size. Required size of hose depends on volume and velocity of fluid flow. If fluid velocity is too high, much of the energy is lost in the form of heat. If hose must be a different size from the accessory ports, jump size adapters should be used.

Operating Pressure
Hose lines are rated for continuous operation at the maximum operating pressures specified for the hose. Generally, the operating pressure is one fourth the hose minimum burst pressure. Never exceed the hose maximum operating pressure.

Operating Temperatures
Operating temperatures specified refer to the maximum temperature of the fluid being conveyed. Continuous operation at or near maximum rated temperatures will materially reduce the service life of the hose. Engineering data concerning critical temperature requirement is available upon request from the manufacturer.

hose selection considerations continued

Pressure Surges
Many hydraulic systems develop pressure shocks which exceed relief valve settings and affect the service life all system components. In systems where shocks are severe, select a hose of higher working pressure.

Spring Guards
Spring guards prolong the life of hose lines that are exposed to sharp and abrasive materials. They distribute bending radii to avoid sharp kinks in hose lines at juncture of couplings and protect the hose from excessive wear and deep cuts.

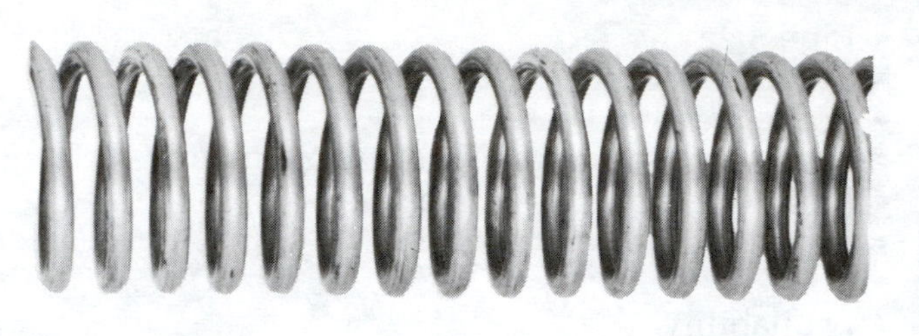

illustration j (c-4)

notes

Hose Application Considerations

1. Hose should be routed to flex in a single plane. If it must be routed through a compound bend it should be divided into two or more bends by clamping to permit each to flex in a single plane.

2. Clamps should also be used to prevent hose cover abrasion. Abrasion may occur when two hose lines cross, or when a hose line rubs against a fixed point.

3. Sometimes when hose is looped or routed outside the contour of the machine it is used as a convenient hand hold or step which leads to premature hose failure.

4. When a hose is used above its rated working pressure it reduces service life and increases the customer costs by increasing down time and requiring unnecessary replacement. It will also cause accidents, unless special precautions are taken.

5. Tension on hose during pressure applications will cause fitting blowoffs. The hose should be the proper length.

6. Protective spring guards should be used whenever a hose is exposed to sharp or abrasive materials.

pressure effects

When flexible hose is subjected to pressure it may change its length as much as +2 to −4%. Installation of lines should compensate for shortening effects by bending the hose or leaving slack in straight runs.

NOTE: *High pressure gaseous systems are very hazardous. Hose lines should be adequately protected from external shock and mechanical or chemical damage. They should also be suitably protected to prevent whiplash action in the event of failure for any reason. Cover should be pinpricked for any gaseous service except ordinary shop air.*

swivel adapters

- Increase Hose Life
- Cut Hose Costs
- Use Parker Swivel Adapters

Parker Swivel adapters feature 360° swiveling action that especially suits them for use in applications where hose moves, bends or twists. Swivel

illustration j (c-6)

adapters connected to hose assemblies relieve twisting, prevent excessive flexing of hose, and eliminate need for long radius bends.

RESULTS: *Longer hose life and reduced equipment "downtime".*

Engineered to operate at pressures up to 3000 p.s.i. for all sizes, the Parker Swivel Adapters can be used with hoses rated up to and including 3000 p.s.i. working pressures.

hose failure

Twisting is one of the primary causes for hose failure. A 7° twist in a large diameter, flexing hose can reduce service life by as much as 90%.

Consideration in routing and installation helps insure efficient system operation and reduces field problems.

vacuum requirements

1. Maximum negative pressures shown for hoses — 16 and larger are suitable only for hose which has suffered no external damage or kinking.

2. If greater negative pressures are required, an internal supporting coil is recommended.

3. No vacuum recommended for multiple wire braid or multiple spiral wrap (4 spiral or more).

application continued

right vs. wrong

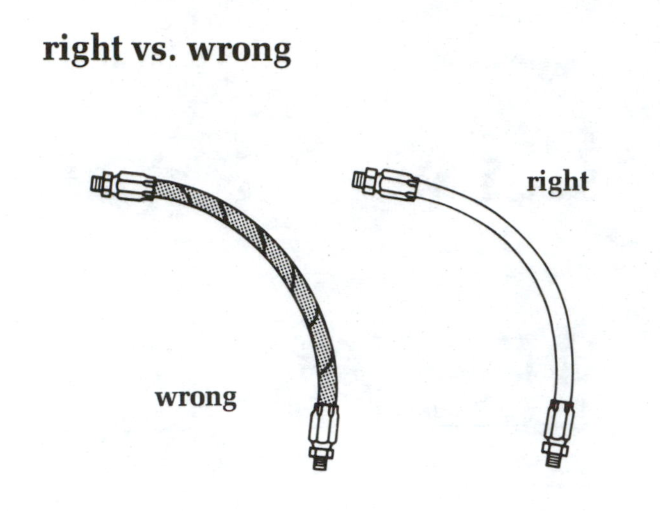

Hose is weakened when installed in twisted position. Also, pressure in twisted hose tends to loosen fitting connections.

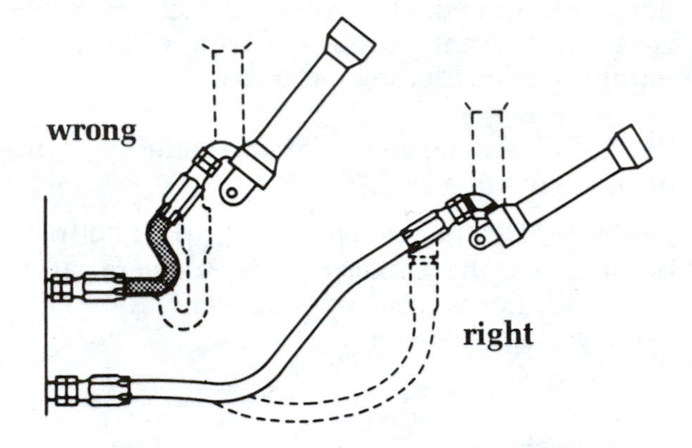

When hose assembly is installed in a flexing application, remember that metal hose fittings are not part of flexible protion. Allow ample free length for flexing.

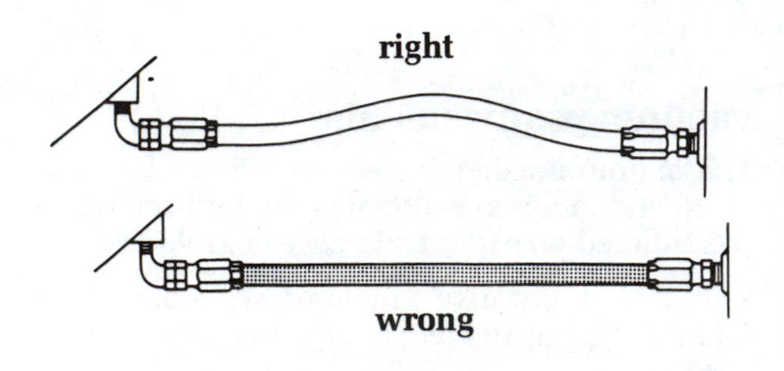

Pressure can change hose length as much as +2% or −4%. Provide slack in line to compensate for hose length changes.

illustration j (c-7)

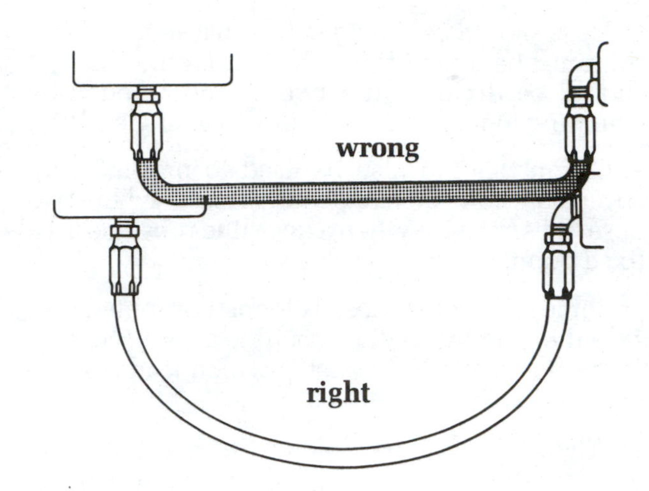

Ample bend radius should be provided to avoid collapsing of line and restriction of flow. Exceeding minimum bend radius will greatly reduce hose assembly life.

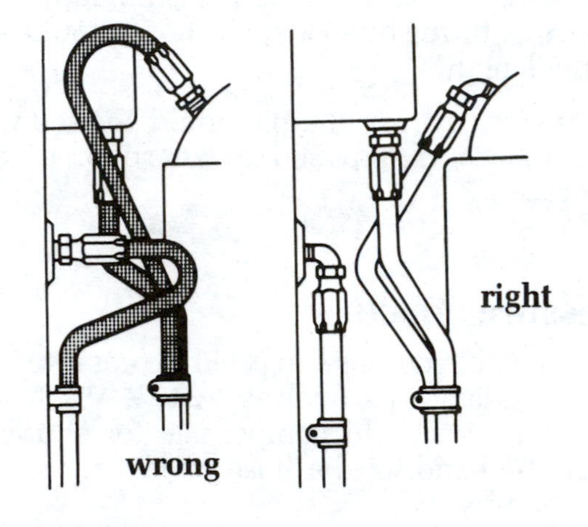

Use elbows or other adapters as necessary to eliminate excess hose length and to insure neater installation for easier maintenance.

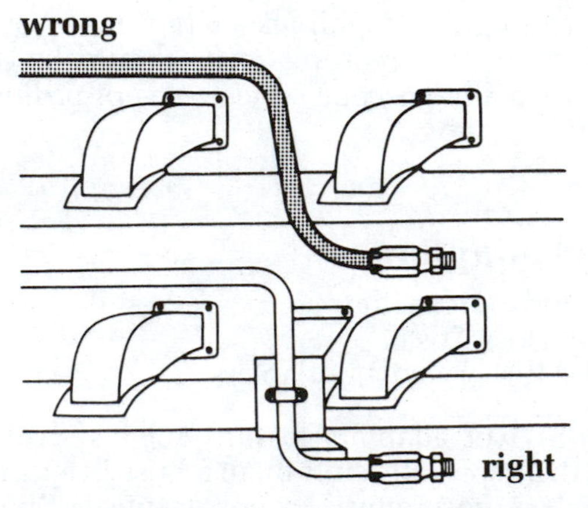

Avoid installing hose line close to exhaust manifold or any other hot section. If impossible, isolate hose with fire proof boot or other protective means.

basic hose construction

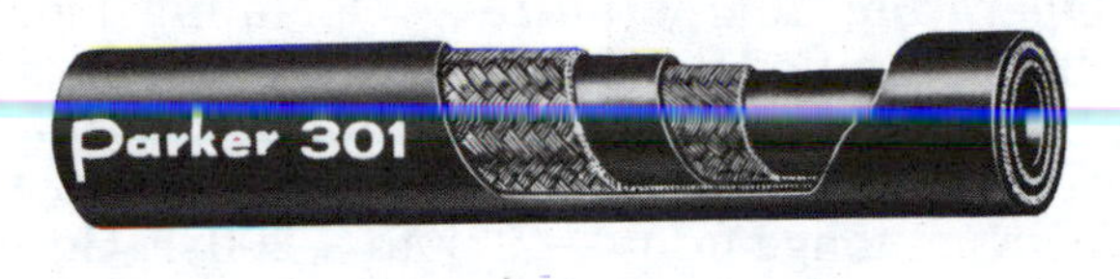

illustration j (c-8)

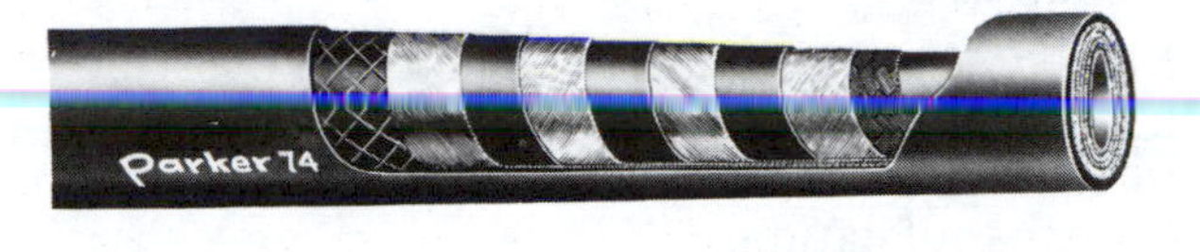

illustration j (c-9)

Most types of hose consist of three basic elements or parts as illustrated above.

- Tube or Inner Lining
- Reinforcement
- Cover

1. INNER TUBE
The tube or liner is the innermost element of the hose. The primary function of the tube or lining is to retain the material-liquids, gases, or combination of these. The type of hose and the service to be encountered determines the materials used and the thickness of the inner tube. This tube must also be compatible with various methods of attaching hose end fittings.

These materials are compounded to operate in various temperature ranges where elevated temperatures or low temperatures are required, in addition to being resistant to the various materials to be conveyed. The materials most commonly used in inner tube construction are;

COMMON NAME	ASTM DESIGNATION	COMPOSITION	GENERAL PROPERTIES
Buna-N	N. B. R.	Nitrile-Butadiene	Excellent oil resistance Good resistance to aromatics Good physical properties
Neoprene	CR	Chloroprene	Very good weathering resistance Good oil resistance Good physical properties
ethylene-propylene	EPR, EPDM	Copolymer of Ethylene, propylene, and (sometimes) another.	Excellent weathering resistance Resistance to certain fluids Fair physical properties
Teflon	TFE	Tetrafluoro-Ethylene	Excellent high temperature resistance Excellent chemical compatibility Low moisture absorption Excellent weather resistance
PKR	—	Proprietary	Good chemical resistance Outstanding temperature range Excellent weather resistance

table j (c-7)

NOTE: *For selection of recommended materials see Compatibility Chart in this Hand Book.*

2. REINFORCEMENT
The reinforcement is the fabric, cord, or metal elements built into the body of the hose for purposes of strength, to withstand internal pressure or external forces or a combination of both.

The materials most commonly used as reinforcement are textile yarns, synthetic yarns, textile fabrics, and wire. The yarn may be cotton, rayon, nylon, dacron, glass or similar materials. The wire may be steel, stainless steel, bronze, aluminum, or other metal. The type and number of layers or reinforcing materials used depends upon the methods of manufacture and the service conditions of the hose.

3. COVER
The cover is the outermost element of the hose and its primary function is to provide protection against damage to the tube and reinforcement. The cover materials are selected to provide resistance to abrasion, sunlight, hot and cold temperature conditions, and to protect the hose from the various oils, solvents, acids, gasoline and other substances encountered in service.

Neoprene, synthetic fabrics, nylon, urethane, and other elastomeric materials are commonly used for hose covers.

Hose Performance Ratings

pressure

Most types of hose are designed and constructed to meet established pressure rating standards, or to meet a working pressure determined by the hose manufacturer.

Pressure ratings of hose are normally divided into three classifications as follows:

A. **Burst Pressure** – The burst pressure rating of a hose is a test pressure value only that can be applied to hose assemblies once. Burst pressure is the pressure at which actual rupture of the hose occurs. The minimum burst pressure rating of a hose assembly is determined from actual bursts of a large number of samples, and normally, the minimum burst rating is approximately 90% of the actual burst obtained from wire reinforced hose.

B. **Proof Pressure** – The proof pressure rating of a hose is a test pressure value only and is normally 50% of the minimum burst pressure rating which is applied to a hose assembly for

C. **Work Pressure** – The working pressure rating of a hose is the actual maximum working or operating pressure and provides a four to one safety factor in relation to the minimum burst pressure rating of the hose.

Do not use hose above its maximum rated working pressure. When a hose is used at a hydraulic system pressure that exceeds the rated working pressure of the hose, a shortened service life and premature failure in the form of either a hose rupture or fitting blow-off will occur. It increases the user costs by increasing the frequency of replacement and downtime.

Almost all hydraulic systems encounter pressure surges and shocks. When these pressure surges exceed the relief valve settings, the service life of the host is affected and when severe surge pressures exist, a higher pressure rated hose should be used.

temperature

Both the operating temperature and ambient temperature conditions can affect the service life of a hose assembly. Consideration must be given to the temperature range rating of the hose and to the installation and routing location of the hose.

A. **Operating Temperature** – Operating temperatures specified for each hose refer to the minimum and maximum temperature of the fluid conveyed. Most hoses are rated to operate in a temperature range from −40°F to +200°F. Hose is not damaged by mere exposure to temperatures below minimum but will crack if pressurized or bent. Continuous operations at or above the maximum rated temperatures will materially reduce the service life of the hose. Hoses with high temperature rubber tubes are available with a temperature range up to +300°F. Temperatures up to +450°F can be handled by Teflon hose. For temperatures higher than +450°F metal tubing must be used.

B. **Ambient Temperature** – The temperature of the surroundings is generally of importance only as it affects the internal temperature. Exceptions occur at both extremes. Hose may be temporarily embrittled by exposure to ambient temperatures of several hundred degrees, or from routing past manifolds and other hot spots. Heat embrittlement from the outside may occur. The hose must be rerouted or shielded to prevent this condition.

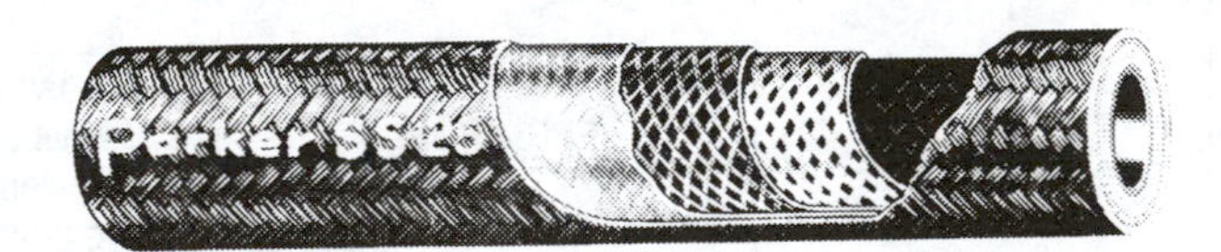

illustration j (c-10)

Minimum Bend Radius

All hose normally has a recommended minimum bend radius established and this radius is based on the hose construction, size, and pressure classification of the hose. It is measured to the inside of the curvature and is specified for non-flexing applications at the rated working pressure of the hose.

On hose installations where flexing occurs, the bend radius must be increased to avoid overstressing the hose and to prevent any tendency to kink or flatten.

NOTE: *Refer to Parker hose catalog data for the recommended minimum bend radius established for each specific hose size and construction.*

hose fitting types

Hose fittings are classified by the method used to attach them to the hose. There are two basic types, permanently attached and reusable. Both have the same components, a nipple and a socket with the hose clamped between them.

reusable

Reusable fittings are normally designed to be screwed or clamped to the hose ends and can be removed from a worn hose and reassembled to a replacement hose.

Three commonly used types of reusable hydraulic hose fittings are shown below.

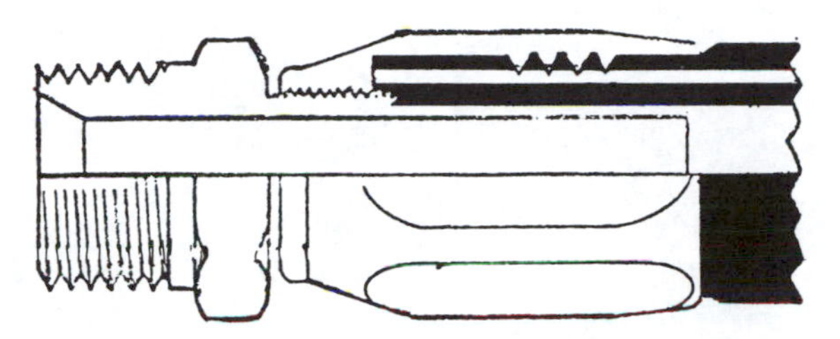

Reusable, Screwed No-Skive

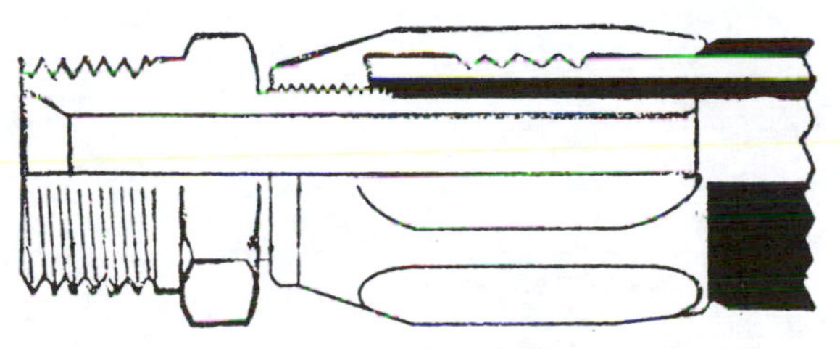

Reusable, Screwed, Skive

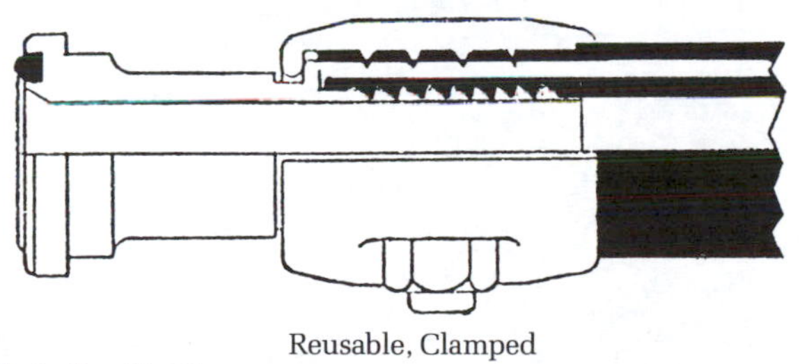

Reusable, Clamped

illustration j (c-11)

The Skive Type fitting is a screw-on design for use on hoses with a thick outer cover and this cover must be removed (skived) from the ends of the hose prior to installing this type of fitting.

The No-Skive Type fitting is a screw-on design for use on hose with a thinner outer cover than the Skive type, but still thick enough to provide adequate protection for the reinforcement. The hose cover does not require removal and provides a supporting cushion of rubber to reduce stress concentration and fills the voids in the gripper thread of the fitting to protect the hose wire reinforcement from moisture and corrision.

The Clamp Type fitting is designed with a barbed nipple which is inserted directly into the hose and two clamp halves are then assembled with bolts and nuts on the outside of the hose, and the halves are then bolted together to provide a positive, leak-proof grip.

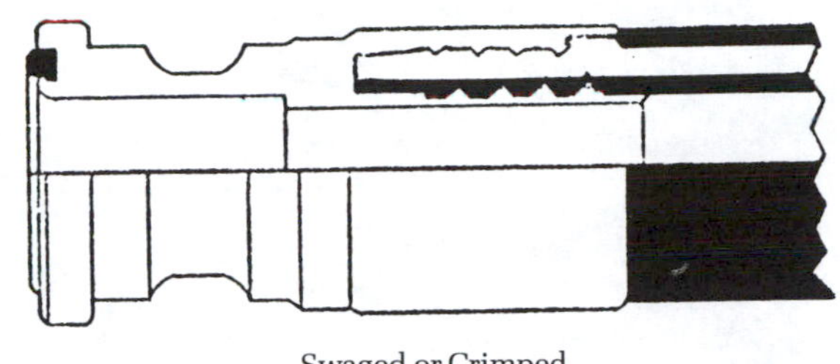

Swaged or Crimped

illustration j (c-12)

Permanent attached type fittings are generally of a one-piece design and the hose is inserted directly into the fitting between the nipple and the socket and the socket is then either swaged or crimped to hold the hose in a vise-like grip. The barbs on the o.d. of the nipple provide additional holding power by being forced into the i.d. of the hose inner tube.

JIC 37° Swivel

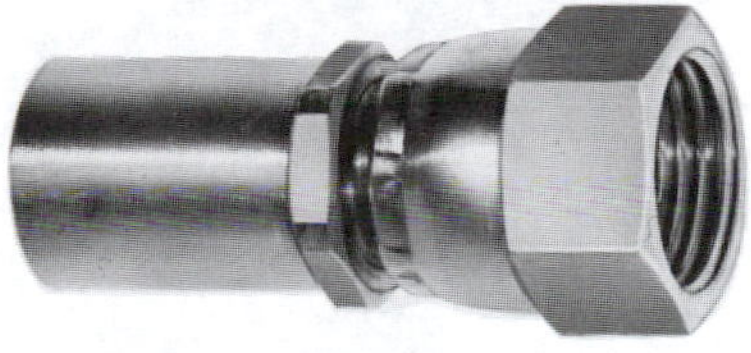

SAE 45° Swivel

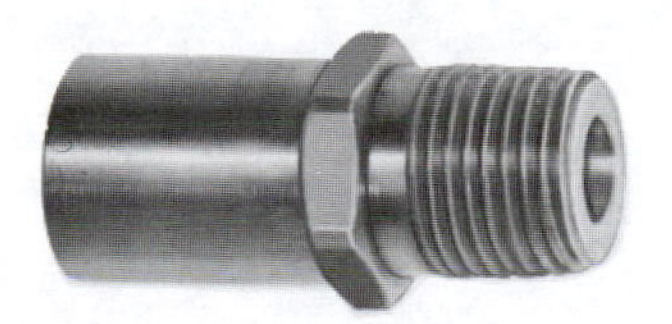

Male Pipe

illus'ration j (c-13)

notes

connectors

quick couplings

coupling seal material selection ..... j (b-2)
quick couplings .... j (b-3)
single shut-off .... j (b-3)
double shut-off .... j (b-3)
straight-thru .... j (b-3)

thermoplastic tubing

fast & tite clear vinyl tubing instrument grade

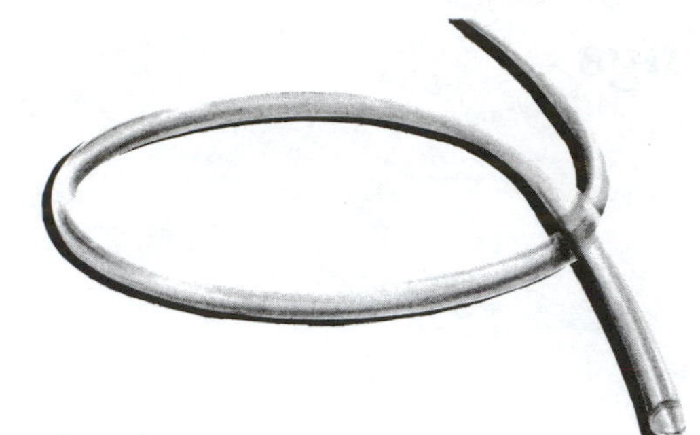

illustration j (d-1)

Formula PV-1 FAST & TITE clear vinyl tubing meets all F.D.A. specifications for use in transmitting fluids for human consumption. It will not impart odor or taste to materials handled and can be steam-cleaned.

Formula PV-2 FAST & TITE clear vinyl tubing is for general industrial and chemical applications.

All Parker FAST & TITE plastic fittings are designed for use with clear vinyl tubing. A "tube support" is recommended when using FAST & TITE fittings.

FORMULA PV-1 Tubing Number	Color	Nominal Tube O.D. (in.)	Tube I.D. (in.)	Average Wall Thickness (in.)	Working Pressure at 73° F (psi)	Coil Length (ft.)
PV42-1	Clear	1/4	.125	.062	65	50
PV43-1	Clear	1/4	.170	.040	55	50
PV53-1	Clear	5/16	.187	.062	55	50
PV64-1	Clear	3/8	.250	.062	55	50
PV86-1	Clear	1/2	.375	.062	45	50
*PV108-1	Clear	5/8	.500	.062	30	50

table j (d-1)

FORMULA PV-2 Tubing Number	Color	Nominal Tube O.D. (in.)	Tube I.D. (in.)	Average Wall Thickness (in.)	Working Pressure at 73° F (psi)	Coil Length (ft.)
PV42-2	Clear	1/4	.125	.062	65	50
PV43-2	Clear	1/4	.170	.040	55	50
PV53-2	Clear	5/16	.187	.062	55	50
PV64-2	Clear	3/8	.250	.062	55	50
PV86-2	Clear	1/2	.375	.062	45	50
*PV108-2	Clear	5/8	.500	.062	30	50

table j (d-2)

fast & tite polyethylene tubing instrument grade

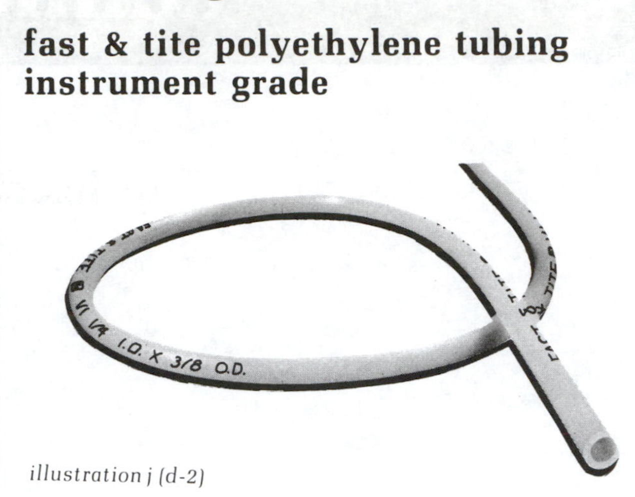

illustration j (d-2)

Black (EB) tubing contains an ultraviolet inhibitor which is recommended for use in sunlit areas. Ingredients of natural color tubing listed below meet F.D.A. specification for use in transmitting fluids for human consumption. All tubing conforms to ASTM standards.

Parker Tubing Number	Color	Nominal Tube O.D. (in.)	Tube I.D. (in.)	Average Wall Thickness (in.)	Working Pressure at 73° F (psi)	Reel Length (ft.)
*E43	Natural					
EB43	Black	1/4	.170	.040	120	1000
E53	Natural					
EB53	Black	5/16	.187	.062	145	500
E64	Natural					
EB64	Black	3/8	.250	.062	125	500
E86	Natural					
EB86	Black	1/2	.375	.062	90	250
E108	Natural	5/8	.500	.062	70	
Color Coded Tubing						
E43-R	Red	1/4	.170	.040	120	500
E43-0	Orange					
E43-Y	Yellow					
E43-P	Purple					
E43-G	Green					
E43-Bl	Blue					

NOTE: These colors also available in size E64, 3/8 x .062.

table j (d-3)

NOTE: Parker Brass "Dubl-Barb" and "Poly-Tite" fittings are also suitable for use with clear vinyl and polyethylene tubing.

NOTE: Polyethylene is not recommended for beverage tubing.

how to measure fast-stor

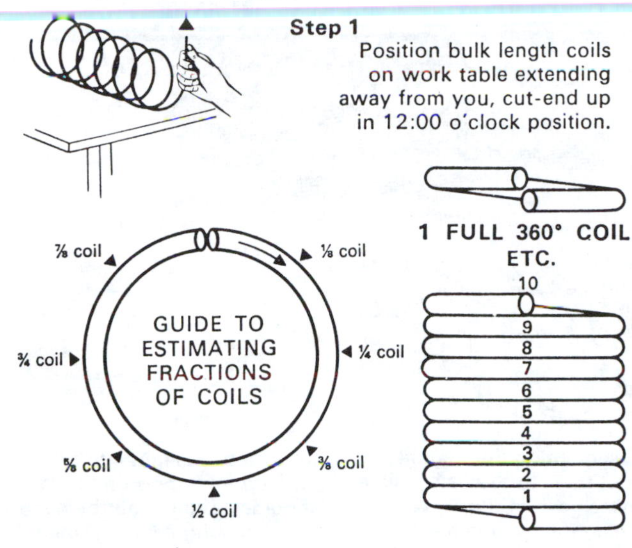

illustration j (d-3)

NET EXTENDED LENGTH REQUIRED		NUMBER OF COILS NEEDED TO OBTAIN REQUIRED NET EXTENDED LENGTH ±3%		
Ft	In	5/16" O.D. Fast-Stor	1/2" O.D. Fast-Stor	5/8" O.D. Fast-Stor
3	36	3-1/2 coils	2-1/4 coils	1-5/8 coils
5	60	5-3/4 coils	3-7/8 coils	2-5/8 coils
7	84	8-1/8 coils	5-3/8 coils	3-3/4 coils
10	120	11-1/2 coils	7-5/8 coils	5-3/8 coils
12	144	13-7/8 coils	9-1/8 coils	6-1/2 coils
15	180	17-3/8 coils	11-1/2 coils	8 coils
16	192	18-1/2 coils	12-1/4 coils	8-5/8 coils
17	204	19-5/8 coils	13 coils	9-1/8 coils
18	216	20-3/4 coils	13-3/4 coils	9-5/8 coils
19	228	22 coils	14-1/2 coils	10-1/4 coils
20	240	23-1/8 coils	15-1/4 coils	10-3/4 coils
25	300	28-7/8 coils	19 coils	13-1/2 coils
30	360	34-5/8 coils	22-7/8 coils	16-1/8 coils
33	396	38-1/8 coils	25-1/4 coils	17-3/4 coils
50	600	57-3/4 coils	38-1/8 coils	27-7/8 coils

table j (d-4)

Alternate Method of Measurement:
Bulk retracted lengths of Fast-Stor are exactly 100′ long. Measuring the tightly retracted length in inches and dividing by 4 produces 4 exact 25′ lenths of tubing.

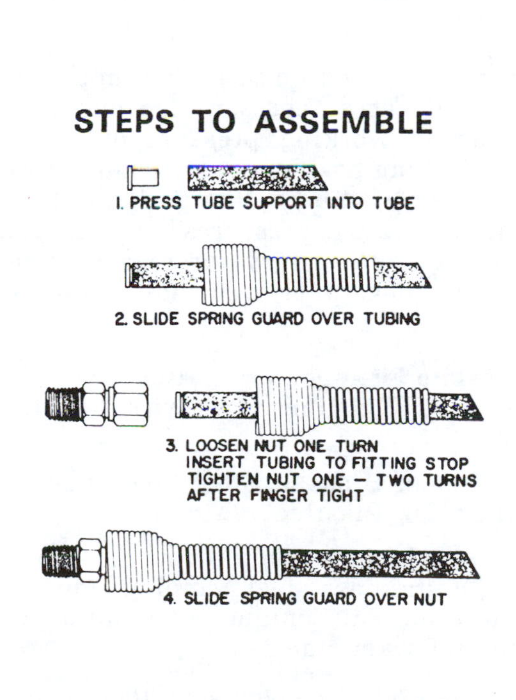

illustration j (d-4)

notes

fast & tite fittings installation

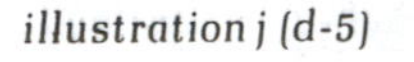

illustration j (d-5)

fast assembly

Fast & Tite Fittings Information

It is not generally necessary to disassemble fitting to connect fitting to tube except when clear vinyl application requires a tube support. Just follow these instructions:

illustration j (d-6)

Cut tube end at right angle for a "square" end. Metal or glass tubing should be de-burred to prevent 'O' ring damage and to promote ease of assembly.

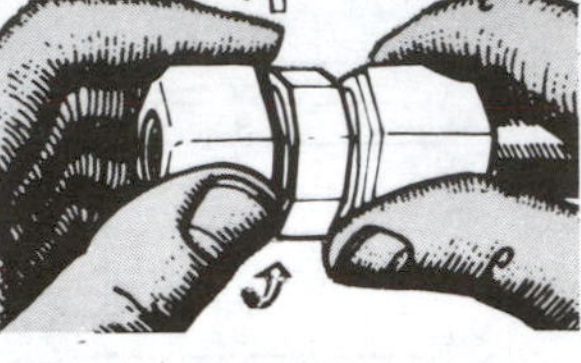

Loosen nut on fitting until about three threads are visible. Moisten end of tubing with water, or other suitable lubricant.

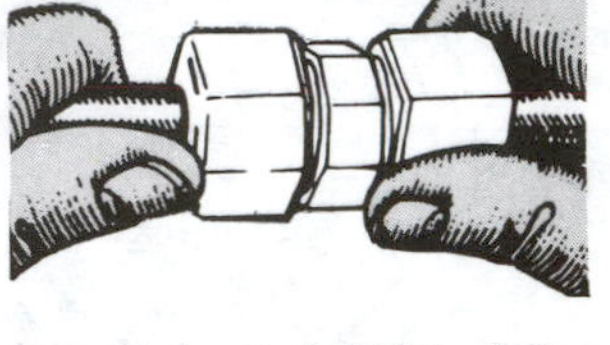

Insert tube straight into fitting until tube bottoms on fitting shoulder. Tighten nut by hand. Additional tightening should not be necessary, but ¼ turn with wrench may be added if desired.

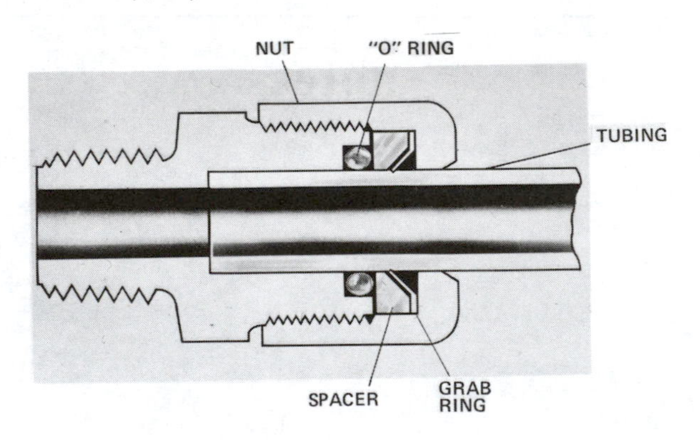

illustration j (d-7)

Fast and Tite Nylon and Polypropylene fittings are approved by the National Sanitation Foundation, Standard 24, Approval #8047, for Mobile Home and Recreational Vehicle hot and cold water plumbing, pressure and non-pressure service.

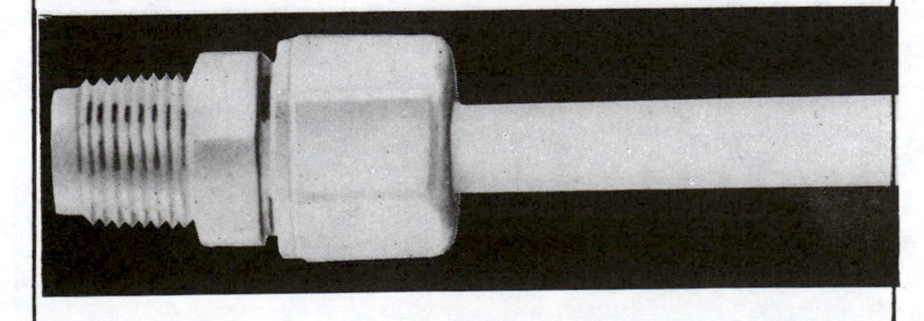

illustration j (d-8)

Working Pressures for Fast & Tite Fittings Air-Oil-Water Pressure in PSI

Tube O.D.(in.)	Up to 75°F	76°F to 125°F	126°F to 175°F
1/4	300	300	300
5/16	300	300	300
3/8	250	250	150
1/2	200	200	150
5/8	150	100	50

table j (d-5)

Ratings are based on use with copper tubing, and in all cases represent the maximum recommended working pressure of the *fitting* only. Working pressures (vs. temperatures) of other types of tubing may limit the tube and fitting assembly to pressures lower than shown above. Consult factory for recommendations on applications other than shown above.

Fast & Tite Fittings Information

FAST & TITE fittings work with the following tubing:

1. Copper 2. Brass 3. Steel 4. Stainless Steel* 5. Glass* 6. Plastic 7. Aluminum

FAST & TITE is the only fitting available combining a stainless steel gripping ring with an O-ring seal. This unique combination is patented by Parker Hannifin.

**Working pressure dependent upon fitting grip allowed by tubing hardness.*

connectors

tube fittings

tables and graphs .... j (e-2)
tubing selection .... j (e-10)
tube fitting selection .... j (e-14)
tubing .... j (e-17)
37° flared fittings .... j (e-19)
bite-type flareless fittings .... j (e-20)
flareless instrumentation fittings .... j (e-21)
installing connectors and tubing ..... j (e-22)
installation equipment table .... j (e-23)
tube fitting troubleshooting .... j (e-24)

tables & graphs

type 304 stainless steel seamless pipe

Iron Pipe Size, In.	Outside Diameter	Inside Diameter	Wall Thickness	Pounds Per Foot
SCHEDULE 5				
3/4	1.050	.920	.065	.6838
1	1.315	1.185	.065	.8678
1-1/4	1.660	1.530	.065	1.107
1-1/2	1.900	1.770	.065	1.274
2	2.375	2.245	.065	1.604
SCHEDULE 10				
1/2	.840	.674	.083	.6710
3/4	1.050	.884	.083	.8572
1	1.315	1.097	.109	1.404
1-1/4	1.660	1.442	.109	1.806
1-1/2	1.900	1.682	.109	2.085
2	2.375	2.157	.109	2.638
SCHEDULE 40				
1/8	.405	.269	.068	2.447
1/4	.540	.364	.088	.4248
3/8	.675	.493	.091	.5676
1/2	.840	.622	.109	.8510
3/4	1.050	.824	.113	1.131
1	1.315	1.049	.133	1.679
1-1/4	1.660	1.380	.140	2.273
1-1/2	1.900	1.610	.145	2.718
2	2.375	2.067	.154	3.653
SCHEDULE 80				
1/8	.405	.215	.095	.3145
1/4	.540	.302	.119	.5351
3/8	.675	.423	.126	.7388
1/2	.840	.546	.147	1.088
3/4	1.050	.742	.154	1.474
1	1.315	.957	.179	2.172
1-1/4	1.660	1.278	.191	2.997
1-1/2	1.900	1.500	.200	3.631
2	2.375	1.939	.218	5.022

table j (e-1)

aluminum pipe

I.P.S. Size, In.	Outside Diameter	Inside Diameter	Wall Thickness	Pounds Per Foot
SCHEDULE 5				
1-1/4	1.660	1.530	.065	.383
1-1/2	1.900	1.770	.065	.441
2	2.375	2.245	.065	.555
SCHEDULE 10				
3/4	1.050	.884	.083	.297
1	1.315	1.097	.109	.486
1-1/4	1.660	1.442	.109	.625
1-1/2	1.900	1.682	.109	.721
2	2.375	2.157	.109	.913
SCHEDULE 40				
* 1/8	.405	.269	.068	.085
* 1/4	.570	.364	.088	.147
* 3/8	.675	.493	.091	.196
1/2	.840	.622	.109	.294
3/4	1.050	.824	.113	.391
1	1.315	1.049	.133	.581
1-1/4	1.660	1.380	.140	.786
1-1/2	1.900	1.610	.145	.940
2	2.375	2.067	.154	1.264
SCHEDULE 80 (Extra Heavy)				
1	1.315	.957	.179	.751
1-1/2	1.900	1.500	.200	1.256
2	2.375	1.939	.218	1.737

table j (e-2)

COPPER PIPE			
Iron Pipe Size, Inch	Outside Diameter, Inch	Wall Thickness, Inch	Lbs. Per Foot
1/8	.405	.062	.259
1/4	.540	.082	.457
3/8	.675	.090	.641
1/2	.840	.107	.955
3/4	1.050	.114	1.30
1	1.315	.126	1.82
1-1/4	1.660	.146	2.69
1-1/2	1.900	.150	3.20
2	2.375	.156	4.22

table j (e-3)

EXTRA HEAVY PIPE			
Iron Pipe Size, Inch	Outside Diameter, Inch	Wall Thickness, Inch	Lbs. Per Foot
1/8	.405	.100	.371
1/4	.540	.123	.625
3/8	.675	.127	.847
1/2	.840	.149	1.25
3/4	1.050	.157	1.71
1	1.315	.182	2.51
1-1/4	1.660	.194	3.46
1-1/2	1.900	.203	4.19
2	2.375	.221	5.80

table j (e-4)

steel pipe

PIPE SIZE	O.D.	WEIGHT CLASSIFICATION						ASA SCHEDULE NUMBER					
		Std.		X.S.		X.X.S.		30		40		80	
		Wall	Wt./Ft.	Wall	Wt./Ft.	Wall	Wt./Ft.	Wall	Wt./Ft.	Wall	Wt./Ft.	Wall	Wt./Ft.
1/8	0.405	.068	.2400	.095	.3100	...	.....	...	.....	.068	.2400	.095	.3100
1/4	0.540	.088	.4200	.119	.5400	...	.....	...	.....	.088	.4200	.119	.5400
3/8	0.675	.091	.5700	.126	.7400	...	.....	...	.....	.091	.5700	.126	.7400
1/2	0.840	.109	.8500	.147	1.090	.294	1.710	...	.....	.109	.8500	.147	1.090
3/4	1.050	.113	1.130	.154	1.470	.308	2.440	...	.....	.113	1.130	.154	1.470
1	1.315	.133	1.680	.179	2.170	.358	3.660	...	.....	.133	1.680	.179	2.170
1-1/4	1.660	.140	2.270	.191	3.000	.382	5.210	...	.....	.140	2.270	.191	3.000
1-1/2	1.900	.145	2.720	.200	3.630	.400	6.410	...	.....	.145	2.720	.200	3.630
2	2.375	.154	3.650	.218	5.020	.436	9.030	...	.....	.154	3.650	.218	5.020

table j (e-5)

tables & graphs continued

flow vs. velocity vs. tubing I.D.

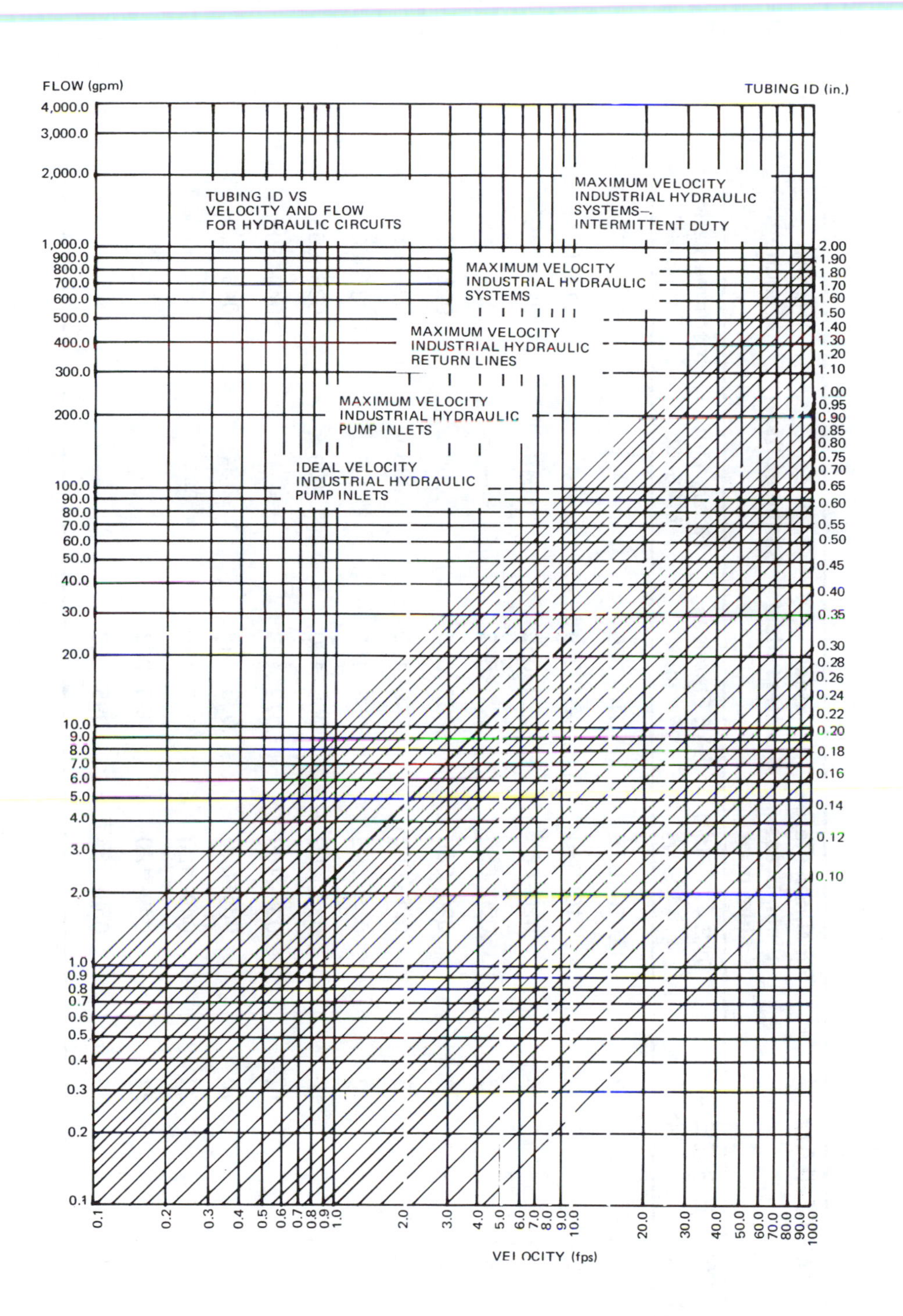

The graph shows relationship among flow, velocity and tubing ID for hydraulic circuits along with maximum recommended velocity for various systems. To determine correct tubing ID, project horizontal line from flow value and vertical line from velocity value. At intersection, follow diagonal upward and to the right to determine tubing ID. If answer falls between two sizes, normally the larger size would be used since it would produce a lower velocity.

graph j (e-1)

connectors j

tables & graphs continued

recommended "min./max." tube wall thickness for common fitting types

MATERIAL:		Steel, Stainless Steel, Brass, Aluminum	Steel Stainless Steel	Steel Stainless Steel	Brass Aluminum
— TUBE SIZE — O.D.-In.	Size No.	SAE 37° Flare "Triple-Lok"	SAE Flareless "Ferulok"	Heavy Duty MA6000	Instrumentation "Intrulok"
1/8	2	.010-.035	.010-.035	---	.012-.028
3/16	3	.010-.035	.020-.049	---	.012-.035
1/4	4	.020-.065	.028-.065	.049	.020-.049
5/16	5	.020-.065	.028-.065	---	.020-.065
3/8	6	.020-.065	.035-.095	.065	.028-.065
1/2	8	.028-.083	.049-.120	.083	.035-.083
5/8	10	.035-.095	.058-.120	.095	.035-.083
3/4	12	.035-.109	.065-.120	.109	.035-.095
7/8	14	.035-.109	.072-.120	---	.049-.095
1	16	.035-.120	.083-.148	.134	.049-.120
1-1/4	20	.049-.120	.095-.188	---	---
1-1/2	24	.049-.120	.095-.203	---	---
2	32	.058-.134	.095-.220	---	---

table j (e-6)

tables & graphs continued

VELOCITY VS. FLOW TABLE

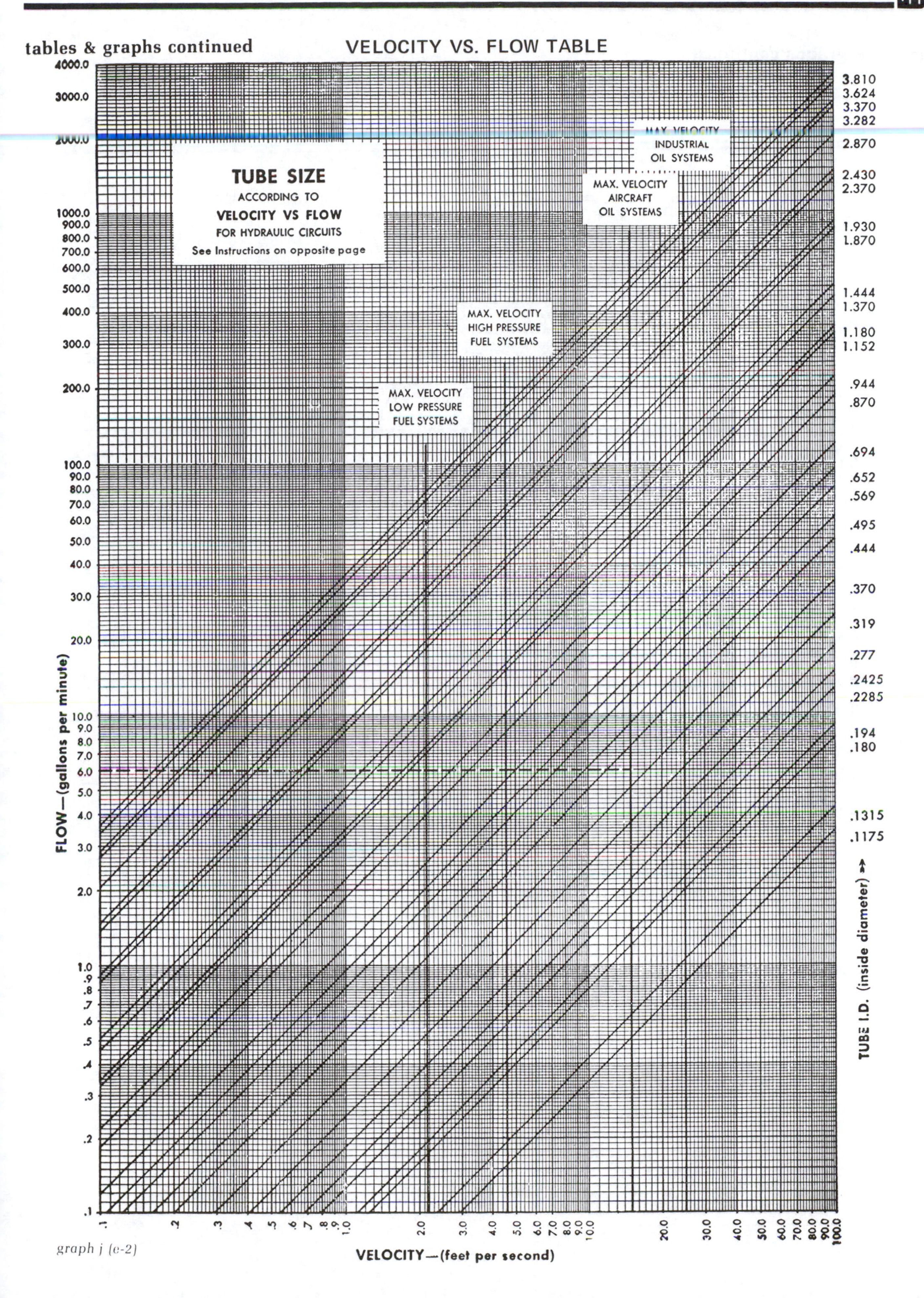

graph j (e-2)

connectors j

tables & graphs continued

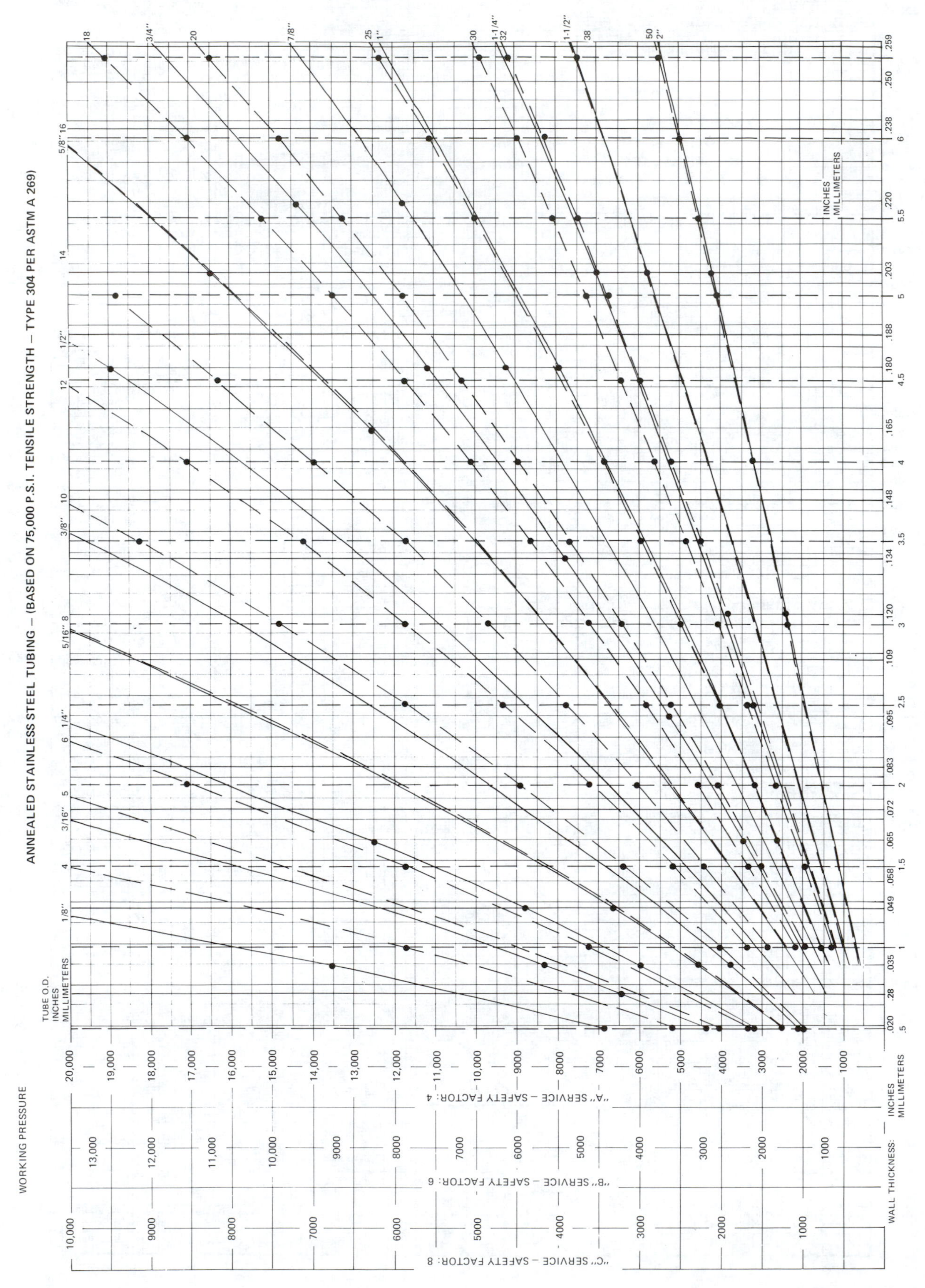

graph j (e-3)

tables & graphs continued

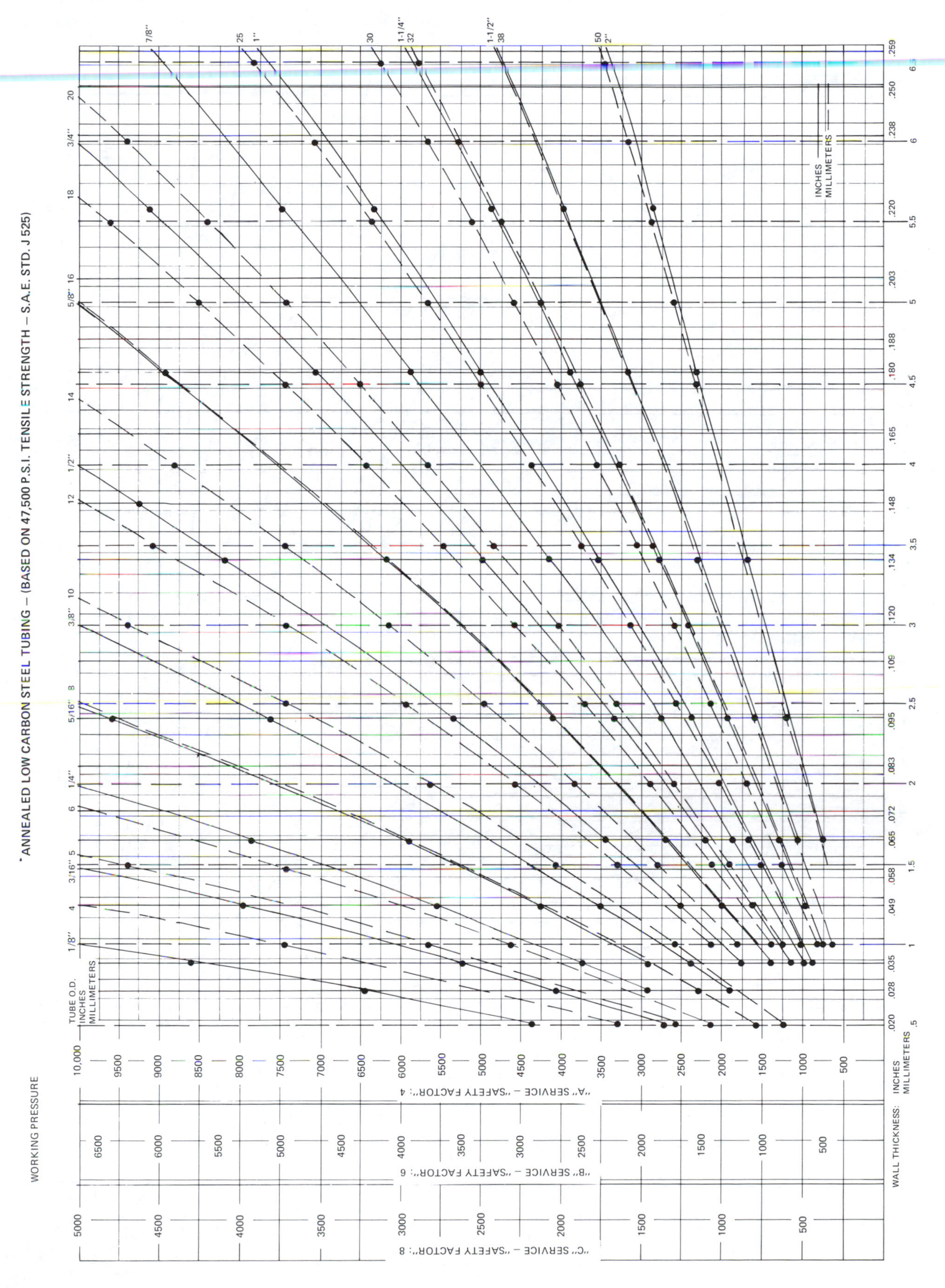

graph j (e-4)

tables & graphs continued

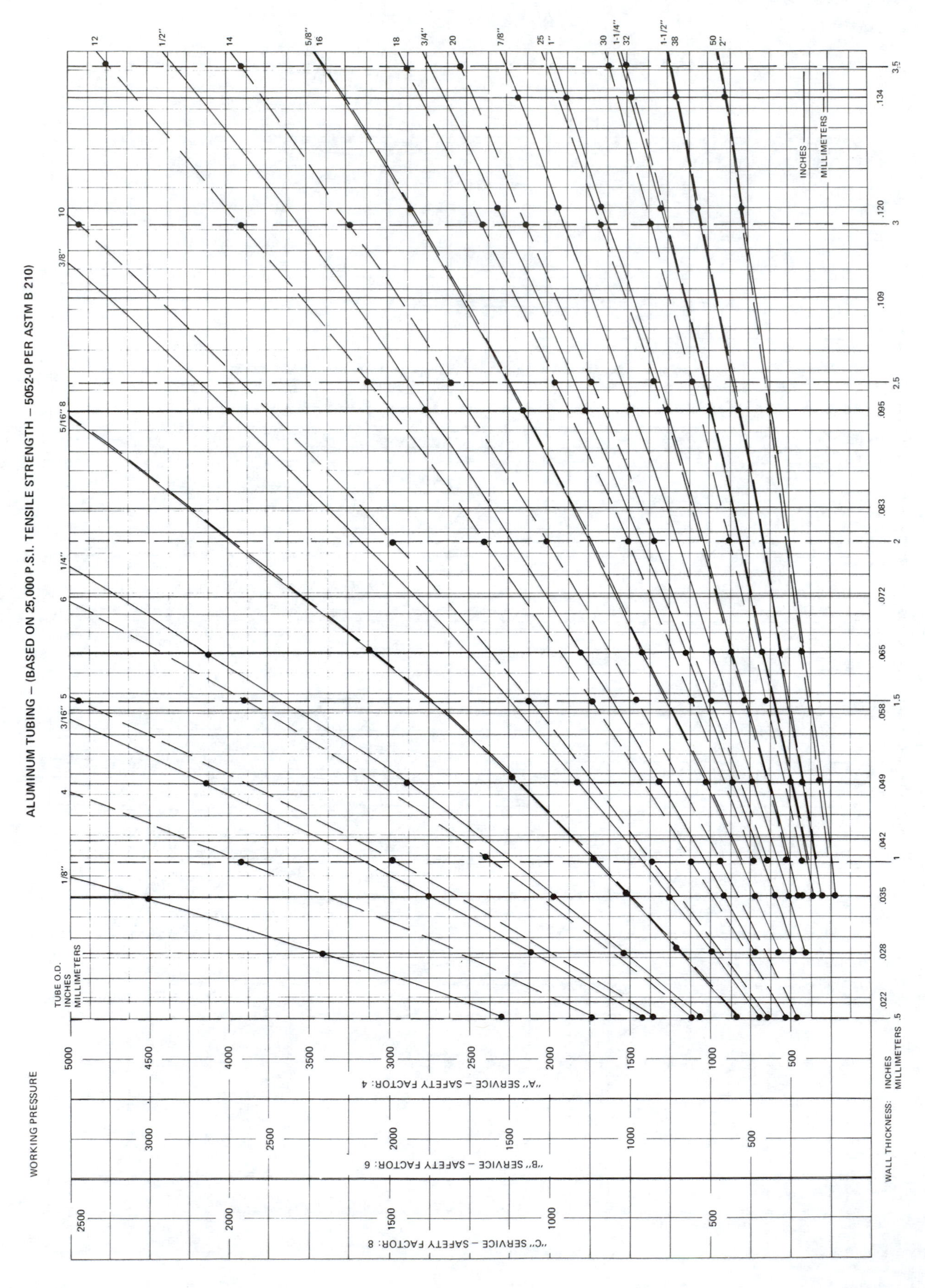

graph j (e-5)

tables & graphs continued

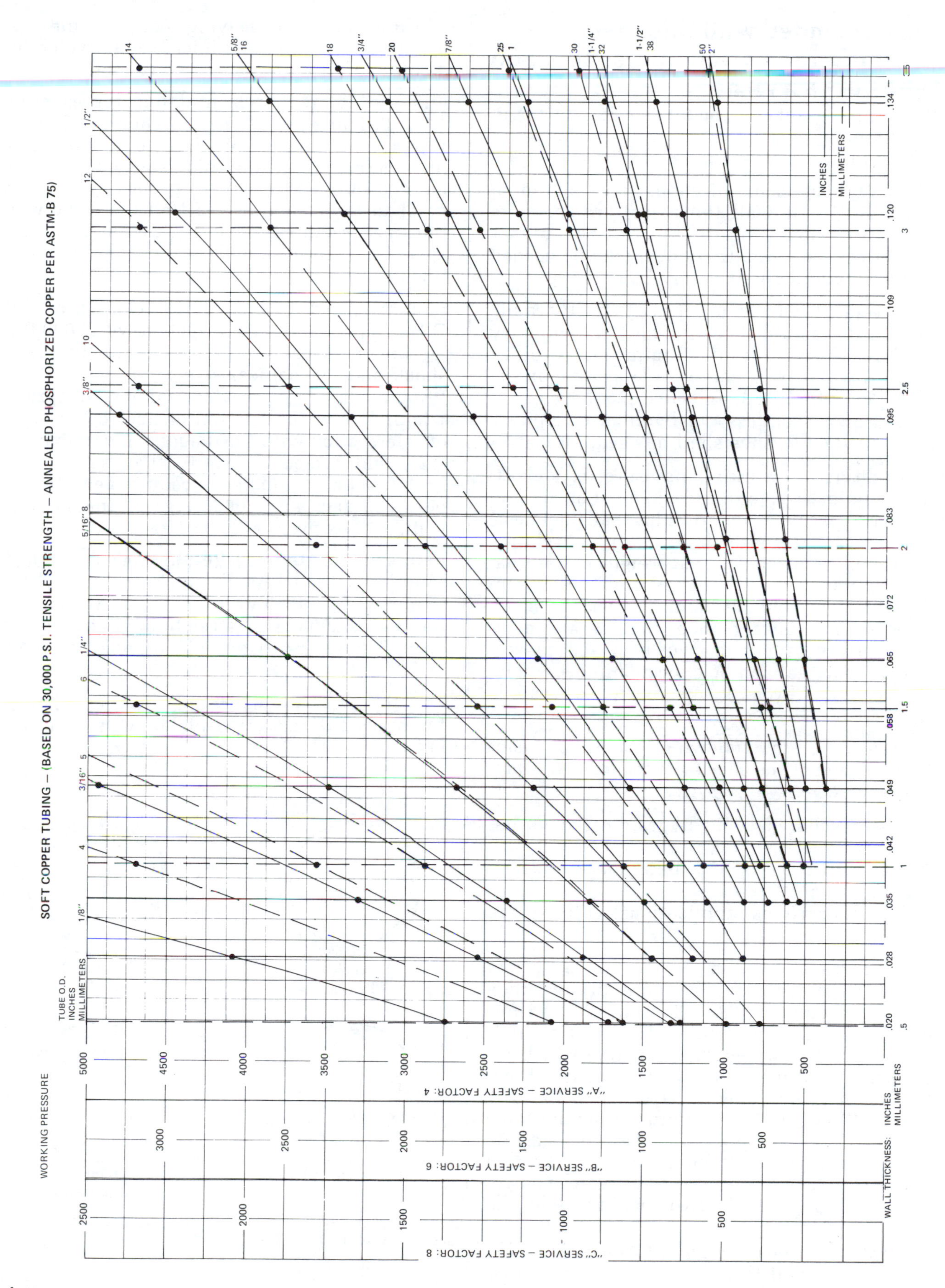

graph j(e-6)

tubing selection

recommended wall thickness of tubing for various pressures and types of service

Working pressure of the system is a factor in determining material and wall thickness of tubing to be used.

The softer materials of copper and aluminum (with sufficient wall thickness) are adequate for lower operating pressures. For higher pressures, the strength of steel may be needed and even stainless steel may be required.

Paragraph H4.1.6.2 of the J.I.C. Hydraulic Standards (established by the Joint Industry Conference for Hydraulic Equipment) specifies that "S.A.E. 1010, conforming to standards; SAE J524, SAE J525, and SAE J356; dead soft, cold drawn seamless steel tubing or equivalent and steel fittings shall be used."

In addition to working pressure, consideration should be given to mechanical strain and service abuse when determining wall thickness. Mechanical strains include vibration, relative motion of connected parts, weight of external contracting parts or unsupported fixtures in the tube line which tend to stretch, bend, shorten or twist the tube. Service abuse involves carelessness of workmen and general rough wear that can be expected in heavy mechanical and process machinery and mobile equipment.

Consequently, the following graphs, by using three different design factors, indicate three grades of severity of service:

"A" — mechanical and hydraulic shocks not excessive
"B" — considerable hydraulic shock and mechanical strain
"C" — hazardous applications with severe service conditions

To use the tables, simply locate the system working pressure in the appropriate grade of service column ("A", "B", or "C") and follow the horizontal line to the right until it intersects the curved diagonal line which represents the tube O.D. size being used. From this point project down vertically to the bottom of the chart to determine minimum table wall thickness required for the intended service. If this projected point lies between two of the standard wall gauge thicknesses shown, then the heavier of the two should be selected.

Examples: Suppose you want to determine recommended wall thickness for 3/4" O.D. steel tubing to be used in a system with 2000 PSI working pressure leaving considerable hydraulic shock and mechanical strain which indicates "B" grade service (safety factor 6).

First, locate the horizontal line representing 2000 PSI in the "B" service column of **graph j (e-4)** and follow this line to the point of intersection with the diagonal curved line marked 3/4". When you project from this point to the base line you find that tube wall thickness requirement lies between .083 and .095. Preference should be for the .095 wall thickness.

Similarly, to determine wall thickness for 1/2" O.D. copper tubing for a system with 2000 PSI working pressure and only moderate hydraulic shock and mechanical strain you locate 2000 PSI in the "A" service column of **graph j (e-6)** follow the horizontal line to the point of intersection with the diagonal curve marked 1/2", and project down to the base line to find tube wall thickness requirement to fall between .058 and .065. Again, the heavier wall, .065, is preferred.

Reference to **table j (e-6),** "Recommended" Min./Max. "Tube Wall Thickness Range for Common Fitting Types" will show that for the 3/4" x .095 wall steel tubing, choices for steel fittings are triple-lok, ferulok, weldlok, or CPI.

For the 1/2" x .065 copper tubing, choices for brass fittings are triplelok, CPI, or intrulok.

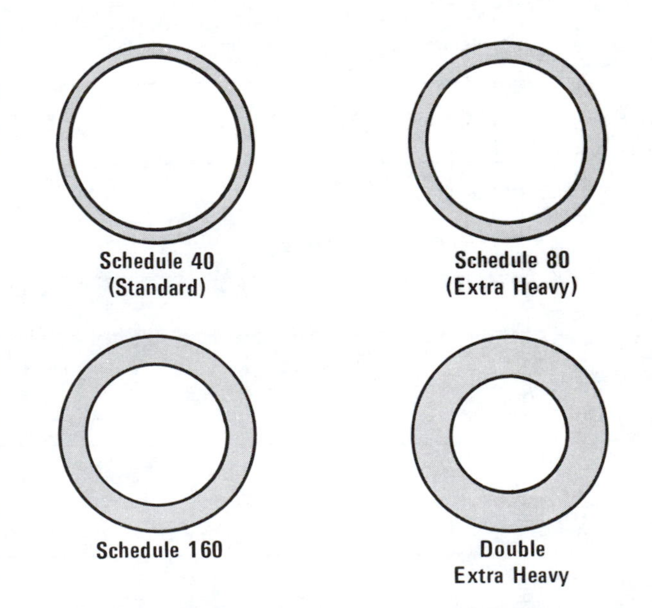

FOR ANY SCHEDULE PIPE, the OD remains the same, but ID decreases as wall thickness increases.

illustration j (e-1)

tubing selection continued

Allowable working pressures for tubing were calculated using formula → $P = \frac{2St}{OD}$ and supplemented with actual burst test data.

Where:

- P = internal burst pressure, psig
- S = allowable stress — lbs/sq. inch
- OD = outside diameter of tube in inches
- t = nominal wall thickness — inches

selecting tubing

This is always the first step. Tube fittings are designed to be used with certain types and size ranges of tubing. Therefore, it is impractical to select a fitting first and then expect to use it with whatever tubing the system may require.

selecting materials and sizes

In general, use —

1. brass fittings with copper tubing
2. steel fittings with steel tubing
3. stainless steel fittings with stainless steel tubing
4. aluminum fittings with aluminum tubing

tube materials

Because of the fabrication procedure normally employed in constructing fluid power transmission lines, minimum cold work in the tube is desirable regardless of the type of material used. Most military and industrial standards which describe tubing intended for power transmission service clearly require a capability for satisfactory flaring and bending and their condition of ductility lends itself well to the attachment of flareless as well as flared tube fittings.

Maximum hardnesses as given in such specifications are:

copper - Rockwell F55
aluminum - not specified - use "0" temperature
carbon steel - Rockwell B65
stainless steel - Rockwell B90

Tube material specifications in common use are:

copper	ASTM - B 68	WW - T - 799
	ASTM - B 75	MIL - T - 873
	ASTM - B 280	MIL - T - 3235
	SAE - U528	
aluminum	ASTM - B210	WW - T - 700
	ASTM - B483	
	ASTM - B491	
carbon steel	SAE - V524	
	SAE - V525	
	SAE - V526	
	SAE - V527	(ASTM - A254 CLI)
	SAE - V356	

NOTE: *All are satisfactory for use with flareless fittings. J526, J527, and J356 should be double flared for reliable sealing in flare type fittings.*

stainless steel	ASTM - A 269	MIL-T-18063
	ASTM - A 249	MIL-T-8504
	ASTM - A 213 (austenitic only)	MIL-T-8506

See Tabulation for Specifications of Various Standard Tube Materials. *(See page j (e-2))*

Corrosion Conditions

Copper tubing and brass fittings are suitable for water, air, and other chemically inactive fluids.

Steel tubing and fittings are used for non-corrosive fluids such as oil. Stainless steel tubing and fittings will handle a great many highly corrosive fluids.

Special alloy tubing and fittings, and special finishes if required for a specific corrosion condition, will be found to be available on special order.

Operating Conditions

Velocity and Flow

The quantity of fluid that must pass through the line in a given period of time is a major determining factor in selecting inside diameter size of tubing to be used, **graph j (e-2),** is provided to indicate proper tubing size for known flow and velocity requirements.

Pressures

Operating pressure of the system is a factor in determining material and wall thickness of tubing to be used.

In general, high pressure installations take steel or stainless steel tubing and fittings rather than heavy wall copper tubing and heavy brass fittings since higher tensile strength of steel permits thinner wall tube and lighter weight

tubing selection continued

fitting without reducing ultimate strength of installation. Tube fittings assembly should not be pressurized beyond the relevant recommended working pressure.

NOTE: *In determining material of tube and fittings on basis of strength requirements, mechanical strain and service abuse must be considered as well as bursting pressures.*

Mechanical strains are those imposed by external sources which — by vibration, relative motion of connected parts, weight of contacting external members or weight of unsupported fixtures within the tube line — tend to stretch, bend, shorten or twist the tube.

Service abuse refers to rough handling by workmen as well as wear and tear normally expected to occur in service, particularly in heavy mechanical and process equipment.

Where installations are subject to considerable mechanical strain or service abuse, it is usually better to increase wall thickness rather than use harder materials.

Tables incorporating factors of pressure and service conditons are provided to indicate selection of tubing as to material and size.

Temperatures
Operating temperature of system is still another factor to consider in determining proper material.

Operating Conditions Summary
Follow these steps to select proper size and material of tubing and fittings:

1. Obtain necessary information:
   a. Flow in gallons per minute
   b. Operating pressure of system
   c. Operating temperature of system

2. Use Flow- Velocity **graph j (e-2),** to determine tube inside diameter which will provide proper flow without excessive velocity.

3. Use Pressure **graphs j (e-3, e-4, e-5, e-6),** choosing A, B or C tables depending on service conditions to determine proper wall thickness for working pressure.

4. If operating temperature is higher than room themperature, determine possible correction of previously selected tube wall thickness and material.

5. Check results by referring again to Flow-Velocity **graph j (e-2),** to determine whether tube outside diameter and wall thickness selected will leave adequate inside diameter to provide sufficient flow without ever exceeding maximum velocity line shown on table. If necessary to adjust, refigure using larger or smaller outside diameter as circumstances dictate.

NOTE: *All tubing outside diameter sizes are shown in 1/16" — for instance: size 4 tubing is 4/16" or 1/4" O.D., size 8 tubing is 8/16" or 1/2" O.D., etc. Fitting sizes adhere to same numbering system. Size 4 fitting is used with size 4 tubing, etc.*

tube size selection by velocity vs. flow

Hydraulic flow involves the volume or quantity of fluid that is required to be pumped through the line to operate the actuating mechanism within the prescribed time cycle. Flow is measured in G.P.M. (gallons per minute).

Velocity involves the rate or speed at which the fluid passes through the line and is measured in FT./SEC. (feet per second). Obviously, a given flow of a certain number of gallons per minute will move through a large line at a slower rate than through a small line. In other words, velocity increases as the inside diameter of the tube decreases.

Shock pressures are undesirable effects because they impair the efficiency of the system.

Consequently, the velocity vs. flow requirements are a determining factor in the selection of tubing with respect to inside diameter as the tube should have sufficient capacity to carry the required flow at a velocity that is not too high.

The Velocity vs. Flow Graph, **graph j (e-1),** indicates the velocity in feet per second for various flows in gallons per minute within various inside diameters of tubing. The inside diameter determines the tube's capacity. (Outside diameter minus twice the wall thickness equals inside diameter.)

Conversely, the table indicates the tube inside diameter when velocity and flow are known, or the resulting flow when tube inside diameter and velocity are known. Following are examples of how to use the table.

Problem No. 1. Find permissible flow in GPM for a velocity of 18 ft./sec. in a .370 I.D. tube.

tubing selection continued

First project 18 ft./sec. point on velocity scale vertically until it intersects diagonal line representing .370 I.D. tube. Then project intersecting point horizontally to flow scale and read 6 GPM as permissible flow.

Problem No. 2. Find permissible tube inside diameter for a velocity of 15 ft./sec. and a desired flow of 12 GPM. First project 15 ft./sec. point on velocity scale vertically until it intersects horizontal projection of 12 GPM point from flow scale. Then from this intersection point, follow diagonal line to read permissible tube I.D. of .569.

In using **graph j (e-1),** if the point for inside diameter falls between two lines, normally the greater inside diameter would be selected since it would mean less velocity.

notes

tube fitting selection

All too often an incorrect design approach is used in selecting a tube fitting for use in a system. Unlike the selection of hydraulic hose where the hose itself is first selected on the basis of its pressure capability and service life expectancy, tube fittings are more often selected by "brand" name first and then adapted to the system.

Seldom is it feasible to use one type of tube fitting for all applications. Requirements vary from system to system and even sections of a system may require different fittings.

In order to select the best possible fitting for any specific application, one must first determine all the parameters of the system or system section plus any peripheral considerations . . . type of fluid, pressure, vibration, shock, leakage limitation, environment, experience of workmen, working conditons, etc. Then these factors must be fed into some kind of organized selection procedure that eliminates unsuitable fittings and compares the possible trade-offs of the remaining fittings. The following step-by-step selection procedure has been found to be very successful.

Relate Hoop Strength and Deformation Capability of Tubing Selected To Fitting Limitations

Hoop strength relates to the tube's collapse resistance when subjected to a uniform external pressure at all points of the O.D. It varies, of course, with tubing material, size and wall thickness. Most flareless fittings exert a uniform stress on the tubing and, in these cases, the tubing must have an adequate hoop strength to withstand this stress. In the area of deformation capability, on the other hand, some fittings, such as the flared fitting, require the tubing to be deformed somewhat in order for the fitting to work properly.

An example of these relationships is shown in **table j (e-6),** "Recommended 'min./max.' Tube Wall Thickness Range for Common Fitting Types" which illustrates that at least two types of fittings can be used with 3/4" O.D. steel tubing.

To use a flared fitting, the tubing must be ductile enough to permit the tube flaring operation. For our 3/4" O.D. steel tube example, the normal wall thickness range is from .035 to .109.

With a bite type flareless hydraulic fitting, the tubing must be strong enough to withstand the high compression force exerted by the fitting biting into the tube O.D. Here, the acceptable wall thickness range for the 3/4" O.D. example, is from .065" to .120".

It follows then, that if a 3/4" steel tubing is selected, only a flared fitting should be used if the wall thickness is .058 or under and only a bite type flareless fitting should be used if the wall thickness is .109" or heavier. Between .035" and .065", the flared type can be used. In the range between .065" and .083", all two types can be used. In actual practice, of course, more than two types of fittings might be considered.

This analysis of the tubing-fitting relationship is most valuable, because in one step you quickly eliminate the unsuitable fittings and reduce the choice to just a few possibilities. These remaining fittings can then be subjected to the following selection steps.

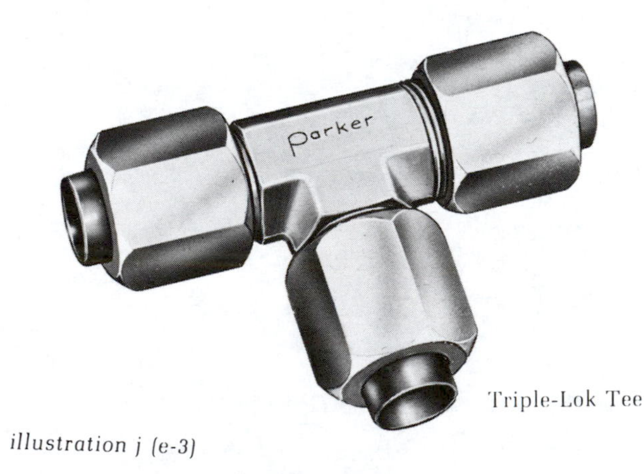

Triple-Lok Tee

illustration j (e-3)

Analyze Holding and Sealing Requirements
All tube fittings perform the functions of both holding and sealing, but some fittings perform one function better than the other. Correspondingly, all applications require both the holding and sealing functions, but, again, on some applications one function is more important or difficult to achieve than the other function. There are three basic possibilities:

- Difficult to hold — easy to seal
- Difficult to seal — easy to hold
- Difficult to both hold and seal

tube fitting selection continued

Mechanical holding power, or the ability to mechanically restrain the tube against pressure blow-out, is a function of a ratio of system pressure and the cross sectional area of the tube plus a safety factor. For example, a 1/4" OD tube has a cross sectional area of 0.0491 in^2. When exposed to a pressure of 1000 psi, the fitting must grip the tube with enough force to resist a 50 lb. blowout force (1000 psi x 0.0491 in^2). Using a 1" OD tube at the same pressure increases the blowout force to 785 lbs. Staying with the 1/4" OD tube, but increasing pressure to 3000 psi, increases blowout force to 150 lbs. It is obvious, then, that as system pressure and/or tube size increase, the required fitting holding power becomes greater. Or, putting it another way, on applications utilizing low pressures and/or small tube sizes, the holding function of the fitting is easy to achieve, while on high pressure systems or where large tube sizes are required it is difficult.

Sealing capability is a function of the relation of the molecular structure of the contained fluid or media to the size of any "hole" or leak path at the fitting seal surfaces.

Molecular structure relates to both molecular size and molecular arrangement. Leakage will result only when the leak path is large enough to permit the molecules of the fluid to pass through it (pressure only affects leak rate). It follows, that as molecular size decreases and/or molecular arrangement becomes more in-line, the potential leak path must be kept smaller to prevent leakage. The smaller the leak path required, the tougher it is to achieve the sealing function at the fitting. Due to it's normal molecular size, helium gas is the most difficult media to seal.

The size of a potential path is determined by the surface finish of the mating parts of the fitting. The smoother the surface finish, the smaller the potential leak path. However, even the best production surface finishes, when viewed under a microscope, will show a series of "hills and valleys" when compared to the size of the molecules to be contained. It is these "hills and valleys" that form a leak path for small molecules and must be eliminated.

Now, going back to the three hold-seal categories, we see that a *difficult to hold-easy to seal application* would involve containing a large molecule media (such as hydraulic oil) at a high system pressure using fairly large tubing. One of the best choices would be a flared fitting with its very strong holding power. If the tube wall was too thick for a flared type, a well designed bite type flareless would work in most cases. For high pressures 6000 psi would be considered. On many applications, a permanent welding type fitting works best.

A difficult to seal-easy to hold application would involve containing a small molecule media (such as helium) at a low system pressure using fairly small tubing. Most good bite type flareless fittings can be used on such applications. The instrumentation type fitting, is particularly suited to such an application because it seals by both a coining out of the "hills and valleys" plus a mild bite that penetrates through the surface layer of the tube into the dense base metal.

A difficult to seal-difficult to hold application consists of both a small molecule media and a high system pressure (i.e., nitrogen at 6000 psi). MA6000 would be recommended.

It can be seen that this fitting selection step is most vital to achieving high reliability. It is imperative that you know the actual holding and sealing requirements of your system and that you match these with a fitting that has the needed capabilities. Any attempt to improvise can only lead to trouble.

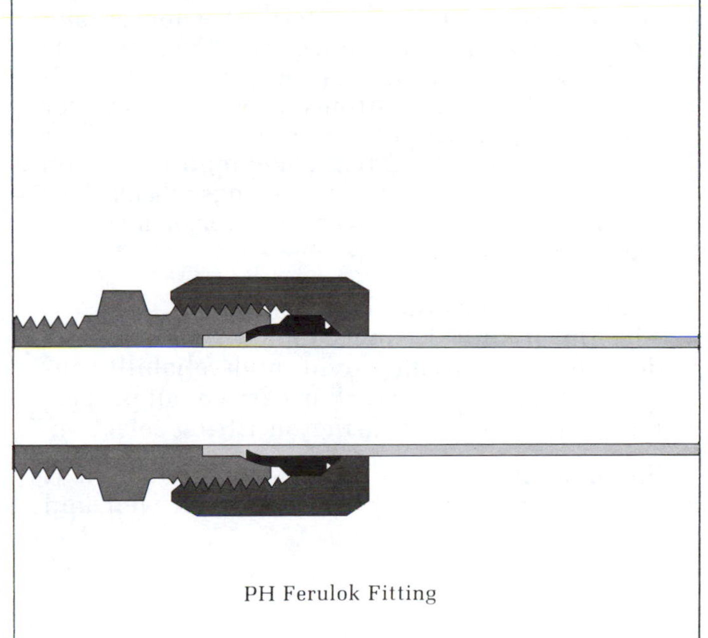

PH Ferulok Fitting

illustration j (e-4)

Other System Parameters

By the time you reach this step, you should have your selection narrowed down to one or two fittings that will do the job. But don't stop yet. It is important to verify that the selected fitting(s) will handle any other critical system

tube fitting selection continued

parameters. These would include such things as shock, vibration and contaminated atmosphere. An instrumentation type fitting, for example, will seal and hold well, but is not designed to stand up to high system shocks. Effects of contaminated atmosphere can be alleviated by proper selection of materials or protective coating.

It is highly desirable to protect each component in a system from shock and vibration which might be imposed by neighboring components. Interconnecting fluid lines should therefore be rigidly supported to prevent the transmission of these forces. Recommended spacing between supports on tube runs is as follows:

TUBE O.D.	APPROX. SPACING (FT)
1/4, 5/16, 3/8, 1/2	3
5/8, 3/4, 7/8	4
1	5
1-1/4, 1-1/2	7
2	8

Prefabricated clamping hardware such as Parker Multi-Clamp makes it easy to plan and to execute an effective, reliable system of support and isolation of extraneous forces from the fittings as well as the components they connect. Well designed hydraulic fittings provide means for damping vibrational forces to prevent their concentration in the critical clamping areas within the fitting. However, the fittings themselves should not be relied upon as vibration dampers for the entire system.

Consider Fabrication Restrictions

The finest tube fitting coupled with the best designed system can provide high reliability only if the fabrication work is carried out properly. This has some bearing on fitting selection.

For one thing, there is the work location. If fabrication is to be performed in a well laid out shop area, then fittings requiring special tooling can be used without problem. On the other hand, if fittings must be made up in hard to get to spots at field locations, then fittings requiring only a couple of wrenches would be the most suitable.

The workmen that are to fabricate the tubing assemblies are a factor. If, for example, you have selected two possible fittings and the workmen have had previous experience with one of them, then that fitting would be the best choice. Even the status of the job is important. If the fitting make-up job is considered to be a good job and the workmen have pride in craftmenship, then a difficult to assemble fitting can be used. If, however, it is considered to be a menial job, favor the foolproof fitting.

Adequate training of fabrication personnel is vital. There are very definite fabrication procedures for every fitting and they cannot be left to chance. Without proper training, the fabrication of reliable tubing assemblies is highly improbable. Then, too, a good training program more than pays for itself. The cost to make up assemblies is reduced because trained personnel can perform the work faster and more efficiently.

Examine Value Factors

Although this step is always important in the purchasing of any equipment or supplies, it should be left to last. The first objective is to select a fitting that will do the job properly. With this accomplished, one can then consider initial cost, availability, servicing and similar value factors. As always, good value analysis includes the consideration of long-range cost in addition to initial cost.

As can be seen, this is a straight-forward, logical fitting selection procedure. It will work well for all applications. There are two prerequisites: 1. know your system parameters and 2. know your fitting capabilities and limitations. All the needed fitting data can be obtained from your fitting manufacturer.

tubing

The standard fluid line systems, whether it be for the simple, everyday household utilities or for the more exacting requirements of industry, was for many years constructed from piping of assorted materials assembled by means of various standard pipe fitting shapes, unions and nipples using the standard NPT pipe threads. Such systems under high pressure are plagued with leakage problems besides being cumbersome, inefficient and costly to assemble and maintain. The trend has been away from the use of pipe to take advantage of these features of superiority offered by tubing.

major advantages of tubing systems

1. **Bending Quality** — Tubing has strong but relatively thin wall, is easy to bend. Tube fabrication is simple.

2. **Greater Strength** — Tubing is relatively stronger. No weakened sections from reduction of wall thickness by threading.

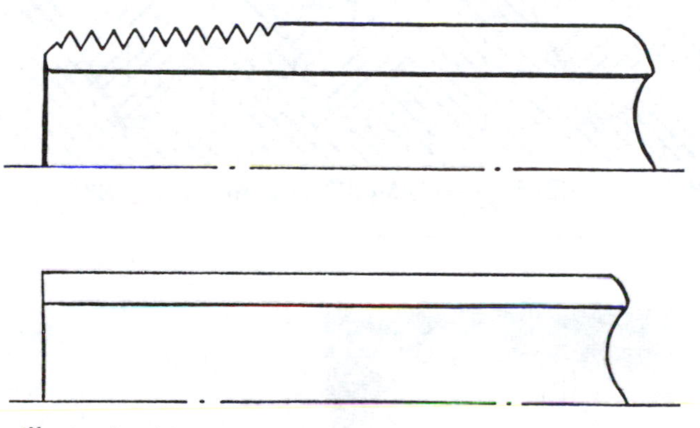

illustration j (e-6)

With no threading necessary, tubing does not require extra wall thickness.

3. **Economy of Weight** — No dead weight in extra wall thickness for threading; tube fittings, being smaller, weigh less; fewer tube fittings needed per installation.

4. **Economy of Space** — Tubing, having better bending qualities and smaller outside diameter, saves space and permits working in closer quarters.

5. **Flexibility** — Tubing, being less rigid, has less tendency to transmit vibration from one connection to another.

6. **Fewer Fittings** — Tubing bends substitute for elbows. Fewer fittings mean fewer joints.

7. **Tighter Joints** — Proper tube fittings, correctly made up, result in leak-proof systems.

8. **Lower Pressure Drop** — Streamlined flow passage of tubing system means less turbulence of fluid.

9. **Better Appearance** — Tubing permits smoother contours, is more adaptable to space limitations.

10. **Cleaner Fabrication** — No sealing compounds on tube connections, no threading; minimum chance of scale, metal chips, foreign particles in system.

11. **Easier Assembly, Disassembly** — Every tube connection serves as a union, can be reassembled repeatedly; easy wrench action.

12. **Less Maintenance** — Foregoing advantages of tubing and tube fittings add up to dependable, trouble-free installations at a lower cost than pipe.

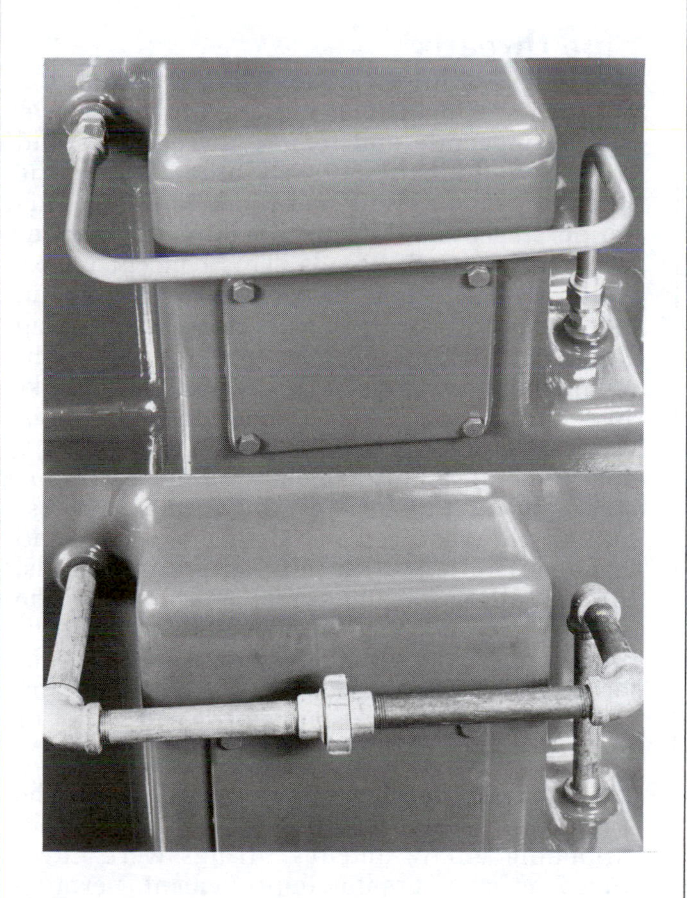

Four joints with tubing; 11 joints with pipe.

illustration j (e-7)

tubing continued

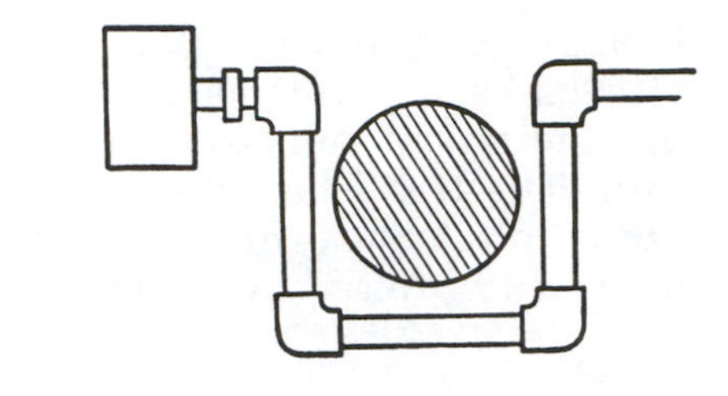

Old Method — Each connection threaded—requires numerous fittings—system not flexible or easy to install and service. Connections not smooth inside—pockets obstruct flow.

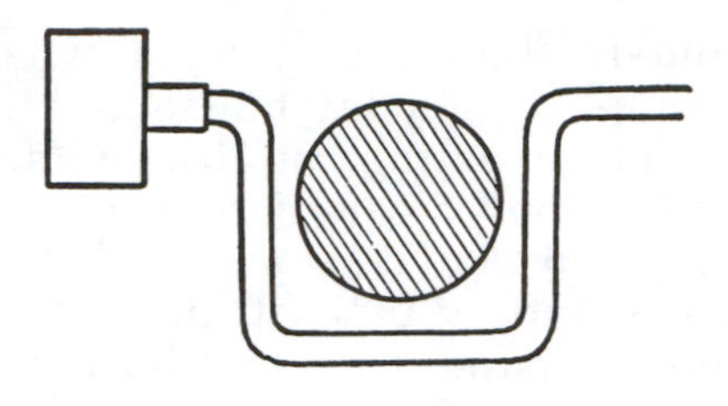

Modern Method — Bendable tubing needs fewer fittings — no "threading on the job"—system light and compact— easy to install or service—no internal pockets or obstructions to free flow.

illustration j (e-8)

Tubing provides simplified, free flow system.

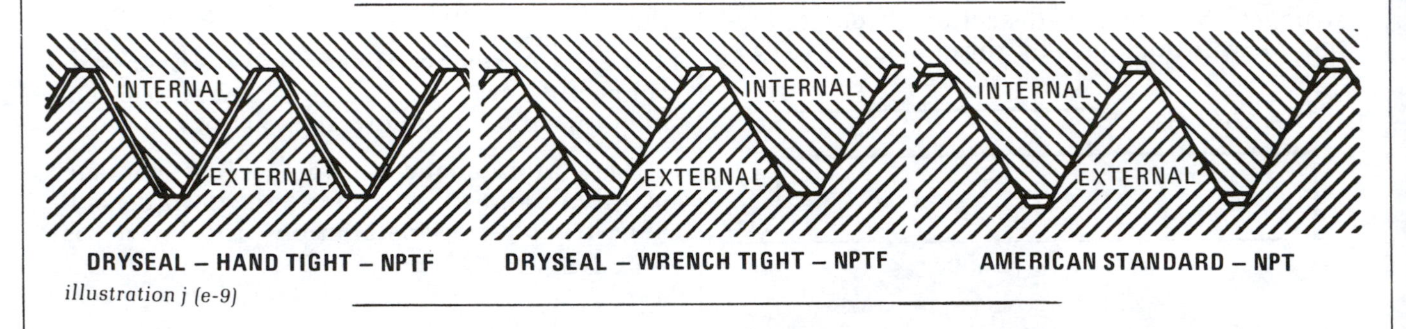

illustration j (e-9)

pipe threads

NPTF dryseal pipe threads were introduced to improve sealing. It was designed to insure that the mating male and female thread roots and crests would contact each other before the threads became tightened flank to flank. This action closes off the spiral leak path commonly found in NPT threads. This concept works well in ductile materials such as brass, and low carbon steel. Problems arose when using NPTF in brittle metals (cast iron, die cast aluminum, and zinc) and self passivaiting metals like austenitic stainless steels and aluminum alloys. With brittle materials it was at best a one time assembly since with no available spring-back, the threads could seldom re-establish the required metal to metal seal and leakage resulted. With self-passivating metals, thread galling became a major problem as the high degree of metal to metal interference promoted local welding or siezing. These problems left us with a connection that in many cases, still required some form of gasket or thread sealant.

The solution to this situation was . . . The straight thread port with "o" ring seal and positionable elbow and tee fittings were introduced as even greater improvement, permitting positive seal without fear of the damage to expensive components, possible in tightening a pipe thread joint.

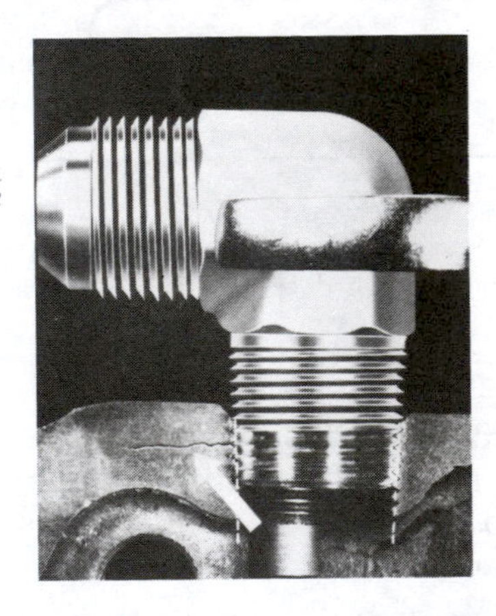

Tapered pipe thread at left in cutaway boss; straight thread at right.

illustration j (e-10)

37° flared fittings

how they work

The flared fitting functions by establishing one circular line contact between the nose of the fitting and the flare on the tube. The flare on the tube also serves as the holding power function in that it is an enlarged section that cannot pass through the nut. As such the flared fitting functions well in high pressure systems because, if properly installed, it is almost blow out proof.

installation

Tube Preparation

1. Prepare end of tube by cutting squarely, removing all burrs from OD and ID of cut tube edge. Also clean all dirt and grit from ID and OD.
2. First place nut on tube, threaded end toward tube end to be flared.
3. Then put tube fitting sleeve on tube so that the large collar end is toward tube end that is to be flared.
4. Place tubing in flaring tool* and flare according to tool manufacturer's directions.

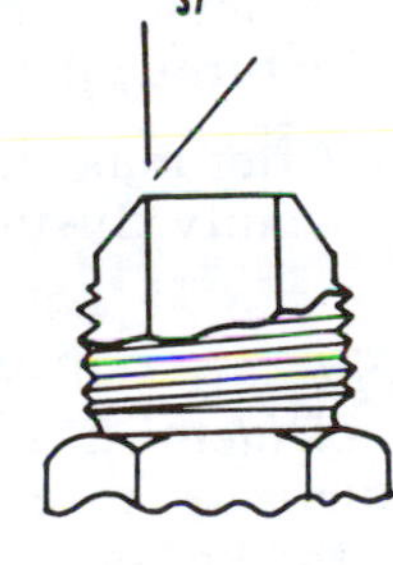

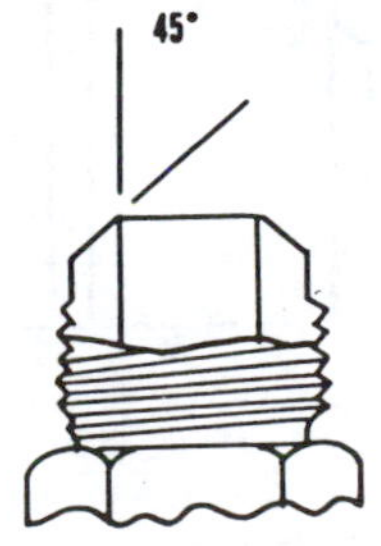

illustration j (e-11)

NOTE: *Be sure that 37° tool is used for 37° fittings and a 45° tool is used for 45° fittings.*

Hammer type, block and pin, power actuated flaring tools are available for flaring any wall tube.

After tubing has been flared, inspect the flare for embedded dirt and out of round distortion. To gauge if tube has been flared enough, use of the sleeve as shown below is a handy quick check gauge. More complete dimensional information is given in the tabulation by size.

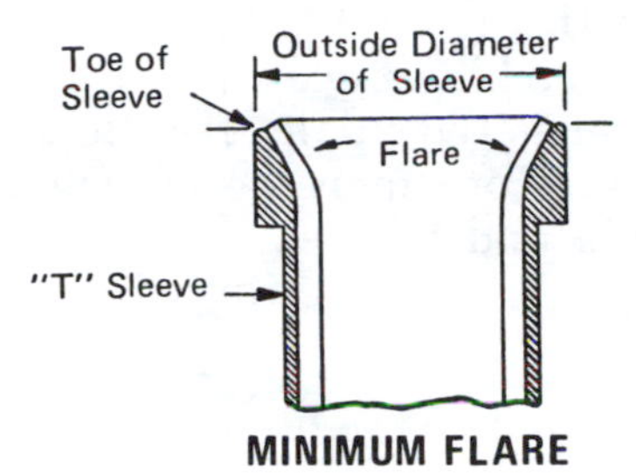

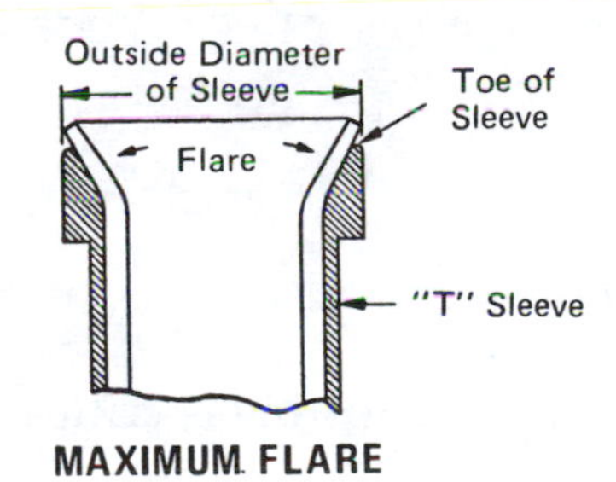

Diameter of maximum flare is equal to outside diameter of sleeve; diameter of minimum flare is equal to maximum inside diameter of sleeve.

illustration j (e-12)

NOTE: *When using stainless steel fittings don't forget to use a thread lubrication such as a high pressure grease. Try ferulube, STP or lubriplate.*

bite-type flareless fittings

how they work

The bite-type flareless hydraulic tube fitting was designed to work on medium to heavy wall tubing which was hard to flare and has three parts:

- Body
- Ferrule
- Nut

The ferrule is driven forward on the tube by the nut. As the ferrule moves forward it meets the seal seat angle of the body which causes the ferrule to cam inward into the tube. The ferrule then bites into the tube causing a small ledge of tube material to be pushed up in front of the biting edge of the ferrule. It is this turned up ledge of material in front of the ferrule that gives the fitting its holding power (i.e., keeps the tube from blowing out). The bite also serves as the seal between the ferrule and the tube. There is also one more necessary seal between the ferrule and the seal seat angle of the fitting body.

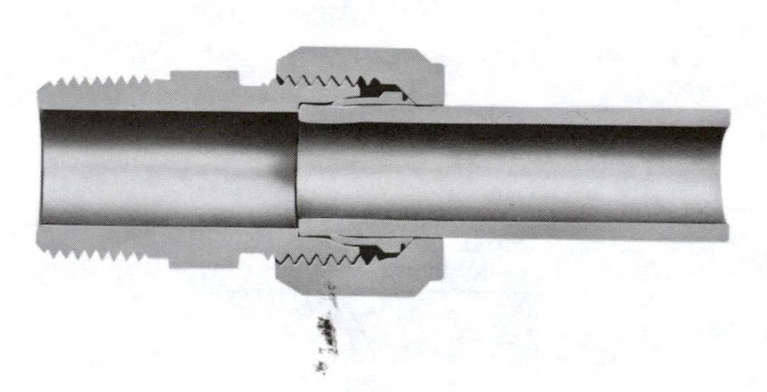

illustration j (e-16)

The make up of the fitting is critical in that the tube must be bottomed against the tube stop in the body on presetting tool during make up. If it is not bottomed, the tube will move forward with the ferrule and a bite will not be achieved. Make sure the tube is bottomed.

A check is possible to see if the tube was bottomed during make up. To make this check look at the end of the tube, and notice if a circular appearing ring has been impressed into the tube end face. This impression will be caused by the tube stop which has a slight reverse angle rather than a flat face.

installation

Tube Preparation

1. Cut the tube squarely by using a tube cutter, hacksaw or an abrasive wheel.

2. Fully deburr the tube end on both the ID and OD using a file and a hand scraper.

3. Clean the tube taking care to remove debris from the inside as this could later contaminate the system.

Make Up

1. Prior to assembly, pre-lubricate the lead edge and back end of the ferrule with Ferulube (lead base, mineral oil compound) or some other anti-galling composition.

2. Place nut and ferrule on the tube and then introduce the tube into the body until it bottoms against the tube stop. Lubricate body threads.

3. While holding tube against the stop, advance the nut and ferrule forward and tighten finger tight. Note that "wiggling" the tube from side to side will insure that the nut is as "finger tight" as possible.

4. Wrench nut 1¾ turns.

NOTE: *Bite type flareless hydraulic fittings in the smaller sizes (3/8" tube and smaller) will make up satisfactorily in the fitting body, but for the larger sizes a hardened preset tool (in essense a tool steel fitting body) should be used. For sizes 3/4" tube and larger, a hydraulic presetting device should be used. Consult your manufacturer for tooling.*

5. **Important —** After make up—loosen nut and remove tube assembly. Inspect for the following:

You should make sure that—

a. A ridge of metal has been raised above the tube surfaces, to a height of at least 50% of the thickness of the ferrule's leading edge, completely around the tube (A).

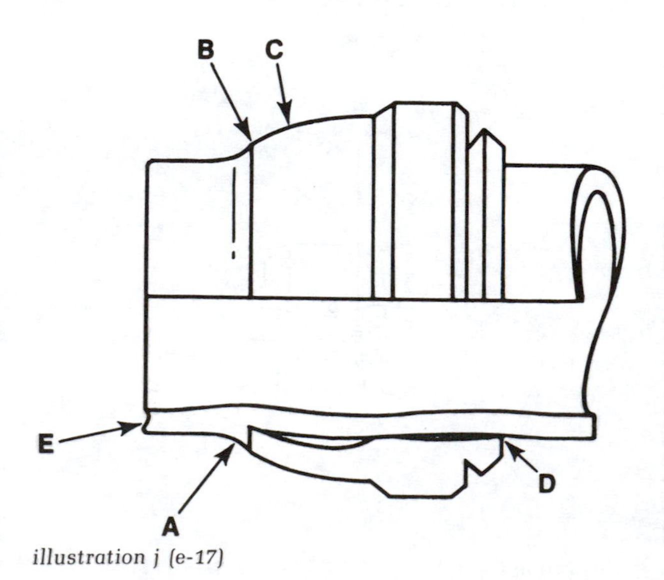

illustration j (e-17)

bite-type flareless fittings continued

b. While the leading edge of the ferrule may be coined flat (B) there is a slight cow to the balance of the pilot section (C).

c. The tail or back end of the ferrule is snug against the tube (D).

d. There is a slight indentation all around the end of the tube (E) that indicates the tube was bottomed in the tool or fitting during presetting (unless evidence of this complete contact is visible the ferrule cannot be considered properly preset).

e. The ferrule may be rotated on the tube but should not be capable of moving back and forth along the tube beyond the bite area. And when rotated 1/2 turn there should never be an uneven gap between ferrule and bite (Stainless Steel will move more than steel because of it's spring back characteristics).

6. **Final Installation** — After inspection re-assemble fitting to finger tight position, wrench until torque build up is felt, then wrench 1/6 turn further.

flareless instrumentation fittings

how they work

Users should note that most instrumentation fittings have a deep internal tube bore and therefore long tube entry. The purpose of this long tube entry is to insure proper tube alignment thus eliminate ferrule cocking during make up. It is then very important that the tube should "fit up" to the fitting properly before make up. If the tubing must be forced into the fitting because of an improper bend or some other form of misalignment, you may end up with a cocked ferrule and a leaking joint. It should also be noted that the use of the fitting as an anchor point for one end of the tubing while it is being hard bent into position is also considered as not being good practice.

installation

Note that most fittings may be used as they come out of the box, they do not need to be disassembled.

1. Cut tube squarely using either a tube cutter or hacksaw. Deburr tube end carefully.

2. Insert tubing into fitting until it bottoms against the built in tube stop. **Note that the tube must be held against the stop during make up. If the tube is allowed to back out slightly, the ferrule may not function properly and a leak may develop.**

3. With the tube held against the stop, tighten the nut finger tight and then wrench 1¼ turns further.

NOTE: *Instrument fittings work by the nut forcing the ferrule forward and inward to a desired point. This is why the 1¼ turns is mentioned. Flareless fittings, unlike flared fittings, should never be made up by torque or "feel" as there are too many tolerances to compensate for (i.e. production, tubing, thread finish). Because of these tolerances two fittings that "feel" that they are made up identically may be miles apart in the amount of actual ferrule action on the tube. Make up by turns, not "feel".*

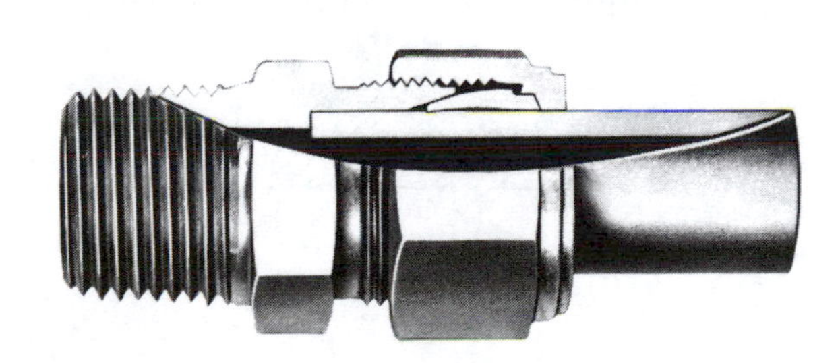

illustration j (e-18)

4. **Remake** — Most flareless instrument fittings will remake many times if the "remake" is made carefully. Almost all manufacturers request that you mark the relationship between the nut and body before disassembly and then return the parts to this relationship during remake. All that is needed then is to tighten the nut a very slight amount further to re-

flareless instrumentation fittings continued

establish the seals. If you can not make the fitting and assuming it is in good condition (i.e. threads not galled or body expanded) then just retighten the nut until you feel a load build up. This will be the point at which the ferrule reseats. Then just tighten a slight additional amount. Remember, do not over do it.

Regardless of the choice of fitting, the techniques and workmanship used in connecting components, running the lines and supporting them are vital to the success of the overall installation. And vital to good workmanship is the proper tooling to provide the correct preparation and fitting assembly technique. Good tooling and care in use of sealants will help minimize introduction of contaminants and undue residual stresses to the system.

Establishment of sound standard shop practices, some of which are shown below, are essential. Your fitting supplier can furnish the proper tooling, both manual and mechanized to suit your particular needs.

installing connectors & tubing

installing connectors

Pipe Threads. When installing pipe thread connectors, apply a good sealing compound such as Unipar or Teflon Tape (see Parker Catalong 4390) to male threads.

These are available from your Parker distributor.

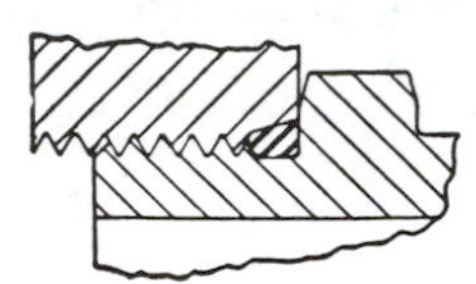

illustration j (e-16)

Straight Threads. Cross section drawing indicates typical installation of straight thread adpater with o-ring into SAE straight thread boss. Parker o-ring provides sealing so pipe joint compound is not used.

Locknut Fittings. Installation of locknut fittings for adjustable positioning follows:

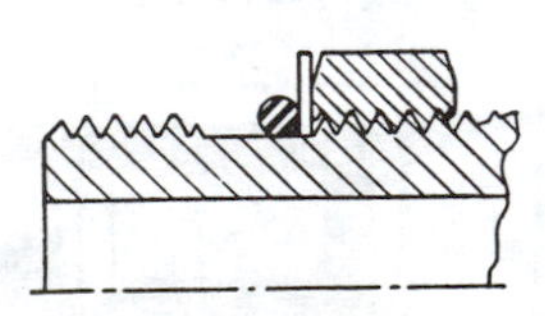

illustration j (e-17)

O-ring and back-up washer should be in proper position on undercut unthreaded section nearest to locknut. (Lubrication of o-ring with Parker o-ring grease is recommended.)

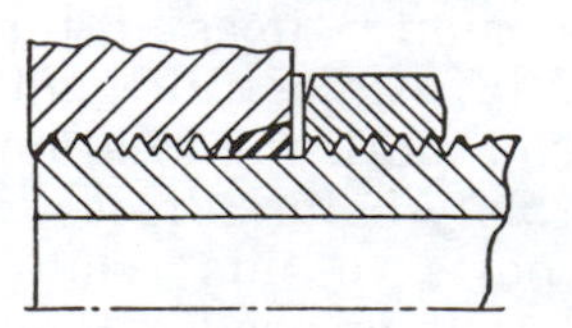

illustration j (e-18)

Step 1. Screw fitting by hand into straight thread boss until back-up washer contacts face of boss.

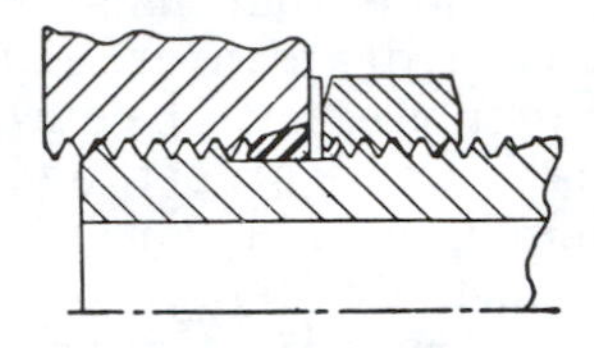

illustration j (e-19)

Step 2. To position fitting, unscrew as far as necessary (up to one full turn), hold fitting in desired position and tighten locknut with wrench so back-up washer contacts face of boss and contains o-ring within boss cavity.

O-Ring Replacement. Do not push o-ring recklessly over threads of fitting which might nick the o-ring surface. A damaged o-ring could lead to leakage trouble.

Use thread protector to replace o-rings on fitting. Place the protector over the threads, then push resilient o-ring over the protector and into the O-ring groove of the fitting.

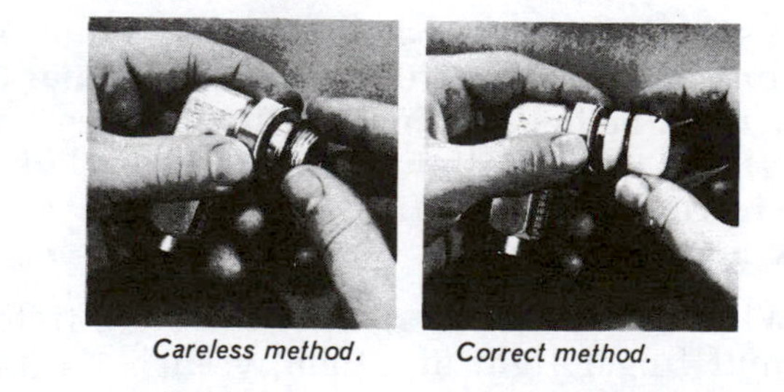

Careless method. *Correct method.*

illustration j (e-20)

Properly installed Parker o-ring will assure leakproof seal.

installing connectors & tubing continued

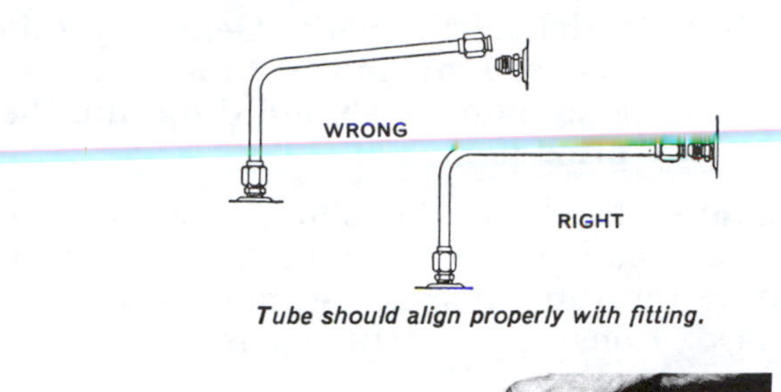

Tube should align properly with fitting.

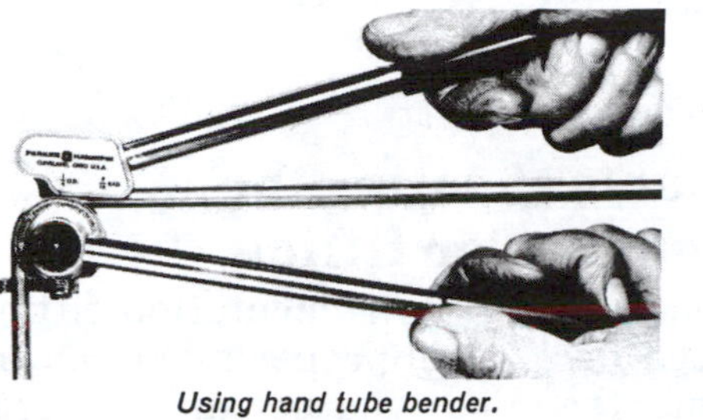

Using hand tube bender.

illustration j (c-24)

tube alignment

Bend the tube into correct radius for proper fit. Springing tube to **force** alignment puts undesirable strain on fitting joint. Correct bends are easily made with Parker benders (Parker Catalog 4390) — without wrinkling, flattening or kinking tube.

CUTTERS & ACCESSORIES	BENDERS	FLARING TOOLS	PRE-SETTING TOOLS	OTHERS
KLOSKUT® Medium (Sizes 2 to 18)	⅛" Hand Tube Bender	Vise Block and Flaring Pin (Sizes 4 to 32)	FERULOK® Ferrule Pre-Setter (Sizes 2 to 32)	Sealers and Lubricants
KLOSKUT® Large (Sizes 12 to 32)	Medium Hand Tube Bender	210A Combination Tool (Sizes 2 to 10)	HY-FER-SET® Ferrule Pre-setter, Hydraulic (Sizes 4 to 32)	Teflon* Tape
IN-EX® De-burring Tool (Sizes 2 to 26)	Ratchet Hand Tube Bender	ROLO-FLAIR® Rotary, Manual (Sizes 2 to 12)		Straight Thread Boss Tap (Sizes 2 to 32)
TRU-KUT SAWING VISE Saw Vise (Sizes 3 to 32)	1" Hand Tube Bender	HYDRA-FLARE Hydraulic Tube Flarer		Straight Thread Boss Counterbore (Sizes 2 to 32)
	EXACTOL® 412 and 420 Medium and Large Crank (Sizes 4 to 20)	232B Electric Power (Sizes 2 to 32)		
	H624 Medium, Bench (Sizes 6 to 24)			
	HB632 Hydraulic (Sizes 6 to 32)			

**These require the use of various accessory items which can be found on the page on which that tool is listed or the following page.*

*Teflon is a DuPont registered trademark.

tube fitting trouble shooting

trouble shooting the flared fitting

If a flare fitting is leaking check the following:

1. If the flared end of tube is flared to proper

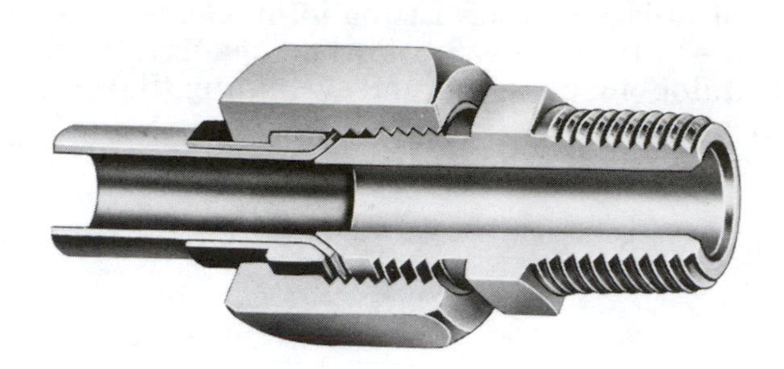

illustration j (e-22)

angle (37° or 45°) — wrong tool might have been used.

2. Check tube flare for embedded burrs, dirt and grit.

3. Check tube flare for severe scratches or indented seams left in ID of tube during its manufacture.

4. Check nose of fitting for deep scratches of surface imperfections.

5. Check nose of fitting for a narrow seal line which should appear as a complete 360° circle line contact. A break in this line may indicate a possible leak path.

6. Check tubing for proper alignment. A flared fitting will accept more misalignment than other fittings but do not ask it to make up for improper tube bending.

trouble shooting bite-type fitting

Problems with bite-type hydraulic fittings are most often traced to faulty make up procedure. Look for the following:

1. **Tube not bottomed.** Check for the indent on the tube end or compare length from end of tube to the front end of the ferrule, of a known good assembly to that of the assembly in question.

2. **Lack of bite.** Inspect for turned up ridge of material. A failure to achieve this ridge can be traced either to the fact that the nut was not tightened enough or that the tube was not bottomed against the stop which allowed the tube to travel forward with the ferrule.

3. **Ferrule cocked on tube.** The ferrule is cocked on the tube when the tube assembly is off axis with that of the body. Generally this condition is caused by faulty tube bending. **All bent tube assemblies should drop into the fitting body prior to make up.**

4. **No bite.** If all of the above checks have been made and the ferrule still shows no sign of biting the tube, it may be that the tube is too hard. Consult your fitting manufacturer in this case.

trouble shooting the compression fitting

Most leaks in instrumentation fittings can be traced back to improper make up procedures. Either the tube was not aligned properly before make up or it was not held in against the stop during make up. To check if it was not held against the stop compare the leaking tube and ferrule assembly against one that you know has been held against the stop. The distance from the front of the ferrule to the end of the tube should be identical. To check for misalignment, notice if the ferrule is cocked on the tube. If it is, then your problem is misalignment and you will have to retube that assembly.

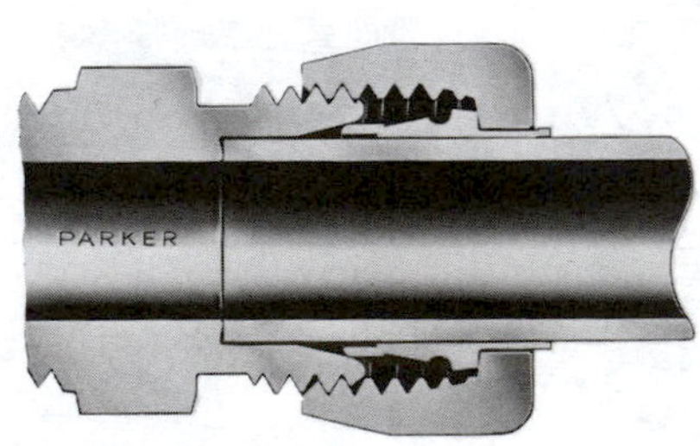

illustration j (e-23)

Another check for leaks can be made by using a soap solution. If the bubbles form at the back of the nut between the nut and the tube, then chances are that you did not get a seal between the ferrule/s and the tube. Again, misalignment may be the cause. However check the tube itself. You could have a scratch or seam running along the tube, allowing a leak to occur. If your soap solution forms bubbles at the front of the nut between it and the body, then the leak is probably between the ferrule/s and the fittings seal seat. Check this area for imbedded dirt. Always remember that if you have a problem that you can not solve, call in the manufacturer. He will probably find a solution and it won't cost you any time or money.

general data

glossary of terms

Abrasion	A wearing, grinding, or rubbing away of material in mechanical elements. The products of abrasion will be introduced into the system as generated particulate contamination.
Absolute Filtration Rating (Largest Particle Passed)	The diameter of the largest hard spherical particle that will pass through a filter under specified conditions. This is an indication of the largest opening in the filter element.
Absorbent (Absorptive)	A filter medium that holds contaminant by mechanical means.
Accumulator	A container in which fluid is stored under pressure as a source of fluid power.
Accumulator, Hydropneumatic	An accumulator in which compressed gas applies force to the stored liquid.
Accumulator, Hydropneumatic, Bladder	A hydropneumatic accumulator in which the liquid and gas are separated by an elastic bag or bladder.
Accumulator, Hydropneumatic, Diaphragm	A hydropneumatic accumulator in which the liquid and gas are separated by a flexible diaphragm.
Accumulator, Hydropneumatic, Piston	A hydropneumatic accumulator in which the liquid and gas are separated by a floating piston
Accumulator, Hydropneumatic, Non-Separator	A hydropneumatic accumulator in which the compressed gas operates directly on liquid within the pressure chamber.
Accumulator, Mechanical	An accumulator incorporating a mechanical device which applies force to the stored fluid.
Accumulator, Mechanical, Spring	A mechanical accumulator in which springs apply force to the stored fluid.
Accumulator, Mechanical, Weighted	A mechanical accumulator in which the gravitational force acting upon weights applies force to the stored fluid.
Adapter (Support Ring)	A seal support shaped to conform with the contour of the seal and the mating element.
Adapter, Female	An adapter with a concave seal support. Referred to as bottom adapter with "V" type sets.
Adapter, Male	An adapter with a convex seal support. Referred to as bottom adapter with "V" type sets.
Adapter, Redestal	An adapter usually used to support a "U" type seal.
Adsorbent (Adsorptive)	A filter medium primarily intended to hold soluble and insoluble contaminants on its surface by molecular adhesion.
Aftercooler	A device which cools a gas after it has been compressed.
Air Bleeder	A device for removal of air.
Air Breather	A device permitting air movement batween atmosphere and the component in which it is installed.
Air Inclusion	The volume of air introduced into a liquid system as a result of servicing a filter.
Air Motor	A device which converts pneumatic fluid power into mechanical force and motion. It usually provides rotary mechanical motion.
Air, Compressed (Pressure)	Air at any pressure greater than atmospheric pressure.
Analog	Of or pertaining to the general class of fluidic devices or circuits whose output varies as a continuous function of its output.
AND Device	A control device which has its output in the logical 1 state if and only if all the control signals assume the logical 1 state.
AND 10050	A United State Air Force-Navy Aeronautical Design Standard in which a straight thread port is used to attach tube fittings to various components. It employs an "0" ring seal compressed in a special cavity.
Angle	A mounting device which is angular in cross section. It is usually made from a 90° structural angle.
Aniline Point	The lowest temperature at which a liquid is completely miscible with an equal volume of freshly distilled aniline (ASTM Designation D611-64).
Artificially Loaded	A filter element that is loaded with a controlled laboratory test contaminant.
Atmosphere, Technical	A unit of pressure used in Germany and equal to 1 Kp/cm^2 (kilopond per square centimeter), approximately equal to 14.2 psig.
Automatic Count	A particle count obtained by an electromechanical or electronic device as opposed to visual microscopic counting technique.

Back Connected	Where connections are made to normally unexposed surfaces of components.
Back-Up Ring (Anti-Extrusion Ring) (Junket Ring) (Bull Ring)	A ring which bridges a clearance to minimize seal extrusion.
Background Contamination	The total of the extraneous particles which are introduced in the process ot obtaining, storing, moving, transferring and analyzing the fluid sample.
Baffle	A device to prevent direct flow or inpimgement.
Bar	A unit of pressure based on 10^5 N/m^2 (Newtons per square meter), approximately equal to 14.5 psig.
Base, Filter	The foundation or support for the filter which may also contain one or more ports.
Bernoulli's Law	If no work is done on or by a flowing frictionless liquid its energy due to pressure and velocity remains constant at all points along the streamline.
Bi-Directional Filter	A filter element designed for flow in both directions.
Bistable	Of or pertaining to the general class of fluidic devices which maintain in either of two position operating states in the presence or absence of the setting input.
Bleed Off	A filter located in a line between a flow control device and reservoir.
Boil Point (Foam All Over) (Mass Bubble Point) (Open Bubble Point)	The differential gas pressure at which gas bubbles are profusely emitted profusely emitted from the entire surface of a wetted filter element under specified test conditions.
Bowl Shell	A case that is closed at one end and mates with the filter head
Boyle's Law	The absolute pressure of a fixed mass of gas varies inversely as the volume, provided the temperature remains constant.
Breathing Capacity	A measure of flow rate through an air breather.
Bridging	A condition of filter element loading in which contaminant pans the space between adjacent sections of a filter element hus blocking a portion of the useful filtration area.
Bubble Point (First Bubble Point) (Initial Bubble Point)	The differential gas pressure at which the first steady stream of gas bubbles is emitted from a wetted filter element under specified test conditions.
Bulk Modulus	The measure of resistance to compressibility of a fluid. It is the reciprocal of the compressibility.
Burst	An outward structural failure of the filer element caused by excessive differential pressure.
Burst Pressure Rating Rating Filter	The maximum specified inside-out differential pressure which can be applied to a filter element without outward structural or filter medium failure
Burst Pressure, Filter	The pressure which causes rupture. Also, the inside-out differential pressure that causes outward structural or filter medium failure of a filter element.
Cap (Back End) (Blind End) (Blind Head) (Rear End) (Rear Head)	A cylinder end closure which completely covers the bore area.
Cap, Filter	An end closure for the filter case.
Case (Shell), Filter	A hollow part that provides a cavity for the filter element.
Case Drain Line	A line conducting fluid from a pump or motor housing to reservoir.
Cavitation	A localized gaseous condition within a liquid stream which occurs where the pressure is reduced to the vapor pressure.
Celsius	A temperature scale. 0 Celsius (or 0 Centigrade) is the freezing point of water. $°C = 5/9\ (°F - 32)$.
Center Tube (Core)	The internal duct and filter media support.
Centipoise	A unit of absolute (dynamic) viscosity. 1 centipoise = 10^{-2} dyne s/cm^2 (dyne-second per square centimeter) = 10^{-3} Ns/m^2 (Newton seconds per square meter).
Centistoke	A unit of kinematic viscosity. 1 centistoke = 10^{-2} cm^2/s (square centimeters per second) = 10^{-6} m^2/s (square meters per second).
Centrifuge Volume	The volume of contaminant (liquid or solid or both) separated from a volume of liquid exposed to centrifugal force.
Channel	A fluid passage, the length of which is large with respect to its cross-sectional area.
Charles' Law	The volume of a fixed mass of gas varies directly with absolute temperature, provided the pressure remains constant.
Circuit	An arrangement of interconnected component parts.
Circuit, Pilot	A circuit used to control a main circuit or component.

Circuit, Pressure Control	Any circuit whose main purpose is to adjust or regulate fluid pressure in the system or any branch of the system.
Circuit, Regenerative	A circuit in which pressurized fluid discharged from a component is returned to the system to reduce power input requirements. On single rod end cylinders the discharge from the rod end is often directed to the opposite end to increase rod extension speed.
Circuit, Safety	A circuit which prevents accidental operation, protects against overloads, or otherwise assures safe operation.
Circuit, Sequence	A circuit which establishes the order in which two or more phases of a circuit occur.
Circuit, Servo	A circuit which is controlled by automatic feed back; i.e., the output of the system is sensed or measured and is compared with the input signal. The difference (error) between the actual output and the input controls the circuit. The controls attempt to minimize the error. The system output may be position, velocity, force, pressure, level, flow rate, or temperature.
Circuit, Speed Control	Any circuit where components are arranged to regulate speed of operation.
Circuit, Synchronizing	A circuit in which multiple operations are controlled to occur at the same time.
Circuit, Unloading	A circuit in which pump volume is returned to reservoir at near zero gage pressure whenever delivery to the system is not required.
Cleanability	The ability of a cleanable filter element to withstand repeated field cleanings and retain adequate dirt capacity and service life.
Cleanliness Level	The analogue of contamination level.
Contamination Level	A quantitative term specifying the degree of contamination.
Clevis (Hinge) (Pendulum)	An "U" shaped mounting device which contains a common pin hole at right angle or normal to the axis of symmetry through each extension. A clevis usually connects with an eye.
Collapse Pressure	The outside-in differential pressure that causes structural or filter medium failure of a filter element.
Collapse Pressure Rating	The maximum specified outside-in differential pressure which can be applied to a filter element without inward structural or filter medium failure.
Combination (Composite)	A filter medium composed of two or more types, grades, or arrangements of filter media to provide properties which are not available in a single filter medium.
Compartment	A space within the base, frame or column of the equipment.
Compressibility	The change in volume of a unit volume of a fluid when subjected to a unit change of pressure.
Compressor	A device which converts mechanical force and motion into pneumatic fluid power.
Compressor, Multiple Stage	A compressor having two or more compressive steps in which the discharge from each supplies the next in series.
Compressor, Single Stage	A compressor having only one compressive step between inlet and outlet.
Conductor	A component whose primary function is to contain and direct fluid.
Conduit	Any confining element employed to transfer fluid.
Contaminant	Any material or substance which is unwanted or adversely affects the fluid power system or components, or both.
Contaminant, Artificial	Contaminant of known composition and particle size distribution which are introduced into fluid systems of fluid system components for test purposes. The most commonly used artificial contaminants include standardized fine air cleaner test dust, standardized coarse air cleaner test dust, carbonyl iron, glass beads, cottonlinters, red iron oxide and black iron oxide.
Contaminant, Built-In	Initial residual contamination in a component, fluid, or system. Typical built-in contaminants are burrs, chips, flash, dirt, dust, fiber, sand, moisture, pipe dope, weld spatter, paints and solvents, flushing solutions, incompatible fluids and operating fluid impurities.
Contaminant, Generated	Contamination created by the operation of a fluid system or component. Generated contaminants are products of erosion, fretting, scoring, wear, corrosion, decomposition, oxidation, and fluid-breakdown. Air bubbles may also be generated under some operating conditions.
Contaminated	A filter element which releases into the effluent foreign particles resulting from handling, storage, and fabrication.
Continuity Equation	Under steady state conditions the mass rate of fluid flow into a fixed space is equal to the mass flow rate out. Hence, the mass flow rate of fluid past all cross sections of a conduit is equal.

Control	A device used to regulate the function of a component or system.
Control, Automatic	A control which actuates equipment in a predetermined manner.
Control, Combination	A combination of more than one basic control.
Control, Cylinder	A control in which a fluid cylinder is the actuating device.
Control, Electric	A control actuated electrically.
Control, Hydraulic	A control actuated by a liquid.
Control, Liquid-Level	A device which controls the liquid level by a float switch or other means.
Control, Manual	A control actuated by the operator.
Control, Mechanical	A control actuated by linkages, gears, screws, cams or other mechanical elements.
Control, Pneumatic	A control actuated by air or other gas pressure.
Control, Pressure	A control in which a pressure signal operates a compensating device.
Control, Pump	A control applied to a positive displacement variable delivery pump to adjust the volumetric output or direction of flow.
Control, Servo	A control actuated by a feed back system which compares the outpu with the reference signal and makes corrections to reduce the difference.
Cooler	A heat exchanger which removes heat from a fluid.
Corrosion	The chemical change in the mechanical elements caused by the interaction of fluid or contaminants, or both. More specifically related to chemical changes in metals. The products of change may be introduced into the system as generated particulate contamination.
Counting Calibration Factor	Ratio of the effective filtration area on the membrane to the area counted, SAE-ARP-598.
Cover	An end closure which provides access to the filter element.
Crest	The outer fold of a pleat.
Cushion	A device which provides controlled resistance to motion.
Cushion, Cylinder	A device to reduce the impact of the piston against the cap or head of the cylinder.
Cushion, Die	A cushion installed with a die on a press to provide controlled resistance against the work. The return motion of the cushion is sometimes used to eject the work.
Cushion, Hydraulic	A cushion in which resistance is developed hydraulically.
Cushion, Hydropneumatic	A cushion in which resistance is developed hydraulically and pneumatically.
Cushion, Pneumatic	A cushion in which resistance is developed pneumatically.
Cycle	A single complete operation consisting of progressive phases starting and ending at the neutral position.
Cycle, Automatic	A cycle of operation which once started is repeated indefinitely until stopped.
Cycle, Manual	A cycle which is manually started and controlled through all phases.
Cycle, Semi-automatic	A cycle which is started upon a given signal, proceeds through a predetermined sequence, and stops with all elements in their initial position.
Cylinder	A device which converts fluid power into linear mechanical force and motion. It usually consists of a movable element such as a piston and such as a piston and piston rod, plunger or ram, operating within a cylindrical bore.
Cylinder Mounting	A device by which a cylinder is fastened to its mating element.
Cylinder Mounting, Centerline	A mounting which permits connection on a plane in line with the piston rod centerline.
Cylinder Mounting, Cylinder, Log	A centerline mounting consisting of two opposite lugs at each end of the cylinder.
Cylinder Mounting, Intermediate	A mounting which provides cylinder connection at an intermediate external position along the piston rod centerline between ends.
Cylinder Mounting, Intermediate Fixed Trunnion	An intermediate mounting consisting of trunnion pins which cannot be repositioned.
Cylinder Mounting, Intermediate, Movable Trunnion	An intermediate mounting consisting of trunnion pins which can be repositioned.
Cylinder Mounting, Universal	A mounting which permits a cylinder to change its alignment in all directions.
Cylinder Mounting, End	A mounting which permits connection at either or both ends of a cylinder.

Cylinder Mounting, End, Both — A mounting at both ends of the cylinder.

Cylinder Mounting, End, Both, Tie Rods Extended — A cylinder mounted at both ends by means of extended tie rods.

Cylinder Mounting, End, Cap, Circular — A direct circular cap mounting.

Cylinder Mounting, End, Cap, Circular Flange — A cap mounting consisting of a supplementary circular flange plate.

Cylinder Mounting, End, Cap, Detachable Eye — A cap mounting consisting of an eye which can be removed or rotated.

Cylinder Mounting, End, Cap, Fixed Clevis — A cap mounting consisting of a clevis integral with the cap to, maintain a fixed clevis-port relationship.

Cylinder Mounting, End, Cap, Fixed Eye — A cap mounting consisting of an eye integral with the cap to maintain a fixed eye-port relationship.

Cylinder Mounting, End, Cap, Rectangular Flange — A cap mounting consisting of a supplementary rectangular flange plate.

Cylinder Mounting, End, Cap, Square — A direct square cap mounting.

Cylinder Mounting, End, Cap, Square Flange — A cap mounting consisting of a supplementary square flange plate.

Cylinder Mounting, End, Cap, Tie Rods Extended — A cap mounting of extended tie rods.

Cylinder Mounting, End, Cap, Trunnion — A cap mounting consisting of trunnion pins near or at the cap end of the cylinder.

Cylinder Mounting, End, Head — A mounting which provides cylinder connection at the head end.

Cylinder Mounting, End, Head, Circular — A direct circular head mounting.

Cylinder Mounting, End, Head, Circular Flange — A head mounting consisting of supplementary circular flange plate.

Cylinder Mounting, End, Head, Female Rabbet — A head mounting consisting of a female pilot recess.

Cylinder Mounting, End, Head, Male Rabbet — A head mounting consisting of a male pilot extension.

Cylinder Mounting, End, Head, Rectangular Flange — A head mounting consisting of a supplementary rectangular flange plate.

Cylinder Mounting, End, Head, Square — A direct square head mounting.

Cylinder Mounting, End, Head, Square Flange — A head mounting consisting of a supplementary square flange plate.

Cylinder Mounting, End, Head, Tie Rods Extended — A head mounting consisting of extended tie rods.

Cylinder Mounting, End, Head, Trunnion — A head mounting consisting of trunnion pins near or at the head end of the cylinder.

Cylinder Mounting, End, Cap — A mounting which permits connection at the cap end.

Cylinder Mounting, Fixed — A mounting which provides rigid connection between the cylinder and the mating element wherein the piston rod reciprocates in a fixed line.

Cylinder Mounting, Pivot — A mounting which permits a cylinder to change its alignment in a plane.

Cylinder Mounting, Side — A mounting which provides cylinder connection at one of its sides.

Cylinder Mounting, Side, End Angles — A side mounting consisting of an angle at each end of the cylinder with the free legs facing a common side.

Cylinder Mounting, Side, End Lugs — A side mounting consisting of one or more lugs at each end of the cylinder and facing a common side.

Cylinder Mounting, Side, End Plates — A side mounting consisting of an extended plate at each cylinder end and facing a common side

Cylinder Mounting, Side, Lug — A side mounting consisting of two opposite lugs at each cylinder end facing a common side

Cylinder Mounting, Side, Tapped — A side mounting consisting of one or more tapped holes at each cylinder end facing a common side

Cylinder Mounting, Side, Through Holes — A side mounting consisting of holes drilled across both ends to a common side.

Cylinder, Adjustable Stroke — A cylinder equipped with adjustable stops at one or both ends to limit piston travel.

Cylinder, Cushioned — A cylinder with a piston-assembly deceleration device at one or both ends of the stroke.

Cylinder, Double Acting — A cylinder in which fluid force can be applied to the movable element in either direction.

Cylinder, Double Rod — A cylinder with a single piston and a piston rod extending from each end.

Cylinder, Dual Stroke — A cylinder combination which provides two working strokes.

Cylinder, Non-Rotating — A cylinder in which relative rotation of the cylinder housing and the piston and piston rod, plunger or ram, is not recommended.

Cylinder, Piston — A cylinder in which the movable element has a greater cross-sectional area than the piston rod.

Cylinder, Plunger (Ram) — A cylinder in which the movable element has a greatar cross-sectional area than the piston rod.

Cylinder, Plunger (Ram) — A cylinder in which the movable element has the same cross-sectional area as the piston rod.

Cylinder, Retractable Stroke — An adjustable-stroke cylinder in which the stop can be temporarily changed to permit full retraction of the piston assembly.

Cylinder, Rotating — A cylinder in which relative rotation of the cylinder housing and the piston and piston rod, plunger or ram, is recommended.

Cylinder, Single Acting — A cylinder in which the fluid force can be applied to the movable element in only one direction.

Cylinder, Single Rod — A cylinder with a piston rod extending from one end.

Cylinder, Spring Return — A cylinder in which a spring returns the piston assembly.

Cylinder, Tandem — Two or more cylinders with interconnected piston assemblies.

Cylinder, Telescoping — A cylinder with nested multiple tubular rod segments which provide a long working stroke in a short retracted envelope.

Damping Parameter — A measure of the time required as a function of the maximum press excursion of the power supply output to attain essentially steady state operation after an abrupt disturbance. Specifically, it is the transient recovery time divided by the maximum excursion.

Darcy's Formula — A formula used to determine the pressure drop due to flow friction through a conduit.

Decay — A falling pressure.

Decay Rate — The ratio of pressure decay to time.

Decomposition — Separation by chemical change into constituent parts, elements, or different compounds. More specifically related fluid and seal chemical changes. The materials affected are primarily organic in nature. The products of change may be introduced into the system as contamination.

Decontamination — The process of removing unwanted material or substance; the reduction of contamination to an acceptable level.

Density — A unit of mass per unit volume usually expressed in kg/1 (kilograms per liter) 1 kg/1 = 0.036 lb/in^3.

Deposited — A filter medium produced by chemical or electrolytic deposit.

Depth — A filter medium which primarily retains contaminant within tortuous passages.

Diagram — A drawing which illustrates pertinent characteristics, element positions, sizes, interconnection, controls and actuation of components and fluid power circuits.

Diagram, Combination — A drawing using a combination of graphical, cut-away and pictorial symbols showing interconnected lines.

Diagram, Cutaway — A drawing showing principle internal parts of all components, controls and actuating mechanisms, all interconnecting lines and function of individual components.

Diagram, Graphical (Schematic) — A drawing or drawings showing each piece of apparatus including all interconnecting lines by means of approved ANSI standard symbols.

Diagram, Pictorial — A drawing showing each component in its actual shape according to the manufacturer's installation.

Diagram, Pressure-Time	A graphical presentation of pressure plotted against time for a complete cycle.
Digital	Of or pertaining to the general class of fluidic devices or circuits whose output varies in discrete steps (i.e., pulses or "on-off" characteristics).
Dirt Capacity (Dust Capacity) (Contaminant Capacity)	The weight of a specified artificial contaminant which must be added to the influent to produce a given differential pressure across a filter at specified conditions. Used as an indication of relative service life.
Displacement, Volumetric	The volume for one revolution or stroke.
Dissolved Air	Air which is dispersed at a molecular level in hydraulic fluid to form a single phase.
Dissolved Water	Water which is dispersed at a molecular level in the hydraulic fluid to form a single phase.
Dither	A low amplitude, relatively high frequency periodic electrical signal, sometimes superimposed on the servovalve input to improve system resolution. Dither is expressed by the dither frequency (Hz) and the peak-to-peak dither current amplitude.
Drift	The percentage above and below the operating pressure at a constant flow rate over a specified length of time.
Droop	The deviation between no flow secondary pressure and secondary pressure at a given flow.
Durometer Hardness	An arbitrary indication of hardness determined by an indentor.
Edge	A filter medium whose passages are formed by the adjacent surfaces of stacked dics, edgewound ribbons, or single-layer filaments.
Effective Area	The total area of the porous medium exposed to flow in a filter element.
Effective Particle Diameter	The diameter of a circle having an area equivalent to the projected area of the particle.
Efficiency, Filter	The ability, expressed as a percent, of a filter to remove specified artificial contaminant at a given contaminant concentration under specified test conditions.
Effluent	The fluid leaving a component.
Electrical Control Power	The power dissipation required for operation of the servovalve. Control power is a maximum with full input signal, and is zero with zero-input signal. It is independent of the coil connection (series, parallel, or differential) for any conventional two-coil operation. For differential operation, the control power is the power consumed in excess of the electrical quiescent power. This power change is a result of the differential current change.
Electrical Quiescent Power	The power dissipation required for differential operation of the servovalve when the current through each coil is equal and opposite in polarity.
Element (Cartridge)	The porous device which performs the actual process of filtration.
Emulsion	A stabilized mixture of two immiscible components, water and oil. It may contain additives.
Emulsion, Oil in Water	A dispersion of oil in a continuous phase of water.
Emulsion, Watar in Oil	A dispersion of water in a continuous phase of oil.
Enlosure	A housing for components.
End Cap	A ported or closed cover for the end of a filter element.
End Load	The axial force applied to the end of a filter element which may cause permanent deformation or seal failure.
End Load Rating	The maximum specified axial force which can be applied to a filter element without permanent deformation or seal failure.
End Seal	The bond between the end cap and the filter medium. Also a sealing device which seals against the end cap by axial contact pressure.
End, Cylinder	Either of two envelope surfaces at right angle or normal to the piston rod centerline.
Entrained Air	A mechanical mixture of air hubbles having a tendency to separate from the liquid phase.
Erosion	The loss of material in mechanical elements caused by the impingement of fluid or fluid suspended particulate matter, or both. The product of erosion will be introduced into the system as generated particulate contamination.
Etched	A filter medium having passages produced by chemical or electrolytic removal.
Extended Area	A filter element whose medium is pleated or otherwise formed to obtain more effective area within a given dimensional envelope.
External Support	A permeable structural enclosure which imparts rigidity to a filter element, and usually protects the filter medium.

Eye (Hinge) (Pendulum) — A mounting device consisting of a single extension which contains counting pin hole at right angle or normal to the axis of symmetry. An eye usually connects with a clevis.

Fan In — The number of operating controls in a single fluidic device which individually and in combination will produce the same output.

Fan In — The number of operating controls in a single fluidic device which individually and in combination will produce the same output.

Fan Out — The number of like devices to which operating controls are supplied by the output of the fluidic device.

Fatigued — A structural failure of the filter medium due to flexing caused by cyclic differential pressure.

Fiber — For the purpose of microscopic particle counting a fiber is a particle whose length is greater than 100 micrometers but at least ten times its width.

Filler Ring — A ring which fills the recess of a "V" or "U" type seal.

Filter Accessories — Auxiliary devices incorporated into a filter to enhance its usefulness.

Filter Components — The parts that make up a filter.

Filter Element Media — The porous materials which perform the actual process filtration.

Filter Performance — Those factors which describe the functions and attributes of a filter or filter element.

Filterable Solids — The solids retained on a membrane for analysis by weight, count or observation as it applies to the section on contamination measurement.

Fitting — A connector or closure for fluid power lines and passages.

Fitting, Bushing — A short externally threaded connector with a smaller size internal thread.

Fitting, Cap — A cover for fluid passages.

Fitting, Closure — A cap or a plug.

Fitting, Compression — A fitting which seals and grips by manual adjustable deformation.

Fitting, Connector — A fitting for joining a conductor to a component port or to one or more other conductors.

Fitting, Coupling — A straight connector for fluid lines.

Fitting, Cross — A fitting with four ports arranged in pairs, each pair on one axis, and the axis at right angles.

Fitting, Elbow — A fitting that makes an angle between mating lines. The angle is always 90 degrees unless another angle is specified.

Fitting, Flange — A fitting which uses a radially extending collar for sealing and connection.

Fitting, Flared — A fitting which seals and grips by a preformed flare at the end of the tube.

Fitting, Flared AN — A United States Air Force-Navy 30° flared tube fitting Design Standard.

Fitting, Flareless — A fitting which seals and grips by means other than a flare.

Fitting, Plug — A closure which fits into a fluid passage.

Fitting, Plug, Dryseal Pipe — A plug made with a thread which conforms to Dryseal Pipe Thread Standards.

Fitting, Plug, Short Pipe Thread — A plug which conforms in all respects to standard pipe threads except that the thread has been shortened one full thread from the small end.

Fitting, Plug, Standard — A plug with American (National) Pipe Thread tapered pipe threads.

Fitting, Plug, Straight — A plug with straight thread conforming Thread to United Thread Standards.

Fitting, Reducer — A fitting having a smaller line size at one end than the other.

Fitting, Tee — A fitting with three ports, a pair on one axis with one side outlet at right angles to this axis.

Fitting, Union — A fitting which permits lines to be joined or separated without requiring the lines to be rotated.

Fitting, Welded — A fitting attached by welding.

Fitting, Wye (Y) — A fitting with three ports, a pair on one axis with one side outlet at any angle other than right angles to this axis. The side outlet is usually 45°, unless another angle is specified.

Flange — A mounting device consisting of a plate or collar extending past the basic cylinder profile to provide clearance area for mounting bolts. A flange is usually at right angle or normal to the piston rod centerline.

Flapper Action — A valve design in which output control pressure is regulated by a pivoted flapper in relation to one or two orifices.

Flash Point	The temperature to which a liquid must be heated under specified conditions of the test method to give off sufficient vapor to form a mixture with air that can be ignited momentarily by a specified flame.
Flip Flop	A digital component or circuit with two stable states and sufficient hystereses so that it has "memory". Its state is changed with a control pulse; a continuous control signal is not necessary for it to remain in a given state.
Flow Characteristic Curve	The change in regulated (secondary) pressure occurring as a result of a change in therate of air flow over the operating range of the regulator.
Flow Characteristic Curve	The change in regulated (secondary) pressure occurring as a result of a change in the rate of air flow over the operating range of the regulator.
Flow Fatigue	The ability of a filter element to resist structural failure of the filter medium due to flexing caused by cyclic differential pressure.
Flow Polarity	The relationship between the direction of control flow and the direction of input current.
Flow Rate	The volume, mass, or weight of a fluid passing through any conductor per unit of time.
Flow, Laminar (Streamline)	A flow situation in which fluid moves in parallel lamina or layers.
Flow, Metered	Flow at a controlled rate.
Flow, Steady State	A flow situation wherein conditions such as pressure, temperature, and velocity at any point in the fluid do not change.
Flow, Turbulent	A flow situation in which the fluid particles move in a random manner.
Flow, Unsteady	A flow situation wherein conditions such as pressure, temperature and velocity change at points in the liquid.
Fluid	A liquid, gas, or combination thereof.
Fluid Breakdown	A change of chemical or mechanical properties of a fluid, or both. Some end products may be insoluble in the fluid.
Fluid Capacity	The liquid volume coincident with the "high" mark of the level indicator.
Fluid Friction	Friction due to the viscosity of fluids.
Fluid Power	Energy transmitted and controlled through use of a pressurized fluid.
Fluid Power System	A system that transmits and controls power through use of a pressurized fluid within an enclosed circuit.
Fluid Stability	Resistance of a fluid to permanent changes in properties.
Fluid, Fatty Oil	A fluid composed of fats derived from animal, marine, or vegetable origin. It may contain additives.
Fluid, Fire Resistant (Non-Flammable)*	A fluid difficult to ignite which shows little tendency to propagate flame. *Deprecated.
Fluid, Hydraulic	A fluid suitable for use in a hydraulic system.
Fluid, Pneumatic	A fluid suitable for use in a pneumatic system.
Fluidic	Of or pertaining to devices, systems, assemblies, etc., using fluidic components.
Fluidic Amplification, Power	The ratio of the change of the power in a specified load impedance connected to a device to the change in the power applied to the controls of the device.
Fluidic Amplification, Pressure	The ratio of the change of the pressure drop across a specific load impedance to the change in the pressure drop applied across the controls of the device.
Fluidic Amplification, Flow	The ratio of the change of the flow in a specified load impedance connected to a device to the change in the flow applied to the controls of the device.
Fluidic Amplifier, Turbulence	A fluidic amplifier in which the power jet is at a pressure such that it is in the transition region of laminar stability and can be caused to become turbulent by a secondary jet or by sound.
Fluidic Amplifier, Closed	A fluidic amplifier which has no vent port.
Fluidic Amplifier, Impact Modulator	A fluidic amplifier in which the impact plane position of two opposed streams is controlled to alter the output.
Fluidic Amplifier, Open	A fluidic amplifier which has a vent port.
Fluidic Amplifier, Stream Deflection	A fluidic amplifier which uses one or more control streams to deflect a power stream, altering the output. It is usually analog.
Fluidic Amplifier, Vortex	A fluidic amplifier in which the angular rate of a vortex is controlled to alter the output.

Fluidic Amplifier, Wall Attachment — A fluidic amplifier in which the control of the attachment of a stream to a wall(s) alters the output. It is usually digital.

Fluidic Device, Active — The general class of fluidic that requires a power supply separate from the controls.

Fluidic Amplifier — A device which enables one or more fluid dynamic signals to control a source of power and thus is capable of delivering at its output an enlarged reproduction of the essential characteristics of the signal.

Fluidic Device, Passive — The general class of fluidic devices that operates on signal power alone.

Fluidics — Engineering science pertaining to the use of fluid dynamic phenomena to sense, control, process information, and/or actuate.

Force Motor — A type of electromechanical transducer having linear motion used at the input stages of servovalves.

Free Air — Air at ambient temperature, pressure, relative humidity, and density.

Free Air — Any compressible gas, air or vapor trapped within a hydraulic system that does not condense or dissolve to form a part of the system fluid.

Free Water — Water droplets or globules in the system fluid that tend to accumulate at the bottom or top of the system fluid depending on the fluid's specific gravity.

Fretting — A type of wear resulting from minute reciprocal sliding motion which products fine particulate contamination without chemical change.

Fretting Corrosion — Oxidation of air and moisture on the surface of a mechanical element.

Front Connected — Where connections are made to normally exposed surfaces of components.

Full System Differential Pressure — A filter element which will withstand a differential pressure at least equal to the maximum system operating pressure without structural or filter medium failure.

Gage — An instrument or device for measuring, indicating, or comparing a physical characteristic.

Gage, Bellows — A gage in which the sensing element is a convoluted closed cylinder. A pressure differential between outside and inside causes the cylinder to expand or contract axially.

Gage Bourdon Tube — A pressure gage in which the sensing element is a curved tube that tends to straighten out when subjected to internal fluid pressure.

Gage, Diaphragm — A gage in which the sensing element is relatively thin and its inner portion is free to deflect with respect to its periphery.

Gage, Fluid Level — A gage which indicates the fluid level at all times.

Gage, Manometer — A differential pressure gage in which pressure is indicated by the height of a liquid column of known density. Pressure is equal to the difference in vertical height between two connected columns multiplied by the density of the manometer liquid. Some forms of manometers are "U" tube, inclined tube, well, and bell types.

Gage, Pilot Tube — A velocity-sensing tubular probe with one end facing fluid flow and the other end connected to a gage.

Gage, Piston — A pressure gage in which the sensing element is a piston operating against a spring.

Gage, Pressurs — A gage which indicates the pressure in the system to which it is connected.

Gage, Vacuum — A pressure gage for pressures less than atmospheric.

Gland — The cavity of a stuffing box.

Gland Follower — The closure for a stuffing box.

Gravimetric Value — The weight of suspended solids per unit volume of fluid. A method employing membrane filters for this determination is outlined in Society of Automotive Engineers Aerospace Recommended Practices #785.

Grooving — Shallow ridges in the filter medium perpendicular to the roots of the pleats.

Hagen Poiseuille Law — The friction factor of Darcy's Formula is a ratio of 64 to the Reynolds Number when flow is laminar.

Head — The height of a column or body of fluid above a given point expressed in linear units. Head is often used to indicate gage pressure. Pressure is equal to the height times the density of the fluid.

Head (Front End) (Front Face) (Front Head) (Rod Head) — The cylinder end closure which covers the differential area between the bore area and the piston rod area.

Head, Filter — An end closure for the filter case or bowl which contains one or more ports.

Head, Friction — The head required to overcome the friction at the interior surface of a conductor and between fluid particles in motion. It varies with flow, size, type and condition of conductors and fittings, and the fluid characteristics.

Head, Static	The height of a column or body of fluid above a given point.
Head, Static Discharge	The static head from the centerline of the pump to the free discharge surface.
Head, Static Suction	The head from the surface of the supply source to the centerline of the pump.
Head, Total Suction	The static head from the surface of the supply source to the free discharge surface.
Head, Velocity	The equivalent head through which the liquid would have to fall to attain a given velocity. Mathematically it is equal to the square of the velocity (in feet) divided by 64.4 feet per second squared.
Heat Exchanger	A device which transfers heat through a conducting wall from one fluid to another.
Hertz	A unit of frequency formerly expressed as cycles per second. 1 hz = 1 cps.
Hose	A flexible line, consisting of a cover, a layer of reinforcement and an inner tube.
Hose, Wire Braided	Hose consisting of a flexible material reinforced with woven wire braid.
Housing	A ported enclosure which directs the flow through the filter element.
Hydraulic Amplifier	A fluid device which enables one or more inputs to control a source of fluid power and thus is capable of delivering at its output an enlarged reproduction of the essential characteristics of the input. Hydraulic amplifiers may utilize sliding spools, no nozzle-flappers, jet pipes, etc.
Hydraulic Filter	A device whose primary function is the retention by porous media of insoluble contaminants from a liquid.
Hydraulic Filter, Disposable	A hydraulic filter which is intended to be discarded and replaced after one service cycle.
Hydraulic Filter, Filtered By-Pass	A hydraulic filter in which by-pass flow is filtered through a reserve filter element.
Hydraulic Filter, Full Flow	A hydraulic filter which filters all influent flow.
Hydraulic Filter, Partial Flow	A hydraulic filter which filters a portion of the influent flow.
Hydraulic Filter, Reservoir (Sump)	A hydraulic filter installed in a reservoir in series with a suction or return line.
Hydraulic Filter, Two-Stage	A hydraulic filter having two filter elements in series.
Hydraulic Filter, By-Pass (Reserve)	A hydraulic filter which provides an alternate unfiltered flow path around the filter element when a preset differential pressure is reached.
Hydraulic Filter, Dual	A hydraulic filter having two filter elements in parallel.
Hydraulic Filter, Duplex	An Assembly of two hydraulic filters with valving for selection of either or both filters.
Hydraulic Filter, Fill Cap	A hydraulic filter which covers the fill opening to the reservoir and filters makeup fluid.
Hydraulic Filter, In-Line	A hydraulic filter in which the inlet, outlet, and filter element axis are in a straight line.
Hydraulic Filter, L-Type	A hydraulic filter in which the inlet and outlet port axis are at right angles, and the filter element axis is parallel to either port axis.
Hydraulic Filter, Manifold	A hydraulic filter containing multiple ports and integral related components which services more than one hydraulic circuit.
Hydraulic Filter, Modular	A hydraulic filter which mounts to or within a manifold or subplate with flow passages at the interface.
Hydraulic Filter, T-Type	A hydraulic filter in which the inlet and outlet ports are located at one end of the filter with port axis in a straight line, and the filter element axis is perpendicular to this line.
Hydraulic Filter, Wash	A hydraulic filter in which a larger unfiltered portion of the fluid flowing parallel to the filter element axis is used to continuously clean the influent surface which filters the lesser flow.
Hydraulic Filter, Y-Type	A hydraulic filter in which the inlet and outlet port axis are in a straight line, and the filter element axis is an acute angle to this line.
Hydraulic Horsepower	Horsepower computed from flow rate and pressure differential.
Hydraulic Motor	A device which converts hydraulic fluid power into mechanical force and motion. It usually provides rotary mechanical motion.
Hydraulic Motor, Fixed Displacement	A hydraulic motor in which the displacement per unit of output motion cannot be varied.
Hydraulic Motor, Linear	A fluid power cylinder providing reciprocating motion.

Hydraulic Motor, Rotary A hydraulic motor capable of continuous rotary motion.

Hydraulic Motor, Rotary, Limited A hydraulic rotary motor having limited motion.

Hydraulic Motor, Variable A hydraulic motor in which the displacement per unit of output motion can be **varied.**

Hydraulic Pump A device which converts mechanical force and motion into hydraulic fluid power.

Hydraulic Pump, Axial Piston A hydraulic pump having multiple pistons disposed with their axis parallel.

Hydraulic Pump, Centrifugal A hydraulic pump which produces fluid velocity and converts it to pressure head.

Hydraulic Pump, Centrifugal, Diffuser (Concentric) A centrifugal hydraulic pump in which fluid enters at the center of the impeller, is accelerated radially, and leaves through vanes arranged to provide a gradually enlarging flow passage.

Hydraulic Pump, Centrifugal, Perpheral A centrifugal hydraulic pump in which fluid enters, follows, and leaves the periphery of the impeller.

Hydraulic Pump, Centrifugal, Volute (Spiral) A centrifugal hydraulic pump in which fluid enters at the center of the impeller, is accelerated radially, and leaves through a gradually enlarging flow passage.

Hydraulic Pump, Reciprocating Duplex A hydraulic pump having two reciprocating pistons.

Hydraulic Pump, Reciprocating Piston A hydraulic pump having two reciprocating pistons.

Hydraulic Pump, Reciprocating Single Piston A hydraulic pump having a single reciprocating piston.

Hydraulic Pump, Fixed Displacement A hydraulic pump in which the displacement per cycle cannot be varied.

Hydraulic Pump, Gear housing. A hydraulic pump having two or more intermeshed rotating members enclosed in a

Hydraulic Pump, Hand A hand operated hydraulic pump.

Hydraulic Pump, Multiple Stage Two or more hydraulic pumps in series.

Hydraulic Pump, Radial Piston A hydraulic pump having multiple pistons disposed radially actuated by an eccentric element.

Hydraulic Pump, Screw A hydraulic pump having one or more screws rotating in a housing.

Hydraulic Pump, Vane A hydraulic pump having multiple radial vanes within a supporting rotor.

Hydraulic Pump, Variable Displacement A hydraulic pump in which the displacement per cycle can be varied.

Hydraulics Engineering science pertaining to liquid pressure and flow.

Hydrodynamics Engineering science pertaining to the energy of liquid flow and pressure.

Hydrokinetics Engineering science pertaining to the energy of liquids in motion.

Hydropneumatics Pertaining to the combination of hydraulic and pneumatic fluid power.

Hydrostatics Engineering science pertaining to the energy of liquids at rest.

Impingement The direct high velocity impact of the fluid flow upon or against any internal portion of the filter.

Incompatible Fluids Fluids which when mixed in a system, will have a deleterious effect on that system, its components, or its operation.

Indicator A device which provides external visual evidence of sensed phenomena.

Indicator, By-Pass An indicator which signals alternate flow.

Indicator, Differential An indicator which signals the difference in pressure at two points.

Indicator, Pressure An indicator which signals pressure conditions.

Influent The fluid entering a component.

Input Current The current to the servovalve which commands control flow.

Inside-Out Flow A filter element designed for normal flow outward and perpendicular to the axis of the filter element.

Intensifier A device which converts low pressure fluid power into higher pressure fluid power.

Intercooler A device which cools a gas between the compressive steps of a multiple stage compressor.

Interface A Point or component where a transition is made between medium, power levels, modes of operation, etc.

Jet Action A valve design in which flow effect is controlled by the relative position of a nozzle and a receiver.

Joint A line positioning connector.

Joint, Rotary A joint connecting lines which have relative operational rotation.

Joint, Swivel A joint which permits variable operational positioning of lines.

Joule A unit of work, energy, or heat. (1 J (joule) = 1 Nm/Newton meter).

Joule Per Second A unit of power, i.e. the rate of doing work. 1 J/s (joule per second) = 1 W (watt). 1 Kw (kilowatt) = 10- J/s = 1.34 HP.

Kelvin Absolute temperature scale in the metric system. See Celsius, the preferred system. °C + 273.15 = °K.

Kilogram A unit of mass. 1 kg = 2.20 pounds.

Kilopond A unit of force used in Germany and equal to 1 kg (f) – (kilogram, force). It is generally expressed per square centimeter to produce a unit of pressure. 1 kp/cm^2 = 1 kg (f) $/cm^2$ = 14.22 psig approximately.

Lantern Ring (Seal Cage) A ring in line with a port in a gland to introduce a lubricant or coolant to the packing and stuffing box.

Lift The height of a column or body of fluid below a given point expressed in linear units. Lift is often used to indicate vacuum or pressure below atmosphere.

Lift, Static Suction The lift from the centerline of the pump to the surface of the supply source. (See Head, Static Suction.)

Line A tube, pipe or hose for conducting fluid.

Line, Drain A line returning leakage fluid independently to the reservoir or vented manifold.

Line, Exhaust A line returning power or control fluid back to the reservoir or atmosphere.

Line, Pilot A line which conducts control fluid.

Line, Suction A supply line at sub-atmospheric pressure to a pump, compressor, or other component.

Line, Working A line which conducts fluid power.

Linear Scale Reticule A reticule having a straight scale marked to permit measurement or estimation of distance or length.

Lines, Joining Lines which connect in a circuit.

Lines, Passing Lines which cross but do not connect in a circuit.

Liter A unit of volume. 1 1 (liter) = 1000 cm^3 (cubic centimeters) = 61 in^3 (cubic inches) approximately = 0.264 U.S. gallons approximately.

Liter per Minute A unit of volumetric flow rate. 1 1/min (liter per minute) = 10^{-3} m^3/min (cubic meters per minute) = 1.02 in^3/s (cubic inches per second) = 0.264 U.S. gal/min (gallons per minute).

Liter per Revolution A unit of volumetric capacity. 1 1/rev (liter per revolution) 10^{-3} m^3/rev (cubic meters per revolution) = 61.0 in^3/rev (cubic inches per revolution).

Logic Devices The general category of components which perform logic functions; for example, AND, NAND, OR, and NOR. This can permit or inhibit signal transmission with certain combinations of control signals.

Logical State Signal levels in logic devices are characterized by two stable states, the logical 1 (one) state and the logical 0 (zero) state. The designation of the two states is chosen arbitrarily. Commonly the logical 1 state represents an "on" signal, and the 0 state represents an "off" signal.

Longest Dimension The greatest dimension of a particle equivalent to the diameter of a sphere enclosing the particle when tangent at a minimum of two points.

Lubricator A device which adds controlled or method amounts of lubricant into a fluid power system.

Lug (Foot) A mounting device consisting of a block extending past the basic cylinder profile. The block usually has a tapped or through mounting hole at right angles to the cylinder axis.

Magnetic A filter element which in addition to its filter medium has a magnet or magnets incorporated into its structure to attract and hold ferromagnetic particles.

Manifold A conductor which provides multiple connection ports.

Manifold, Vented A manifold which is open to the atmosphere and returns fluid to the reservoir.

Maximum Excursion	The maximum pressure deviation from the operating pressure after an abrupt disturbance.
Maximum Inlet Pressure	The maximum rated gage pressure applied to the inlet port of the regulator.
Mean Filtration Rating	A measurement of the average size of the pores of the filter medium.
Medium	The porous material that performs the actual process of filtration.
Meter	A unit of length. 1 m (meter) = 39.37 in (inch) = 25.4 mm (millimeters) exactly.
Meter, Flow	A device which indicates either flow rate, total flow, or a combination of both.
Metric System	A decimal system based on the meter and kilogram, which varies somewhat between Countries. See SI System.
Micrometer (Micron) Deprecated**	Unit of measurement one millionth of a meter long, or approximately 0.00003937 inch expressed in English units.
Microscope Gating	A particle counting and sizing technique in which the image of the particles are moved through the linear scale of a reticule for counting and sizing.
Microscopic	Particles whose diameter is below the threshold of normal vision, below forty micrometers for most individuals.
Microscopic Filar Eyepiece	A micrometer eyepiece containing a movable hairline connected to an external micrometer scale used to measure distance.
Migration	Contaminant released downstream.
Migration, Abrasion	Migration generated by parts that rub together and wear during vibration or shock induced by flow or other stimuli.
Migration, Built-In Dirt	Migration composed of foreign materials introduced during handling, storage, and fabrication.
Migration, Contaminant	Migration due to unloading.
Migration, Media	Migration composed of the materials making up the filter medium.
Modular (Plug-In)	A filter element which has not separate housing of its own, but whose housing is incorporated into the equipment which it services. It may also incorporated a suitable closure for the filter cavity.
Muffler	A device for reducing gas flow noise. Noise is decreased by back pressure control of gas expansion.
NAND Device	A control device which has its output in the logical 0 state if and only if all the control signals assume the logical 1 state.
Neutralization Number	A measure of the total acidity or basicity of an oil; this includes organic or inorganic acids, or bases, a combination thereof (ASTM Designation D974-64).
Newt	The standard unit of kinematic viscosity in the English system. It is expressed in square inches per second.
Newton	A unit of force based on the unit of mass, Kg (kilogram), multiplied by the acceleration, m/s^2 (meters per second per second) which produces Kgm/s^2 called the Newton. 1 N = 1 Kgm/s^2 = 0.225 lb (f).
Nipple	A short length of pipe or tube.
Noise, Fluidic	RMS of random pressure variations with respect to the operating pressure defined in terms of a signal-to-noise ratio.
Nominal Filtration Rating	An arbitrary micro-meter value indicated by the filter manufacturer. Due to lack of reproducibility this rating is deprecated.
Non-Combustible Residue	Matter not changed to gaseous state when laboratory membranes are ashed at 1500°F.
Non-Volatile Residue	The residue remaining on laboratory ware after the solvent or fluid has evaporated.
Non-Woven	A filter medium composed of a mat of fibers.
NCR Device	A control device which has its output in the logical state 1 state if and only if all the control signals assume the logical 0 state.
NOT Device	A control device which has its output in the logical 1 state if and only if the control signal assumes the logical 0 state. The NOT device is a single input NOR device.
Open Area	The pore area of a filter medium often expressed as a percent of total area.
Open Area Ratio	The ratio of pore area to total area of a filter medium expressed as a percent of total area.
Operating Band	The range of pressure above and below the operating pressure within which it is desired to keep the supply output.
Optical Density	A method of expressing degree of contamination of a fluid by removal of contaminant by filtration and measuring change in optical transmission of the filter disc or fluid, or both.

OR Device — A control device which has its output in the logical 0 state if and only if all the control signals assume the logical 0 state.

Outer Wrapper — A permeable enclosure which protects the filter medium.

Output Stage — The final state of hydraulic amplification used in a servovalve.

Outside-In Flow — A filter element designed for normal flow perpendicular and toward the axis of the filter element.

Oxidation — The interaction of air and moisture on the surface of a mechanical element.

Packing — A sealing device consisting of bulk deformable material or one or more mating deformable elements, reshaped by manually adjustable compression to obtain and maintain effectiveness. It usually uses axial compression to obtain radial sealing.

Packing, "U" — A packing in which the deformable element has a "U" shaped cross-section.

Packing, "V" — A packing in which the deformable element has a "V" shaped cross-section.

Packing, "W" — A packing in which the deformable element has a "W" shaped cross-section.

Packing, Coil — Packing in coil form.

Panel — A plate or a surface for mounting components.

Panel Mounting — A panel on which a number of components may be mounted.

Particle — A minute piece of matter with observable length, width, and thickness; usually measured in micrometers.

Particle Count Blank — An allowance for the determinable background contamination.

Particle Size Distribution — The tabular or graphical listing of the number particles according to particle size ranges.

Pascal's Law — A pressure applied to a confined fluid at rest is transmitted with equal intensity throughout the fluid.

Passage — A machined or cored fluid-conducting path which lives within or passes through a component.

Patch Test — Any method of evaluating fluid contamination wherein the sample is passed through a standardized laboratory filter, and the change in color, reflectivity, etc., of the laboratory filter is compared with previously established standards.

Permeability — The relationship of flow per unit area to differential pressure across a filter medium.

Petroleum Fluid — A fluid composed of petroleum oil. It may contain additives.

Phase — A distinct functional operation during a cycle. Some typical sequential phases are: neutral, rapid advance, feed or pressure stroke, dwell and rapid return.

Phase, Dwell — The phase of cycle where a specified motion is stopped for a pre-determined length of time.

Phase, Feed — The phase of a cycle where work is performed on the workpiece.

Phase, Neutral — The phase of a cycle from which the work sequence begins.

Phase, Rapid Advance — The phase of a cycle where tools or workpiece approach at high speed to the feed position.

Phase, Rapid Return — The phase of a cycle where tools or workpiece return at high speed to the cycle starting position.

Pilot Line — A filter located in a line conducting fluid to a control device or devices.

Pinched Pleat — A pleat closed off by excessive differential pressure or crowding, thus reducing the effective area of the filter element.

Pipe — A line whose outside diameter is standardized for threading. Pipe is available in Standard, Extra Strong, Double Extra Strong or Schedule wall thickness.

Pipe Thread — Screw threads for joining pipe.

Pipe Thread, Dryseal — Pipe threads in which sealing is a function of root and crest interference.

Pipe Thread, Tapered — Pipe threads in which the pitch diameter follows a helical cone to provide interference in tightening.

Plain — A filter element whose medium is not pleated or otherwise extended and has the geometric form of a cylinder, cone, disc, plate, etc.

Pleated (Corrugated) — A filter element whose medium consists of a series of uniform folds and has the geometric form of a cylinder, cone, disc, plate, etc.

Pleats (Corrugations) — A series of folds in the filter medium usually of uniform height and spacing.

Pneumatics — Engineering science pertaining to gaseous pressure and flow.

Poise — The standard unit of absolute viscosity in the c.g.s. (centimeter-gram-second) system. It is the ratio of the shearing stress to the shear rate of a fluid and is expressed in dyne seconds per square centimeter; 1 centipoise equals. 01 poise.

Pore Size Distribution	The ratio of the number of holes of a given size to the total number of holes per unit area expressed as a percent and as a function of hole size.
Porosity (Void Fraction)	The ratio of pore volume to total volume of a filter medium expressed as a percent.
Port	An internal or external terminus of a passage in a component.
Port-to-Port Dimension	The distance between two ports measured from face to face or between center lines.
Port, Bias	The port at which a biasing signal is applied.
Port, Bleed	A port which provides a passage for the purging of gas from a system or component.
Port, Control	A port which provides passage for a control signal.
Port, Cylinder	A port which provides a passage to or from an actuator.
Port, Differential Pressure	A port(s) which provides a passage to the upstream and downstream sides of a component.
Port, Discharge	A port which provides a passage for fluid power to the system.
Port, Drain	A port for removal of fluid from a component, open to atmosphere, or connected to an unrestricted line.
Port, Exhaust	A port which provides a passage to the atmosphere.
Port, Fill	A port which provides a passage for filling purposes.
Port, Inlet	A port which provides a passage for the influent.
Port, Outlet (Output)	A port which provides a passage for the effluent.
Port, Pipe	A port which conforms to pipe thread standards.
Port, Plain "0" Ring	A port which usea an "0" ring in a groove located on the port face.
Port, Pressure	A port which provides a passage from the source of fluid.
Port, SAE	A straight thread port used to attach tube and hose fittings. It employs an "0" ring compressed in a wedge-shaped cavity. A standard of the Society of Automotive Engineers J514 and ANSI/B116.1.
Port, Suction	A port which provides a passage for atmospheric charging of a pump or compressor.
Port, Supply	The port at which power is provided to an active device.
Port, Tank (Reservoir) (Return)	A port which provides a passage to the fluid source.
Port, Vent	A port which provides a passage to ambient conditions.
Pour Point	The lowest temperature at which a liquid will flow under specified conditions (ASTM Designation D97-66).
Power Capacity	The total volume of gas available at the operating pressure (applies to compressed gas storage supply source).
Power Supply	That component or group of components which supplies and processes the fluid for operating fluidic systems.
Power Unit	A combination of pump, pump drive, reservoir, controls and conditioning components which may be required for its application.
Precipitate	Particles separated from a fluid as a result of a chemical or physical change.
Precipitation Number	The number of milliliters of precipitate formed when 10 ml. of lubricating oil are mixed with 90 ml. of ASTM Precipitation naphtha and centrifuged under prescribed conditions (ASTM Designation D91-61).
Precoat	A filter medium in loose powder form (such as Fuller's or diatomaceous earth) introduced into the upstream fluid to condition a filter element.
Precooler	A device which cools a gas before it is compressed.
Pressure	Force per unit area, usually expressed in pounds per square inch.
Pressure Line	A filter located in a line conducting working fluid to a working device or devices.
Pressure Vessel	A container which holds fluid under pressure.
Pressure, Absolute	The pressure above zero absolute, i.e., the sum of atmospheric and gage pressure. In vacuum related work it is usually expressed in millimeters of mercury (mm Hg), or inches of mercury (in Hg).
Pressure, Atmospheric	Pressure exerted by the atmosphere at any specific location. (Sea level pressure is approximately 14.7 pounds per square inch absolute).
Pressure, Back	The pressure encountered on the return side of a system.
Pressure, Breakloose	The minimum pressure which initiates (Backout) movement.
Pressure, Burst	The pressure which creates loss of fluid through the component envelope.
Pressure, Charge	The pressure at which replenishing fluid is forced into a fluid power system.

Pressure, Cracking	The pressure at which a pressure operated valve begins to pass fluid.
Pressure, Differential	The difference in pressure between any two points of a system or a component.
Pressure, Gage	Pressure differential above or below atmospheric pressure.
Pressure, Head	The pressure due to the height of a column or body of fluid. It is usually expressed in feet.
Pressure, Maximum Inlet	The maximum rated gage pressure applied to the inlet.
Pressure, Operating	The pressure at which a system is operated.
Pressure, Override	The difference between the cracking pressure of a valve and the pressure reached when the valve is passing full flow.
Pressure, Peak	The maximum pressure encountered in the operation of a component.
Pressure, Pilot	The pressure in the pilot circuit.
Pressure, Precharge	The pressure of compressed gas in an accumulator prior to the admission of a liquid.
Pressure, Proof	The non-destructive test pressure in excess of the maximum rated operating pressure, which causes no permanent pressure, which causes no permanent deformation, excessive external leakage, or other malfunction.
Pressure, Rated	The qualified operating pressure which is recommended for a component or a system by the manufacturer.
Pressure, Shock	The pressure existing in a wave moving at sonic velocity.
Pressure, Static	The pressure in a fluid at rest.
Pressure, Suction	The absolute pressure of the fluid at the inlet of a pump.
Pressure, Surge	The pressure existing from surge conditions.
Pressure, System	The pressure which overcomes the total resistances in a system. It includes all losses as well as useful work.
Pressure, Vapor	The pressure at a given fluid temperature, in which the liquid and gaseous phases are in equilibrium.
Pressure, Working	The pressure which overcomes the resistance of the working device.
Primary	The first filter element in a series, or the main filter element of a filtered by-pass filter assembly.
Pump Intake Line	A filter located in a line conducting fluid to the pump inlet.
Pump Intake Line, Supercharge	A pump intake line filter in which the fluid is above atmospheric pressure.
Pump Intake Line, Suction	A pump intake line filter in which the fluid is below atmospheric pressure.
Quick Disconnect	A coupling which can quickly join or separate a fluid line without the use of tools or special devices.
Quick Disconnect, Break-Away	A quick disconnect which provides automatic separation of the coupling halves when a predetermined axial force is applied.
Quick Disconnect, Un-Valves* *Deprecated	A quick disconnect with no shut-off valves.
Quick Disconnect, Valved	A quick disconnect with a shut-off valve in each half.
Quick, Disconnect, One Valve	A quick disconnect with a shut-off valve in one half only.
Quiescent Current	A direct current that is present in each servovalve coil when using a differential coil connection, the polarity of the current in the coils being in opposition such that no electrical control power exists.
Rabbet	A mounting device which uses matching male and female forms (usually coaxial circular) between the cylinder and its mating element.
Rated Current	The specified servovalve input current of either polarity to produce rated flow. Rated current must be specified for a particular coil connection differential, series, or parallel, and does not include null bias current.
Rated Flow	The maximum flow that the power supply system is capable of maintaining at a specific operating pressure.
Raw Count	The actual number counted in each particle size range in a given sample.
Reduced Pressure Range	The adjustment range of the regulator.
Regulator Characteristic Curve	The change in regulated (secondary) pressure occurring as result of a change in the supply (primary) pressure to a regulator.
Regulator, Air Line Pressure	A regulator which tranaforms a fluctuating air pressure supply to provide a constant lower pressure output.

Regulator, Constant Bleed Air Line Pressure — An air line pressure regulator that depends upon a bleed to atmosphere for proper operation.

Regulator, Pressure Relieving Air Line — An air line pressure regulator which automatically vents over-pressures applied to the regulated (secondary) pressure.

Relief Characteristic Curve — The change in the relief flow rate in a relieving type air line regulator which occurs as a result of an increase in regulated (secondary) pressure over set pressure.

Reserve Capacity — The volume of air above the "high" mark of the level indicator.

Reservoir — A container for storage of liquid in a fluid power system.

Reservoir, Atmospheric — A reservoir for storage of fluid media at atmospheric pressure.

Reservoir, Hydraulic — A reservoir for storing and conditioning liquid in a hydraulic system.

Reservoir, Non-Integral — An independent or removable reservoir.

Reservoir, Pressure Sealed — A sealed reservoir for storage of fluids under pressure.

Reservoir, Sealed — A reservoir for storage of fluids isolated from atmospheric conditions.

Reservoir, Top Mounted — A reservoir with provisions for mounting the pump and components on top.

Residual Dirt Capacity — The dirt capacity remaining in a service loaded filter element after use, but before cleaning, measured under the same conditions as the dirt capacity of new filter element.

Response Time — The time required for effective transition.

Restrictor — A device which reduces the cross-sectional flow area.

Restrictor, Choke — A restrictor, the length of which is relatively large with respect to its cross-sectional area.

Restrictor, Orifice — A restrictor, the length of which is relatively small with respect to its cross-section area. The orifice may be fixed or variable. Variable types are non-compensated, pressure compensated, or pressure and temperature compensated.

Return Line — A filter located in a line conducting fluid from working devices to reservoir.

Reversal, Position — A reversal of direction of movement initiated by a signal given at some predetermined point of movement.

Reversal, Pressure — A reversal of direction of movement initiated by a signal responsive to rise in pressure.

Reyn — The standard unit of absolute viscosity in the English system. It is expressed in pound-seconds per square inch.

Reynolds Number — A numerical ratio of the dynamic forces of mass flow to the shear stress due to viscosity. Flow usually changes from laminar to turbulent between Reynolds Numbers 2,000 and 4,000.

Ring, "0" — A ring which has a round cross-section.

Ring, "U" — A ring which has a "U" shaped cross-section.

Ring, "V" — A ring which has a "V" shaped cross-section.

Ring, "W" — A ring which has a "W" shaped cross-section.

Ring, Wiper — A ring which removes material by a wiping action.

Ring, Piston — A piston sealing ring. It is usually one of a series and is often split to facilitate expansion or contraction.

Ring, Scraper — A ring which removes material by a scraping action.

Ripple — A periodic variation of the pressure above and below the operating pressure. It is defined as a percentage of the operating pressure in terms of the maximum peak-to-peak value obtained at the point of rating.

Rise Rate — The ration of pressure rise to time.

Root — The inner fold of a pleat.

Scoring — Scratches in the direction of motion of mechanical parts caused by abrasive contaminants.

Scoring Size — A particle whose dimensions are such that it is capable of entering a working clearance.

Seal — (See Sealing Device)

Seal, "0" Ring — A sealing ring which has a round cross-section.

Seal, Axial (End) (Face) (Shoulder) — A sealing device which seals by axial contact pressure.

Seal, Cup — A sealing device with a radial base integral with an axial cylindrical projection at its outer diameter.

Seal, Diaphragm (Flat Diaphragm) A relatively thin, flat or molded sealing device fastened and sealed at its periphery with its inner portion free to move.

Seal, Dished Diaphragm A diaphragm in which the central area is depressed in a free state. It permits longer travel than a flat comparable diaphragm.

Seal, Dynamic A sealing device used between parts that have relative motion.

Seal, Flange (Hat) A sealing device with a radial base integral with an axial projection at its inner diameter.

Seal, Lip A sealing device which has a flexible sealing projection.

Seal, Lubricant A sealing device which uses lubricant as a sealing barrier.

Seal, Mechanical A sealing device in which sealing action is aided by mechanical force.

Seal, Oil A sealing device which retains oil.

Seal, Piston A sealing device installed on a piston to maintain a sealing fit with a cylinder bore.

Seal, Pressured Actuated A sealing device in which sealing action is aided by fluid pressure.

Seal, Radial A sealing device which seals by radial contact pressure.

Seal, Rod (Shaft) (Stem) A sealing device which seals the periphery of a piston rod.

Seal, Rotary (Shaft) A sealing device used between parts that have relative rotary motion.

Seal, Sliding A sealing device used between parts that have relative reciprocating motion.

Seal, Static (Gasket) A sealing device used between parts that have no relative motion.

Seal, Water A sealing device which used water as a sealing barrier.

Seal, Wiper A sealing device which operates by a wiping action.

Sealing Device (Seal) A device which prevents or controls the escape of a fluid or entry of a foreign material.

Seating Action A valve design in which flow is stopped by a seated obstruction in a flow path.

Seating Action, Ball A seating action valve design which uses a solid ball to obstruct the flow path.

Seating Action, Diaphragm A seating action valve design which uses a diaphragm to obstruct the flow path.

Seating Action, Disc (Globe) A seating action valve design which uses a disc to obstruct the flow path.

Seating Action, Disc, Swing A seating action valve design which uses a hinged disc to obstruct the flow path.

Seating Action, Gate A seating action valve design which uses a solid gate to obstruct the flow path.

Seating Action, Gate, Spreader A gate valve which uses two companion discs which are positively seated by common spreaders to obstruct the flow path.

Seating Action, Gate, Wedge A gate valve which uses a solid wedge shaped gate to obstruct the flow path.

Seating Action, Needle A seating action valve design which uses an externally adjustable tapered closure to obstruct the flow path.

Seating Action, Plug A seating action valve design which uses a plug to obstruct the flow path.

Seating Action, Poppet A seating action valve design in which the seating element pops open to obtain free flow in one direction and immediately reseats when flow reverses.

Second Longest Dimension The greatest dimensions perpendicular to the particle's longest dimension.

Secondary The second of two filter elements in series.

Self Cleaning A filter element designed to be cleaned without removing it from the filter assembly.

Separator A device whose primary function is to isolate contaminants by physical properties other than size.

Separator, Adsorbent A separator that retains certain soluble and insoluble contaminates by molecular adhesion.

Separator, Centrifugal A separator that removes nonmiscible fluid and solid contaminants that have a different specific gravity than the fluid being purified by accelerating the fluid in a circular path and using the radial acceleration component to isolate these contaminants.

Separator, Coalescing A separator that divides a mixture or emulsion of two nonmiscible liquids using the interfacial tension between the two liquids and the difference in wetting of the liquids on a particular porous medium.

Separator, Electrostatic A separator that removes contaminant from dielectric fluids by applying an electrical charge to the contaminant which is then attracted to a collection device of different electrical charge.

Separator, Magnetic A separator that uses a magnetic field to attract and hold ferromagnetic particles.

Separator, Two Phase	A separator that is capable of dividing a liquid and gas mixture.
Separator, Vacuum	A separator that uses subatmospheric pressure to remove certain gases and liquids from another liquid because of their difference in vapor pressure.
Servovalve	A valve which modulates output as a function of an input command.
Servovalve Control Flow	The flow through the servovalve control ports. Conventional test equipment normally measures no-load flow.
Servovalve Phase Lag	The instantaneous time by which the servovalve sinusoidal flow follows the sinusoidal input current, measured at a specified frequency and expressed in degrees.
Servovalve, Electrohydraulic	A servovalve which is capable of continuously controlling hydraulic output as a function of an electrical input.
Servovalve, Electrohydraulic Flow Control	An electrohydraulic servovalve whose primary function is controll of output flow.
Servovalve, Four-Way	A multi-orifice flow control valve with supply, return and two control ports arranged so that the valve action in one direction opens supply to control port #1 and opens control port#2 to return. Reversed valve action opens supply to control port #2 and opens control port #1 to return.
Servovalve, Three-Way	A multi-orifice flow control valve with supply, return and one return and one control port arranged so that valve action in one direction opens supply to control port and reversed valve action opens the control port to return.
Servovalve, Two-Way	A single orifice flow control valve with supply, and one control port arranged so that action is in one direction only, from supply to control port.
Shear Action	A valve design in which flow is modulated by an element which slides across the flow path.
Shear Action, Ball	A shear action valve design which uses a ported ball that rotates on an axis normal to the flow path.
Shear Action, Plug	A shear action valve design which uses a ported plug that rotates on an axis normal to the flow path.
Shear Action, Sliding Plate, Linear	A sliding plate shear action valve design in which the motion of the plate is linear.
Shear Action, Sliding Plate, Rotary	A sliding plate shear action valve design in which the motion of the plate is rotary.
Shear Action, Sliding Plate Plate	A shear action valve design which uses a plate that slides across the flow path.
Shear Action, Spool (Plunger)	A shear action valve design which uses a spool that slides through the flow path.
Shell	A structural form to which a sealing element is assembled or bounded.
Shock Wave	A pressure wave front which moves at sonic velocity.
Side (Base)	An envelope surface which is parallel to the piston rod centerline.
Side Seal	The longitudinal seam of the filter medium in a filter element.
Silencer	A device for reducing gas flow noise. Noise is decreased by tuned resonant control of gas expansion.
Silt	Fine particulate matter, generally less than five micrometers in size.
Silting	An accumulation of fine particles at a specific location in a fluid system.
Sintered	A metallic or non-metallic filter medium processed to cause diffusion bonds at all contacting points.
Sloughing Off	The release of contaminant from the upstream surface of a filter element to the upstream side of the filter enclosure.
Sludge	Particulate contaminant or a mixture of a particulate and liquid contaminant separated from the fluid in an unconsolidated state.
Specific Gravity	The ratio of the weight of a given volume of material or liquid to the weight of an equal volume of water.
Spring, Expander	A spring which produces outward radial force.
Spring, Finger (Lug)	A spring with flexible fingers which produce force.
Spring, Garter	A compression or tension ring formed from helical wire spring with connected ends to produce force.
Spring, Spreader	A spring which produces sealing force against both lips of "U" or "V" seals.
Spring, Wave (Marcel) (Wave Washer)	A compression spring of waved configuration which produces force.
Stability, Chemical	Resistance to chemical change.

Stability, Hydrolytic	Resistance to permanent changes in properties caused by chemical reaction with water.
Stability, Oxidation	Resistance to permanent changes caused by chemical reaction with oxygen.
Stability, Thermal	Resistance to permanent changes caused solely by heat.
Stage	A hydraulic amplifier used in a servovalve. Servovalve may be single stage, two stage, three stage, etc.
Standard	A document, or an object for physical comparison, for defining product characteristics, products, or processes: prepared by a consensus of a properly constituted group of those substantially affected and having the qualifications to prepare the standard for voluntary use.
Standard Air	Air at a temperature of 68°F, a pressure of 14.70 pounds per square inch absolute, and a relative humidity of 36% (0.0750 pounds per cubic foot). In gas industries, the temperature of "standard air" is usually given as 60°F.
Start-Up Time	The period of time needed to reach a steady state condition within the operating band starting from a long term off condition.
Starvation	Insufficient filter effluent to allow proper functioning of downstream components.
Steady State Pressure	A band indicating maximum and minimum deviation indicated in percent of operating pressure, all
Regulation	as a function of flow.
Stoke	The standard unit of kinematic viscosity in the c.g.s. (centimeter-gram-second) system. It is expressed in square centimeters per second; 1 centistoke equals .01 stoke.
Stop, Positive Position	A structural member which accurately stops motion.
Stop, Positive Safety	A structural member which confines maximum travel to safe limits.
Stuffing Box	A cavity and closure for sealing device.
Subplate (Back Plate)	An auxiliary ported plate for mounting components.
Surface	A filter medium which primarily retains contaminent on the influent face.
Surface Tension	The centractile surface force of a liquid in contact with a fluid by which it tends to assume a spherical form and to present the least possible surface. It is expressed in pounds per foot or dynes per centimeter.
Surge	A transient rise of pressure or flow.
Switch, Float	An electric switch which is responsive to liquid level.
Switch, Flow	An electric switch operated by fluid flow.
Switch, Pressure	An electric switch operated by fluid pressure.
Switch, Pressure	An electric switch operated by a difference in pressure.
Symbol, Fluid Power	A representation of the characteristics of a fluid power component by means of lines on a flat surface.
Symbol, Graphical (Schematic)	A simplified symbol which indicates essential characteristics applicable to all similar components.
Symbol, Pictorial	A symbol showing the actual shape of a component according to the manufacturer's description.
Symbol, Combination	A symbol which combines graphical, cutaway, and pictorial representations.
Symbol, Cutaway	A symbol showing principal internal parts, controls and actuating mechanisms, interconnecting lines, and functions of a component.
Synthetic Fluid	Fluid which has been artificially compounded for use in a fluid power system.
Synthetic Fluid	A fluid composed of esters which are compounds of carbon, hydrogen, and oxygen only. It may contain additives.
Synthetic Fluid Halogentated	A fluid composed of halogenated organic materials. It may contain additives.
Synthetic Fluid Phosphate Ester	A fluid composed of phosphate esters. It may contain additives.
Synthetic Fluid Phosphate Ester Base	A fluid which contains a phosphate ester as one of the major components.
Synthetic Fluid Silicate Ester	A fluid composed of organic silicates. It may contain additives.
Synthetic Fluid, Polyglycol	A non-aqueous fluid composed of polyglycol derivatives. It may contain additives.
Synthetic Fluid, Silicone	A fluid composed of silicones. It may contain additives.
Tank	A container for the storage of fluid in a fluid power system.
Tank, Air-Oil	A tank in which pressurized air is used to force oil into the outlet port.

Tank, Vacuum — A tank for gas at less than atmospheric pressure.

Tie Rod — An axial external cylinder element which traverse the length of the cylinder. It is prestressed at assembly to hold the ends of the cylinder against the tubing. Tie rod extensions can be a mounting device.

Toricelli's Theorem — The liquid velocity at an outlet discharging into the free atmosphere is proportional to the square root of the head.

Torque Motor — A type of electromechanical transducer having rotary motion used in the input stages of servovalves.

Torr — A unit of absolute pressure less than atmospheric pressure, equivalent to 1 mm. Hg. It is used primarily to describe a vacuum 25 mm. Hg absolute or less.

Tortuosity — The ratio of the average effective flow path length to minimum theoretical flow path length (thickness) of a filter medium.

Total Area — The entire area of a porous medium, whether effective or not, in a filter element.

Total Statistical Count — The raw count multiplied by a counting calibration factor.

Transient Recovery Time — The period of time required for an abrupt change in the power supply output pressure to dampen out to within the operating band.

Trunnion — A mounting device consisting of a pair of opposite projecting cylindrical pivots. The cylindrical pivot pins are at right angle or normal to the piston rod centerline to permit the cylinder to swing in a plane.

Tube — A line whose size is its outside diameter. Tube is available in varied wall thicknesses.

Two-Stage — A filter element assembly composed of two filter elements or media in series.

Unloading — The release of contaminant that was initially captured by the filter medium.

Vacuum — Pressure less than atmospheric pressure. It is usually expressed in inches of mercury (in. Hg) as referred to the existing atmospheric pressure.

Vacuum Pump — A device which uses mechanical force and motion to evacuate gas from a connected chamber to create sub-atmospheric pressure.

Valve — A device which controls fluid flow direction, pressure, or flow rate.

Valve Actuator — The valve part(s) through which force is applied to move or position flow-directing elements.

Valve Actuator, Manual — A valve actuator consisting of a hand lever, palm button, foot treadle, or other manual energizing devices.

Valve Actuator, Mechanical — A valve actuator consisting of a cam, lever, roller, screw, spring, stem, or other mechanical energizing devices.

Valve Actuator, Pilot — A valve actuator which uses pilot fluid.

Valve Actuator, Pilot Differential Area — A pilot valve actuator wherein pilot fluids act on unequal areas.

Valve Actuator, Pilot Barrier — A pilot valve actuator wherein the working fluid is isolated from the actuator.

Valve Actuator, Pilot Differential Pressure — A pilot actuator wherein pilot fluid acts at unequal pressure.

Valve Actuator, Pilot External — A pilot valve actuator wherein fluid is received from an external source.

Valve Actuator, Pilot Internal — A pilot valve actuator wherein pilot fluid is received from within the valve.

Valve Actuator, Pilot Solenoid Controlled — A pilot valve actuator wherein pilot fluid is controlled by the action of one or more solenoids.

Valve Actuator, Solenoid — A valve actuator which uses one or more solenoids.

Valve Flow Condition — A flow pattern in a directional control valve.

Valve Flow Condition, Closed — All ports are closed.

Valve Flow Condition, Regenerative — Working ports are connected to supply.

Valve Flow Condition, Tandem — Working ports are blocked and supply is connected to the reservoir port.

Valve Flow Condition, Float — Working ports are connected to exhaust or reservoir.

Valve Flow Condition, Hold — Working ports are blocked to hold a powered device in a fixed position.

Valve Flow Condition, Open — All ports are open.

Valve Mounting — The mounting characteristics of a valve.

Valve Mounting, Base	The valve is mounted to a plate which has top and side ports.
Valve Mounting, Line	The valve is mounted directly to system lines.
Valve Mounting, Manifold	The valve is mounted to a plate which provides multiple connection ports for two or more valves.
Valve Mounting, Sub-plate	The valve is mounted to a plate which provides straight-through top and bottom ports.
Valve Position	The point at which flow directing elements provide a specific flow condition in a valve.
Valve Position, Center	The selective mid-position in a directional control valve.
Valve Position, Detent	A predetermined position maintained by a holding device acting on the flow-direction elements of a directional control valve.
Valve Position, Normal	The valve position when signal or actuating force is not being applied.
Valve Position, Offset	An off-center position in a directional control valve.
Valve Position, Return	The initial valve position.
Valve, Air	A valve for controlling air.
Valve, Decompression	A pressure control valve that controls the rate at which the contained energy of the compressed is released.
Valve, Directional Control, Check	A directional control valve which permits flow of fluid in only one direction.
Valve, Directional Control, Four Way	A directional control valve whose primary function is to pressurize and exhaust two ports.
Valve, Directional Control, Selecter (Diversion)	A directional control valve whose primary function is to selectively interconnect two or more ports.
Valve, Directional Control, Straightway	A two port directional control valve.
Valve, Directional Control, Three Way	A directional control valve whose primary function is to pressurize and exhaust a port.
Valve, Directional Control	A valve whose primary function is to direct or prevent flow through selected passages.
Valve, Flow Control (Flow Metering)	A valve whose primary function is to control flow rate.
Valve, Flow Control Deceleration	A flow control valve which gradually reduces flow rate to provide deceleration.
Valve, Flow Control Pressure Compensated	A flow control valve which controls the rate of flow independent of system pressure.
Valve, Flow Control Pressure Compensated	A pressure compensated flow control valve which controls the rate of flow independent of system pressure.
Valve, Flow Control Pressure-Temperature Compensated	A pressure compensated flow control valve which controls the rate of flow independent of fluid temperature.
Valve, Flow Dividing	A valve which divides the flow from a single source into two or more branches.
Valve, Flow Dividing Pressure Compensated	A flow dividing valve which divides the flow at constant ratio regardless of the difference in the resistances of the branches.
Valve, Four Positions	A directional control valve having four positions to give four selections of flow conditions.
Valve, Hydraulic	A valve for controlling liquid.
Valve, Load Dividing	A pressure control valve used to proportion pressure between two pumps in series.
Valve, Pilot	A valve applied to operate another valve or control.
Valve, Pneumatic	A valve for controlling gas.
Valve, Prefill	A valve which permits full flow from a tank to a "working" cylinder during the advance portion of a cycle, permits the operating pressure to be applied during the working portion of the cycle, and permits free flow from the cylinder to the tank during the return portion of the cycle.
Valve, Pressure Control	A valve whose primary function is to control pressure.
Valve, Counterbalance	A pressure control valve which maintains back pressure to prevent a load from falling.
Valve, Pressure Reducing	A pressure control valve whose primary function is to limit outlet pressure.

Valve, Priority A valve which directs flow to one operating circuit at a fixed rate and directs excess flow to another operating circuit.

Valve, Relief A pressure control valve whose primary function is to limit system pressure.

Valve, Relief, Safety A relief valve whose primary function is to provide pressure limitation after malfunction.

Valve, Sequence A valve whose primary function is to direct flow in a pre-determined sequence.

Valve, Shutoff A valve which operates fully open or fully closed.

Valve, Shuttle A connective valve which selects one of two or more circuits because flow or pressure changes between the circuits.

Valve, Surge Damping A valve which reduces shock by limiting the rate of acceleration of fluid flow.

Valve, Three Positions A directional control valve having three positions to give three selections of flow conditions.

Valve, Time Delay A valve which the change of flow occurs only after a desired time interval has elapsed.

Valve, Two Positions A directional control valve having two positions to give two selections of flow conditions.

Valve, Unloading A pressure control valve whose primary function is to permit a pump or compressor to operate at minimum load.

Varnish Materials generated by the hydraulic fluid due to oxidation, Thermal instability, hydrolytic instability, or other reactions. The materials are insoluble in the hydraulic fluid are generally found as brownish deposits on the work surface.

Viscosity A measure of the internal friction or the resistance of a fluid to flow.

Viscosity Index A measure of the viscosity-temperature characteristics of a fluid as referred to that of two arbitrary reference fluids (ASTM Designation D2270-64).

Viscosity, Absolute The ratio of the shearing stress to the shear rate of a fluid. It is usually expressed in centipoise.

Viscosity, Kinematic The absolute viscosity divided by the density of the fluid. It is usually expressed in centistokes.

Viscosity, SAE Number The Society of Automotive Engineers arbitrary numbers for classifying fluids according to their viscosities. The numbers in no way indicate the viscosity index of fluids.

Viscosity, SUS Saybolt Universal Seconds (SUS), which is the time in seconds for 60 milliliters of oil to flow through a standard orifice at a given temperature (ASTM Designation D88-56).

Wash A filter element in which a larger unfiltered portion of the fluid flowing parallel to the filter element axis is used to continuously clean the influent surface which filters the lesser flow.

Water Glycol Fluid A fluid whose major constituents are water and one or more glycols or polyglycols.

Wound A filter medium comprised of layers of helical wraps of a continuous strand or filament in a predetermined pattern.

Woven A filter medium made from strands of fiber, thread, or wire interlaced into a cloth on a loom.

general data

the "how to requirements" in designing
a hydraulic circuit .... k (b-2)
noise measurement techniques ...... k (b-5)
noise measurement glossary of terms k (b-14
advantages of using cv factor .... k (b-16)
orifice pressure drop .... k (b-17)
viscosity vs. temperature .... k (b-19)
formulas, graphs, & tables .... k (b-19)

the "how to requirements" in designing a hydraulic circuit

1. What is required —

a. Is it a straight line or rotary drive, or combination of both?

b. At what speed is the actuator to operate?

C. Does the speed vary during the forward and reverse strokes?

d. Is the speed influenced by external forces?

e. What forces have to be exerted?

f. Do the forces vary during the forward and reverse strokes?

g. Are the forces positive or negative?

2. From this basic information the size of the actuators can be determined. This is achieved by —

a. Assuming a maximum working pressure.

b. Determining the size of the actuator.

c. Determining the pressure reduction in the circuit on certain parts of the circuit if more than one actuator is used.

d. The maximum and minimum flows in the circuit can be determined.

e. Individual actuators should be considered and selected from manufacturers' list.

3. The pump capacity can then be calculated.
a. The overall circuit time cycle should be studied to determine how long pressure is required.

b. In some cases it may be beneficial to consider an accumulator system by —

A. Calculating the total demand of the circuit in the overall time cycle, together with the maximum and minimum pressures that can be tolerated in the circuit.

B. $\frac{\text{The total demand (in gallons)} \times 60}{\text{The overall time cycle (in sec.)}}$

= the maximum pump capacity in GPM

C. From the circuit demand, maximum and minimum pressure, the size of the accumulator can be calculated.

D. This should be checked against manufacturers' lists and a suitable unit, larger if necessary, selected. (Not forgetting the use of auxiliary gas bottles).

F. If necessary the differential pressures can be adjusted to give the best results.

c. An accumulator - time - p.s.i. chart should be compiled and compared with the overall cycle time.

d. If a direct pump system is to be used then the maximum rate of flow or combined rate of flow is the pump capacity.

e. If constant pressure has to be ensured, then a pump having an automatic pressure device or a small secondary pump, will have to be considered and selected.

f. It may well be that the pump of the exact size or rating is not available. Therefore it may be that the actuator size or speed may have to be adjusted to suit.

g. The pump Hp. can be calculated or obtained from the manufacturers' catalog.

PH Manifolded Directional Control Valve
illustration k (b-1)

4. The actuator controls can be considered —

a. They can be one or any combination of the following:

Hand controlled
Solenoid controlled
Pilot controlled
Pressure controlled
Mechanically controlled

b. The sequence of operations must be laid out.

c. The required interlocks determined.

d. The starting and fail safe conditions made clear.

e. From this information the type of control valves can be selected:

Spring centered
Spring offset
Detent
Free spool

f. Where solenoid valves are used the operational sequence can be compiled.

g. Where pilot and pressure operated schemes are being designed, it is important to determine:

how to design a hydraulic circuit continued

a. if sufficient pressure is available,

b. that any operating signal to a valve has an opposing low pressure passage to enable the piston, spool or actuator to move.

h. Where necessary pressure reducing valves should be included.

1. If these are positioned between the directional valve and actuator then check valve must be included for free flow.

2. If they are fitted on the inlet side of a directional valve then the circuit should be considered to decide if reduced pressure is to be required on both movements of the actuator.
It is sometimes possible to arrange the circuit so that a reducing valve only operates when one position of the directional valve is selected.

i. Flow control valve should be included in the circuit. Careful consideration be given to —

a. position of valve - meter in, meter out or bleed off.

b. to ensure that a check valve is included to permit free flow when required if the flow control valve is positioned between the actuator and directional valve.

1. If a high flow condition is required in the same direction as the low flow then a pilot operated check valve or an additional directional valve is required in parallel with the flow control valve.

j. Check valves should be included in the circuit where —

a. flow is to be restricted to one direction,

b. where a minimum pressure is to be maintained while flow is taking place,

c. where anti-cavitation is required,

d. to prevent an actuator from moving if pressure in the circuit falls below that required to operator or support the actuator.

1. Pilot operated check valve should be included to hold an actuator positively in a given position while the circuit is in a certain condition and where the actuator is to be reversed when the certain condition has been cancelled and replaced by an alternative condition.

2. Care must be taken to ensure that the pilot operated check valve has a large enough differential or is of such a design to allow the valve to be opened with the pressure available.

3. It is important to consider if there is any intensification in the circuit acting on the pilot check valve.

4. If a pilot operated check valve is used to support a vertical load it is important to support the load with a counterbalance valve unless the actuator is connected up in such a manner as to not require such a precaution.

k. If sequence valves are to be fitted to the circuit care must be taken to ensure that sufficient pressure is available all times the sequence valve is required to operate or transmit signals. Any sequence valve is pressure sensitive and will reseat as soon as pressure is removed.

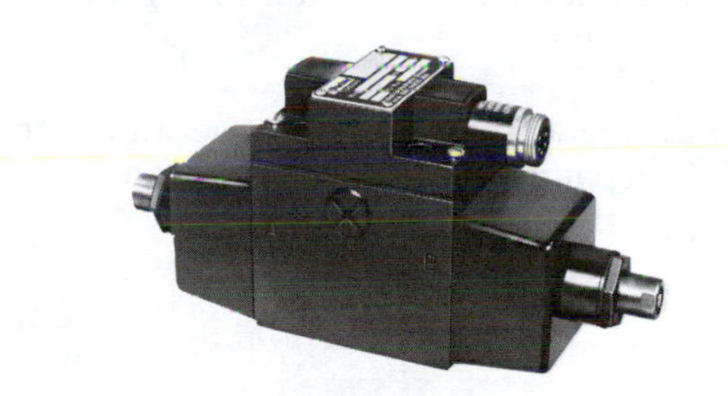

PH Directional Control Valve
illustration k (b-2)

5. The power unit arrangement can be determined.

a. The position of the power pack may influence the type of power unit.

1. It can be incorporated as part of the machine.

2. It can be incorporated in a control cabinet.

3. It can be free standing.

b. The pump motor unit can be considered for mounting on top of the oil reservoir.

c. The pump motor unit can be considered for mounting under the oil reservoir. Shut off valve should be incorporated in the suction line. Check valves, counterbalance valves, or additional shut off valves to be

how to design a bydraulic circuit continued

positioned on the return lines to prevent the system from draining if maintenance is required.

d. The pump unit can be placed inside the oil reservoir in which case the problems of shut off valves can be eliminated.

e. The size of the oil reservoir has to be determined. This may be dictated by —

a. the heat to be dissipated

b. the equipment to be incorporated

c. the floor space

f. The accessories to be incorporated on the power pack should be determined. These should include —

a. Suction filters

b. Filler cap and strainer

c. Air breather

d. Level gauge

e. Drain top

f. Access to the inside of the reservoir

g. Bulkhead connection for drain and return lines.

6. Additional equipment can be added to the circuit, dependent upon the requirements.

a. Pressure gauge

b. Shut off valves

c. Filters — high pressure or low pressure

d. Temperature controls

e. Oil level safety switches

f. Oil coolers

g. Thermostats

h. Heaters

i. Filling devices

j. Flexible hoses

k. Signal lights

l. Electrical controls

7. Pipe sizes and identification can be added to the circuit drawings.

a. Definition between pressure lines, suction and return lines to be made.

b. The type of pipe, hose, tubing and fitting should be decided.

Also material:

a. Steel, stainless steel, aluminum, brass Buna-N®, Vitron®, nylon, plastic

Plastic

Fittings

1. flared
2. bite type
3. compression
4. welded
5. flange
6. threaded

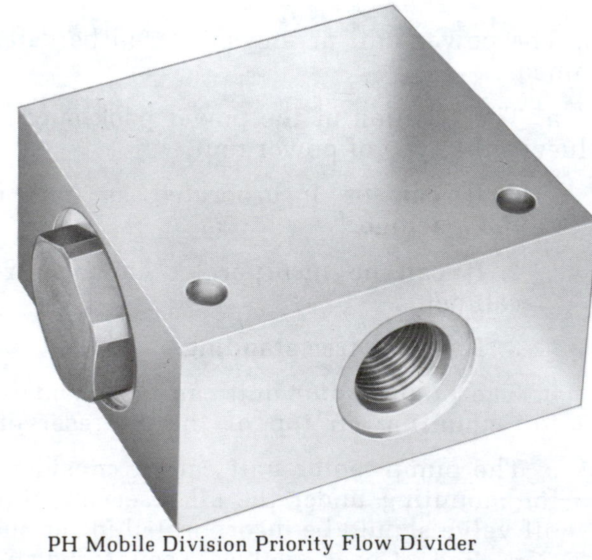

PH Mobile Division Priority Flow Divider

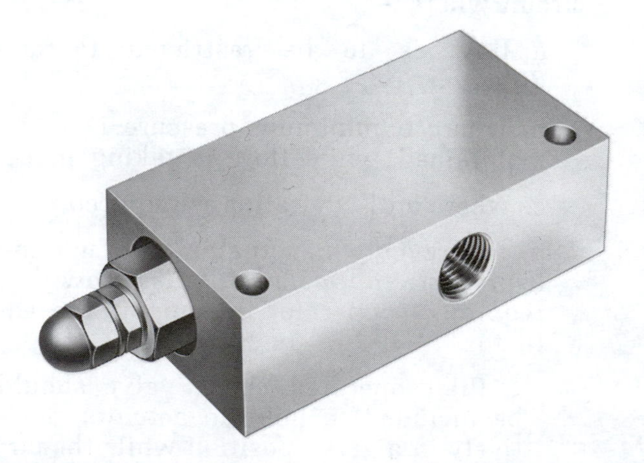

PH Mobile Division Differential Relief Valve

illustration k (b-3)

noise measurement techniques

Prepared by National Machine Tool Builders Association

Forward

This outline of Noise Measurement Techniques reflects the concern of the National Machine Tool Builders' Association (NMTBA) and its members for the emerging problem of noise control in the industrial environment.

All industries are involved and concerned with this problem. Recently it has been established that excessive noise exposures can in some cases result in a loss of hearing. These findings have resulted in the adaption of amendments to Title 41-Public Contracts and Property Management, Part 50-204, Safety and Health Standards for Federal Supply Contracts (Walsh-Healey Act). These amendments specifically establish noise level exposure limits for those employees working on Federal Supply Contracts. It is anticipated that additional Federal and State legislation may extend coverage of noise regulations to all employees.

Reducing the amount of noise generated by machine tools, therefore, is expected to become another important design parameter for the machine tool builder.

To achieve this objective, it is readily apparent that all of the pricipals involved must communicate, using the same terms and procedures in the definition of this most complex problem. The principals are the machine tool builder, the component manufacturer, the ultimate user, and their consultants.

To this end, the NMTBA established a committee in 1969 representing the principals and assigned it the task of preparing a document which would delineate suggested measuring techniques and procedures for the determination of noise emanating from machine tools.

This document is divided into three basic sections: (1) Measurement, (2) Instrumentation, (3) Reporting. Considerable explanatory material is included in these sections and a glossary of terms is presented in the appendix to assist in the use of this document.

Since certain aspects of the noise problem lack definition, this document should be used as a guide and modified to suit the particular test situation. It will be amended to incorporate improved techniques and procedures as they are developed.

Noise Measurement Techniques

Noise Measurement Techniques Committee (An Ad Hoc committee formed to work in co-operation with the NMTBA Research and Development Committee).

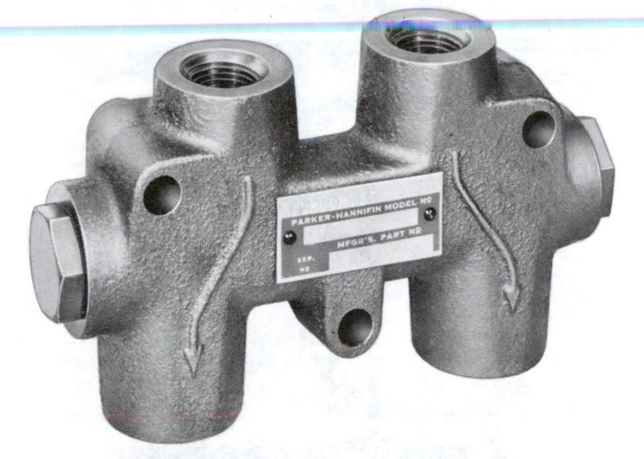

PH Mobile Division Cylinder Lock Valve
illustration k (b-4)

1. **Measurement** — These procedures apply to measurements made in facilities under the control of the machine tool builder. As such it is assumed that the builder will provide a suitable test space so that reasonably accurate noise level data may be obtained and possibly repeated at a later date. Therefore, ambient noise and reverberation correction factors are not included.

1.1 **General Provisions.**

1.1.1 *Ambient Noise Levels* — To obtain an accurate measure of the noise produced by a machine, the ambient noise level should meet the following conditions:

a. The ambient level of the frequency band being measured should preferably be at least 10 dB lower than the band level generated by the machine.

b. The ambient level must remain steady for the duration of the test, or if varying, should not exceed a level 10 dB below that of the machine under test.

1.1.2 *Test Space* — A plot plan shall be included on the data sheet, **illustration k (b-6).** It shall describe the test space, including all major reflecting surfaces such as walls, cabinets, control panels, etc., within 30 feet of the machine envelope.

A brief description of the major reflecting surface materials shall be provided.

1.1.3 *Machine Operating Conditions* — Measurements shall be obtained with the machine in the unloaded mode of operation at minimum and/or maximum speed conditions. In addition, data may be obtained in the loaded mode,

noise measurement techniques continued

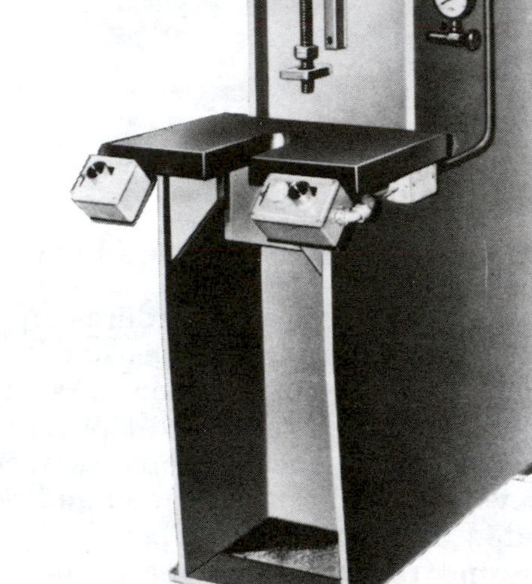

illustration k (b-5)

in which case the specific loads (actual or simulated) and conditions shall be defined.

1.1.4 *Measurement Locations* — The measurement locations shall be shown in **illustration k (b-6).** The machine envelope selected shall include all active surfaces and noise sources. No microphone position shall be within three (3) feet from a major machine or other reflecting surface. The microphone shall be oriented to have the incidence, either normal or grazing, that provides the flattest frequency response for the microphone being used. No microphone frequency response conditions factors shall be used.

NOTE: *A Microphone providing the flattest response to a noise source located normal to it, in a free field, is preferable because it reduces the probability of errors due to extraneous noise sources.*

1.1.5 *Calibration of Instruments* — Measurements shall be initiated each day by performing the instrument field calibrations recommended by their manufacturer. These shall be verified at the end of each day. If this check indicated a deviation of 1 dB from the established calibration, the data acquired that day shall be invalid.

1.1.6 *Tape Recording* — When noises are tape recorded for record or measurement purposes, a comparison between the original and recorded noise shall be made to certify that clipping does not occur and that recording noises do not distort the desired signal.

1.2 STEADY STATE NOISE — Noises that are continuous or that consist of impulses spaced less than one second apart shall be considered to be steady state noises.

The following measurements should be obtained at each location:

a. Sound pressure level, dB(A), with meter on slow response.

b. Overall sound pressure level, dB(C), with meter on slow response.

c. Octave band spectral analysis. The octave bands shall be those specified in ANS S1.11 whose center frequencies range from 63 to 8000 Hz. Where 1/3 octave filters are used, the octave band levels shall be determined by adding the appropriate 1/3 octave band levels on an energy basis.

d. The average level shall be used when the noise level fluctuates randomly over a range of 5 dB or less. Where such fluctuations exceed 5 dB the maximum and minimum levels shall be reported.

1.3 **Cyclical Noise —** Cyclical noise levels are those that change repetitiously in excess of 5 dB during machine duty cycle. This is in recognition that within machine operational events noise levels do vary. The various levels and their duration shall be recorded along with a listing of their causative events.

The following measurements should be obtained at each location:

a. Sound pressure level, dB(A), with meter on slow response.

b. Overall sound pressure level, dB(C), with meter on slow response.

NOTE: *For engineering evaluations, it may be desirable to use fast meter response when noise levels with durations less than three seconds are encountered.*

c. Octave band spectral analysis. The octave bands shall be those specified in NAS S1.11 whose center frequencies range from 63 to 8000 Hz. Where 1/3 octave filters are used, the octave band levels shall be determined by adding the appropriate 1/3 octave band levels on an energy basis.

noise measurement techniques continued

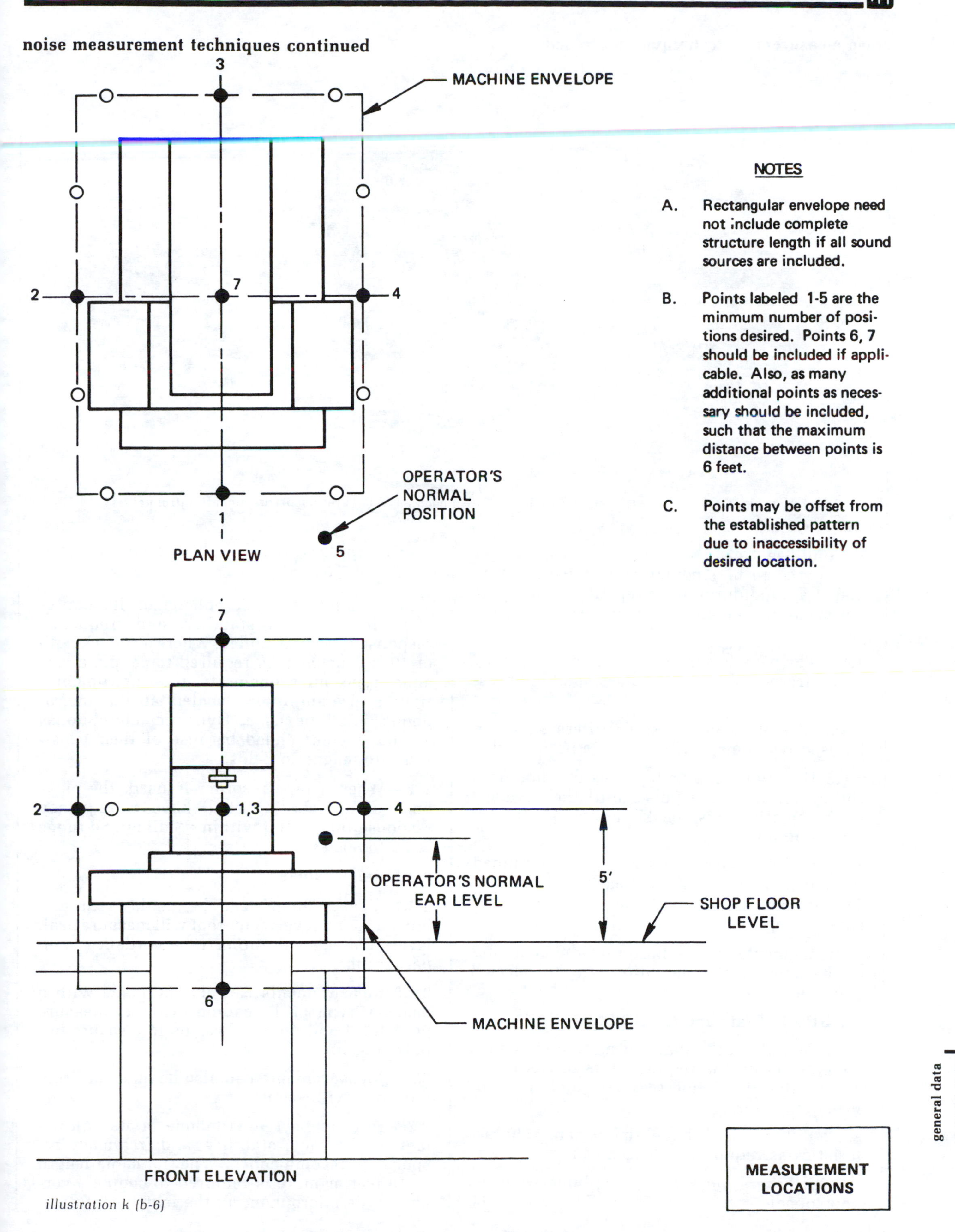

illustration k (b-6)

MEASUREMENT LOCATIONS

noise measurement techniques continued

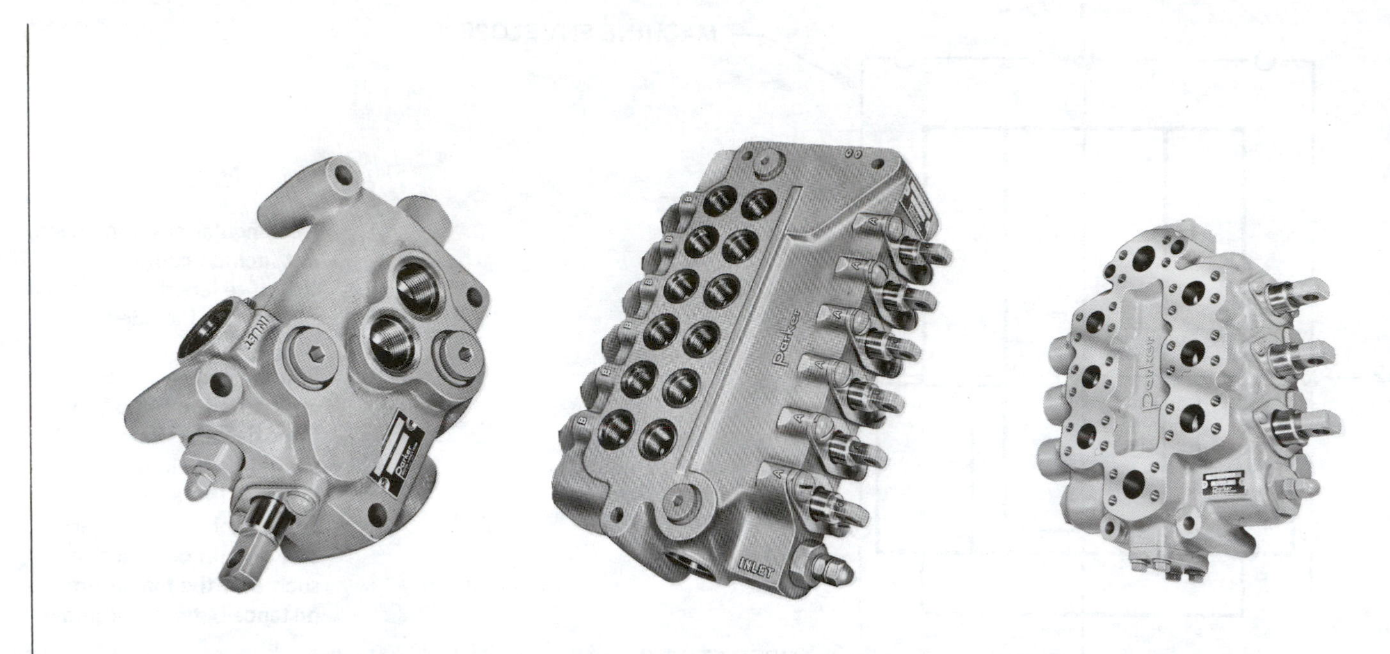

PH Mobile Division V.D.P. Single Spool, V.D.P. Six Spool, V.D.S.P. Three Spool Directional Control Valves

illustration k (b-7)

1.4 **Impulse Noise —** Impulse noises shall be considered to be singular noise pulses of less than 1 second duration or repetitive noise pulses occurring at greater than 1 second intervals.

1.4.1 The peak levels, durations and frequency of occurrence of noise impulses shall be measured.

1.4.2 Weighting networks or filters shall not be used when measuring impulse peak levels.

1.4.3 Duration shall be taken as the time from the initiation of the pulse until the envelope of the positive noise peaks decays 8.7dB from its maximum.

1.4.4 Impulse parameters shall be determined from measurements of at least 10 impulses. The average values and the range of values shall be reported.

2. **Instrumentation —** Based on the character of the noise, instrumentation will be selected from the following:

2.1 **Steady State and Cyclical Noise**

2.1.1 Sound levels shall be measured with a meter meeting the requirements of ANS S1.4 except that it is not required to have a B-weighted filter.

2.1.2 Filters shall have the general characteristics as required in ANS S1.11.

NOTE: *Filters marked as complying with the now obsolete ASA Standard Z24.10 may be sued by applying the procedures given in the appendix of ANS S1.11.*

2.1.3 A condenser microphone or its equivalent in accuracy, stability, and frequency response is recommended. Where a cable length of 10 feet or more is required to couple a ceramic type microphone to the instrumentation, a pre-amplifier, located at the microphone, shall be used. Dynamic microphones are not recommended because of their sensitivity to magnetic fields.

2.1.4 When a tape recorder is used, the playback of a recording shall have a frequency response that is flat within + 3 dB over a range of 50 to 8000 Hz.

2.2 **Impulse Noise**

2.2.1 Measurements can be made with any impact sound level meter that will measure peak levels and also duration in accordance with paragraph 1.4.3.

2.2.2 Measurements can also be made with a memory type peak reading meter to measure peak level and an oscilloscope to measure impulse durations.

2.2.3 Measurements can also be made entirely with an oscilloscope.

2.2.4 A condenser microphone or its equivalent in accuracy, stability, and frequency response is recommended. Where a cable length of 10 feet or more is required to couple a ceramic type microphone to the instrumentation,

noise measurement techniques continued

a pre-amplifier, located at the microphone, shall be used. Dynamic microphones are not recommended because of their sensitivity to magnetic fields.

2.2.5 When a tape recorder is used, the playback of a recording shall have a frequency response that is flat with + 3 dB over a range of 50 to 8000 Hz.

3. **Reporting**

3.1 DATA FORM — The data form shown in this document, **illustration k (b-10-1)** is only intended as a guide to insure adequate data is recorded. Because of the wide range of differences in machine tools, a standardized data form is not practical for the industry, therefore one should be tailored to meet individual requirements. The following discussion will assist in this process.

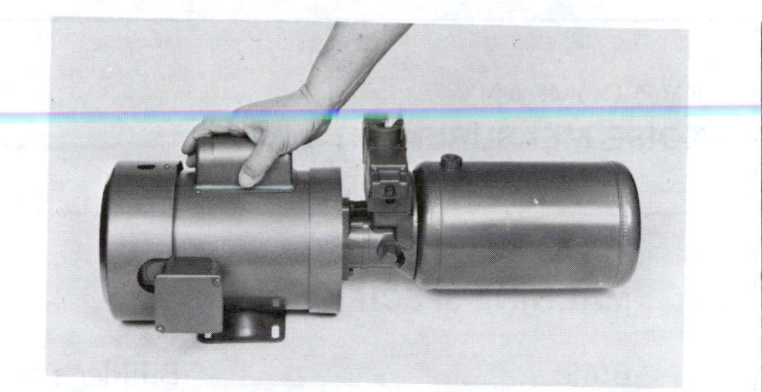

PH Mobile Division — Power Unit

illustration k (b-8)

It will be noted that the sample data form is designed to record all required data in a four page booklet to avoid possible detachment or loss of portions of the data. It is recommended that this principle be followed whenever possible.

Example

MACHINE SPECIFICATIONS

BUILDER: ABC Machine Co. BUILDER'S NO. 60843 BUYER'S P.O. 1234

TYPE: 6 Station transfer machine for finishing gizmo heads.

MODEL: 6YS SERIAL NO. 0001 OUTPUT: 400/hr.

Example

MACHINE SPECIFICATIONS

BUILDER: DEF Lathe Builders BUILDER'S NO. 12345 BUYER'S P.O. 5678

TYPE: Lathe

MODEL: 14-4800 SERIAL NO. 1002 SWING: 14"

CENTER DIST. 48" HORSE POWER: 30 SPEED CONTROL: Hydrostatic Drive

illustration k (b-9)

3.2 **Explanation of Sample Data Form Layout**

3.2.1 *Heading* — Identify the firm making the noise evaluation; either builder, consultant, university or commercial test laboratory.

3.2.2 *Machine Specifications, Item A* — Use items that adequately describe the machine tool under test so that it shall not be construed to be any other machine and also provides a brief description of the machine, if necessary.

a. In case of special machines, the type description may be adequate so that operating parameters are not necessary.

b. For general purpose machines, the operating parameters that differentiate the machine from others made by the same manufacturer shall be used.

3.2.3 *Instrumentation, Item B* — A comprehensive list of instruments is given on the sample data form. Those not used by the testing activity should be omitted.

3.2.4 *Certification, Item D* — Certification should be made with the signature of the person responsible to the builder for performance of the test.

noise measurement techniques continued

XYZ COMPANY
NOISE MEASUREMENT DATA

A. MACHINE SPECIFICATIONS

Builder ____________ Builder's No. ____________ Buyer's P.O. No. ____________
Equipment Specification: Type ____________ Model ____________
Serial No. ____________ Size ____________ Capacity ____________
Speed ____________ Horsepower ____________ Auxiliaries ____________

B. INSTRUMENTATION

INSTRUMENT	MODEL	SERIAL NO.	CALIBRATION DATE
Sound Level Meter			
Microphone			
Calibrator			
Tape Recorder			
Analyzer			
Impact Meter			
Oscilloscope			

C. CONCLUSIONS

D. CERTIFICATION

Certified by: ____________
Name: ____________
Position: ____________
Company: ____________

illustration k (b-10-1)

noise measurement techniques continued

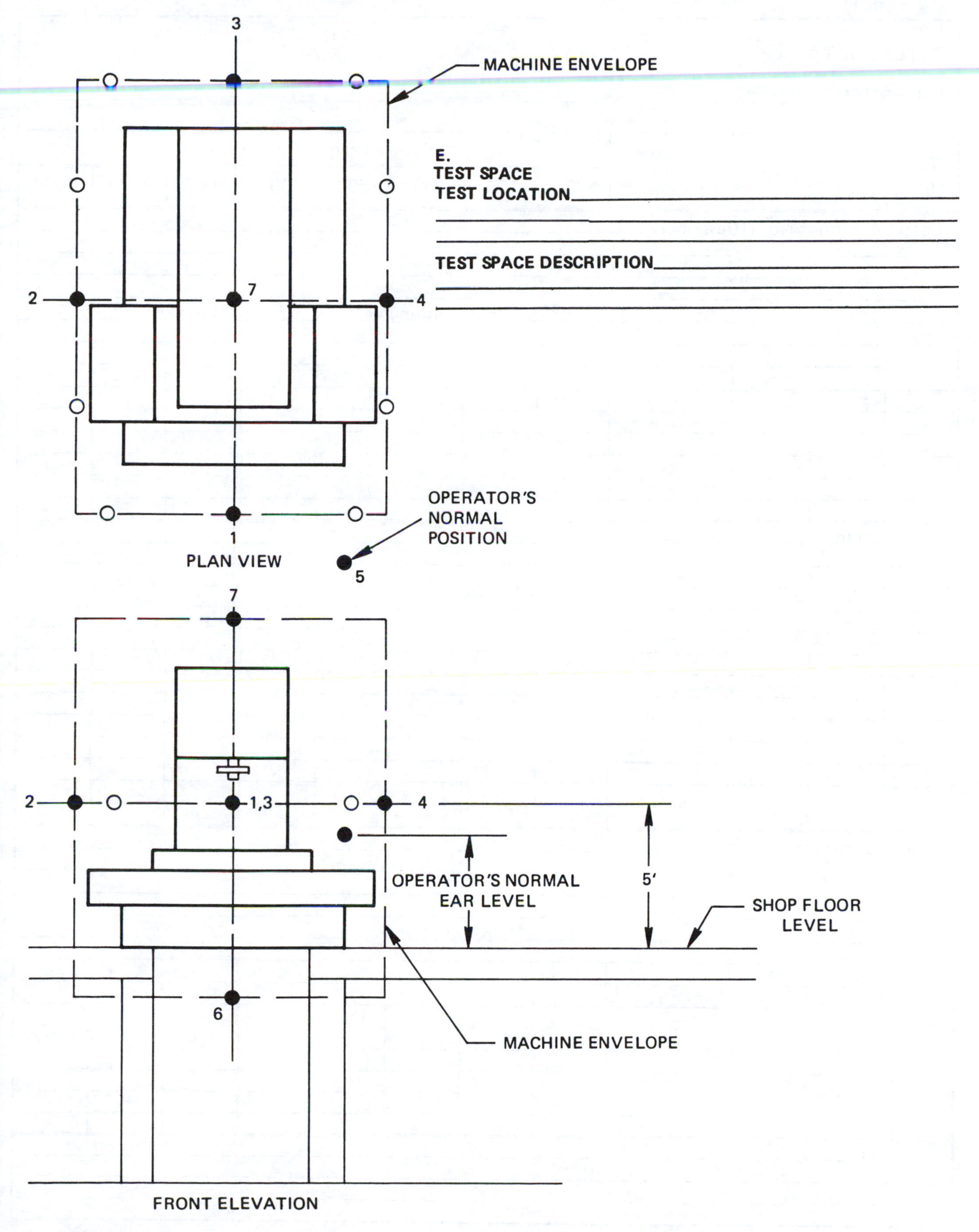

illustration k (b-10-2)

noise measurement techniques continued

F. TEST DATA

OBSERVER: ______________________ DATE: ______________

Test Point	dB(A)	dB(C)	Peak dB of Impulse	Center Frequency — Hertz							
				63	125	250	500	1000	2000	4000	8000

Load Conditions No. 1(Describe) ______________________

______________ Min. Duration Per ______________ Min. Cycle ______________

Load Conditions No. 2(Describe) ______________________

______________ Min. Duration Per ______________ Min. Cycle ______________

Load Conditions No. 3(Describe) ______________________

______________ Min. Duration Per ______________ Min. Cycle ______________

Remarks ______________________

illustration k (b-10-3)

noise measurement techniques continued

OBSERVER: ______________________ DATE: ______________

Test Point	dB(A)	dB(C)	Peak dB of Impulse	Center Frequency — Hertz							
				63	125	250	500	1000	2000	4000	8000
Load Conditions No. 4(Describe)											
Min. Duration Per ______ Min. Cycle ______											
Load Conditions No. 5(Describe)											
Min. Duration Per ______ Min. Cycle ______											
Load Conditions No. 6(Describe)											
Min. Duration Per ______ Min. Cycle ______											
Remarks											

illustration k (b-10-4)

noise measurement techniques continued

3.2.5 *Test Space, Item E* — Indicate with an area sketch, **illustration k (b-10-2),** the position of the machine as placed in the test area and orient by some identifying features. Indicate microphone locations including elevations. State method of mounting or supporting the machine.

3.2.6 *Test Data, Item F*

a. Provide a sufficient number of lines to accommodate the anticipated maximum number of microphone locations for each load condition.

b. Each period where the noise level changes 5 dB or more at any test point shall be considered to be a change in load condition. The description of each load condition must indicate overall machine operating conditions or performance as well as the specific operation or duty cycle event.

c. The duration serves as a basis for computing employee exposure; therefore, it can be recorded in any customary time units.

GLOSSARY OF TERMS

Ambient Noise Level (background noise). The noise level in the area surrounding the machine or component to be tested with machine being tested not operating.

Clipping (recording). Where the reproduced signal amplitude varies from the input signal effective amplitude by more than one dB, due to electronic compression or magnetic saturation.

Cyclical Noise. Cyclical Noise levels are those that change repetitiously in excess of 5 dB during machine duty cycle.

Decibel (dB). A non-dimensional number used to express sound pressure and sound power. It is logarithmic expression of the ratio of a measured quantity to a reference quantity.

For example, two machines side by side, each separately producing a sound pressure level of 80 dB, do not produce a combined sound pressure level of 160 dB when both are operating. They produce a combined sound pressure level of approximately 83 dB.

For sound **pressure** the reference quantity is 20 micronewtons/meter squared* so that: Sound Pressure in dB = 20 $\log_{10}$ (measured pressure $/20\text{x}10^{-6}$).

For sound **power** the reference quantity is 10^{-12} watts so that: Sound Power in dB = 10 $\log_{10}$ (measured power/10^{-12}).

The reference quantities and the coefficients in the basic equations were chosen to make the acoustic power per meter squared **numerically** equal to the sound pressure at that point.

Where a weighted network filter is employed in making sound pressure measurements this is indicated by a suffix added to the unit symbol, i.e., when a A-weighted network is used the symbol dB (A) is used.

*Confusion exists because the sound pressure reference is stated is several different forms in acoustic literature. The forms that follow are all equivalent.

20 micronewtons/meter squared
.0002 dynes/cm squared
.0002 microbar (micro atmospheres)

dB (A). A sound level reading in decibels made on the A-weighted network of a sound level meter.

dB (C). A sound level reading in decibels made on the C-weighted network of a sound level meter.

Fast Response. A selectable mode of operation of a sound level meter or analyzer in which the indicator has minimum damping and can therefore rapidly respond to changes in level.

Flat Response. Is the characterization of microphone, instrument, or recorder having a sensitivity or response that is constant regardless of frequency.

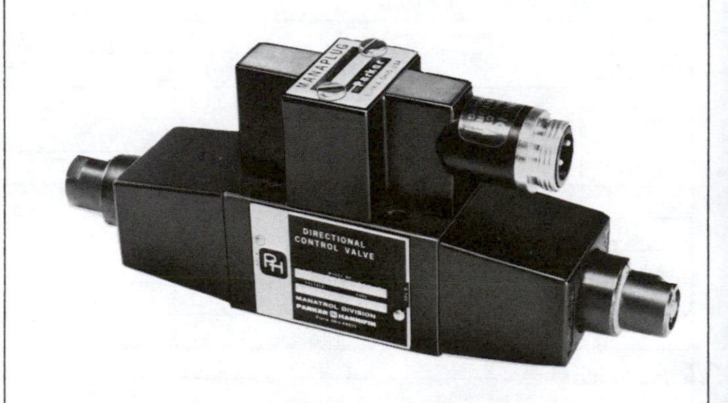

PH Electro-Hydraulic Valve

illustration k (b-11)

noise measurement techniques continued

Frequency Bands. A division of the audible range of frequencies into sub-groups for detailed analysis of sound.

Grazing Incidence. Microphone positioned so that its axis is perpendicular to a line from the microphone to the noise source.

Hertz (Hz). Synonymous term for "cycles per second." Most standardizing agencies have adopted "hertz" as the preferred unit of frequency.

Impact Noise. Same as impulse noise.

Impulse Noise. Impulse noises shall be considered to be singular noise pulses of less than 1 second duration or repetive noise pulses occurring at greater than 1 second intervals.

Machine Envelope. The space occupied by a machine tool. The boundary of the machine envelope is a reasonably uniform line around the periphery of the machine approximately 3 feet out from machine components. The machine envelope extends up and over the top and down and under the bottom of machines. For machines mounted on long runways, the machine envelope shall contain only "active" portions of the machine, excluding vacant portions of the runway.

Machine Tool. A power-driven machine, not portable by hand, used to shape or form metal by cutting, impact, pressure, electrical techniques, or a combination of these processes.

Machine Tool Builder. Any individual, partnership, corporation or other form of enterprise which is engaged in the development, manufacture, or assembly of a machine tool.

Microphone Directivity. The variation in response of a microphone dependent on the direction of arrival of the sound wave.

Normal Incidence. Microphone positioned so it is pointing toward sound source.

Octave Band. A division of the audible range of frequencies into sub-groups such that in each division the upper limit is twice the lower limit. Because of this ratio the center frequencies used to designate the octaves will each be twice the center frequency of the preceding octave band.

One-Third-Octave Bands. A split of an octave band into three equal parts for more detailed analysis of distribution and sound energy.

Operator's Normal Position. The location of the operator's head with respect to the machine tool during the time the machine is operating. The position should be recorded for the particular machine installation. More than one position may be necessary.

Reflecting Surface. A surface oriented to the sound source so that sound waves are modified by or reflected from the surface.

Reverberation. Reverberation is the sound that persists in an enclosed space, as a result of repeated reflection or scattering, after the source of sound has stopped.

Slow Response. A selectable mode of operation of a sound level meter or analyzer in which the indicator has high damping and therefore slowly responds to change in level. This mode tends to provide an average reading.

Steady State Noise. Noises that are continuous or that consist of impulses spaced less than one second apart shall be considered to be steady state noises.

Test Space. An area in the machine tool builder's facility occupied by the machine tool when tests are made.

advantages of using c_v factor

1. The Cv factor reduces the work involved in making a flow calculation.

2. The Cv factor eliminates the difficult problem of converting physical units which many times arises when the flow equation is used.

3. The Cv factor provides a very convenient method for sizing and comparing valve flow capacities.

4. The Cv factor is very convenient for analyzing experimental flow-pressure drop data on a valve.

5. The Cv factor provides a method of presenting the maximum amount of flow-pressure drop data on a valve.

Formal Definition of the Cv Factor:
The Cv factor is the number of U.S. gallons of water that pass through a given orifice area in one minute at a pressure drop of one psi.

Example Calculation Using the Flow Equation:

$$Q = C_d A_o \sqrt{\frac{2\Delta P}{\rho}}$$

Where:
Q = Flow (in³/sec)
C_d = Orifice Discharge Coefficient (0.611 for a sharp edged orifice)
A_o = Area of Orifice (in²)
ΔP = Pressure Drop Across the Orifice (lb./in²)
ρ = Density of the Fluid Passing Through the Orifice (lb-sec²/in⁴)

Given:

$$C_d = 0.611$$

$$A_o = (\pi/4)\, D_o^2 = (\pi/4)\,(0.25)^2 = 0.0491 \text{ in}^2$$

$$P = 50 \text{ lb./in}^2$$

ρ =

$$\frac{62.4 \text{ lb.}}{1 \text{ ft}^3} \times \frac{1 \text{ ft}^3}{1728 \text{ in}^3} \times \frac{1 \text{ sec}^2}{386 \text{ in}} = \underset{\text{(water)}}{0.000094} \frac{\text{lb-sec}^2}{\text{in}^4}$$

$$Q = (0.611)\,(0.0491) \sqrt{\frac{2(50)}{0.000094}} = 30.6 \text{ in}^3/\text{sec}$$

To Convert to GPM:

$$Q = \frac{30.6 \text{ in}^3}{1 \text{ sec}} \times \frac{60 \text{ sec}}{1 \text{ min}} \times \frac{1 \text{ gal}}{231 \text{ in}^3} = 7.95 \text{ GPM}$$

Mathematical Definition of the Cv Factor:

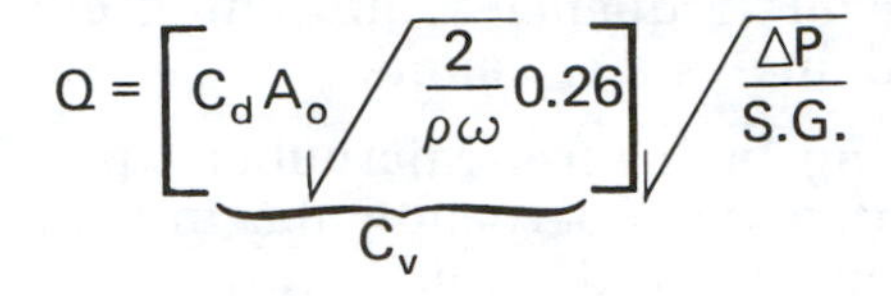

$$Q = \underbrace{\left[C_d A_o \sqrt{\frac{2}{\rho\omega} 0.26} \right]}_{C_v} \sqrt{\frac{\Delta P}{S.G.}}$$

Where:
$\rho\omega$ = Density of Water
S.G. = Specific Gravity of Fluid

Now making the same calculation as above if it is known that Cv = 1.13:

$$Q = C_v \sqrt{\frac{\Delta P}{S.G.}}$$

$$Q = (1.13) \sqrt{\frac{50}{1.0}} = 7.95 \text{ GPM}$$

Now suppose that the fluid is changed to hydraulic oil with a S.G. of 0.85.

$$Q = (1.13) \sqrt{\frac{50}{0.85}} = 8.65 \text{ GPM}$$

The Cv factor shown in the Parker Hannifin catalogs can be used in metric calculations if it is converted properly. We have designated Cv as the metric equivalent.

Metric Flow Factor:

$$Q = C_E \sqrt{\frac{P}{S.G.}}$$

Where:
Q = Liters/Minute
P = Kilograms/Centimeter Squared
S.G. = Specific Gravity

$$C_E = 14.27\, C_v$$

Definition:
The Cv factor is the number of liters of water that pass through a given orifice area in one minute at a pressure drop of one kilogram per square centimeter.

Relationship to U.S. Cv Factor:
The Cv factor is 14.27 times the U.S. Cv factor.

orifice pressure drop

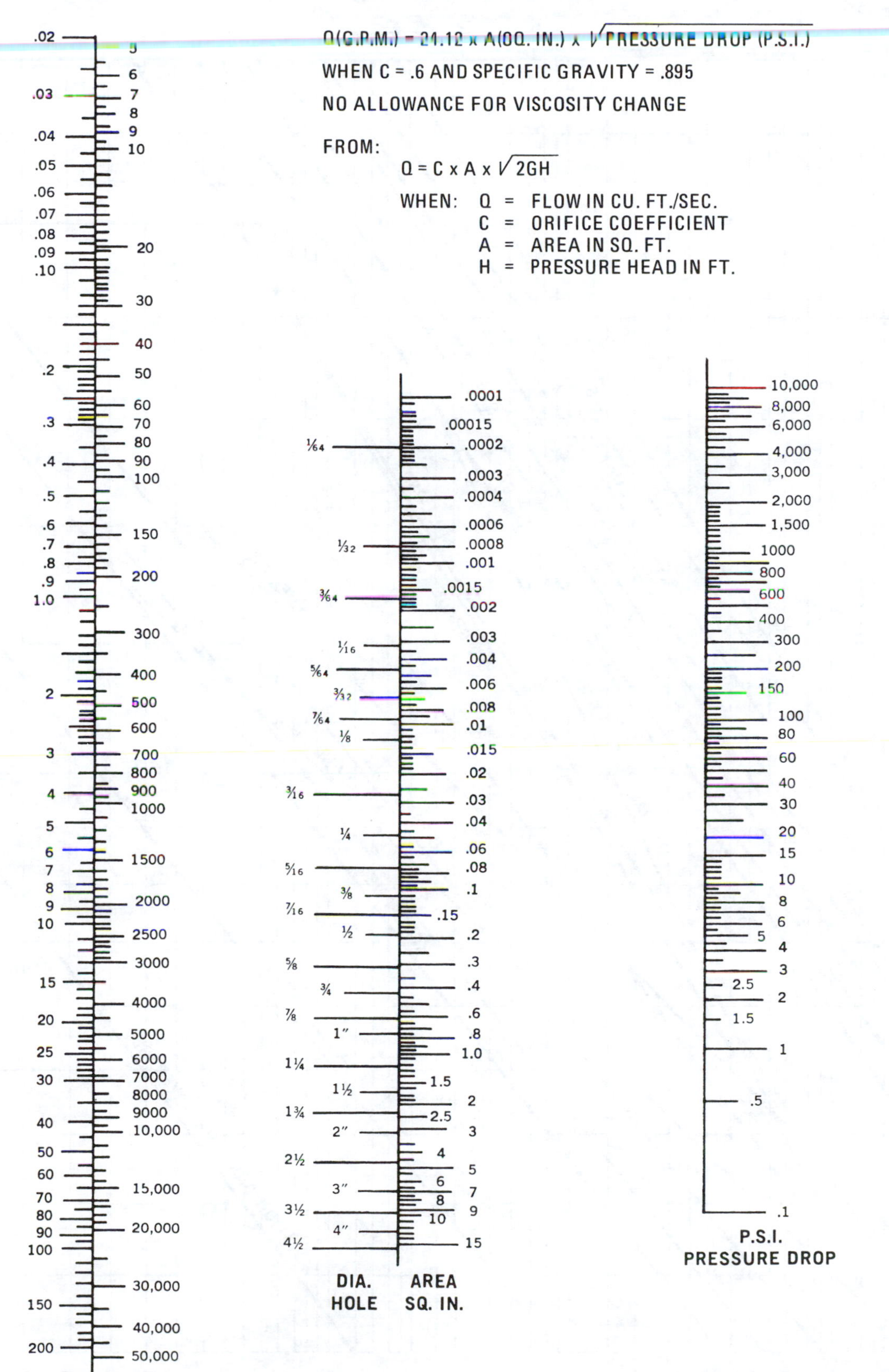

graph k (b-1)

viscosity vs. temperature

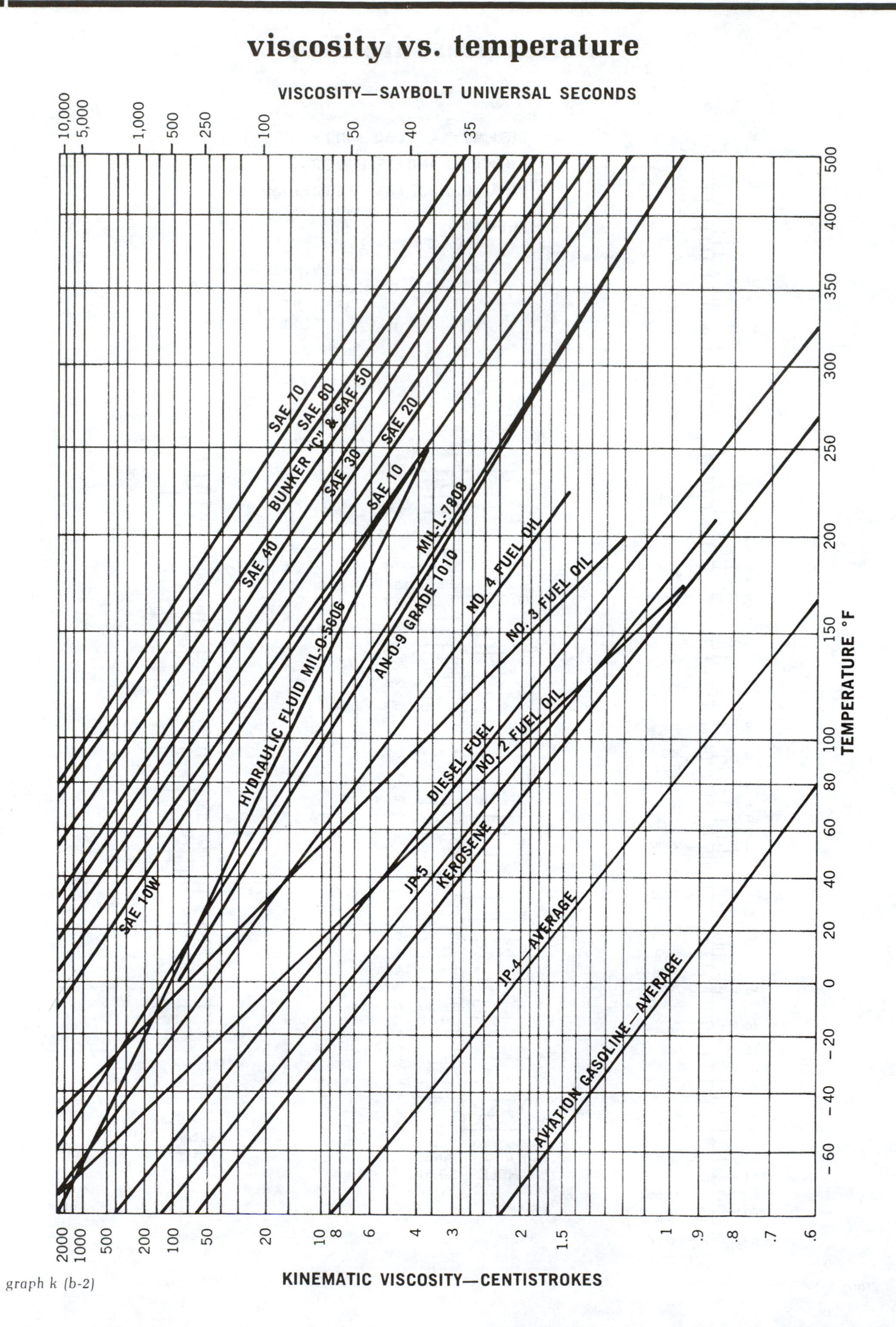

graph k (b-2)

formulas, graphs & tables

Metric Conversion Table

ORIGINAL VALUE	DESIRED VALUE							
	Mega	Kilo	Units	Deci	Centi	Milli	Micro	Micromicro
Mega	← 3	3 →	6 →	7 →	8 →	9 →	12 →	18 →
Kilo	← 3		3 →	4 →	5 →	6 →	9 →	15 →
Units	← 6	← 3		1 →	2 →	3 →	6 →	12 →
Deci	← 7	← 4	← 1		1 →	2 →	5 →	11 →
Centi	← 8	← 5	← 2	← 1		1 →	4 →	10 →
Milli	← 9	← 6	← 3	← 2	← 1		3 →	9 →
Micro	←12	← 9	← 6	← 5	← 4	← 3		6 →
Micromicro	←18	←15	←12	←11	←10	← 9	← 6	

table k (b-1)

The above metric conversion table provides a fast and automatic means of conversion from one metric notation to another. The notation "Unit" represents the basic units of measurement, such as amperes, volts, ohms, watts, cycles, meters, grams, etc. To use **table k (b-1),** first locate the original or given value in the left-hand column. Now follow this line horizontally to the vertical column headed by the prefix of the desired value. The figure and arrow at this point indicates number of places and direction decimal point is to be moved.

Example: *Convert 0.15 ampere to milliamperes. Starting at the "Units" box in the left-hand column (since ampere is a basic unit of measurement), move horizontally to the column headed by the prefix "Milli", and read 3 ⟶. Thus 0.15 ampere is the equivalent of 150 milliamperes.*

MILLIMETERS TO FRACTIONS TO DECIMALS

MM.	INCHES		MM.	INCHES	
	FRAC.	DEC.		FRAC.	DEC.
.3969	1/64	.0156	13.0969	33/64	.5156
.7938	1/32	.0312	13.4938	17/32	.5312
1.1906	3/64	.0468	13.8906	35/64	.5468
1.5875	1/16	.0625	14.2875	9/16	.5625
1.9844	5/64	.0781	14.6844	37/64	.5781
2.3812	3/32	.0937	15.0812	19/32	.5937
2.7781	7/64	.1093	15.4781	39/64	.6093
3.1750	1/8	.1250	15.8750	5/8	.6250
3.5719	9/64	.1406	16.2719	41/64	.6406
3.9688	5/32	.1562	16.6688	21/32	.6562
4.3656	11/64	.1718	17.0656	43/64	.6718
4.7625	3/16	.1875	17.4625	11/16	.6875
5.1594	13/64	.2031	17.8594	45/64	.7031
5.5562	7/32	.2187	18.2562	23/32	.7187
5.9531	15/64	.2343	18.6531	47/64	.7343
6.3500	1/4	.2500	19.0500	3/4	.7500
6.7469	17/64	.2656	19.4469	49/64	.7656
7.1438	9/32	.2812	19.8438	25/32	.7812
7.5406	19/64	.2968	20.2406	51/64	.7968
7.9375	5/16	.3125	20.6375	13/16	.8125
8.3344	21/64	.3281	21.0344	53/64	.8281
8.7312	11/32	.3437	21.4312	27/32	.8437
9.1281	23/64	.3593	21.8281	55/64	.8593
9.5250	3/8	.3750	22.2250	7/8	.8750
9.9219	25/64	.3906	22.6219	57/64	.8906
10.3188	13/32	.4062	23.0188	29/32	.9062
10.7156	27/64	.4218	23.4156	59/64	.9218
11.1125	7/16	.4375	23.8125	15/16	.9375
11.5094	29/64	.4531	24.2094	61/64	.9531
11.9062	15/32	.4687	24.6062	31/32	.9687
12.3031	31/64	.4843	25.0031	63/64	.9843
12.7000	1/2	.5000	25.4000	1	1.0000

table k (b-2)

Metric Relationships

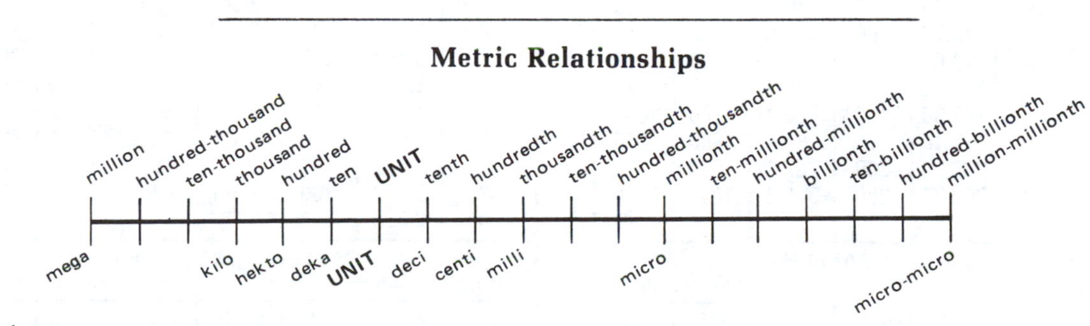

graph k (b-3)

Graph k (b-3) above, shows the relation between the American and the metric systems of notation.

Example: *Convert 5.0 milliwatts to watts. Place the finger on milli and count the number of steps from there to units (since the term watt is a basic unit). The number of steps so counted is three, and the direction was to the left. Therefore, 5.0 milliwatts is the equivalent of .005 watts.*

Example: *Convert 0.00035 microfarads to micro-microfarads. Here the number of steps counted will be size to the right. Therefore 0.00035 microfarads is the equivalent of 350 micro-microfarads.*

formulas, graphs & tables continued

NUMERICAL CONVERSION TABLE

INCHES INTO MILLIMETERS

1 inch = 25.40005 mm.

in.	mm.	in.	mm.	in.	mm.	in.	mm.
1/64	0.397	1-1/2	38.10	15	381.0	36	914.4
1/32	0.794	1-3/4	44.45	16	406.4	38	965.2
3/64	1.191	2	50.80	17	431.8	40	1016.0
1/16	1.588	2-1/2	63.50	18	457.2	42	1067.0
3/32	2.381	3	76.20	19	482.6	44	1118.0
1/8	3.175	3-1/2	88.90	20	508.0	46	1168.0
5/32	3.969	4	101.60	21	533.4	48	1219.0
3/16	4.763	4-1/2	114.30	22	558.8	50	1270.0
1/4	6.350	5	127.00	23	584.2	55	1397.0
5/16	7.938	6	152.40	24	609.6	60	1524.0
3/8	9.525	7	177.80	25	635.0	65	1651.0
7/16	11.110	8	203.20	26	660.4	70	1778.0
1/2	12.700	9	228.60	27	685.8	75	1905.0
5/8	15.880	10	254.00	28	711.2	80	2032.0
3/4	19.050	11	279.40	29	736.6	85	2159.0
7/8	22.230	12	304.80	30	762.0	90	2286.0
1	25.400	13	330.20	32	812.8	95	2413.0
1-1/4	31.750	14	355.60	34	863.6	100	2540.0

MILLIMETERS INTO INCHES

1 mm. = 0.03937000 in.

mm.	in.	mm.	in.	mm.	in.	mm.	in.
1	0.039	28	1.102	130	5.118	750	29.53
2	0.079	30	1.181	140	5.512	800	31.50
3	0.118	35	1.378	150	5.906	850	33.46
4	0.158	40	1.575	160	6.299	900	35.43
5	0.197	45	1.772	170	6.693	950	37.40
6	0.236	50	1.969	180	7.087	1000	39.37
7	0.276	55	2.165	190	7.480	1250	49.21
8	0.315	60	2.362	200	7.874	1500	59.05
9	0.354	65	2.559	250	9.842	1750	68.90
10	0.394	70	2.756	300	11.810	2000	78.74
12	0.472	75	2.953	350	13.780	2500	98.43
14	0.551	80	3.150	400	15.750	3000	118.10
16	0.630	85	3.346	450	17.720	3500	137.80
18	0.709	90	3.543	500	19.690	4000	157.50
20	0.787	95	3.740	550	21.650	4500	177.20
22	0.866	100	3.937	600	23.620	5000	196.90
24	0.945	110	4.331	650	25.590	7500	295.30
26	1.024	120	4.724	700	27.560	10000	393.70

SQUARE IN. INTO SQUARE CM.

1 sq. in. = 6.45163 sq. cm.

sq. in.	sq. cm.	sq. in.	sq. cm.	sq. in.	sq. cm.	sq. in.	sq. cm.
1/4	1.613	7	45.16	45	290.3	200	1290
1/2	3.226	8	51.61	50	322.6	250	1613
3/4	4.839	9	58.06	55	354.8	300	1935
1	6.452	10	64.52	60	387.1	350	2258
1-1/2	9.677	12	77.42	65	419.4	400	2581
2	12.900	14	90.32	70	451.6	450	2903
2-1/2	16.130	16	103.20	75	483.9	500	3226
3	19.350	18	116.10	80	516.1	600	3871
3-1/2	22.580	20	129.00	85	548.4	700	4516
4	25.810	25	161.30	90	580.6	800	5161
4-1/2	29.030	30	193.50	95	612.9	900	5806
5	32.260	35	225.80	100	645.2	1000	6452
6	38.710	40	258.10	150	967.7		

SQUARE CM. INTO SQUARE IN.

1 sq. cm. = 0.155000 sq. in.

sq. cm.	sq. in.	sq. cm.	sq. in.	sq. cm.	sq. in.	sq. cm.	sq. in.
1	0.155	18	2.790	80	12.40	600	93.0
2	0.310	20	3.100	85	13.18	700	108.5
3	0.465	25	3.875	90	13.95	800	124.0
4	0.620	30	4.650	95	14.73	900	139.5
5	0.775	35	5.425	100	15.50	1000	155.0
6	0.930	40	6.200	150	23.25	1500	232.5
7	1.085	45	6.975	200	31.00	2000	310.0
8	1.240	50	7.750	250	38.75	2500	387.5
9	1.395	55	8.525	300	46.50	3000	465.0
10	1.550	60	9.300	350	54.25	3500	542.5
12	1.860	65	10.080	400	62.00	4000	620.0
14	2.170	70	10.850	450	69.75	5000	775.0
16	2.480	75	11.630	500	77.50		

GALLONS* INTO LITERS

1 liter = 0.2641775 gal.* *U.S. liquid gallon (231 in.^{3}).

gal.	liters	gal.	liters	gal.	liters	gal.	liters
1	3.785	10	37.85	90	340.7	300	1136
2	7.571	20	75.71	100	378.5	400	1514
3	11.360	30	113.60	120	454.2	500	1893
4	15.140	40	151.40	140	529.9	600	2271
5	18.930	50	189.30	160	605.7	700	2650
6	22.710	60	227.10	180	681.4	800	3028
7	26.500	70	265.00	200	757.1	900	3407
8	30.280	80	302.80	250	946.3	1000	3785
9	34.070						

LITERS INTO GALLONS*

1 gal.* = 3.785334 liter *U.S. liquid gallon (231 in.^{3}).

liters	gal.	liters	gal.	liters	gal.	liters	gal.
1	0.264	10	2.642	250	66.04	1000	264.2
2	0.528	20	5.284	300	79.25	2000	528.4
3	0.793	30	7.925	400	105.70	3000	792.5
4	1.057	40	10.570	500	132.10	4000	1057.0
5	1.321	50	13.210	600	158.50	5000	1321.0
6	1.585	100	26.420	700	184.90	6000	1585.0
7	1.849	150	39.630	800	211.30	8000	2113.0
8	2.113	200	52.840	900	237.80	10000	2642.0
9	2.378						

CUBIC FEET INTO GALLONS*

1 cu. ft. = 7.48052 gal.* *U.S. liquid gallon (231 in.^{3}).

cu. ft.	gal.	cu. ft.	gal.	cu. ft.	gal.	cu. ft.	gal.
1	7.481	10	74.81	90	673.2	300	2244
2	14.960	20	149.60	100	748.1	400	2992
3	22.440	30	224.40	120	897.7	500	3740
4	29.920	40	299.20	140	1047.0	600	4488
5	37.400	50	374.00	160	1197.0	700	5236
6	44.880	60	448.80	180	1346.0	800	5984
7	52.360	70	523.60	200	1496.0	900	6732
8	59.840	80	598.40	250	1870.0	1000	7481
9	67.320						

GALLONS* INTO CUBIC FEET

1 gal.* = 0.133685 cu. ft. *U.S. liquid gallon (231 in.^{3}).

gal.	cu. ft.	gal.	cu. ft.	gal.	cu. ft.	gal.	cu. ft.
1	0.134	10	1.337	250	33.42	1000	133.7
2	0.267	20	2.674	300	40.11	2000	267.4
3	0.401	30	4.011	400	53.47	3000	401.1
4	0.535	40	5.347	500	66.84	4000	534.7
5	0.668	50	6.684	600	80.21	5000	668.4
6	0.802	100	13.370	700	93.58	6000	802.1
7	0.934	150	20.050	800	106.90	8000	1069.0
8	1.069	200	26.740	900	120.30	10000	1337.0
9	1.203						

table k (b-3)

References on the following page.

formulas, graphs & tables continued

FEET PER SEC. INTO METERS PER SEC.
1 ft. per sec. = 0.3048006 m. per sec.

ft./sec.	m./sec.	ft./sec.	m./sec.	ft./sec.	m./sec.	ft./sec.	m./sec.
1	0.305	9	2.743	24	7.315	50	15.24
2	0.610	10	3.048	26	7.925	60	18.29
3	0.914	12	3.658	28	8.534	70	21.34
4	1.219	14	4.267	30	9.144	75	22.86
5	1.524	16	4.877	35	10.670	80	24.38
6	1.829	18	5.486	40	12.190	90	27.43
7	2.134	20	6.096	45	13.720	100	30.48
8	2.438	22	6.706				

POUNDS* INTO KILOGRAMS
1 lb.* = 0.45359243 kg. *Avoirdupois

lb.	kg.	lb.	kg.	lb.	kg.	lb.	kg.
1	0.454	10	4.536	250	113.4	1000	453.6
2	0.907	20	9.072	300	136.1	2000	907.2
3	1.361	30	13.610	400	181.4	3000	1361.0
4	1.814	40	18.140	500	226.8	4000	1814.0
5	2.268	50	22.680	600	272.2	5000	2268.0
6	2.722	100	45.360	700	317.5	6000	2722.0
7	3.175	150	68.040	800	362.9	8000	3629.0
8	3.629	200	90.720	900	408.2	10000	4536.0
9	4.082						

POUNDS PER SQ. IN. INTO KG. PER SQ. CM.
1 lb. per sq. in. = 0.070307 kg. per cm.^{2}

lb./in.^{2}	kg./cm.^{2}	lb./in.^{2}	kg./cm.^{2}	lb./in.^{2}	kg./cm.^{2}	lb./in.^{2}	kg./cm.^{2}
1	0.070	20	1.406	140	9.843	600	42.18
2	0.141	30	2.109	160	11.250	700	49.21
3	0.211	40	2.812	180	12.660	800	56.25
4	0.281	50	3.515	200	14.060	900	63.28
5	0.352	60	4.218	250	17.580	1000	70.31
6	0.422	70	4.921	300	21.090	2500	175.80
7	0.492	80	5.625	350	24.610	5000	351.50
8	0.562	90	6.328	400	28.120	7500	527.30
9	0.633	100	7.031	450	31.640	10000	703.10
10	0.703	120	8.437	500	35.150		

HORSEPOWER INTO KW.
1 horsepower = 0.74571 kw.

hp.	kw.	hp.	kw.	hp.	kw.	hp.	kw.
1	0.746	9	6.711	24	17.90	50	37.29
2	1.491	10	7.457	26	19.39	60	44.74
3	2.237	12	8.948	28	20.88	70	52.20
4	2.983	14	10.440	30	22.37	75	55.93
5	3.729	16	11.930	35	26.10	80	59.66
6	4.474	18	13.420	40	29.83	90	67.11
7	5.220	20	14.910	45	33.56	100	74.57
8	5.966	22	16.410				

METERS PER SEC. INTO FEET PER SEC.
1 m. per sec. = 3.280833 ft. per sec.

m./sec.	ft./sec.	m./sec.	ft./sec.	m./sec.	ft./sec.	m./sec.	ft./sec.
0.5	1.640	8	26.25	16	52.49	26	85.30
1	3.281	9	29.53	17	55.77	28	91.86
2	6.562	10	32.81	18	59.05	30	98.42
3	9.842	11	36.09	19	62.34	35	114.80
4	13.120	12	39.37	20	65.62	40	131.20
5	16.400	13	42.65	22	72.18	45	147.60
6	19.680	14	45.93	24	78.74	50	164.00
7	22.970	15	49.21				

KILOGRAMS INTO POUNDS*
1 kg. = 2.2046223 lb.* *Avoirdupois

kg.	lb.	kg.	lb.	kg.	lb.	kg.	lb.
1	2.205	10	22.05	90	198.4	300	661.4
2	4.409	20	44.09	100	220.5	400	881.8
3	6.614	30	66.14	120	264.5	500	1102.0
4	8.818	40	88.18	140	308.6	600	1323.0
5	11.020	50	110.20	160	352.7	700	1543.0
6	13.230	60	132.30	180	396.8	800	1764.0
7	15.430	70	154.30	200	440.9	900	1984.0
8	17.640	80	176.40	250	551.2	1000	2205.0
9	19.840						

KG. PER SQ. CM. INTO POUNDS PER SQ. IN.
1 kg. per cm.^{2} = 14.2233 lb. per sq. in.

kg./cm.^{2}	lb./in.^{2}	kg./cm.^{2}	lb./in.^{2}	kg./cm.^{2}	lb./in.^{2}	kg./cm.^{2}	lb./in.^{2}
0.25	3.556	8	113.8	35	497.8	85	1209
0.50	7.112	9	128.0	40	568.9	90	1280
0.75	10.670	10	142.2	45	640.0	95	1351
1	14.220	12	170.7	50	711.2	100	1422
2	28.450	14	199.1	55	782.3	200	2845
3	42.670	16	227.6	60	853.4	300	4267
4	56.890	18	256.0	65	924.5	400	5689
5	71.120	20	284.5	70	995.6	500	7112
6	85.340	25	355.6	75	1067.0	600	8534
7	99.560	30	426.7	80	1138.0		

KW. INTO HORSEPOWER
1 kw. = 1.3410 hp.

kw.	hp.	kw.	hp.	kw.	hp.	kw.	hp.
1	1.341	9	12.07	24	32.18	50	67.05
2	2.682	10	13.41	26	34.87	60	80.46
3	4.023	12	16.09	28	37.55	70	93.87
4	5.364	14	18.77	30	40.23	75	100.60
5	6.705	16	21.46	35	46.94	80	107.30
6	8.046	18	24.14	40	53.64	90	120.70
7	9.387	20	26.82	45	60.35	100	134.10
8	10.730	22	29.50				

table k (b-4)

Pressure

- One atmosphere = 14.70 lb. per sq. in.
- One lb. per sq. in. = 0.06804 atmosphere
- One cm. mercury = 0.1934 lb. per sq. in.
- One in. mercury = 0.4912 lb. per sq. in.
- One meter water * = 1.421 lb. per sq. in.
- One foot water * = 0.4332 lb. per sq. in.

Density

- One lb. per cu. in. = 27.68 gram per cu. cm.
- One gram per cu. cm. = 0.03613 lb. per cu. in.

* 15°C and g = 32.1740 ft. per sec. per sec.

Energy or Work

- One British Thermal Unit = 777.52 ft. lb.
- One kilogram calorie = 3086 ft. lb.
- One foot lb. = 0.1383 kg. m.
- One kg. m. = 7.233 ft. lb.

Length

- One mile (statute) = 1.609 kilometer
- One kilometer = 0.6214 mile
- One mile (statute) = 5280 ft.

formulas, graphs & tables continued

METRIC EQUIVALENTS

LENGTH			
1 centimeter	.3937 inches	1 inch	2.5400 centimeters
1 meter	3.2800 feet	1 foot	.3048 meters
1 meter	1.0940 yards	1 yard	.9144 meters
1 kilometer	.6210 miles	1 mile	1.6100 kilometers
AREA			
1 sq. cm.	.1550 sq. in.	1 sq. in.	6.4500 sq. cm.
1 sq. meter	10.7600 sq. ft.	1 sq. ft.	.0929 sq. meters
1 sq. meter	1.1960 sq. yd.	1 sq. yd.	.8360 sq. meters
1 sq. kilo.	.3860 sq. miles	1 sq. mi.	2.5900 sq. kilo.
VOLUME			
1 cubic cm.	.0610 cu. in.	1 cu. in.	16.3800 cu. cm.
1 cubic m.	35.3150 cu. ft.	1 cu. ft.	.0280 cu. m.
1 cubic m.	1.3080 cu. yds.	1 cu. yd.	.7645 cu. m.
CAPACITY			
1 liter	.0353 cu. ft.	1 cu. ft.	28.3200 liters
1 liter	.2643 gal.	1 gallon	3.7850 liters
1 liter	61.0230 cu. in.	1 cu. in.	.0164 liters
1 liter	2.2020 lbs. of fresh water at 62 degrees F.		

table k (b-5)

CONDUCTIVE HEAT TRANSFER FORMULA

$$Q = \frac{K\Delta TA}{\Delta X}$$

Where

Q = Conductive heat transfer
A = Cross sectional area
ΔX = Length of heat path
ΔT = Temperature difference
Use consistent units

LINEAR EXPANSION

$$L = L_O\ (1 + \alpha\Delta T)$$

$$\alpha = \frac{\Delta \ell}{L_O}\frac{1}{\Delta T}$$

L = New length
L_O = Original length
α = Coefficient of linear expansion
ΔT = Change in temperature

TEMPERATURE CONVERSION TABLE

-60 to 43			44 to 93			94 to 510		
°C		°F	°C		°F	°C		°F
-51	**-60**	-76	6.7	**44**	111.2	34.4	**94**	201.2
-46	**-50**	-58	7.2	**45**	113.0	35.0	**95**	203.0
-40	**-40**	-40	7.8	**46**	114.3	35.6	**96**	204.8
-34	**-30**	-22	8.3	**47**	116.6	36.1	**97**	206.6
-29	**-20**	- 4	8.9	**48**	118.4	36.7	**98**	208.4
-23	**-10**	14	9.4	**49**	120.2	37.2	**99**	210.2
-17.8	**0**	32	10.0	**50**	122.0	37.8	**100**	212.0
-17.2	**1**	33.8	10.6	**51**	123.8	38	**100**	212
-16.7	**2**	35.6	11.1	**52**	125.6	43	**110**	230
-16.1	**3**	37.4	11.7	**53**	127.4	49	**120**	248
-15.6	**4**	39.2	12.2	**54**	129.2	54	**130**	266
-15.0	**5**	41.0	12.8	**55**	131.0	60	**140**	284
-14.4	**6**	42.8	13.3	**56**	132.8	66	**150**	302
-13.9	**7**	44.6	13.9	**57**	134.6	71	**160**	320
-13.3	**8**	46.4	13.4	**58**	136.4	77	**170**	338
-12.8	**9**	48.2	15.0	**59**	138.2	82	**180**	356
-12.2	**10**	50.0	15.6	**60**	140.0	88	**190**	374
-11.7	**11**	51.8	16.1	**61**	141.8	93	**200**	392
-11.1	**12**	53.6	16.7	**62**	143.6	99	**210**	410
-10.6	**13**	55.4	17.2	**63**	145.4	100	**212**	413.6
-10.0	**14**	57.2	17.8	**64**	147.2	104	**220**	428
- 9.4	**15**	59.0	18.3	**65**	149.0	110	**230**	446
- 8.9	**16**	60.8	18.9	**66**	150.8	116	**240**	464
- 8.3	**17**	62.6	19.4	**67**	152.6	121	**250**	482
- 7.8	**18**	64.4	20.0	**68**	154.4	127	**260**	500
- 7.2	**19**	66.2	20.6	**69**	156.2	132	**270**	518
- 6.7	**20**	68.0	21.1	**70**	158.0	138	**280**	536
- 6.1	**21**	69.8	21.7	**71**	159.8	143	**290**	554
- 5.6	**22**	71.6	22.2	**72**	161.6	149	**300**	572
- 5.0	**23**	73.4	22.8	**73**	163.4	154	**310**	590
- 4.4	**24**	75.2	23.3	**74**	165.2	160	**320**	608
- 3.9	**25**	77.0	23.9	**75**	167.0	166	**330**	626
- 3.3	**26**	78.8	24.4	**76**	168.8	171	**340**	644
- 2.8	**27**	80.6	25.0	**77**	170.6	177	**350**	662
- 2.3	**28**	82.4	25.6	**78**	172.4	182	**360**	680
- 1.7	**29**	84.2	26.1	**79**	174.3	188	**370**	698
- 1.1	**30**	86.0	26.7	**80**	176.0	193	**380**	716
- 0.6	**31**	87.8	27.2	**81**	177.8	199	**390**	734
0.0	**32**	89.6	27.8	**82**	179.6	204	**400**	752
0.6	**33**	91.4	28.3	**83**	181.4	210	**410**	770
1.1	**34**	93.2	28.9	**84**	183.2	216	**420**	788
1.7	**35**	95.0	28.4	**85**	185.0	221	**430**	806
2.2	**36**	96.8	30.0	**86**	186.8	227	**440**	824
2.8	**37**	98.6	30.6	**87**	188.6	232	**450**	842
3.3	**38**	100.4	31.1	**88**	190.4	238	**460**	860
3.9	**39**	102.2	31.7	**89**	192.2	243	**470**	878
4.4	**40**	104.0	32.2	**90**	194.0	249	**480**	896
5.0	**41**	105.8	32.8	**91**	195.8	254	**490**	914
5.6	**42**	107.6	33.3	**92**	197.6	260	**500**	932
6.1	**43**	109.4	33.9	**93**	199.4	266	**510**	950

LOOK UP READING IN MIDDLE COLUMN
FIND °C AT LEFT OR °F AT RIGHT

table k (b-6)

formulas, graphs & tables continued

CONVERSION FACTOR

$$T_C = \frac{5}{9}(T_F - 32) \qquad T_F = \frac{9}{5}T_C + 32$$

Temperature Range °K	300-4	300-20	300-77
Aluminum (99% pure)	728	700	505
Copper (Elect TP)	1,614	1,480	929
Stainless Steel	30.6	30.4	27.3
Titanium Alloy	16.5	16.4	14.5
Teflon	.702	.686	.570

table k (b-7)

HEAT COLORS

	°F
Faint Red	930
Blood Red	1075
Dark Cherry	1175
Medium Cherry	1275
Cherry	1375
Bright Cherry	1450
Salmon	1550
Dark Orange	1650
Orange	1725
Lemon	1830
Light Yellow	1975
White	2200
Blue White	2350

table k (b-8)

Velocity

Velocity should be kept approximately 15 ft./sec. after pump outlet.

Velocity should be kept approximately 4 ft./sec. before pump inlet.

Pressure loss/foot length of pipe:

$$P = \frac{V \times G.P.M.}{18{,}300 \times D^4}$$

Head Pressure

PSIA = 14.7 + (0.40 x h) for oil

PSIA = 14.7 + (0.433 x h) for water

PSIG = 0.40 x h for oil

inches of mercury =

suction lift (ft.) x .883 x SP. GR.

Specific Gravity

$$SP.\ GR. = \frac{\rho \text{ of liquid}}{\rho \text{ of water}} = \frac{\text{weight of liquid}}{\text{weight of equal volume of water}}$$

SP. GR. of water = 1.0

density (of water) = 62.4 lb./ft.^{3}

$$\text{density} = \frac{\text{weight of liquid}}{\text{equal volume water}}$$

Reynolds Number:

$$R = \frac{\rho V L}{\mu}$$

where:
ρ = density
V = free stream velocity
L = plate length
μ = coefficient of viscosity

Darcy's Formula:

$$h_f = \frac{flv^2}{2Dg}$$

where:
f = friction coefficient
l = pipe length
V = velocity
D = diameter
g = 32.2

Bernoulli's Equation

$$h + \frac{144p}{\rho} + \frac{V^2}{2g} = Z$$

where:
h = height
p = pressure
ρ = density
V = velocity
g = 32.2

Spring Rate

$$\text{spring rate} = \frac{\text{force}}{\text{distance compressed}}$$

Kinematic Viscosity Formula:

$$\gamma = \frac{\mu}{\rho}$$

where:
μ = coefficient or viscosity
ρ = density

formulas, graphs & tables continued

KINEMATIC VISCOSITY CONVERSION FACTORS

	m^2 sec.	m^2 hr.	cm^2 sec.	ft^2 sec.	ft^2 hr.
m^2 sec.	1.0	3,600	1×10^4	10.7639	3.875×10^4
m^2 hr.	277.8×10^{-6}	1.0	2.778	299.9×10^{-4}	10.7639
cm^2 sec. (Stokes)	1×10^{-4}	0.36	1.0	10.7639×10^{-4}	3.875
ft^2 sec.	0.092903	334.45	929.03	1.0	3,600
ft^2 hr.	25.806×10^{-6}	0.092903	0.25806	277.8×10^{-6}	1.0

table k (b-9)

ABSOLUTE VISCOSITY CONVERSION FACTORS

	Kg. sec. m^2	Kg. Hr. m^2	g* cm. sec.	Kg* m. hr.	lb. sec. ft^2	lb. hr. ft^2	lb* ft. sec.
Kg. Sec m^2	1.0	277.8×10^{-6}	98.1	3.5316×10^4	0.2048	56.89×10^{-6}	6.5919
Kg Hr m^2	3,600	1.0	0.35316×10^6	127.1×10^6	737.28	0.2048	2.373×10^4
g* cm. sec (Poise)	0.01019	2.833×10^{-6}	1.0	360	2.088×10^{-3}	0.58×10^{-6}	0.06721
Kg.* m. hr.	2.831×10^{-5}	7.8655×10^{-9}	2.788×10^{-6}	1.0	5.798×10^{-6}	1.6107×10^{-9}	0.1866×10^{-3}
lb. sec. $ft.^2$	4.882	1.356×10^{-3}	478.96	0.1724×10^6	1.0	277.7×10^{-6}	32.185
lb. hr. $ft.^2$	1.7578×10^4	4.882	1.7244×10^6	620.8×10^6	3,600	1.0	11.587×10^4
lb* ft. sec.	0.1517	42.139×10^{-6}	14.882	5.358×10^3	0.03107	8.631×10^{-6}	1.0

* mass

table k (b-10)

SELECTED "SI" UNITS FOR GENERAL PURPOSE – FLUID POWER USAGE

QUANTITY	SI UNIT FOR FLUID POWER	"CUSTOMARY US" UNIT FOR FLUID POWER	CONVERSION
Length	Millimeter (mm)	Inch (in)	1 in = 25.4 mm
Pressure (Note 1)	Bar (assumed to be "gauge" unless otherwise stated)	Pounds per square inch (psig or psia)	1 bar = 14.5 psi
Pressure (Note 2)	Bar (a value less than 1.0 For example 0.95 bar)	Inches of mercury (in Hg)	1 in Hg (@ 60°F) = 0.034 bar
Flow (Note 3)	Liters per minute (ℓ/min)	Gallons per minute (USGPM)	1 USGPM = 3.79 ℓ/min
Flow (Note 4)	Cubic decimeters per second (dm_n^3/s)	Cubic feet per minute (cfm)	1 dm_n^3/s = 2.12 scfm
Force	Newton (N)	Pound(f) lb(f)	1 lb(f) = 4.44 N
Mass	Kilogram (kg)	Pound(m) lb(m)	1 kg = 2.20 lb(m)
Time	Second (s)	Second (s)	–
Volume (Note 3)	Liter (ℓ)	Gallon (US gal)	1 US gal = 3.79 ℓ
Temperature	Degrees Celcius (°C)	Degrees Fahrenheit (°F)	°C = 5/9 (°F – 32)
Torque	Newton-meters (N•m)	Pounds(f)-inches lb(f)-in	1 N•m = 8.88 lb(f)-in
Power	Kilowatt (kW)	Horsepower (HP)	1 kW = 1.34 HP
Shaft speed	Revolutions per minute (rev/min)	Revolutions per minute (RPM)	–
Frequency	Hertz (Hz)	Cycles per second (cps)	1 Hz = 1 cps
Displacement (Note 3)	Milliliters per revolution (mℓ/rev)	Cubic inches per revolution (cip)	1 mℓ/rev = .061 cipr
Kinematic viscosity	Centistokes (cSt)	Saybolt (SUS)	cSt = (4.635) (SUS) (Note 5)
Velocity	Meter per second (m/s)	Feet per second (fps)	1 m/s = 3.28 fps
Material stress	Deka Newtons per square millimeter (da N/mm^2)	Pounds per square inch (psi)	1 da N/mm^2 = 1,450 psi

table k (b-11)

Note 1 – Pressures above atmospheric
Note 2 – Pressures below atmospheric
Note 3 – Liquid
Note 4 – Gas – under standard temperature, humidity, and pressure conditions per ISO/R 554-1967
Note 5 – @ 38°C; factor is 4.667 @ 99°C

CONVENTIONAL U.S. CONVERSION FACTORS

TO CONVERT → INTO → MULTIPLY BY INTO ← TO CONVERT ← DIVIDE BY		
abcoulomb	statcoulombs	2.998×10^{10}
acre	sq. chain (Gunters)	10
acre	sq. rods	160
acre	sq. links (Gunters)	10^{5}
acre	hectare or sq. hectometer	.4047
acres	sq. feet	43,560
acres	sq. meters	4,047
acres	sq. miles	1.562×10^{-3}
acres	sq. yards	4,840
acre-feet	cu. feet	43,560
acre-feet	gallons	3.259×10^{5}
amperes/sq. cm.	amps./sq. in.	6.452
amperes/sq. cm.	amps./sq. meter	10^{4}
amperes/sq. in.	amps./sq. cm.	.155
amperes/sq. in.	amps./sq. meter	1,550
amperes/sq. meter	amps./sq. cm.	10^{-4}
amperes/sq. meter	amps./sq. in.	6.452×10^{-4}
ampere-hours	coulombs	3,600
ampere-hours	faradays	.03731
ampere-turns	gilberts	1.257
ampere-turns/cm.	amp.-turns/in.	2.54
ampere-turns/cm.	amp.-turns/meter	100
ampere-turns/cm.	gilberts/cm.	1.257
ampere-turns/in.	amp.-turns/cm.	.3937
ampere-turns/in.	amp.-turns/meter	39.37
ampere-turns/in.	gilberts/cm.	.495
ampere-turns/meter	amp.-turns/cm.	.01
ampere-turns/meter	amp.-turns/in.	.0254
ampere-turns/meter	gilberts/cm.	.01257
angstrom unit	inch	3.937×10^{-6}
angstrom unit	meter	10^{-10}
angstrom unit	micron or (Mu)	10^{-4}
ares	sq. yards	119.6
ares	acres	.02471
ares	sq. meters	100
astronomical unit	kilometers	1.495×10^{8}
atmospheres	ton/sq. inch	.007348
atmospheres	cms. of mercury	76
atmospheres	ft. of water (at 4°C)	33.9
atmospheres	in. of mercury (at 0°C)	29.92
atmospheres	kgs./sq. cm.	1.0332
atmospheres	kgs./sq. meter	10,332
atmospheres	pounds/sq. in.	14.7
atmospheres	tons/sq. ft.	1.058
atmospheres	bars	1.0332
barrels (U.S., dry)	cu. inches	7056
barrels (U.S., dry)	quarts (dry)	105
barrels (U.S., liquid)	gallons	31.5
barrels (oil)	gallons (oil)	42
bars	atmospheres	.9869
bars	dynes/sq. cm.	10^{6}
bars	kgs./sq. meter	1.02×10^{4}
bars	pounds/sq. ft.	2,089
bars	pounds/sq. in.	14.5
bars	cm. Hg.	75.01
baryl	dyne/sq. cm.	1
bolt (U.S. Cloth)	meters	36.576

TO CONVERT → INTO → MULTIPLY BY INTO ← TO CONVERT ← DIVIDE BY		
Btu	liter-atmosphere	10.409
Btu	ergs	1.055×10^{10}
Btu	foot-lbs.	778.3
Btu	gram-calories	252
Btu	horsepower-hrs.	3.931×10^{-4}
Btu	joules	1,054.8
Btu	kilogram-calories	.252
Btu	kilogram-meters	107.5
Btu	kilowatt-hrs.	2.928×10^{-4}
Btu/hr.	foot-pounds/sec.	.2162
Btu/hr.	gram-cal./sec.	.07
Btu/hr.	horsepower-hrs.	3.929×10^{-4}
Btu/hr.	watts	.2931
Btu/min.	foot-lbs./sec.	12.96
Btu/min.	horsepower	.02356
Btu/min.	kilowatts	.01757
Btu/min.	watts	17.57
Btu/sq. ft./min.	watts/sq. in.	.1221
bucket (Br. dry)	cu. cm.	1.818×10^{4}
bushels	cu. ft.	1.2445
bushels	cu. in.	2,150.4
bushels	cu. meters	.03524
bushels	liters	35.24
bushels	pecks	4
bushels	pints (dry)	64
bushels	quarts (dry)	32
calories, gram (mean)	joules	4.183
calories, gram (mean)	Btu (mean)	3.9685×10^{-3}
candle/sq. cm.	lamberts	3.142
candle/sq. inch	lamberts	.487
centares (centiares)	sq. meters	1
centigrade	fahrenheit	(C° x 9/5) + 32
centigrams	grams	.01
centiliter	ounce fluid (U.S.)	.3382
centiliter	cubic inch	.6103
centiliter	drams	2.705
centiliter	liters	.01
centimeters	feet	3.281×10^{-2}
centimeters	inches	.3937
centimeters	kilometers	10^{-5}
centimeters	meters	.01
centimeters	miles	6.214×10^{-6}
centimeters	millimeters	10
centimeters	mils	393.7
centimeters	yards	1.094×10^{-2}
centimeter-dynes	cm.-grams	1.020×10^{-3}
centimeter-dynes	meter-kgs.	1.020×10^{-8}
centimeter-dynes	pound-feet	7.376×10^{-8}
centimeter-grams	cm.-dynes	980.7
centimeter-grams	meter-kgs.	10^{-5}
centimeter-grams	pound-feet	7.233×10^{-5}
centimeters of mercury	atmospheres	.01316
centimeters of mercury	feet of water	.4461
centimeters of mercury	kgs./sq. meter	136
centimeters of mercury	pounds/sq. ft.	27.85
centimeters of mercury	pounds/sq. in.	.1934

TO CONVERT →	INTO →	MULTIPLY BY
INTO ←	TO CONVERT ←	DIVIDE BY
centimeters/sec.	feet/min.	1.1969
centimeters/sec.	feet/sec.	.03281
centimeters/sec.	kilometers/hr.	.036
centimeters/sec.	knots	.1943
centimeters/sec.	meters/min.	.6
centimeters/sec.	miles/hr.	.02237
centimeters/sec.	miles/min.	3.728×10^{-4}
centimeters/sec./sec.	feet/sec./sec.	.03281
centimeters/sec./sec.	kms./hr./sec.	.036
centimeters/sec./sec.	meters/sec./sec.	.01
centimeters/sec./sec.	miles/hr./sec.	.02237
chain (surveyors)	inches	792
chain	meters	20.12
chain	yards	22
chain	feet	66
circular mils	sq. cms.	5.067×10^{-6}
circular mils	sq. mils	.7854
circular mils	sq. inches	7.854×10^{-7}
circumference	radians	6.283
cords	cord feet	8
cord feet	cu. feet	16
coulomb	statcoulombs	2.998×10^{9}
coulomb	faradays	1.036×10^{-5}
coulombs/sq. cm.	coulombs/sq. in.	64.52
coulombs/sq. cm.	coulombs/sq. meter	10^{4}
coulombs/sq. in.	coulombs/sq. cm.	.155
coulombs/sq. in.	coulombs/sq. meter	1,550
coulombs/sq. meter	coulombs/sq. cm.	10^{-4}
coulombs/sq. meter	coulombs/sq. in.	6.452×10^{-4}
cubic centimeters	cu. feet	3.531×10^{-5}
cubic centimeters	cu. inches	.06102
cubic centimeters	cu. meters	10^{-6}
cubic centimeters	cu. yards	1.308×10^{-6}
cubic centimeters	gallons (U.S. liquid)	2.642×10^{-4}
cubic centimeters	liters	.001
cubic centimeters	pints (U.S. liquid)	2.113×10^{-3}
cubic centimeters	quarts (U.S. liquid)	1.057×10^{-3}
cubic feet	bushels (dry)	.8036
cubic feet	cu. cms.	28,317
cubic feet	cu. inches	1,728
cubic feet	cu. meters	.02832
cubic feet	cu. yards	.03704
cubic feet	gallons (U.S. liquid)	7.48052
cubic feet	liters	28.32
cubic feet	pints (U.S. liquid)	59.84
cubic feet	quarts (U.S. liquid)	29.92
cubic feet of water	pounds	62.428
cubic feet/min.	cu. cms./sec.	472
cubic feet/min.	gallons/sec.	.1247
cubic feet/min.	liters/sec.	.4720
cubic feet/min.	pounds of water/min.	62.43
cubic feet/sec.	million gals./day	.646317
cubic feet/sec.	gallons/min.	448.831
cubic inches	cu. cms.	16.39
cubic inches	cu. feet	5.787×10^{-4}
cubic inches	cu. meters	1.639×10^{-5}
cubic inches	cu. yards	2.143×10^{-5}
cubic inches	gallons	4.329×10^{-3}
cubic inches	liters	.01639
cubic inches	mil-feet	1.061×10^{5}
cubic inches	pints (U.S. liquid)	.03463
cubic inches	quarts (U.S. liquid)	.01732

TO CONVERT →	INTO →	MULTIPLY BY
INTO ←	TO CONVERT ←	DIVIDE BY
cubic meters	bushels (dry)	28.38
cubic meters	cu. cms.	10^{6}
cubic meters	cu. feet	35.31
cubic meters	cu. inches	61,023
cubic meters	cu. yards	1.308
cubic meters	gallons (U.S. liquid)	264.2
cubic meters	liters	1,000
cubic meters	pints (U.S. liquid)	2,113
cubic meters	quarts (U.S. liquid)	1,057
cubic yards	cu. cms.	7.646×10^{5}
cubic yards	cu. feet	27
cubic yards	cu. inches	46,656
cubic yards	cu. meters	.7646
cubic yards	gallons (U.S. liquid)	202
cubic yards	liters	764.6
cubic yards	pints (U.S. liquid)	1,615.9
cubic yards	quarts (U.S. liquid)	807.9
cubic yards/min.	cubic ft./sec.	.45
cubic yards/min.	gallons/sec.	3.367
cubic yards/min.	liters/sec.	12.74
dalton	gram	1.65×10^{-24}
days	seconds	86,400
decigrams	grams	.1
deciliters	liters	.1
decimeters	meters	.1
degrees (angle)	quadrants	.01111
degrees (angle)	radians	.01745
degrees (angle)	seconds	3,600
degrees/sec.	radians/sec.	.01745
degrees/sec.	revolutions/min.	.1667
degrees/sec.	revolutions/sec.	2.778×10^{-3}
dekagrams	grams	10
dekaliters	liters	10
dekameters	meters	10
diameter of circle	circumference	3.1416
diameter of circle	area	$.7854\ D^{2}$
diameter of circle	volume of sphere	$.5236\ D^{3}$
drams (apoth)	ounces (avdp)	.1371
drams (apoth)	ounces (apoth)	.125
drams (apoth)	grains	60
drams (apoth)	grams	3.888
drams (fluid)	cu. cms.	3.697
drams (avdp)	ounces (avdp)	.0625
drams (avdp)	grains	27.3437
drams (avdp)	grams	1.7718
drams (avdp)	dram (apoth)	1.4623
dynes	grams	1.02×10^{-3}
dynes	joules/cm.	10^{-7}
dynes	joules/meter (newtons)	10^{-5}
dynes	kilograms	1.02×10^{-6}
dynes	poundals	7.233×10^{-5}
dynes	pounds	2.248×10^{-6}
dyne-cm.	erg	1
dyne/sq. cm.	atmospheres	9.869×10^{-7}
dyne/sq. cm.	inch of mercury at 0°C	2.953×10^{-5}
dyne/sq. cm.	inch of water at 4°C	4.015×10^{-4}
dyne/sq. cm.	gram (weight)	.01573
dyne/sq. cm.	bars	10^{-6}

TO CONVERT → / INTO ←	INTO → / TO CONVERT ←	MULTIPLY BY → / DIVIDE BY ←
ell	cm.	114.3
ell	inches	45
em, pica	inch	.167
em, pica	cm.	.4233
ergs	Btu	9.480×10^{-11}
ergs	dyne-centimeters	1
ergs	foot-pounds	7.367×10^{-8}
ergs	gram-calories	2.389×10^{-8}
ergs	gram-cms.	1.02×10^{-3}
ergs	horsepower-hrs.	3.725×10^{-14}
ergs	joules	10^{-7}
ergs	kg.-calories	2.389×10^{-11}
ergs	kg.-meters	1.02×10^{-8}
ergs	kilowatt-hrs.	2.778×10^{-14}
ergs	watt-hours	2.778×10^{-11}
ergs/sec.	dyne-cm./sec.	1
ergs/sec.	Btu./min.	5.688×10^{-6}
ergs/sec.	ft.-lbs./min.	4.427×10^{-6}
ergs/sec.	ft.-lbs./sec.	7.376×10^{-8}
ergs/sec.	horsepower	1.341×10^{-10}
ergs/sec.	kg.-calories/min.	1.433×10^{-9}
ergs/sec.	kilowatts	10^{-10}
farads	microfarads	10^{6}
faraday/sec.	ampere (absolute)	9.65×10^{4}
faradays	ampere-hours	26.8
faradays	coulombs	9.649×10^{4}
fathoms	meter	1.828804
fathoms	feet	6
feet	centimeters	30.48
feet	kilometers	3.048×10^{-4}
feet	meters	.3048
feet	miles (naut.)	1.645×10^{-4}
feet	miles (stat.)	1.894×10^{-4}
feet	millimeters	304.8
feet of water	atmospheres	.0295
feet of water	in. of mercury	.8826
feet of water	kgs./sq. cm.	.03048
feet of water	kgs./sq. meter	304.8
feet of water	pounds/sq. ft.	62.43
feet of water	pounds/sq. in.	.4335
feet/min.	cms./sec.	.508
feet/min.	feet/sec.	.01667
feet/min.	kms./hr.	.01829
feet/min.	meters/min.	.3048
feet/min.	miles/hr.	.01136
feet/sec.	cms./sec.	30.48
feet/sec.	kms./hr.	1.0973
feet/sec.	knots	.5921
feet/sec.	meters/min.	18.29
feet/sec.	miles/hr.	.6818
feet/sec.	miles/min.	.01136
feet/sec./sec.	cms./sec./sec.	30.48
feet/sec./sec.	kms./hr./sec.	1.097
feet/sec./sec.	meters/sec./sec.	.3048
feet/sec./sec.	miles/hr./sec.	.6818
feet/100 feet	per cent grade	1
foot-candle	lumen/sq. meter	10.764

TO CONVERT → / INTO ←	INTO → / TO CONVERT ←	MULTIPLY BY → / DIVIDE BY ←
foot-pounds	Btu	1.286×10^{-3}
foot-pounds	ergs	1.356×10^{7}
foot-pounds	gram-calories	.3238
foot-pounds	hp.-hrs.	5.05×10^{-7}
foot-pounds	joules	1.356
foot-pounds	kg.-calories	3.24×10^{-4}
foot-pounds	kg.-meters	.1383
foot-pounds	kilowatt-hrs.	3.766×10^{-7}
foot-pounds/min.	Btu/min.	1.286×10^{-3}
foot-pounds/min.	foot-pounds/sec.	.01667
foot-pounds/min.	horsepower	3.03×10^{-5}
foot-pounds/min.	kg.-calories/min.	3.24×10^{-4}
foot-pounds/min.	kilowatts	2.26×10^{-5}
foot-pounds/sec.	Btu/hr.	4.6263
foot-pounds/sec.	Btu/min.	.07717
foot-pounds/sec.	horsepower	1.818×10^{-3}
foot-pounds/sec.	kg.-calories/min.	.01945
foot-pounds/sec.	kilowatts	1.356×10^{-3}
furlongs	miles (U.S.)	.125
furlongs	rods	40
furlongs	feet	660
gallons	cu. cms.	3,785
gallons	cu. feet	.1337
gallons	cu. inches	231
gallons	cu. meters	3.785×10^{-3}
gallons	cu. yards	4.951×10^{-3}
gallons	liters	3.785
gallons (liq. Br. Imp.)	gallons (U.S. liquid)	1.20095
gallons (U.S.)	gallons (Imp.)	.83267
gallons of water	pounds of water	8.3453
gallons/min.	cu. ft./sec.	2.228×10^{-3}
gallons/min.	liters/sec.	.06308
gallons/min.	cu. ft./hr.	8.0208
gausses	lines/sq. in.	6.452
gausses	webers/sq. cm.	10^{-8}
gausses	webers/sq. in.	6.452×10^{-8}
gausses	webers/sq. meter	10^{-4}
gilberts	ampere-turns	.7958
gilberts/cm.	amp.-turns/cm.	.7958
gilberts/cm.	amp.-turns/in.	2.021
gilberts/cm.	amp.-turns/meter	79.58
gills (British)	cubic cm.	142.07
gills	liters	.1183
gills	pints (liquid)	.25
grade	radian	.01571
grains	drams (avoirdupois)	.03657143
grains (troy)	grains (avdp)	1
grains (troy)	grams	.0648
grains (troy)	ounces (avdp)	2.0833×10^{-3}
grains (troy)	pennyweight (troy)	.04167
grains/U.S. gal.	parts/million	17.118
grains/U.S. gal.	pounds/million gal.	142.86
grains/Imp. gal.	parts/million	14.286
grams	dynes	980.7
grams	grains	15.43
grams	joules/cm.	9.807×10^{-5}
grams	joules/meter (newtons)	9.807×10^{-3}
grams	kilograms	.001
grams	milligrams	1,000
grams	ounces (avdp)	.03527
grams	ounces (troy)	.03215
grams	poundals	.07093
grams	pounds	2.205×10^{-3}

TO CONVERT → INTO	INTO → TO CONVERT	MULTIPLY BY / DIVIDE BY
grams/cm.	pounds/inch	5.6×10^{-3}
grams/cu. cm.	pounds/cu. ft.	62.43
grams/cu. cm.	pounds/cu. in.	.03613
grams/liter	grains/gal.	58.417
grams/liter	pounds/1,000 gal.	8.345
grams/liter	pounds/cu. ft.	.062427
grams/liter	parts/million	1,000
grams/sq. cm.	pounds/sq. ft.	2.0481
gram-calories	Btu	3.9683×10^{-3}
gram-calories	ergs	4.1868×10^{7}
gram-calories	foot-pounds	3.088
gram-calories	horsepower-hrs.	1.5596×10^{-6}
gram-calories	kilowatt-hrs.	1.163×10^{-6}
gram-calories	watt-hrs.	1.163×10^{-3}
gram-calories/sec.	Btu/hr.	14.286
gram-centimeters	Btu	9.297×10^{-8}
gram-centimeters	ergs	980.7
gram-centimeters	joules	9.807×10^{-5}
gram-centimeters	kg.-cal.	2.343×10^{-8}
gram-centimeters	kg.-meters	10^{-5}
hand	cm.	10.16
hectares	acres	2.471
hectares	sq. feet	1.076×10^{5}
hectograms	grams	100
hectoliters	liters	100
hectometers	meters	100
hectowatts	watts	100
hefner	Intl. candles	.9
henries	millihenries	1,000
hogsheads (British)	cubic ft.	10.114
hogsheads (U.S.)	cubic ft.	8.42184
hogsheads (U.S.)	gallons (U.S.)	63
horsepower	Btu/min.	42.44
horsepower	foot-lbs./min.	33,000
horsepower	foot-lbs./sec.	550
horsepower	kg.-calories/min.	10.68
horsepower	kilowatts	.7457
horsepower	watts	745.7
horsepower (metric) (542.5 ft. lb./sec.)	horsepower (550 ft. lb./sec.)	.9863
horsepower (550 ft. lb./sec.)	horsepower (metric) (542.5 ft. lb./sec.)	1.014
horsepower (boiler)	Btu/hr.	33,479
horsepower (boiler)	kilowatt-hrs.	9.803
horsepower-hrs.	Btu	2,547
horsepower-hrs.	ergs	2.6845×10^{13}
horsepower-hrs.	foot-lbs.	1.98×10^{6}
horsepower-hrs.	gram-calories	641,190
horsepower-hrs.	joules	2.684×10^{6}
horsepower-hrs.	kg.-calories	641.2
horsepower	kg-meters	2.737×10^{5}
horsepower-hrs.	kilowatt-hrs.	.7457
hours	days	4.167×10^{-2}
hours	weeks	5.952×10^{-3}
hundredweights (long)	pounds	112
hundredweights (long)	tons (long)	.05
hundredweights (short)	ounces (avoirdupois)	1,600
hundredweights (short)	pounds	100
hundredweights (short)	tons (metric)	.0453592
hundredweights (short)	tons (long)	.0446429

TO CONVERT → INTO	INTO → TO CONVERT	MULTIPLY BY / DIVIDE BY
inches	centimeters	2.54
inches	meters	2.54×10^{-2}
inches	miles	1.578×10^{-5}
inches	millimeters	25.40
inches	mils	1,000
inches	yards	2.778×10^{-2}
inches of mercury	atmospheres	.03342
inches of mercury	feet of water	1.133
inches of mercury	kgs./sq. cm.	.03453
inches of mercury	kgs./sq. meter	345.3
inches of mercury	pounds/sq. ft.	70.73
inches of mercury	pounds/sq. in.	.4912
inches of water (at 4°C)	atmospheres	2.458×10^{-3}
inches of water (at 4°C)	inches of mercury	.07355
inches of water (at 4°C)	kgs./sq. cm.	2.54×10^{-3}
inches of water (at 4°C)	ounces/sq. in.	.5781
inches of water (at 4°C)	pounds/sq. ft.	5.204
inches of water (at 4°C)	pounds/sq. in.	.03613
International ampere	ampere (absolute)	.9998
International volt	volts (absolute)	1.0003
joules	Btu	9.48×10^{-4}
joules	ergs	10^{7}
joules	foot-pounds	.7376
joules	kg.-calories	2.389×10^{-4}
joules	kg.-meters	.102
joules	watt-hrs.	2.778×10^{-4}
joules/cm.	grams	1.02×10^{4}
joules/cm.	dynes	10^{7}
joules/cm.	joules/meter (newtons)	100
joules/cm.	poundals	723.3
joules/cm.	pounds	22.48
kilograms	dynes	980,665
kilograms	grams	1,000
kilograms	joules/cm.	.09807
kilograms	joules/meter (newtons)	9.807
kilograms	poundals	70.93
kilograms	pounds	2.205
kilograms	tons (long)	9.842×10^{-4}
kilograms	tons (short)	1.102×10^{-3}
kilograms/cu. meter	grams/cu. cm.	.001
kilograms/cu. meter	pounds/cu. ft.	.06243
kilograms/cu. meter	pounds/cu. in.	3.613×10^{-5}
kilograms/cu. meter	pounds/mil.-foot	3.405×10^{-10}
kilograms/meter	pounds/ft.	.672
kilograms/sq. cm.	dynes	980,665
kilograms/sq. cm.	atmospheres	.9678
kilograms/sq. cm.	feet of water	32.81
kilograms/sq. cm.	inches of mercury	28.96
kilograms/sq. cm.	pounds/sq. ft.	2,048
kilograms/sq. cm.	pounds/sq. in.	14.22
kilograms/sq. meter	atmospheres	9.678×10^{-5}
kilograms/sq. meter	bars	98.07×10^{-6}
kilograms/sq. meter	feet of water	3.281×10^{-3}
kilograms/sq. meter	inches of mercury	2.896×10^{-3}
kilograms/sq. meter	pounds/sq. ft.	.2048
kilograms/sq. meter	pounds/sq. in.	1.422×10^{-3}
kilograms/sq. mm.	kgs./sq. meter	10^{6}
kilogram-calories	Btu	3.968
kilogram-calories	foot-pounds	3,088
kilogram-calories	hp.-hrs.	1.56×10^{-3}
kilogram-calories	joules	4,186
kilogram-calories	kg.-meters	426.9
kilogram-calories	kilojoules	4.186
kilogram-calories	kilowatt-hrs.	1.163×10^{-3}

TO CONVERT → INTO	→ INTO ← TO CONVERT	→ MULTIPLY BY ← DIVIDE BY
kilogram meters	Btu	9.294×10^{-3}
kilogram meters	ergs	9.804×10^{7}
kilogram meters	foot-pounds	7.233
kilogram meters	joules	9.804
kilogram meters	kg.-calories	2.342×10^{-3}
kilogram meters	kilowatt-hrs.	2.723×10^{-6}
kilolines	maxwells	1,000
kiloliters	liters	1,000
kilometers	centimeters	10^{5}
kilometers	feet	3,281
kilometers	inches	3.937×10^{4}
kilometers	meters	1,000
kilometers	miles	.6214
kilometers	millimeters	10^{6}
kilometers	yards	1,094
kilometers/hr.	cms./sec.	27.78
kilometers/hr.	feet/min.	54.68
kilometers/hr.	feet/sec.	.9113
kilometers/hr.	knots	.5396
kilometers/hr.	meters/min.	16.67
kilometers/hr.	miles/hr.	.6214
kilometers/hr./sec.	cms./sec./sec.	27.78
kilometers/hr./sec.	ft./sec./sec.	.9113
kilometers/hr./sec.	meters/sec./sec.	.2778
kilometers/hr./sec.	miles/hr./sec.	.6214
kilowatts	Btu/min.	56.92
kilowatts	foot-lbs./min.	4.426×10^{4}
kilowatts	foot-lbs./sec.	737.6
kilowatts	horsepower	1.341
kilowatts	kg.-calories/min.	14.34
kilowatts	watts	1,000
kilowatt-hrs.	Btu	3,413
kilowatt-hrs.	ergs	3.6×10^{13}
kilowatt-hrs.	foot-lbs.	2.655×10^{6}
kilowatt-hrs.	gram-calories	859,850
kilowatt-hrs.	horsepower-hrs.	1.341
kilowatt-hrs.	joules	3.6×10^{6}
kilowatt-hrs.	kg.-calories	860.5
kilowatt-hrs.	kg.-meters	3.671×10^{5}
kilowatt-hrs.	pounds of water evaporated from and at 212°F.	3.53
kilowatt-hrs.	pounds of water raised from 62° to 212°F.	22.75
knots	feet/hr.	6,080
knots	kilometers/hr.	1.8532
knots	nautical miles/hr.	1
knots	statute miles/hr.	1.151
knots	yards/hr.	2,027
knots	feet/sec.	1.689
lambert	candles/sq. cm.	.3183
lambert	candles/sq. in.	2.054
league	miles (approx.)	3
light year	miles	5.9×10^{12}
light year	kilometers	9.49×10^{12}
lines/sq. cm.	gausses	1
lines/sq. in.	gausses	.155
lines/sq. in.	webers/sq. cm.	1.55×10^{-9}
lines/sq. in.	webers/sq. in.	10^{-8}
lines/sq. in.	webers/sq. meter	1.55×10^{-5}

TO CONVERT → INTO	→ INTO ← TO CONVERT	→ MULTIPLY BY ← DIVIDE BY
links (engineer's)	inches	12
links (surveyor's)	inches	7.92
liters	bushels (U.S. dry)	.02838
liters	cu. cm.	1,000
liters	cu. feet	.03531
liters	cu. inches	61.02
liters	cu. meters	.001
liters	cu. yards	1.308×10^{-3}
liters	gallons (U.S. liquid)	.2642
liters	pints (U.S. liquid)	2.113
liters	quarts (U.S. liquid)	1.057
liters/min.	cu. ft./sec.	5.886×10^{-4}
liters/min.	gals./sec.	4.403×10^{-3}
lumen	spherical candle power	.07958
lumen	watt	.001496
lumen/sq. ft.	lumen/sq. meter	10.76
lumen/sq. ft.	foot-candles	1
lumen/sq. cm.	lamberts	1
lumen/sq. cm.	phot (incident)	1
lumen/sq. meter	lux (incident)	1
lux	foot-candles	.0929
lux	lumen/sq. meter	1
maxwells	kilolines	.001
maxwells	webers	10^{-8}
megalines	maxwells	10^{6}
megohms	microhms	10^{12}
megohms	ohms	10^{6}
meters	centimeters	100
meters	feet	3.281
meters	inches	39.37
meters	kilometers	.001
meters	miles (naut.)	5.396×10^{-4}
meters	miles (stat.)	6.214×10^{-4}
meters	millimeters	1,000
meters	yards	1.094
meters	varas	1.179
meters/min.	cms./sec.	1.667
meters/min.	feet/min.	3.281
meters/min.	feet/sec.	.05468
meters/min.	kms./hr.	.06
meters/min.	knots	.03238
meters/min.	miles/hr.	.03728
meters/sec.	feet/min.	196.8
meters/sec.	feet/sec.	3.281
meters/sec.	kilometers/hr.	3.6
meters/sec.	kilometers/min.	.06
meters/sec.	miles/hr.	2.2369
meters/sec.	miles/min.	.03728
meters/sec./sec.	cms./sec./sec.	100
meters/sec./sec.	ft./sec./sec.	3.281
meters/sec./sec.	kms./hr./sec.	3.6
meters/sec./sec.	miles/hr./sec.	2.237
meter-kilograms	cm.-dynes	9.807×10^{7}
meter-kilograms	cm.-grams	10^{5}
meter-kilograms	pound-feet	7.233
microfarad	farads	10^{-6}
micrograms	grams	10^{-6}
microhms	megohms	10^{-12}
microhms	ohms	10^{-6}

TO CONVERT → INTO → MULTIPLY BY INTO ← TO CONVERT ← DIVIDE BY		
microliters	liters	10^{-6}
microns	meters	10^{-6}
miles (naut.)	feet	6,080.27
miles (naut.)	kilometers	1.853
miles (naut.)	meters	1,853
miles (naut.)	miles (stat.)	1.1516
miles (naut.)	yards	2,027
miles (stat.)	centimeters	1.609×10^{5}
miles (stat.)	feet	5,280
miles (stat.)	inches	6.336×10^{4}
miles (stat.)	kilometers	1.609
miles (stat.)	meters	1,609
miles (stat.)	miles (naut.)	.8684
miles (stat.)	yards	1,760
miles/hr.	cms./sec.	44.7
miles/hr.	feet/min.	88
miles/hr.	feet/sec.	1.467
miles/hr.	kms./hr.	1.6093
miles/hr.	kms./min.	.02682
miles/hr.	knots	.8684
miles/hr.	meters/min.	26.82
miles/hr.	miles/min.	.1667
miles/hr./sec.	cms./sec./sec.	44.7
miles/hr./sec.	feet/sec./sec.	1.467
miles/hr./sec.	kms./hr./sec.	1.609
miles/hr./sec.	meters/sec./sec.	.447
miles/min.	cms./sec.	2,682
miles/min.	feet/sec.	88
miles/min.	kms./min.	1.609
miles/min.	knots	.8684
miles/min.	miles/hr.	60
mil.-feet	cu. inches	9.425×10^{-6}
millibars	lbs./sq. in.	.0145
milliers	kilograms	1,000
millimicrons	meters	10^{-9}
milligrams	grains	.01543236
milligrams	grams	.001
milligrams/liter	parts/million	1
millihenries	henries	.001
milliliters	liters	.001
millimeters	centimeters	.1
millimeters	feet	3.281×10^{-3}
millimeters	inches	.03937
millimeters	kilometers	10^{-6}
millimeters	meters	.001
millimeters	miles	6.214×10^{-7}
millimeters	mils	39.37
millimeters	yards	1.094×10^{-3}
million gals./day	cu. ft./sec.	1.54723
mils	centimeters	2.54×10^{-3}
mils	feet	8.333×10^{-5}
mils	inches	.001
mils	kilometers	2.54×10^{-8}
mils	yards	2.778×10^{-5}
miner's inches	cu. ft./min.	1.5
minims (British)	cubic cm.	.059192
minims (U.S. fluid)	cubic cm.	.061612

TO CONVERT → INTO → MULTIPLY BY INTO ← TO CONVERT ← DIVIDE BY		
minutes (angles)	degrees	.01667
minutes (angles)	quadrants	1.852×10^{-4}
minutes (angles)	radians	2.909×10^{-4}
minutes (angles)	seconds	60
myriagrams	kilograms	10
myriameters	kilometers	10
myriawatts	kilowatts	10
nepers	decibels	8.686
ohm (International)	ohm (absolute)	1.0005
ohm	megohms	10^{-6}
ohm	microhms	10^{6}
ounces (avdp)	drams	16
ounces (avdp)	grains	437.5
ounces (avdp)	grams	28.349527
ounces (avdp)	pounds	.0625
ounces (avdp)	ounces (troy)	.9115
ounces (avdp)	tons (long)	2.790×10^{-5}
ounces (avdp)	tons (metric)	2.835×10^{-5}
ounces (fluid)	cu. inches	1.805
ounces (fluid)	liters	.02957
ounces (troy)	grains	480
ounces (troy)	grams	31.103481
ounces (troy)	ounces (avdp.)	1.09714
ounces (troy)	pennyweights (troy)	20
ounces (troy)	pounds (troy)	.08333
ounces/sq. inch	dynes/sq. cm.	4309
ounces/sq. inch	pounds/sq. in.	.0625
parsec	miles	19×10^{12}
parsec	kilometers	3.084×10^{13}
parts/million	grains/U.S. gal.	.0584
parts/million	grains/Imp. gal.	.07016
parts/million	pounds/million gal.	8.345
pecks (British)	cubic inches	554.6
pecks (British)	liters	9.091901
pecks (U.S.)	bushels	.25
pecks (U.S.)	cubic inches	537.605
pecks (U.S.)	liters	8.809582
pecks (U.S.)	quarts (dry)	8
pennyweights (troy)	grains	24
pennyweights (troy)	ounces (troy)	.05
pennyweights (troy)	grams	1.55517
pennywieghts (troy)	pounds (troy)	4.1667×10^{-3}
phots	lumen/sq. cm.	1
phots	lux	10^{4}
pints (dry)	cu. inches	33.6
pints (liquid)	cu. cms.	473.2
pints (liquid)	cu. feet	.01671
pints (liquid)	cu. inches	28.87
pints (liquid)	cu. meters	4.732×10^{-4}
pints (liquid)	cu. yards	6.189×10^{-4}
pints (liquid)	gallons	.125
pints (liquid)	liters	.4732
pints (liquid)	quarts (liquid)	.5
Planck's quantum	erg.-second	6.624×10^{-27}
poise	gram/cm. sec.	1

TO CONVERT → INTO →	INTO ← TO CONVERT ←	MULTIPLY BY / DIVIDE BY
poundals	dynes	13,825
poundals	grams	14.1
poundals	joules/cm.	1.383×10^{-3}
poundals	joules/meter (newtons)	.1383
poundals	kilograms	.0141
poundals	pounds	.03108
pounds	drams	256
pounds	dynes	44.4823×10^{4}
pounds	grains	7,000
pounds	grams	453.5924
pounds	joules/cm.	.04448
pounds	joules/meter (newtons)	4.448
pounds	kilograms	.4536
pounds	ounces	16
pounds	ounces (troy)	14.5833
pounds	poundals	32.17
pounds	pounds (troy)	1.21528
pounds	tons (short)	.0005
pounds (troy)	grains	5,760
pounds (troy)	grams	373.24177
pounds (troy)	ounces (avdp)	13.1657
pounds (troy)	ounces (troy)	12
pounds (troy)	pennyweights (troy)	240
pounds (troy)	pounds (avdp)	.822857
pounds (troy)	tons (long)	3.6735×10^{-4}
pounds (troy)	tons (metric)	3.7324×10^{-4}
pounds (troy)	tons (short)	4.1143×10^{-4}
pounds of water	cu. feet	.01602
pounds of water	cu. inches	27.68
pounds of water	gallons	.1198
pounds of water/min.	cu. ft./sec.	2.67×10^{-4}
pound-feet	cm.-dynes	1.356×10^{7}
pound-feet	cm.-grams	13,825
pound-feet	meter-kgs.	.1383
pounds/cu. ft.	grams/cu. cm.	.01602
pounds/cu. ft.	kgs./cu. meter	16.02
pounds/cu. ft.	pounds/cu. in.	5.787×10^{-4}
pounds/cu. ft.	pounds/mil.-foot	5.456×10^{-9}
pounds/cu. in.	gms./cu. cm.	27.68
pounds/cu. in.	kgs./cu. meter	2.768×10^{4}
pounds/cu. in.	pounds/cu. ft.	1,728
pounds/cu. in.	pounds/mil.-foot	9.425×10^{-6}
pounds/ft.	kgs./meter	1.488
pounds/in.	gms./cm.	178.6
pounds/mil.-foot	gms./cu. cm.	2.306×10^{6}
pounds/sq. ft.	atmospheres	4.725×10^{-4}
pounds/sq. ft.	feet of water	.01602
pounds/sq. ft.	inches of mercury	.01414
pounds/sq. ft.	kgs./sq. meter	4.882
pounds/sq. ft.	pounds/sq. in.	6.944×10^{-3}
pounds/sq. in.	atmospheres	.06804
pounds/sq. in.	feet of water	1.307
pounds/sq. in.	inches of mercury	2.036
pounds/sq. in.	kgs./sq. meter	703.1
pounds/sq. in.	pounds/sq. ft.	144
pounds/sq. in.	millibars	68.9
quadrants (angle)	degrees	90
quadrants (angle)	minutes	5,400
quadrants (angle)	radians	1.571
quadrants (angle)	seconds	3.24×10^{5}
quarts (dry)	cu. inches	67.2
quarts (liquid)	cu. cms.	946.4
quarts (liquid)	cu. feet	.03342
quarts (liquid)	cu. inches	57.75
quarts (liquid)	cu. meters	9.464×10^{-4}
quarts (liquid)	cu. yards	1.238×10^{-3}
quarts (liquid)	gallons	.25
quarts (liquid)	liters	.9463
radians	degrees	57.3
radians	minutes	3,438
radians	quadrants	.6366
radians	seconds	2.063×10^{5}
radians/sec.	degrees/sec.	57.3
radians/sec.	revolutions /min.	9.549
radians/sec.	revolutions/sec.	.1592
radians/sec./sec.	revs./min./min.	573
radians/sec./sec.	revs./min./sec.	9.549
radians/sec./sec.	revs./sec./sec.	.1592
revolutions	degrees	360
revolutions	quadrants	4
revolutions	radians	6.283
revolutions/min.	degrees/sec.	6
revolutions/min.	radians/sec.	.1047
revolutions/min.	revs./sec.	.01667
revolutions/min./min.	radians/sec./sec.	1.745×10^{-3}
revolutions/min./min.	revs./min./sec.	.01667
revolutions/min./min.	revs./sec./sec.	2.778×10^{-4}
revolutions/sec.	degrees/sec.	360
revolutions/sec.	radians/sec.	6.283
revolutions/sec.	revs./min.	60
revolutions/sec./sec.	radians/sec./sec.	6.283
revolutions/sec./sec.	revs./min./min.	3,600
revolutions/sec./sec.	revs./min./sec.	60
rod	chain (Gunters)	.25
rod	meters	5.029
rod	yards	5.5
rod	feet	16.5
scruples	grains	20
seconds (angle)	degrees	2.778×10^{-4}
seconds (angle)	minutes	.01667
seconds (angle)	quadrants	3.087×10^{-6}
seconds (angle)	radians	4.848×10^{-6}
slug	kilogram	14.59
slug	pounds	32.17
sphere	steradians	12.57
square centimeters	circular mils	1.973×10^{5}
square centimeters	sq. feet	1.076×10^{-3}
square centimeters	sq. inches	.155
square centimeters	sq. meters	10^{-4}
square centimeters	sq. miles	3.861×10^{-11}
square centimeters	sq. millimeters	100
square centimeters	sq. yards	1.196×10^{-4}
square feet	acres	2.296×10^{-5}
square feet	circular mils	1.833×10^{8}
square feet	sq. cms.	929
square feet	sq. inches	144
square feet	sq. meters	.0929
square feet	sq. miles	3.587×10^{-8}
square feet	sq. millimeters	9.29×10^{4}
square feet	sq. yards	.1111

TO CONVERT → INTO ←	INTO → TO CONVERT ←	MULTIPLY BY / DIVIDE BY
square inches	circular mils	1.273×10^{6}
square inches	sq. cms.	6.452
square inches	sq. feet	6.944×10^{-3}
square inches	sq. millimeters	645.2
square inches	sq. mils	10^{6}
square inches	sq. yards	7.716×10^{-4}
square kilometers	acres	247.1
square kilometers	sq. cms.	10^{10}
square kilometers	sq. ft.	10.76×10^{6}
square kilometers	sq. inches	1.55×10^{9}
square kilometers	sq. meters	10^{6}
square kilometers	sq. miles	.3861
square kilometers	sq. yards	1.196×10^{6}
square meters	acres	2.471×10^{-4}
square meters	sq. cms.	10^{4}
square meters	sq. feet	10.76
square meters	sq. inches	1,550
square meters	sq. miles	3.861×10^{-7}
square meters	sq. millimeters	10^{6}
square meters	sq. yards	1.196
square miles	acres	640
square miles	sq. feet	27.88×10^{6}
square miles	sq. kms.	2.59
square miles	sq. meters	2.59×10^{6}
square miles	sq. yards	3.098×10^{6}
square millimeters	circular mils	1,973
square millimeters	sq. cms.	.01
square millimeters	sq. feet	1.076×10^{-5}
square millimeters	sq. inches	1.55×10^{-3}
square mils	circular mils	1.273
square mils	sq. cms.	6.452×10^{-6}
square mils	sq. inches	10^{-6}
square yards	acres	2.066×10^{-4}
square yards	sq. cms.	8,361
square yards	sq. feet	9
square yards	sq. inches	1,296
square yards	sq. meters	.8361
square yards	sq. miles	3.228×10^{-7}
square yards	sq. millimeters	8.361×10^{5}
temperature (°C) + 273	absolute temperature (Kelvin)	1
temperature (°C) + 17.78	temperature (°F)	1.8
temperature (°F) + 460	absolute temperature (Rankine)	1
temperature (°F) -32	temperature (°C)	5/9
tons (long)	kilograms	1,016
tons (long)	pounds	2,240
tons (long)	tons (short)	1.12
tons (metric)	kilograms	1,000
tons (metric)	pounds	2,205
tons (short)	kilograms	907.1848
tons (short)	ounces	32,000
tons (short)	ounces (troy)	29,166.66
tons (short)	pounds	2,000
tons (short)	pounds (troy)	2,430.56
tons (short)	tons (long)	.89287
tons (short)	tons (metric)	.9078

table k (b-12)

TO CONVERT → INTO ←	INTO → TO CONVERT ←	MULTIPLY BY / DIVIDE BY
tons (short)/sq. ft.	kgs./sq. meter	9,765
tons (short)/sq. ft.	pounds/sq. in.	13.9
tons of water/24 hrs.	pounds of water/hr.	83.333
tons of water/24 hrs.	gallons/min.	.16643
tons of water/24 hrs.	cu. ft./hr.	1.3349
volt/inch	volt/cm.	.3937
volt (absolute)	statvolts	.003336
watt (International)	watt (absolute)	1.0002
watts	Btu/hr.	3.4192
watts	Btu/min.	.05688
watts	ergs./sec.	107
watts	foot-lbs./min.	44.27
watts	foot-lbs./sec.	.7378
watts	horsepower	1.341×10^{-3}
watts	horsepower (metric)	1.36×10^{-3}
watts	kg.-calories/min.	.01433
watts	kilowatts	.001
watts	lumens	668
watts (Abs.)	Btu (mean)/min.	.056884
watts (Abs.)	joules/sec.	1
watt-hours	Btu	3.413
watt-hours	ergs.	3.6×10^{10}
watt-hours	foot-pounds	2,656
watt-hours	gram-calories	859.85
watt-hours	horsepower-hrs.	1.341×10^{-3}
watt-hours	kilogram-calories	.8605
watt-hours	kilogram-meters	367.2
watt-hours	kilowatt-hrs.	.001
webers	maxwells	10^{8}
webers	kilolines	10^{5}
webers/sq. in.	gausses	1.55×10^{7}
webers/sq. in.	lines/sq. in.	10^{8}
webers/sq. in.	webers/sq. cm.	.155
webers/sq. in.	webers/sq. meter	1,550
webers/sq. meter	gausses	10^{4}
webers/sq. meter	lines/sq. in.	6.452×10^{4}
webers/sq. meter	webers/sq. cm.	10^{-4}
webers/sq. meter	webers/sq. in.	6.452×10^{-4}
yards	centimeters	91.44
yards	kilometers	9.144×10^{-4}
yards	meters	.9144
yards	miles (naut.)	4.934×10^{-4}
yards	miles (stat.)	5.682×10^{-4}
yards	millimeters	914.4

general data

symbol

introduction .... k (c-2)
symbol rules .... k (c-2)
fluid conductor .... k (c-3)
energy storage and fluid storage ..... k (c-4)
fluid conditioners .... k (c-5)
linear devices .... k (c-6)
actuators and controls .... k (c-6)
rotary devices .... k (c-8)
instruments and accessories .... k (c-9)
valves .... k (c-10)
representative composite symbols ... k (c-13)
pneumatic logic symbols .... k (c-18)
parker mobile symbols .... k (c-20)

USAS Y32.10 graphic symbols

1. Introduction

1.1 General

Fluid power systems are those that transmit and control power through use of a pressurized fluid (liquid or gas) within an enclosed circuit.

Types of symbols commonly used in drawing circuit diagrams for fluid power systems are Pictorial, Cutaway, and Graphic. These symbols are fully explained in the USA Standard Drafting Manual (Ref. 2).

1.1.1 *Pictorial symbols* are very useful for showing the interconnection of components. They are difficult to standardize from a functional basis.

1.1.2 *Cutaway symbols* emphasize construction. These symbols are complex to draw and the functions are not readily apparent.

1.1.3 *Graphic symbols* emphasize the function and methods of operation of components. These symbols are simple to draw. Component functions and methods of operation are obvious. Graphic symbols are capable of crossing language barriers, and can promote a universal understanding of fluid power systems.

Graphic symbols for fluid power systems should be used in conjunction with the graphic symbols for other systems published by the USA Standards Institute (Ref. 3-7 inclusive).

1.1.3.1 Complete graphic symbols are those which give symbolic representation of the component and all of its features pertinent to the circuit diagram.

1.1.3.2 Simplified graphic symbols are stylized versions of the complete symbols.

1.1.3.3 Composite graphic symbols are an organization of simplified or complete symbols. Composite symbols usually represent a complex component.

1.2 Scope and Purpose

1.2.1 *Scope* — This standard presents a system of graphic symbols for fluid power diagrams.

1.2.1.1 Elementary forms of symbols are:

Circles	Rectangles	Arcs	Dots
Squares	Triangles	Arrows	Crosses

1.2.1.2 Symbols using words or their abbreviations are avoided. Symbols capable of crossing language barriers are presented herein.

1.2.1.3 Component function rather than construction is emphasized by the symbol.

1.2.1.4 The means of operating fluid power components are shown as part of the symbol (where applicable).

1.2.1.5 This standard shows the basic symbols, describes the principles on which the symbols are based, and illustrates some representative composite symbols. Composite symbols can be devised for any fluid power component by combining basic symbols.

Simplified symbols are shown for commonly used components.

1.2.1.6 This standard provides basic symbols which differentiate between hydraulic and pneumatic fluid power media.

1.2.2 *Purpose*

1.2.2.1 The purpose of this standard is to provide a system of fluid power graphic symbols for industrial and educational purposes.

1.2.2.2 The purpose of this standard is to simplify design, fabrication, analysis, and service of fluid power circuits.

1.2.2.3 The purpose of this standard is to provide fluid power graphic symbols which are internationally recognized.

1.2.2.4 The purpose of this standard is to promote universal understanding of fluid power systems.

1.3 Terms and Definitions

Terms and corresponding definitions found in this standard are listed in **Section k-a.**

2. Symbol Rules

2.1 Symbols show connections, flow paths, and functions of components represented. They can indicate conditions occurring during transition from one flow path arrangement to another. Symbols do not indicate construction, nor do they indicate values, such as pressure, flow rate, and other component settings.

2.2 Symbols do not indicate locations of ports, direction of shifting of spools, or positions of actuators on actual component.

2.3 Symbols may be rotated or reversed without altering their meaning except in the cases of: a.) Lines to Reservoir, 4.1.1;b.) Vented Manifold, 4.1.2.3; c.) Accumulator, 4.2.

2.4 Line Technique

Keep line widths approximately equal. Line width does not alter meaning of symbols.

2.4.1 Solid Line — Main

(Main line conductor, outline, and shaft)

graphic symbols continued

2.4.2 Dash Line — Pilot

(Pilot line for control)

2.4.3 Dotted Line

(Exhaust or Drain line)

2.4.4 Center Line

(Enclosure outline)

2.4.5 Sensing Line — Same as line to which it connects.

2.4.6 Lines Crossing (The intersection is not necessarily at a 90° angle.)

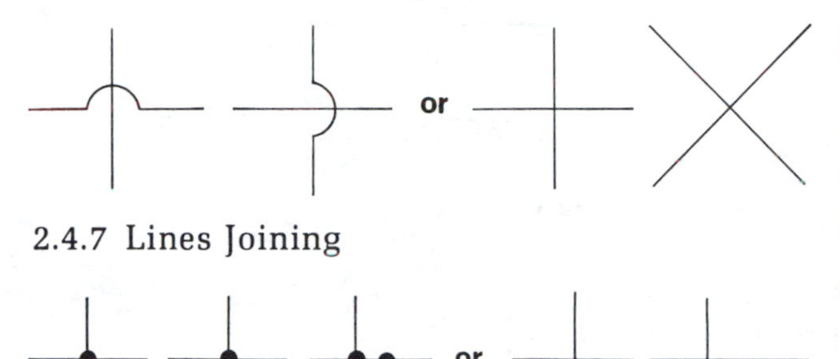

2.4.7 Lines Joining

2.5 Basic symbols may be shown any suitable size. Size may be varied for emphasis or clarity. Relative sizes should be maintained. (As in the following example.)

2.5.1 Circle and Semi-Circle

2.5.1.1 Large and small circles may be used to signify that one component is the "main" and the other the auxiliary.

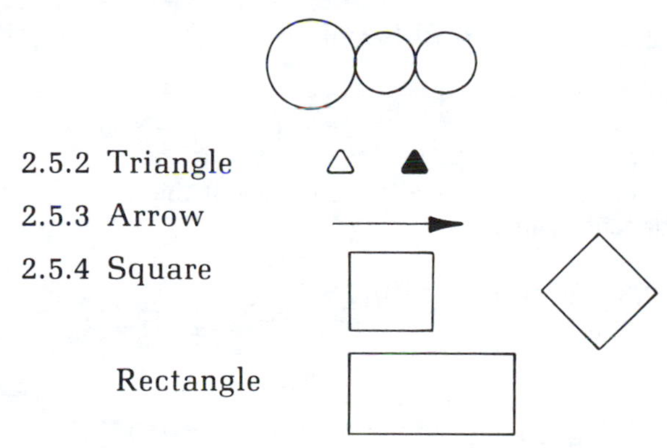

2.5.2 Triangle

2.5.3 Arrow

2.5.4 Square

Rectangle

2.6 Letter combinations used as parts of graphic symbols are not necessarily abbreviations.

2.7 In multiple envelope symbols, the flow condition shown nearest an actuator symbol takes place when that control is caused or permitted to actuate.

2.8 Each symbol is drawn to show normal, at-rest, or neutral conditon of component unelss multiple diagrams are furnished showing various phases of circuit operation. Show an actuator symbol for each flow path conditon possessed by the component.

2.9 An arrow through a symbol at approximately 45 degrees indicates that the component can be adjusted or varied.

2.10 An arrow parallel to the short side of a symbol, within the symbol, indicates that the component is pressure compensated.

2.11 A line terminating in a dot to represent a thermometer is the symbol for temperature cause or effect.

See Temperature Controls 7.9, Temperature Indicators and Recorders 9.1.2, and Temperature Compensation 10.16.3 and 4.

2.12 External ports are located where flow lines connect to basic symbol, except where component enclosure symbol is used.

External ports are located at intersections of flow lines and component enclosure symbol when enclosure is used, see Section 11.

2.13 Rotating shafts are symbolized by an arrow which indicates direction of rotation (assume arrow on near side of shaft).

3. Conductor, Fluid

3.1 Line, Working (main)

3.2 Line, Pilot (for control)

3.3 Line, Exhaust and Liquid Drain

3.4 Line, sensing, etc. such as gauge lines

graphic symbols continued

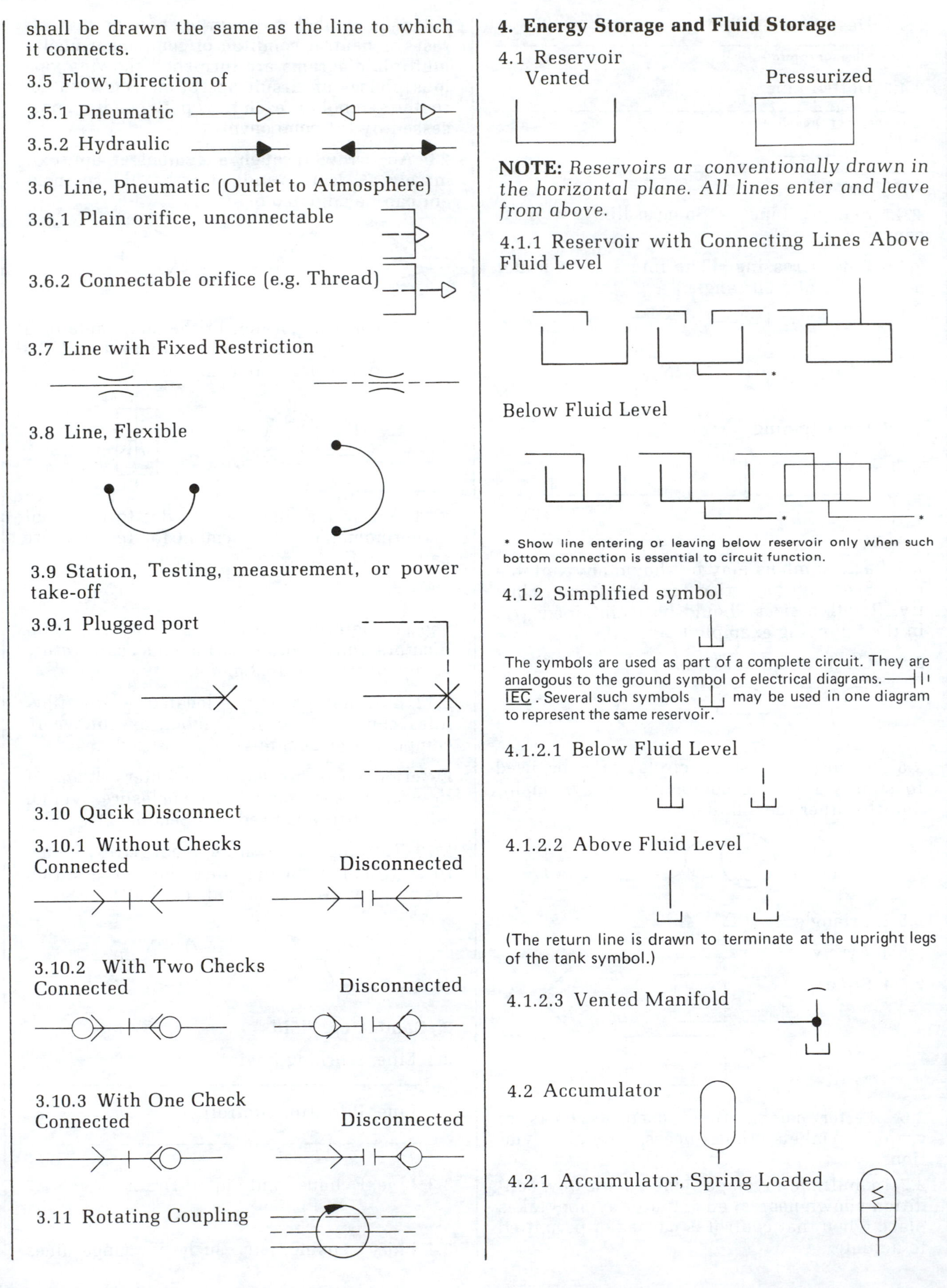

shall be drawn the same as the line to which it connects.

3.5 Flow, Direction of

3.5.1 Pneumatic

3.5.2 Hydraulic

3.6 Line, Pneumatic (Outlet to Atmosphere)

3.6.1 Plain orifice, unconnectable

3.6.2 Connectable orifice (e.g. Thread)

3.7 Line with Fixed Restriction

3.8 Line, Flexible

3.9 Station, Testing, measurement, or power take-off

3.9.1 Plugged port

3.10 Qucik Disconnect

3.10.1 Without Checks
Connected — Disconnected

3.10.2 With Two Checks
Connected — Disconnected

3.10.3 With One Check
Connected — Disconnected

3.11 Rotating Coupling

4. Energy Storage and Fluid Storage

4.1 Reservoir
Vented — Pressurized

NOTE: *Reservoirs ar conventionally drawn in the horizontal plane. All lines enter and leave from above.*

4.1.1 Reservoir with Connecting Lines Above Fluid Level

Below Fluid Level

* Show line entering or leaving below reservoir only when such bottom connection is essential to circuit function.

4.1.2 Simplified symbol

The symbols are used as part of a complete circuit. They are analogous to the ground symbol of electrical diagrams. IEC. Several such symbols may be used in one diagram to represent the same reservoir.

4.1.2.1 Below Fluid Level

4.1.2.2 Above Fluid Level

(The return line is drawn to terminate at the upright legs of the tank symbol.)

4.1.2.3 Vented Manifold

4.2 Accumulator

4.2.1 Accumulator, Spring Loaded

graphic symbols continued

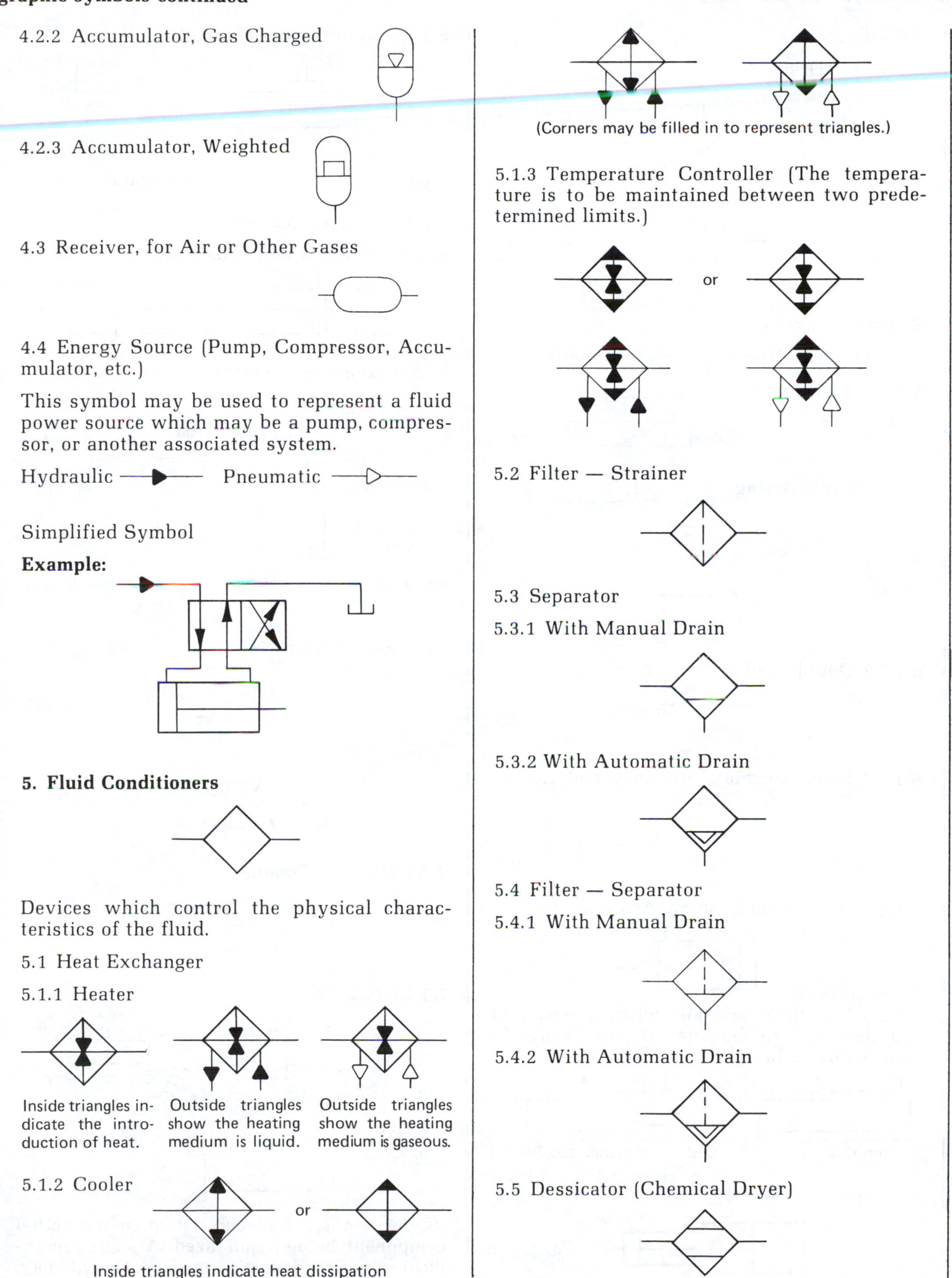

4.2.2 Accumulator, Gas Charged

4.2.3 Accumulator, Weighted

4.3 Receiver, for Air or Other Gases

4.4 Energy Source (Pump, Compressor, Accumulator, etc.)

This symbol may be used to represent a fluid power source which may be a pump, compressor, or another associated system.

Hydraulic Pneumatic

Simplified Symbol

Example:

5. Fluid Conditioners

Devices which control the physical characteristics of the fluid.

5.1 Heat Exchanger

5.1.1 Heater

Inside triangles indicate the introduction of heat.

Outside triangles show the heating medium is liquid.

Outside triangles show the heating medium is gaseous.

5.1.2 Cooler

or

Inside triangles indicate heat dissipation

(Corners may be filled in to represent triangles.)

5.1.3 Temperature Controller (The temperature is to be maintained between two predetermined limits.)

or

5.2 Filter — Strainer

5.3 Separator

5.3.1 With Manual Drain

5.3.2 With Automatic Drain

5.4 Filter — Separator

5.4.1 With Manual Drain

5.4.2 With Automatic Drain

5.5 Dessicator (Chemical Dryer)

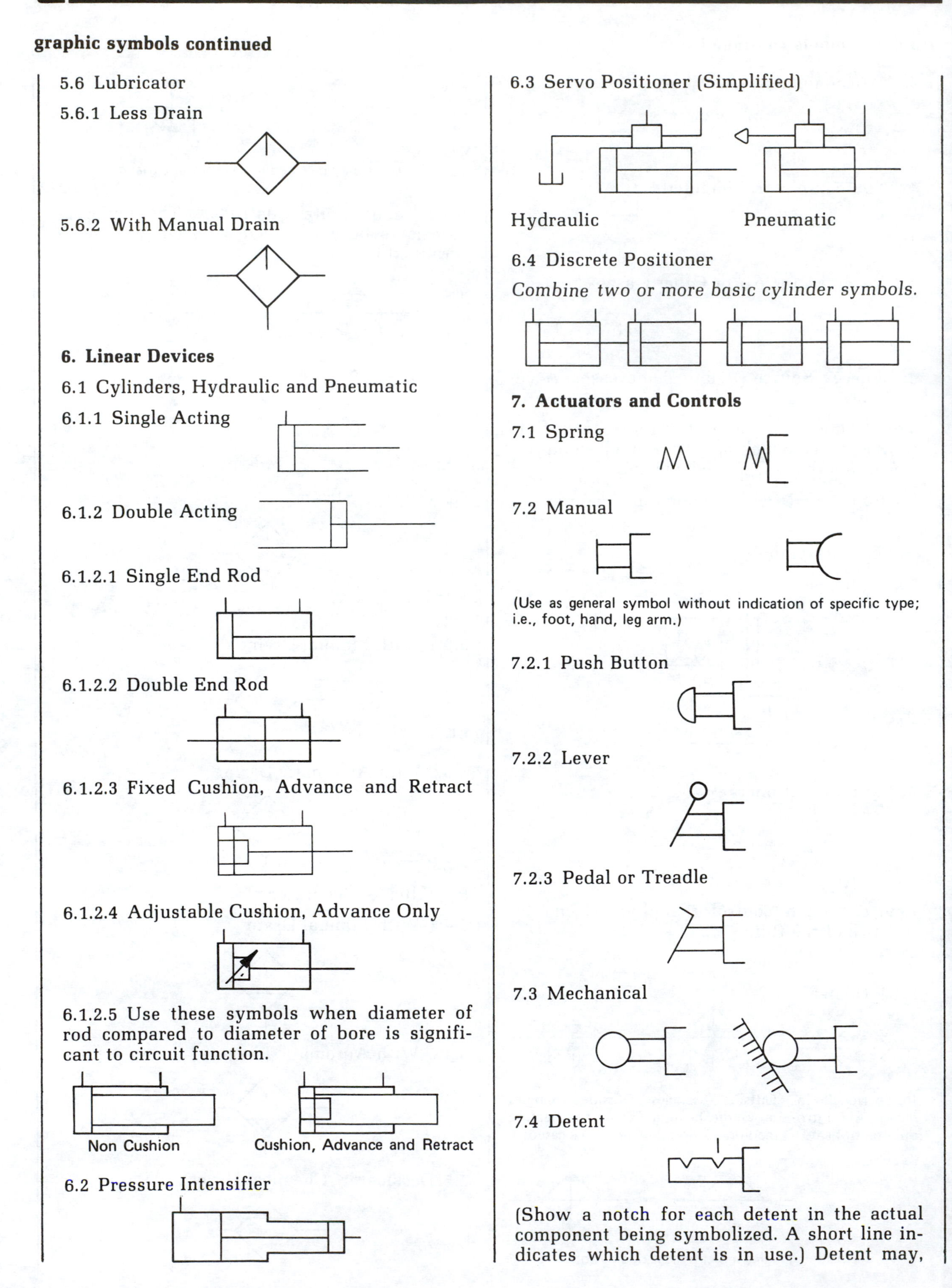

graphic symbols continued

5.6 Lubricator

5.6.1 Less Drain

5.6.2 With Manual Drain

6. Linear Devices

6.1 Cylinders, Hydraulic and Pneumatic

6.1.1 Single Acting

6.1.2 Double Acting

6.1.2.1 Single End Rod

6.1.2.2 Double End Rod

6.1.2.3 Fixed Cushion, Advance and Retract

6.1.2.4 Adjustable Cushion, Advance Only

6.1.2.5 Use these symbols when diameter of rod compared to diameter of bore is significant to circuit function.

Non Cushion

Cushion, Advance and Retract

6.2 Pressure Intensifier

6.3 Servo Positioner (Simplified)

Hydraulic

Pneumatic

6.4 Discrete Positioner

Combine two or more basic cylinder symbols.

7. Actuators and Controls

7.1 Spring

7.2 Manual

(Use as general symbol without indication of specific type; i.e., foot, hand, leg arm.)

7.2.1 Push Button

7.2.2 Lever

7.2.3 Pedal or Treadle

7.3 Mechanical

7.4 Detent

(Show a notch for each detent in the actual component being symbolized. A short line indicates which detent is in use.) Detent may,

graphic symbols continued

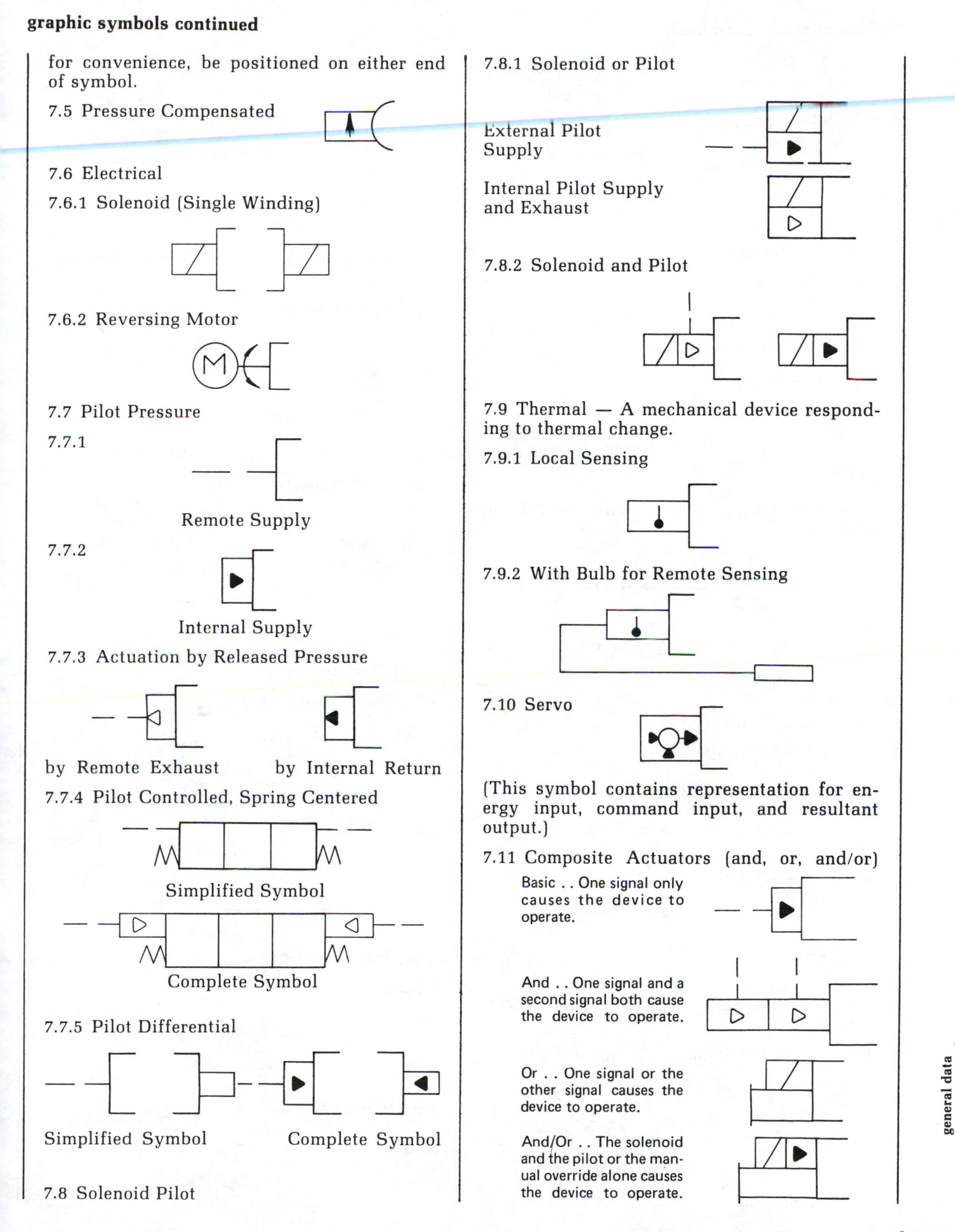

for convenience, be positioned on either end of symbol.

7.5 Pressure Compensated

7.6 Electrical

7.6.1 Solenoid (Single Winding)

7.6.2 Reversing Motor

7.7 Pilot Pressure

7.7.1

Remote Supply

7.7.2

Internal Supply

7.7.3 Actuation by Released Pressure

by Remote Exhaust | by Internal Return

7.7.4 Pilot Controlled, Spring Centered

Simplified Symbol

Complete Symbol

7.7.5 Pilot Differential

Simplified Symbol | Complete Symbol

7.8 Solenoid Pilot

7.8.1 Solenoid or Pilot

External Pilot Supply

Internal Pilot Supply and Exhaust

7.8.2 Solenoid and Pilot

7.9 Thermal — A mechanical device responding to thermal change.

7.9.1 Local Sensing

7.9.2 With Bulb for Remote Sensing

7.10 Servo

(This symbol contains representation for energy input, command input, and resultant output.)

7.11 Composite Actuators (and, or, and/or)

Basic . . One signal only causes the device to operate.

And . . One signal and a second signal both cause the device to operate.

Or . . One signal or the other signal causes the device to operate.

And/Or . . The solenoid and the pilot or the manual override alone causes the device to operate.

graphic symbols continued

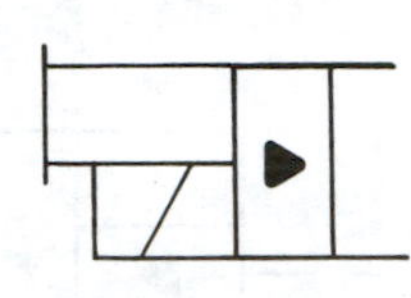

The solenoid and the pilot or the manual override and the pilot.

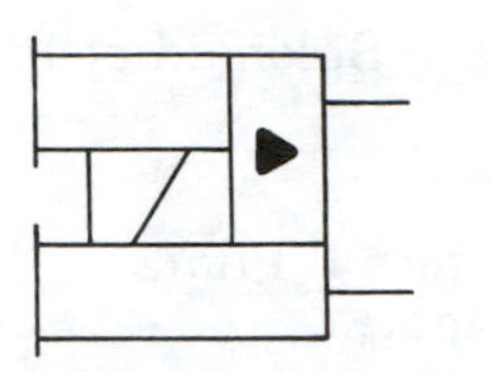

The solenoid and the pilot or a manual override and the pilot or a manual override alone.

8. Rotary Devices

8.1 Basic Symbol

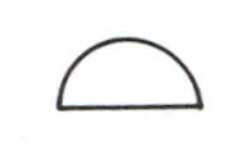

8.1.1 With Ports

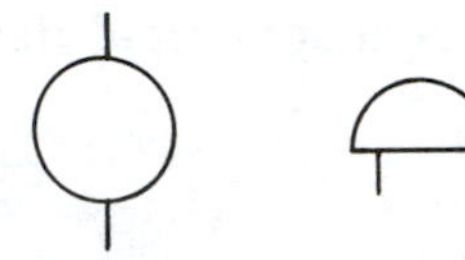

8.1.2 With Rotating Shaft, with control and with Drain.

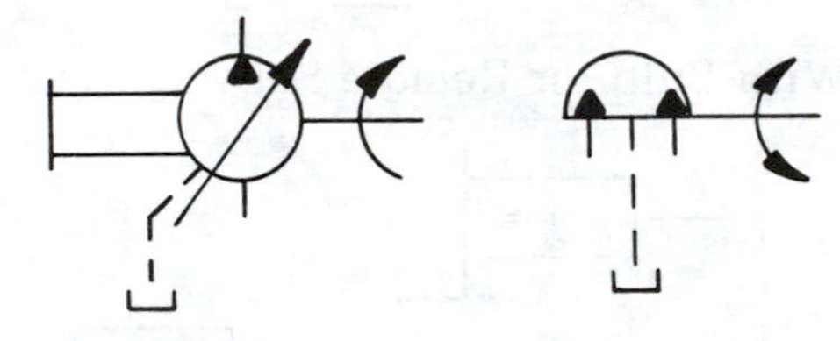

8.2 Hydraulic Pump

8.2.1 Fixed Displacement

8.2.1.1 Unidirectional

8.2.1.2 Bidirectional

8.2.2 Variable Displacement, Non-compensated

8.2.2.1 Unidirectional

Simplified

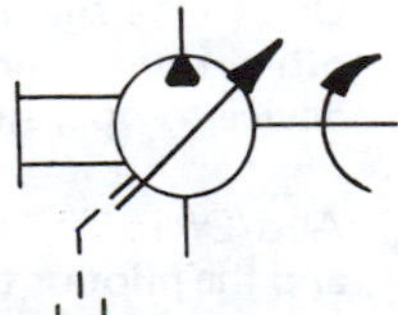

Complete

8.2.2.2 Bidirectional

Simplified

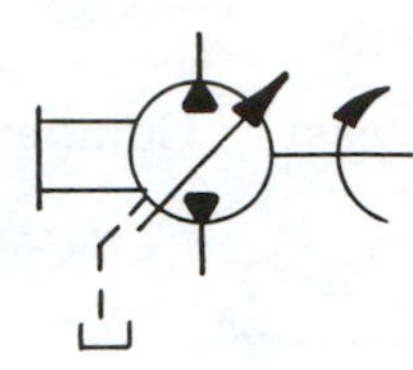

Complete

8.2.3 Variable Displacement, Pressure Compensated

8.2.3.1 Unidirectional

Simplified

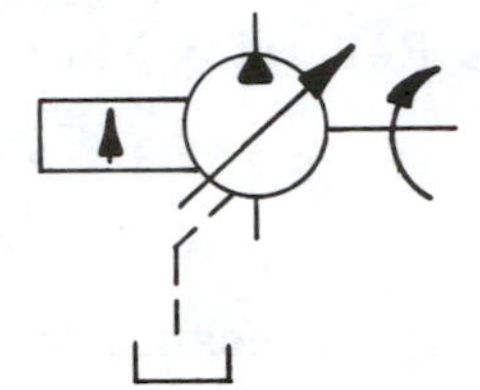

Complete

8.2.3.2 Bidirectional

Simplified

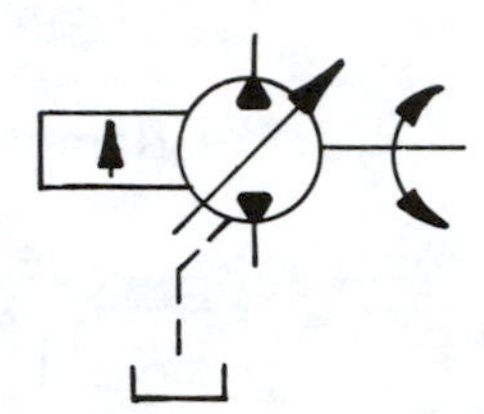

Complete

8.3 Hydraulic Motor

8.3.1 Fixed Displacement

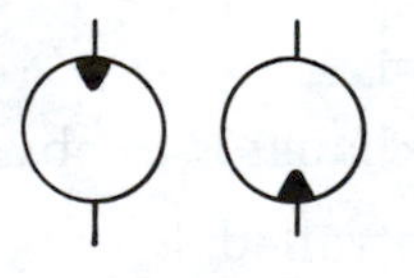

8.3.1.2 Bidirectional

8.3.2 Variable Displacement

8.3.2.1 Unidirectional

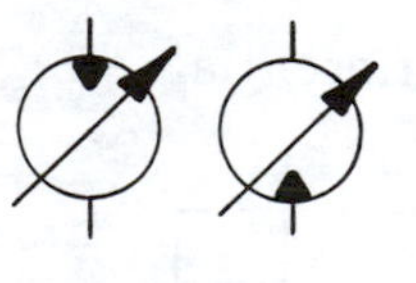

8.3.2.2 Bidirectional

graphic symbols continued

8.4 Pump-Motor, Hydraulic

8.4.1 Operating in one direction as a pump. Operating in the other direction as a motor.

8.4.1.1 Complete Symbol

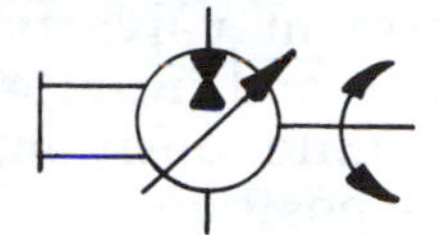

8.4.1.2 Simplified Symbol

8.4.2 Operating one direction of flow as either a pump or as a motor.

8.4.2.1 Complete Symbol

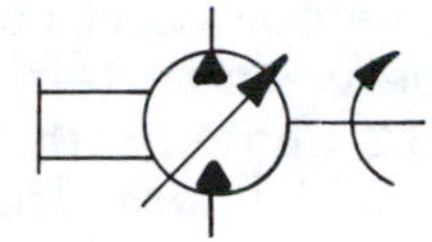

8.4.2.2 Simplified Symbol

8.4.3 Operating in both directions of flow either as a pump or as a motor. (Variable displacement, pressure compensated shown.)

8.4.3.1 Complete Symbol

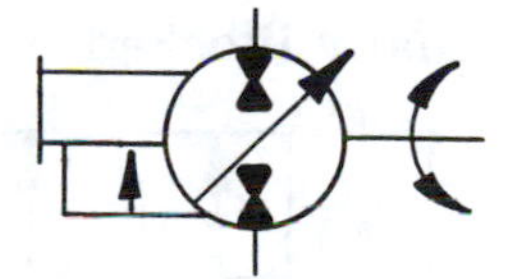

8.4.3.2 Simplified Symbol

8.5 Pump, Pneumatic

8.5.1 Compressor, Fixed Displacement

8.5.2 Vacuum Pump, Fixed Displacement

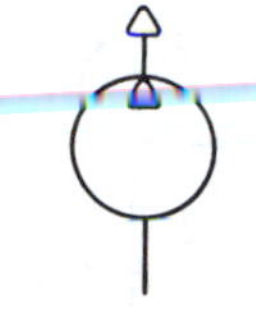

8.6 Motor, Pneumatic

8.6.1 Unidirectional

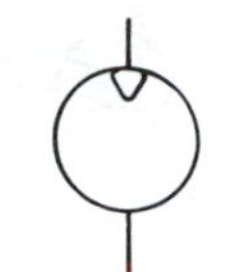

8.6.2 Bidirectional

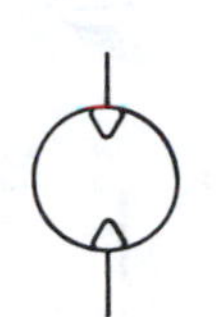

8.7 Oscillator

8.7.1 Hydraulic

8.7.2 Pneumatic

8.8 Motors, Engines

8.8.1 Electric Motor

IEC

8.8.2 Heat Engine (E.G. internal combustion engine.)

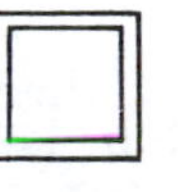

9. Instruments and Accessories

9.1 Indicating and Recording

9.1.1 Pressure

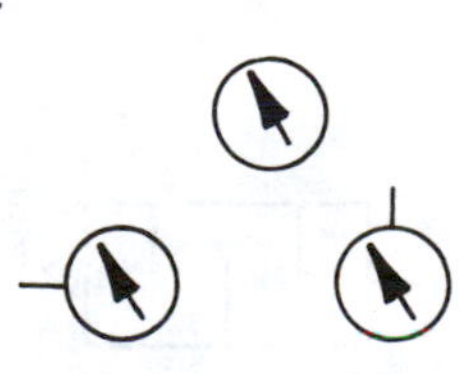

graphic symbols continued

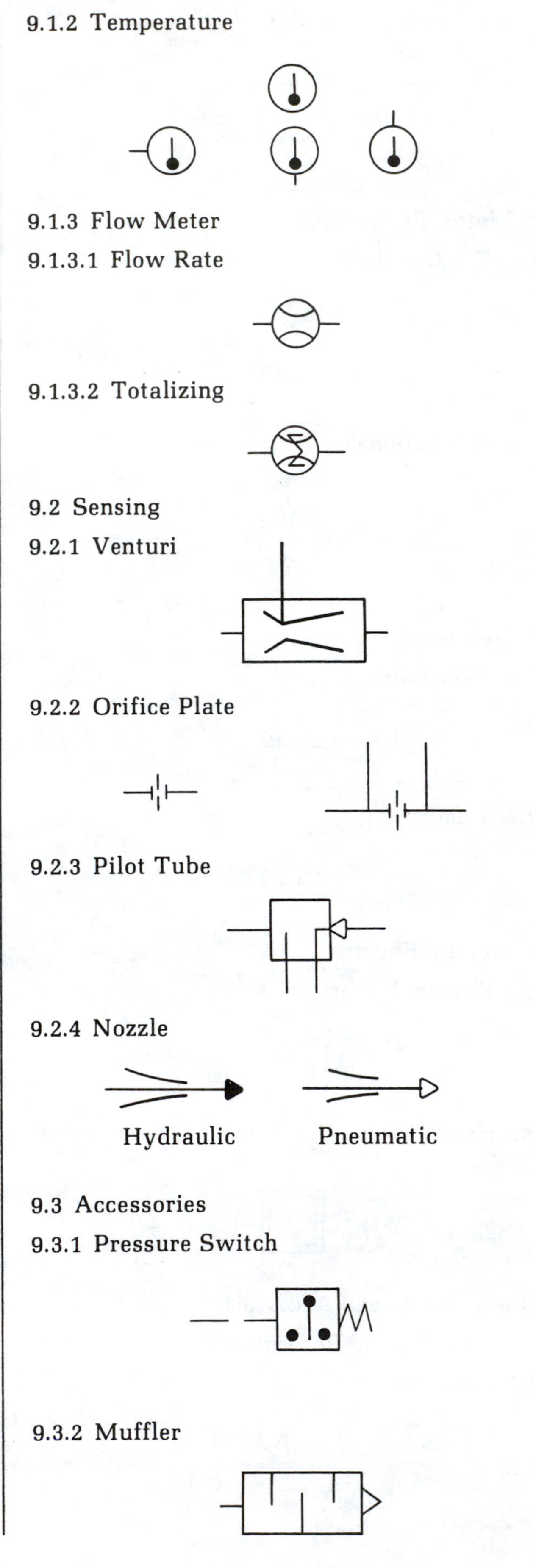

10. Valves

A basic valve symbol is composed of one or more envelopes with lines inside the envelope to represent flow paths and flow conditioners between ports. Three symbol systems are used to represent valve types: single envelope, both finite and infinite position; multiple envelope, finite position; and multiple envelope, infinite position.

10.1 In infinite position single envelope valves, the envelope is imagined to move to illustrate how pressure or flow conditions are controlled as the valve is actuated.

10.2 Multiple envelopes symbolize valves providing more than one finite flow path option for the fluid. The multiple envelope moves to represent how flow paths change when the the valving element within the component is shifted to its finite positions.

10.3 Multiple envelope valves capable of infinite positioning between certain limits are symbolized as in 10.2 above with the addition of horizontal bars which are drawn parallel to the envelope. The horizontal bars are the clues to the infinite positioning function possessed by the valve re-represented.

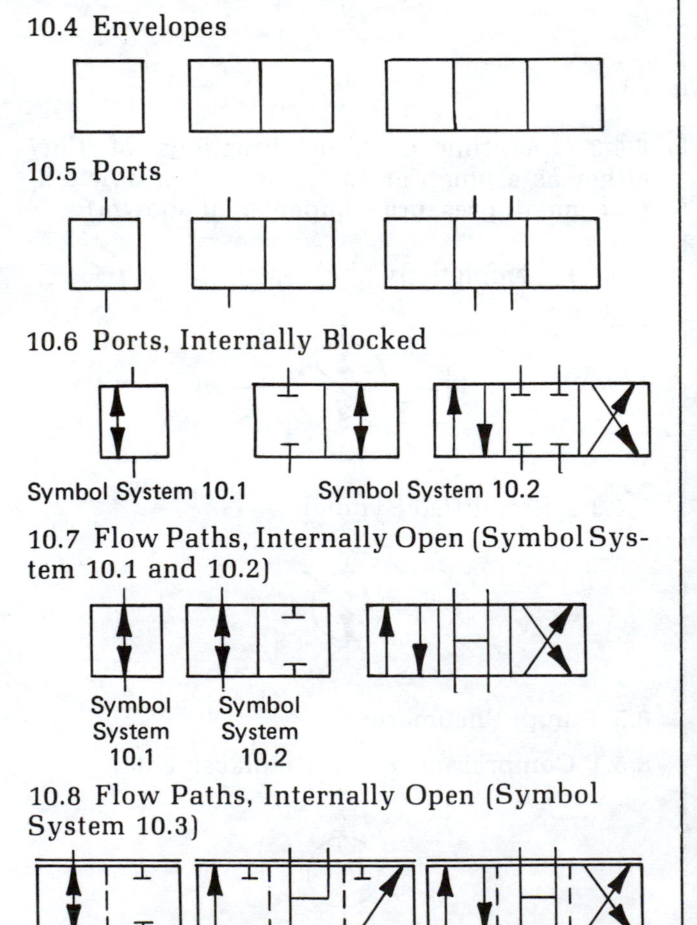

graphic symbols continued

10.9 Two-Way Valves (2 Ported Valves)

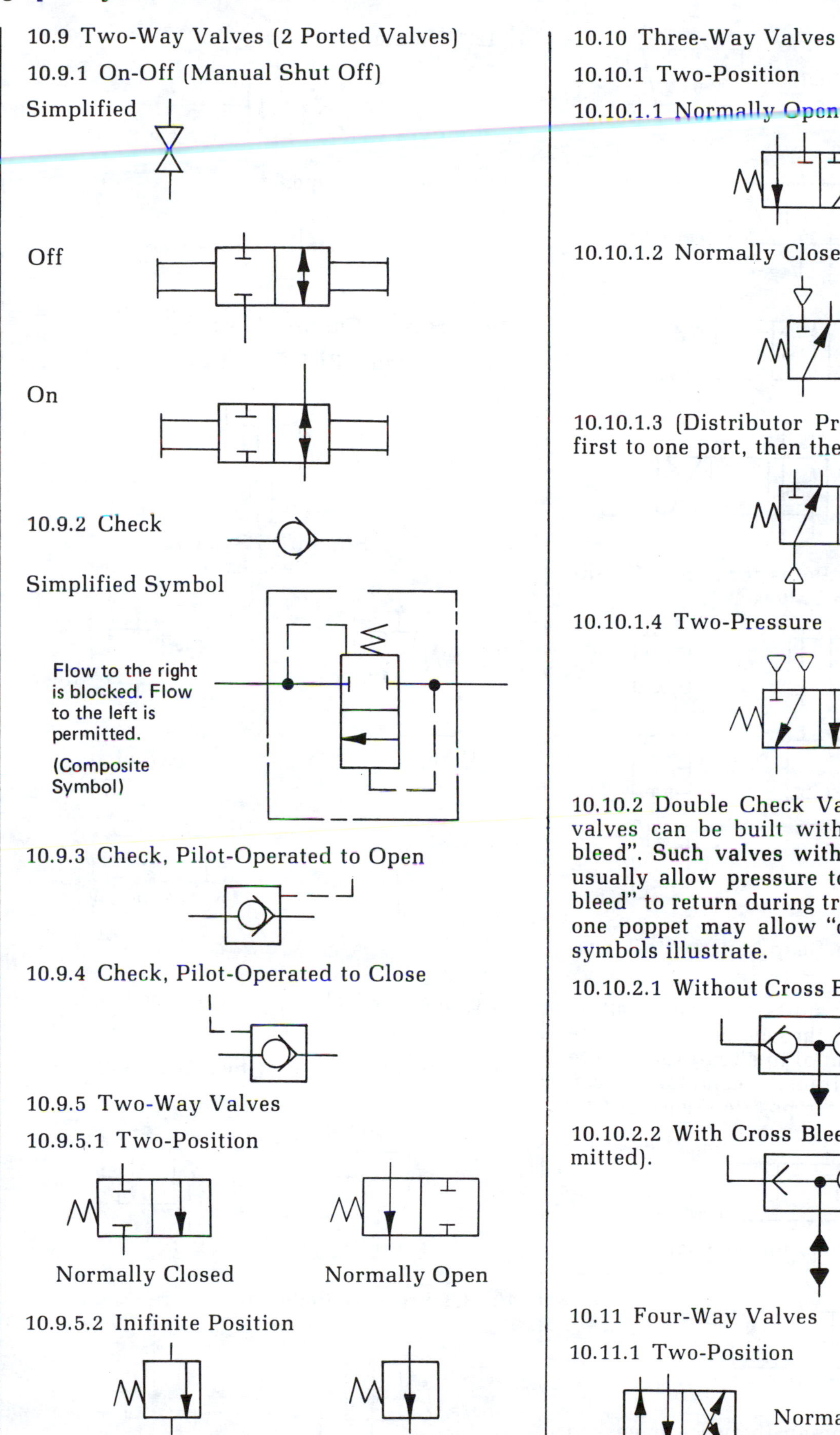

10.10 Three-Way Valves

10.10.1 Two-Position

10.10.1.1 Normally Open

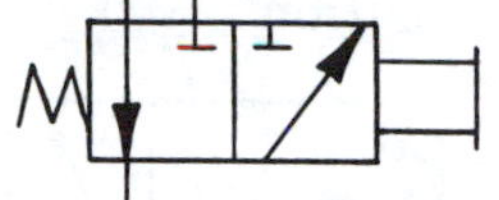

10.10.1.2 Normally Closed

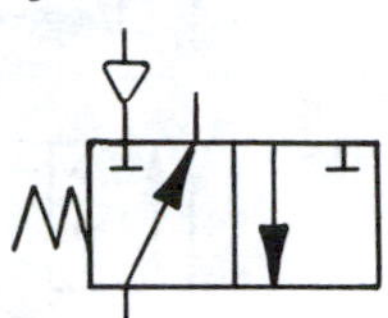

10.10.1.3 (Distributor Pressure is distributed first to one port, then the other).

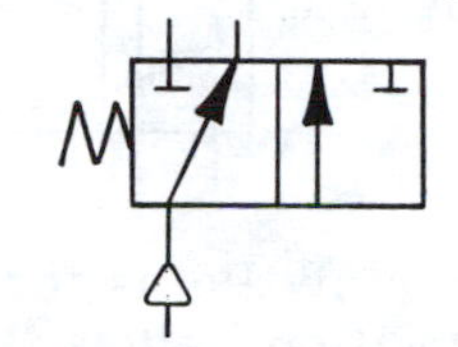

10.10.1.4 Two-Pressure

10.10.2 Double Check Valve — Double check valves can be built with and without "cross bleed". Such valves with two poppets do not usually allow pressure to momentarily "cross bleed" to return during transition. Valves with one poppet may allow "cross bleed" as these symbols illustrate.

10.10.2.1 Without Cross Bleed (One way flow).

10.10.2.2 With Cross Bleed (Reverse flow permitted).

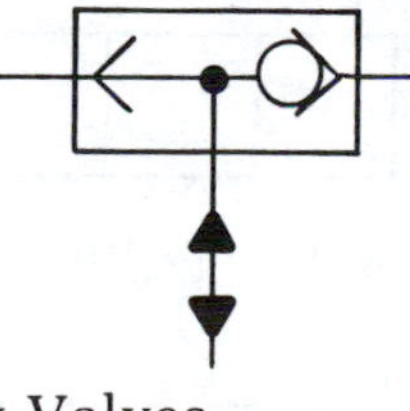

10.11 Four-Way Valves

10.11.1 Two-Position

graphic symbols continued

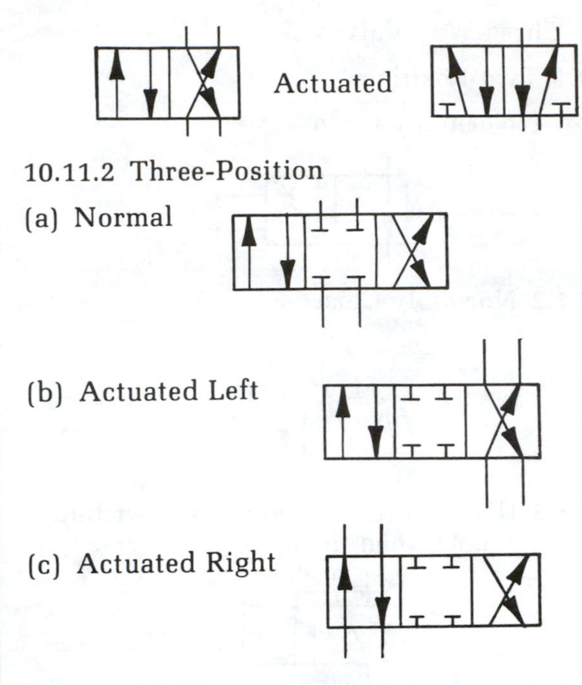

Actuated

10.11.2 Three-Position

(a) Normal

(b) Actuated Left

(c) Actuated Right

10.11.3 Typical Flow Paths for Center Condition of Three-Position Valves

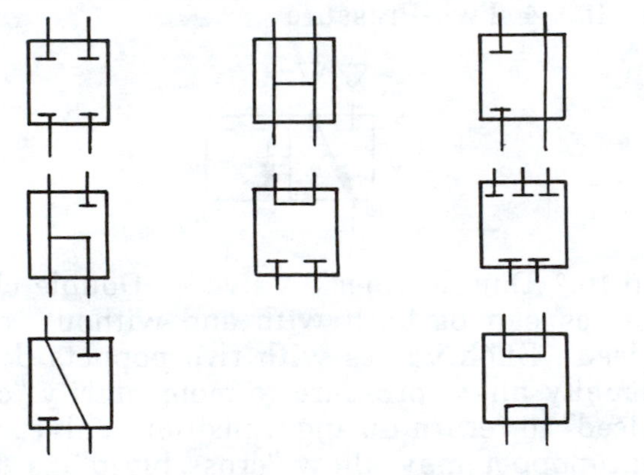

10.11.4 Two-Position, Snap Action with Transition.

As the valve element shifts from one position to the other, it passes through an intermediate position. If it is essential to circuit function to symbolize this "in transit" conditon, it can be shown in the center position, enclosed by dashed lines.

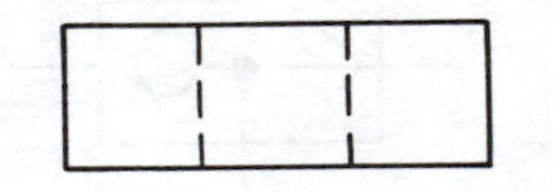

Typical Transition Symbol

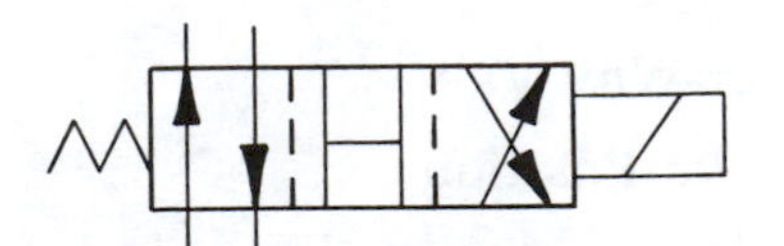

10.12 Infinite Positioning (between open and closed)

10.12.1 Normally Closed

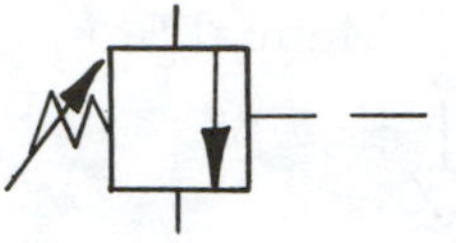

10.12.2 Normally Open

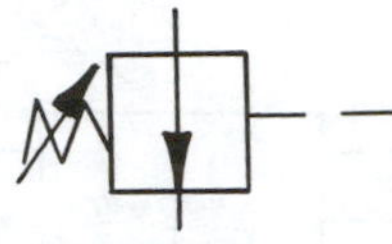

10.13 Pressure Control Valves

10.13.1 Pressure Relief

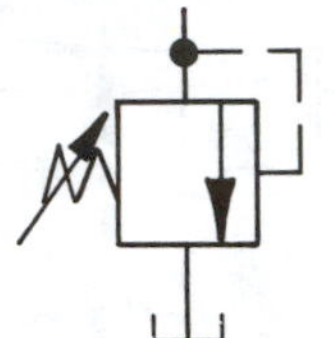

Simplified Symbol Denotes

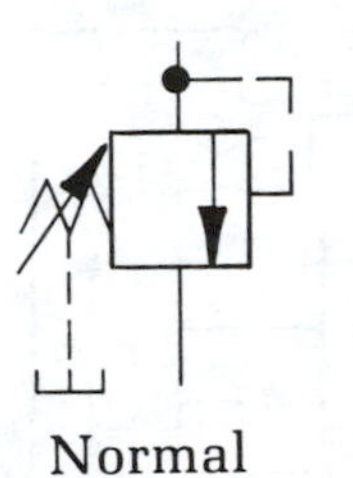

Normal

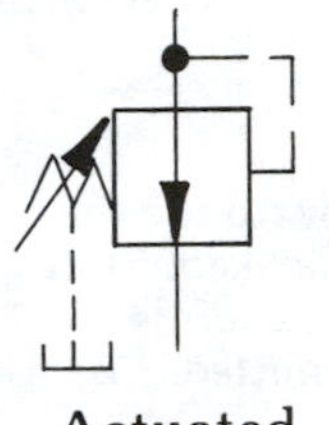

Actuated (Relieving)

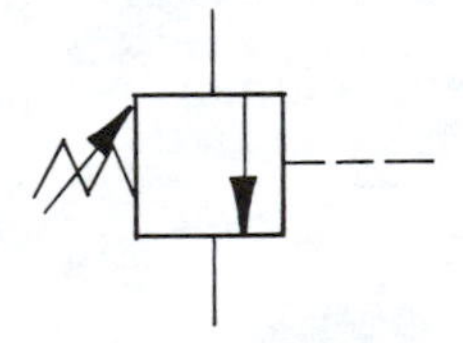

Unloading Valve

Counter Balance

10.13.2 Sequence

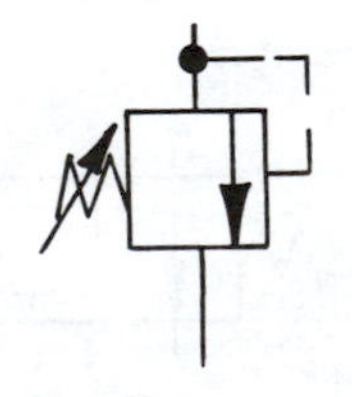

Sequence Actuated

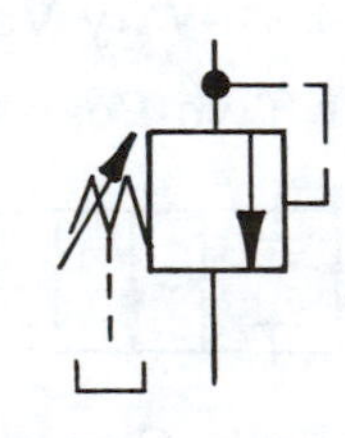

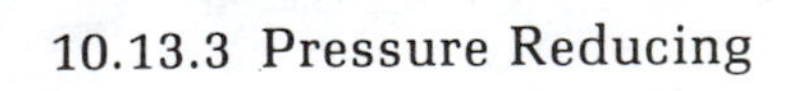

10.13.3 Pressure Reducing

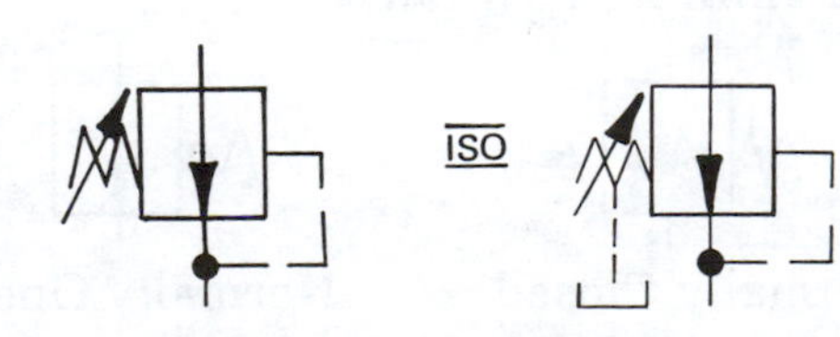

graphic symbols continued

10.13.4 Pressure Reducing and Relieving

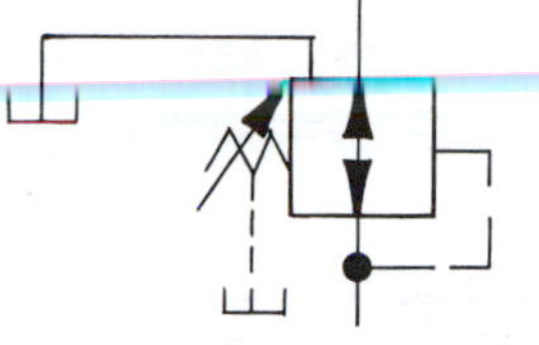

10.13.5 Airline Pressure Regulator (Adjustable, Relieving)

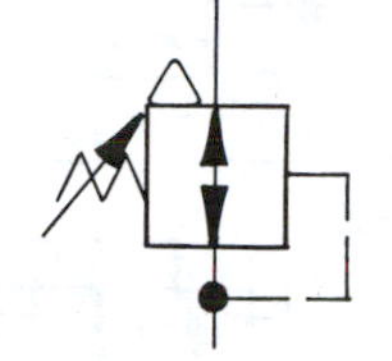

10.14 Infinite Positioning Three-Way Valves

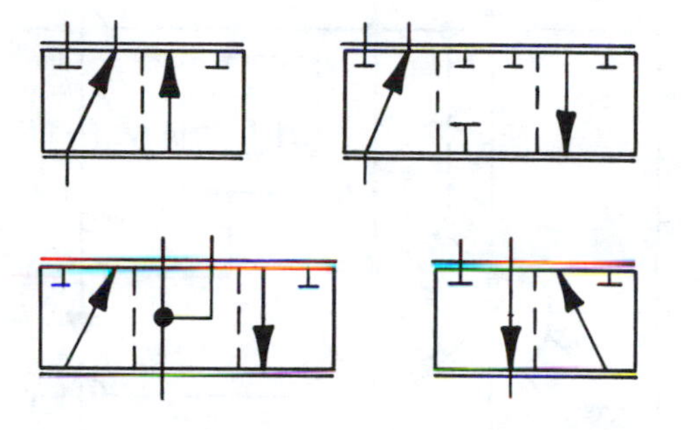

10.15 Infinite Positioning Four-Way Valves

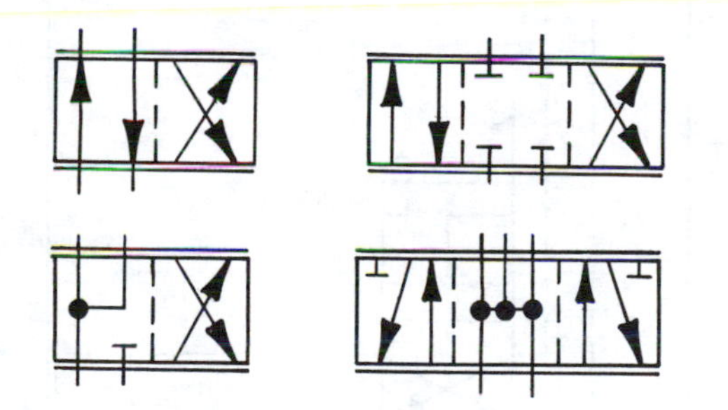

10.16 Flow Control Valves (See 3.7)

10.16.1 Adjustable, Non-Compensated (Flow control in each direction).

10.16.2 Adjustable with Bypass

Flow is controlled to the right. Flow to the left bypasses control.

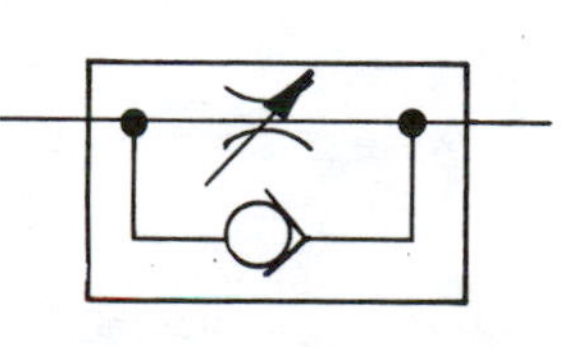

10.16.3 Adjustable and Pressure Compensated with Bypass

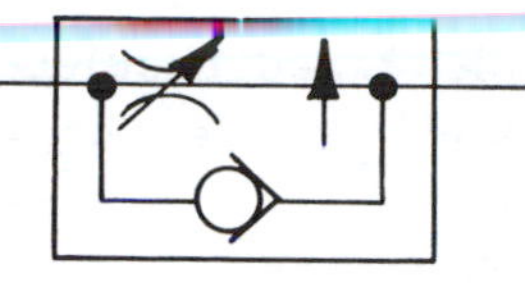

10.16.4 Adjustable, Temperature and Pressure Compensated.

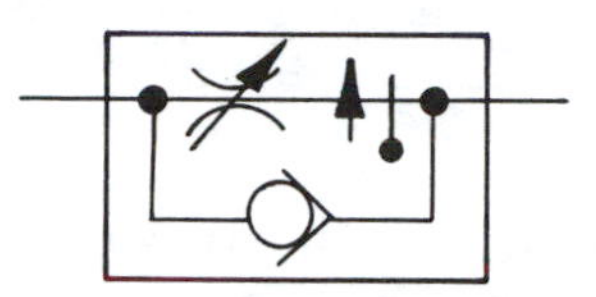

11. Representative Composite Symbols

11.1 Component Enclosure

Component enclosure may surround a complete symbol or a group of symbols to represent an assembly. It is used to convey more information about component connections and functions. Enclosure indicates extremity of component or assembly. External ports are assumed to be on enclosure line and indicate connections to component.

Flow lines shall cross enclosure line without loops or dots.

11.2 Airline Accessories (Filter, Regulator and Lubricator).

Composite

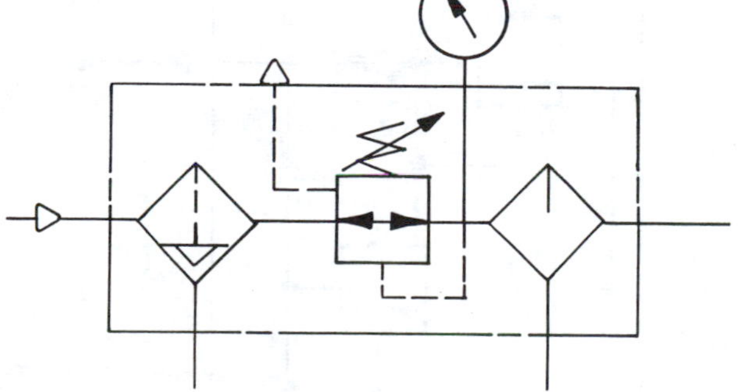

Simplified

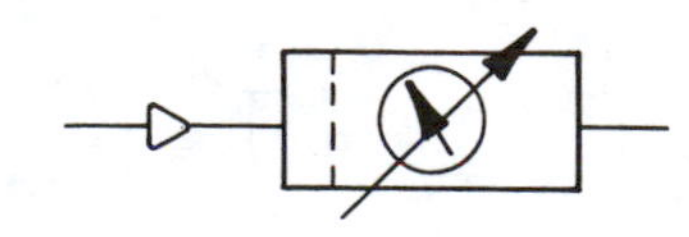

graphic symbols continued

11.3 Pumps and Motors

11.3.1 Pumps

11.3.1.1 Double, Fixed Displacement, One Inlet and Two Outlets.

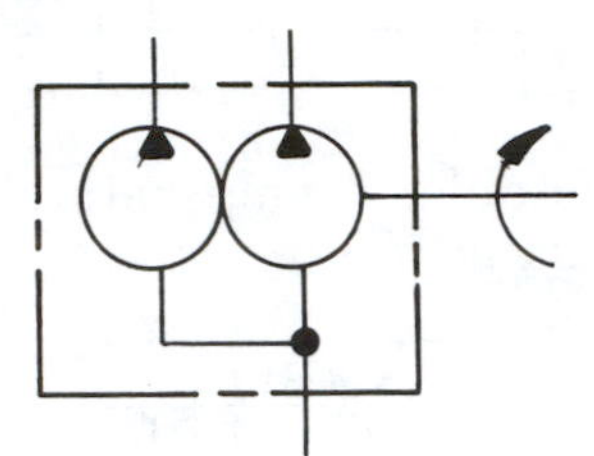

11.3.1.2 Double, with Integral Check Unloading and Two Outlets.

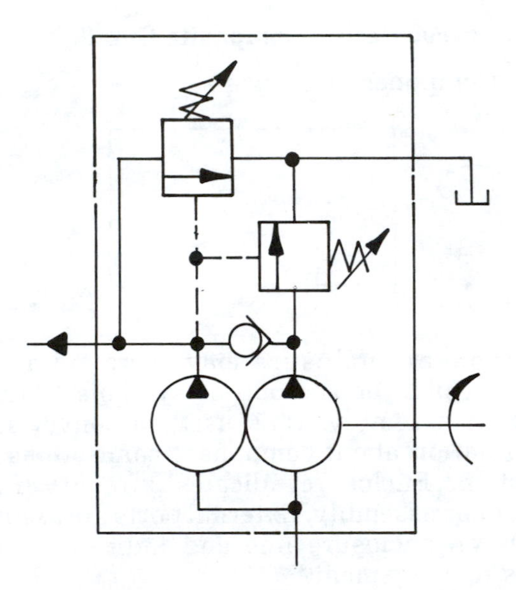

11.3.1.3 Integral Variable Flow Rate Control with Overload Relief.

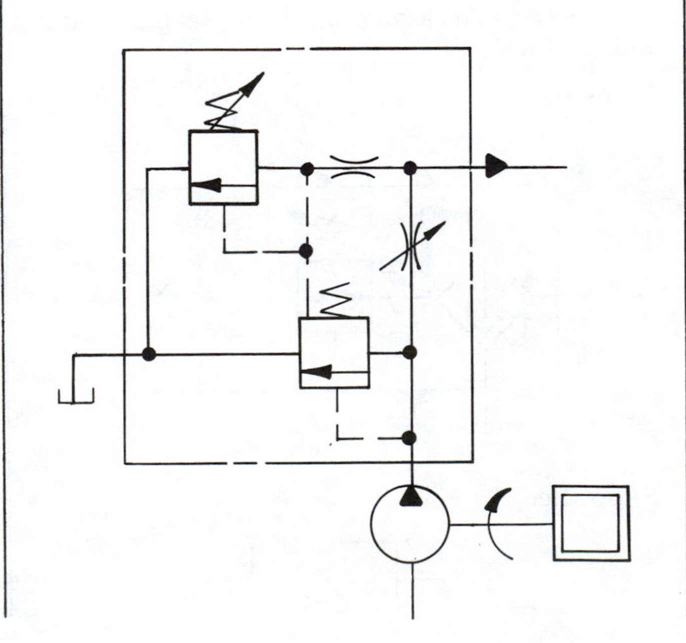

11.3.1.4 Variable Displacement with Integral Replenishing Pump and Control Valves.

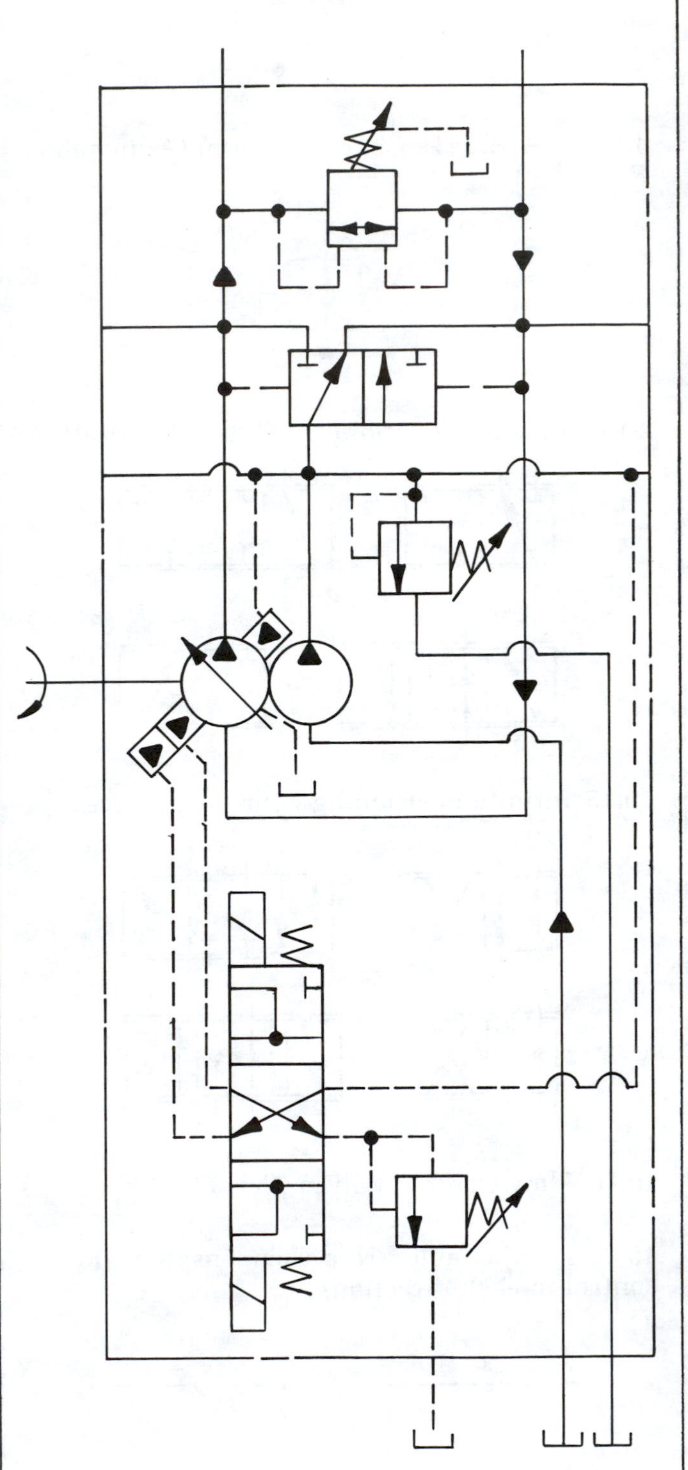

graphic symbols continued

11.4.6 Two-Positions, Four Connection Solenoid and Pilot Actuated, with Manual Pilot Override.

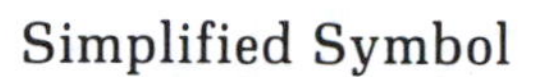

Simplified Symbol

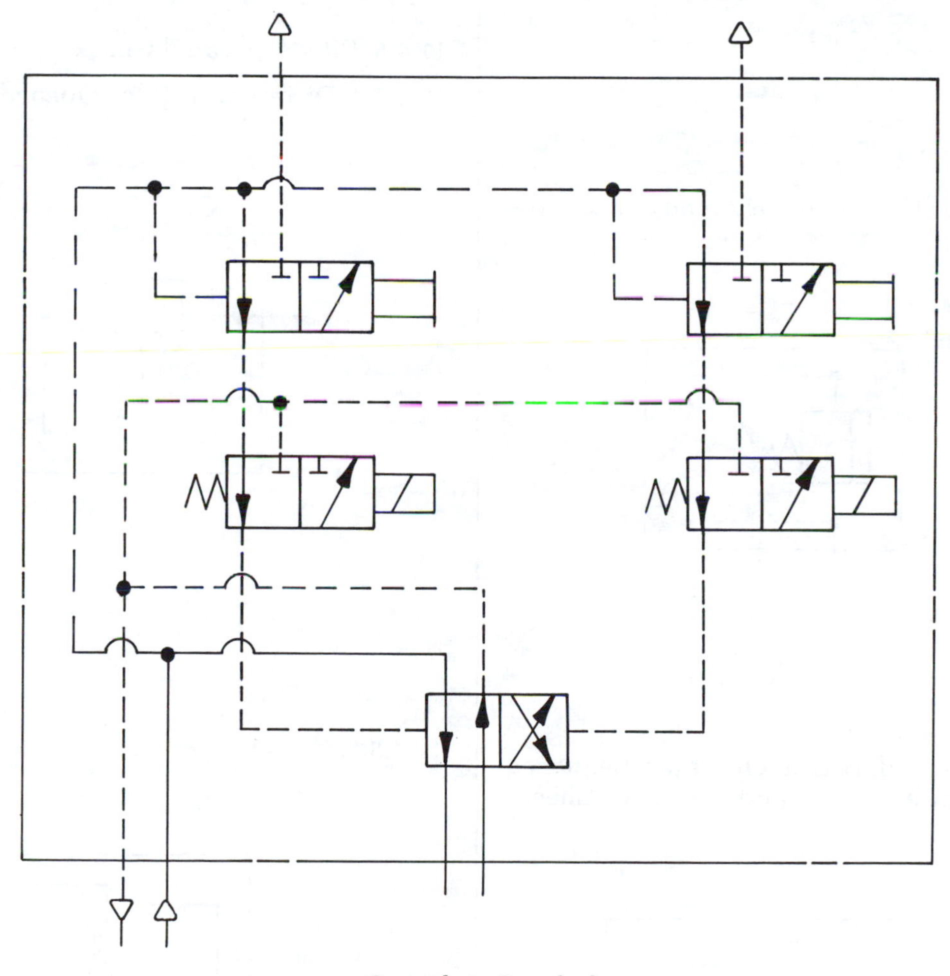

Complete Symbol

graphic symbols continued

11.4 Valves

11.4.1 Relief, Balanced Type

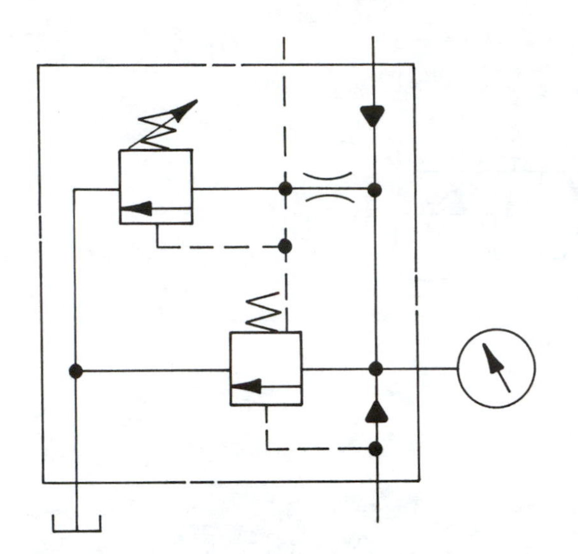

11.4.2 Remote Operated Sequency with Integral Check.

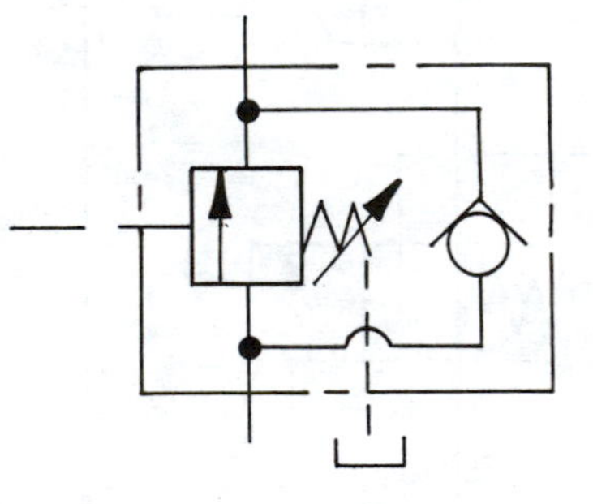

11.4.3 Remote and Direct Operated Sequence with Differential areas and Integral Check.

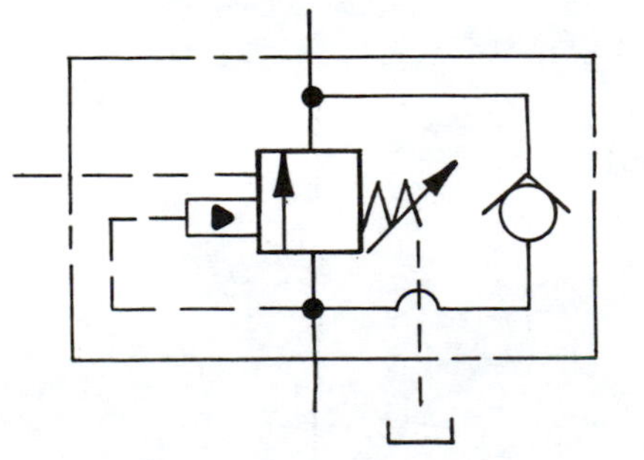

11.4.4 Pressure Reducing with Integral Check.

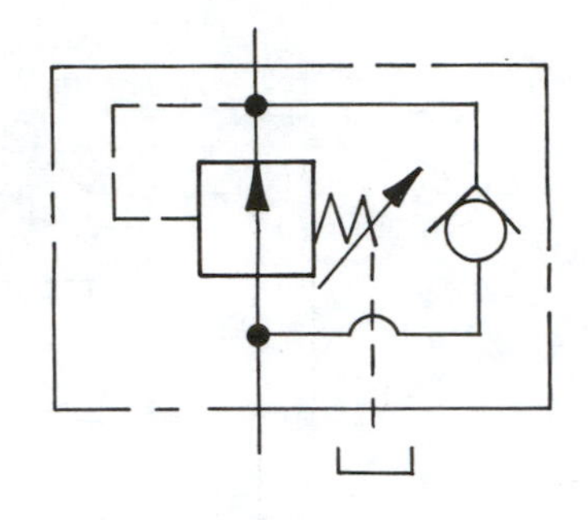

11.4.5 Pilot Operated Check

11.4.5.1 Differential Pilot Opened.

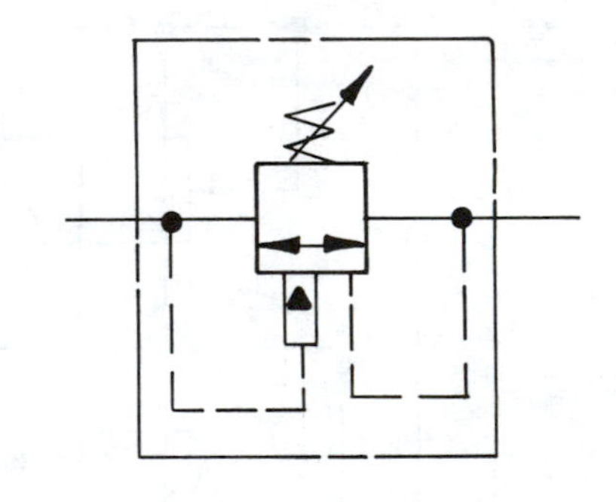

11.4.5.2

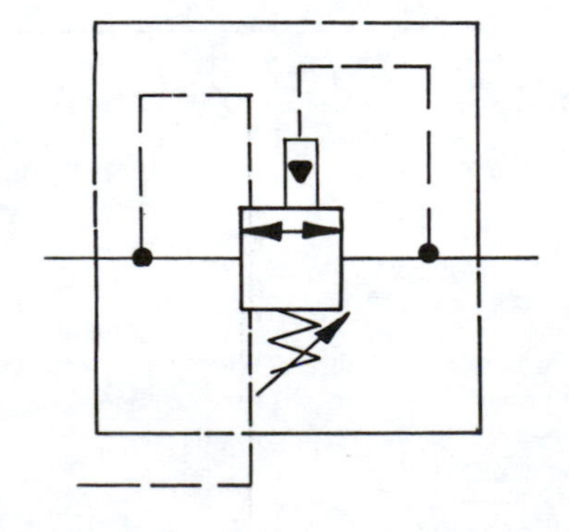

graphic symbols continued

11.4.7 Two-Position, Five Connection, Solenoid Control Pilot Actuated with Detents and Throttle Exhaust.

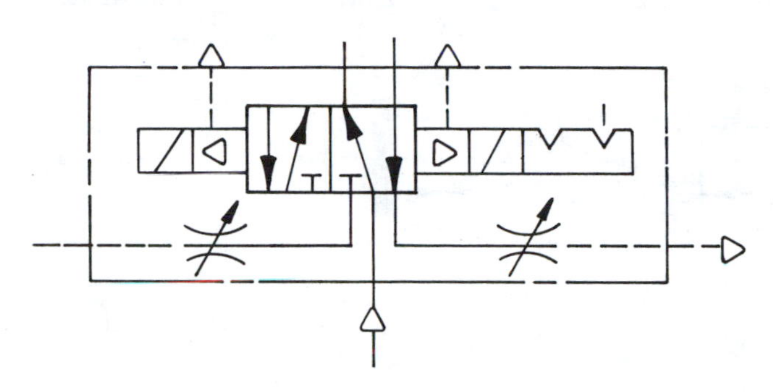

Simplified Symbol

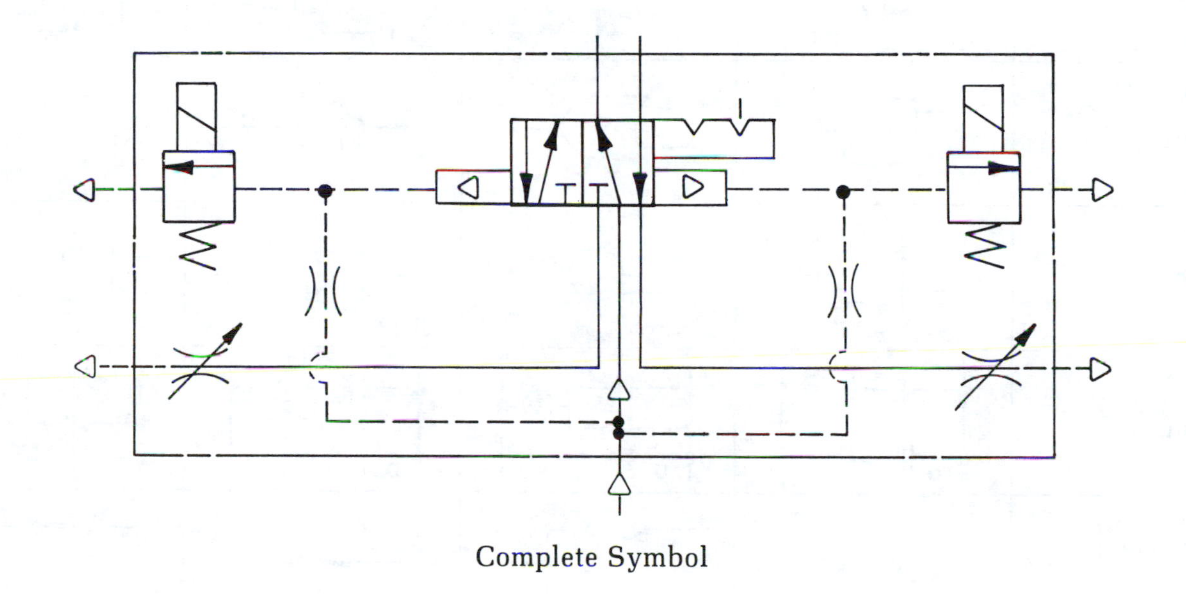

Complete Symbol

11.4.8 Variable Pressure Compensated Flow Control and Overload Relief.

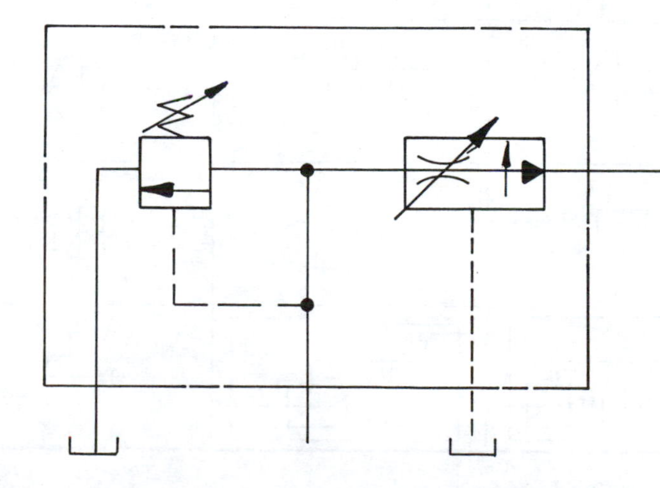

pneumatic logic symbols

LOGIC ELEMENT	AND	OR	NOT	NAND
LOGIC ELEMENT FUNCTION	OUTPUT IF ALL CONTROL INPUT SIGNALS ARE OFF	OUTPUT IF ANY ONE OF THE CONTROL INPUTS IS ON	OUTPUT IF SINGLE CONTROL INPUT SIGNAL IS OFF	NO OUTPUT IF ALL CONTROL INPUT SIGNALS ARE ON
STANDARD LOGIC SYMBOL	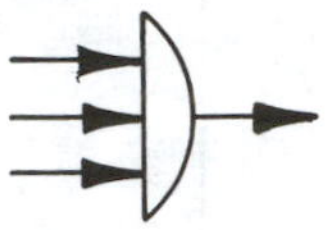	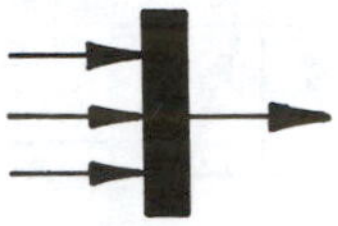	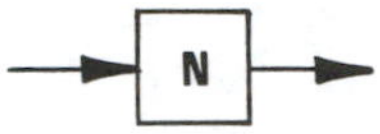 N	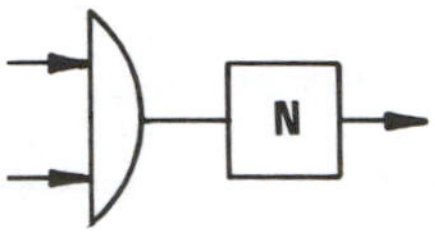 N
BOOLEAN ALGEBRA SYMBOL	( )•( )	( )+( )	$\overline{(\ )}$	$\overline{(\ )\bullet(\ )}$
PNEUMATIC LOGIC SYMBOL	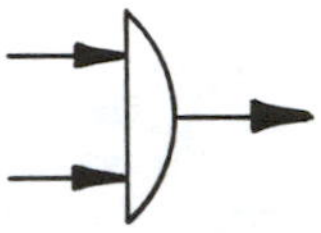	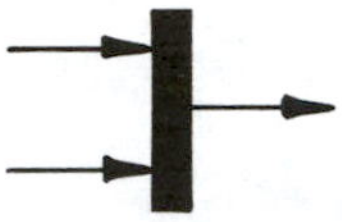	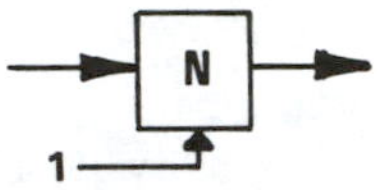 N 1	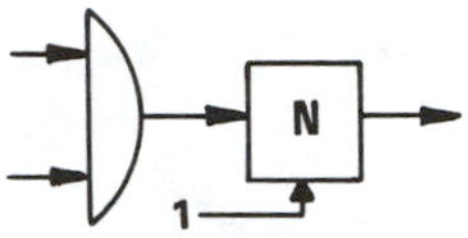 N 1
MIL-STD-806B LOGIC SYMBOL	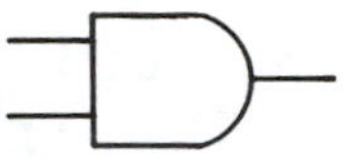	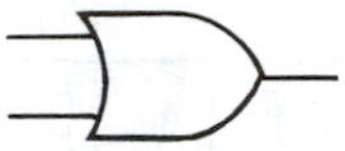	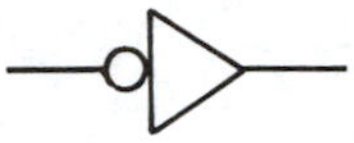	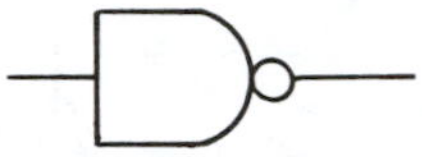
NEMA LOGIC SYMBOL	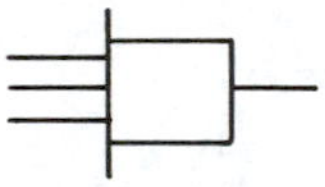			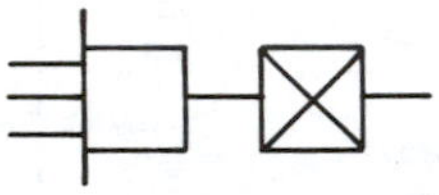
ELECTRICAL RELAY LOGIC SYMBOL	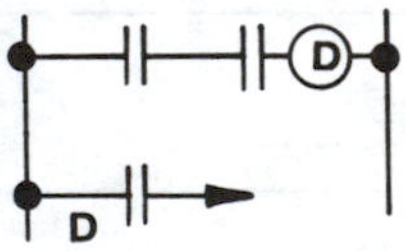 D D	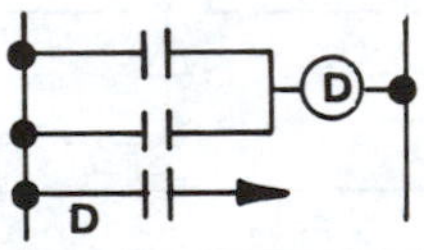 D D	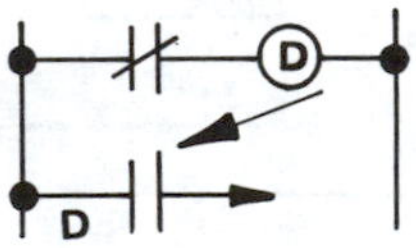 D D	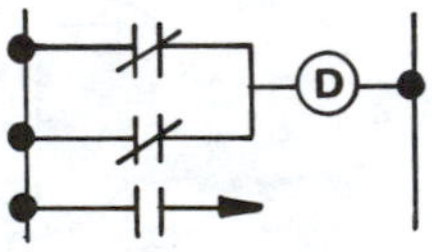 D
ELECTRICAL SWITCH LOGIC SYMBOL		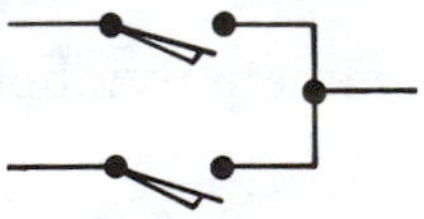		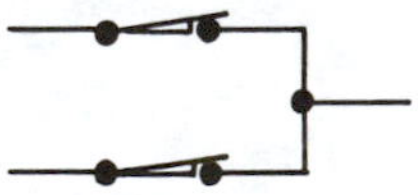
A.S.A. (J.I.C.) VALVING SYMBOL	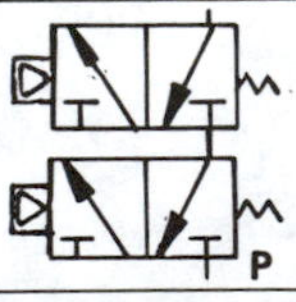 P	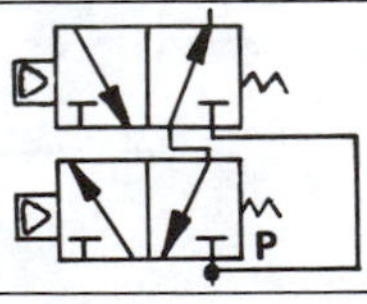 P	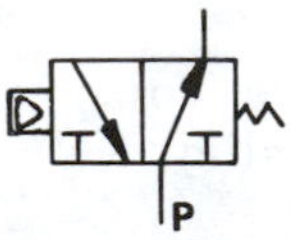 P	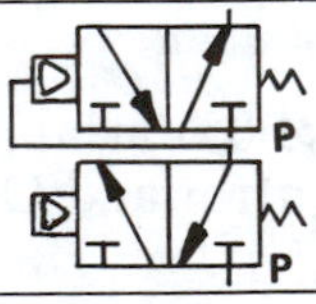 P P
FLUIDIC DEVICE TURBULENCE AMPLIFIER	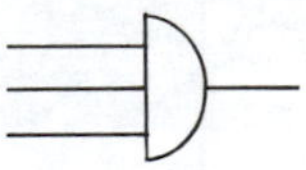	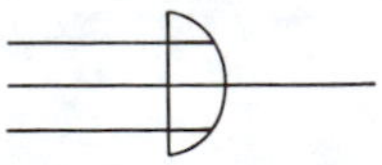	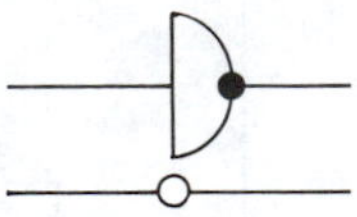	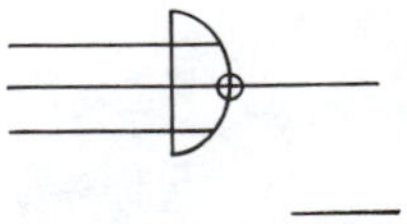
PROPOSED N.F.P.A./A.S.A. SYMBOL	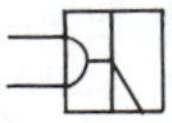	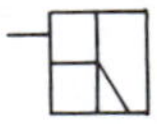		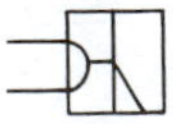
NOR LOGIC EQUIVALENT OF PROPOSED N.F.P.A./A.S.A. SYM.	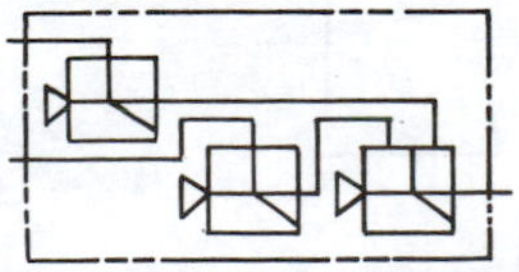	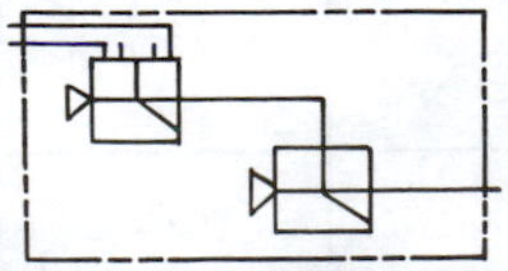	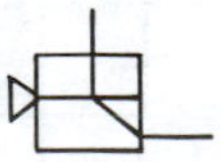	

pneumatic logic symbols continued

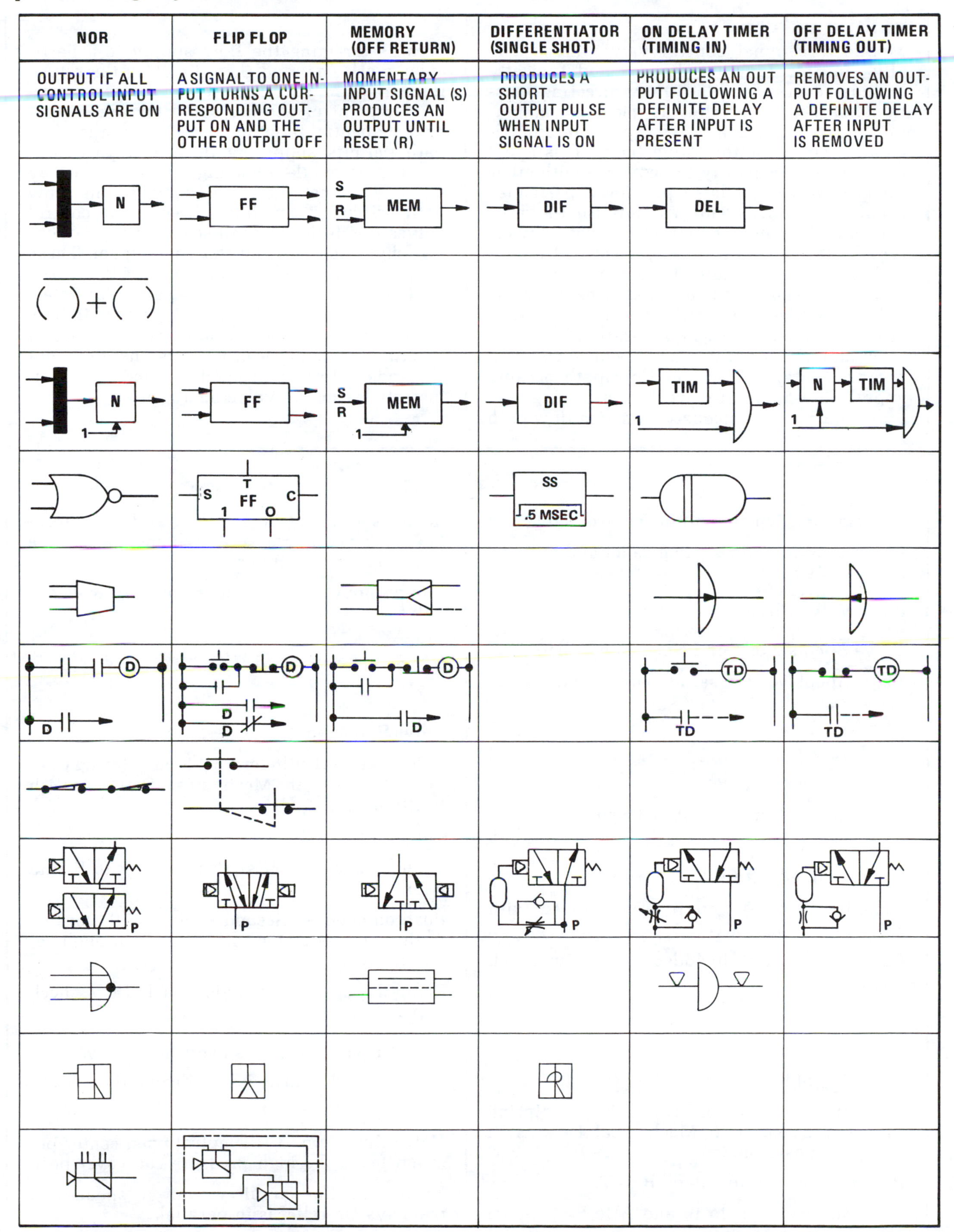

NOR	FLIP FLOP	MEMORY (OFF RETURN)	DIFFERENTIATOR (SINGLE SHOT)	ON DELAY TIMER (TIMING IN)	OFF DELAY TIMER (TIMING OUT)
OUTPUT IF ALL CONTROL INPUT SIGNALS ARE ON	A SIGNAL TO ONE INPUT TURNS A CORRESPONDING OUTPUT ON AND THE OTHER OUTPUT OFF	MOMENTARY INPUT SIGNAL (S) PRODUCES AN OUTPUT UNTIL RESET (R)	PRODUCES A SHORT OUTPUT PULSE WHEN INPUT SIGNAL IS ON	PRODUCES AN OUTPUT FOLLOWING A DEFINITE DELAY AFTER INPUT IS PRESENT	REMOVES AN OUTPUT FOLLOWING A DEFINITE DELAY AFTER INPUT IS REMOVED

Parker mobile symbols

mobile valves

Mobile directional control valves are generally multiple function devices and typically include several "en-bloc" flow directing functions. Directing the flow is generally accomplished by moving a valve "spool" to a particular position in the valve body. This spool movement is generally in sequence with other integral valve spools to provide a particular set of output flow events. Actuation of the valve spools is accomplished either manually, by direct hydraulic pilots, or by electrically controlled hydraulic pilots. To achieve these various flow events, the flow paths between the valve spools are generally designed to permit either parallel, series, or series-parallel, (parallel, and series-parallel are the most common arrangements). A parallel flow path permits supply flow to several work ports, using several spools simultaneously. A series flow path permits directing the flow supply to a particular work port and then directing the return flow from the power device at the port to the supply point of another spool for use on the next sequential function. A series-parallel flow path permits supplying flow to a particular work port while blocking all spool supply points "downstream". To achieve this flow path circuitry, mobile directional control valves are designed with several supply point switching positions on the valve spool. These supply point switching positions are then operated in combination with a particular internal valve geometry to yield a particular flow path when the spool is actuated. The circuit diagram for this type of valve is, therefore, not a standard form and must be modified to illustrate the actual valve operation.

Description: Typical en-bloc control valve

Pump flow to main system relief and to valve spool system.

Basic Valve Circuitry

Valve Spools No. 1 and No. 2 in parallel.

Valve Spool No. 3 in series-parallel with No. 2.

Valve Spool No. 1:

3-Position — Spring Centered Spool.

Pull on spool = P to A, and B to T.

Push on spool = P to B, and A to T.

Cylinder Ports A and B are equipped with cross-over relief valves.

Pump flow input includes back flow check valve.

Valve Spool No. 2:

3-Position — (Regenerative Make-Up) Spring Centered Spool with Mechanical Detents in both End Positions.

Pull on spool = P to A, and B to T.

Push on spool = P to B, and A to B. — Spool includes A to B check valve, and A to T relief valve set at pressure high enough to force A oil to return to B port.

Pump input flow includes back flow check valve.

Valve Spool No. 3:

4-Position — (Cylinder Port Float(Spring Centered Spool with Mechanical Detent in 4th Position.

Pull on spool = P to A, and B to T.

Push on spool — First Position = P to B, and A to T.

Push on spool — Second Position = P blocked, A through restrictor to T, and B through restrictor to T.

Pump input flow includes back flow check valve.

Cylinder port A includes overload relief valve.

Cylinder port B includes anti-cavitation check valve.

Valve Spool No. 3 — Neutral open center position feeds the high pressure carryover port.

Valve tank line circuitry from all spools and main system relief is in parallel.

parker mobile symbols continued

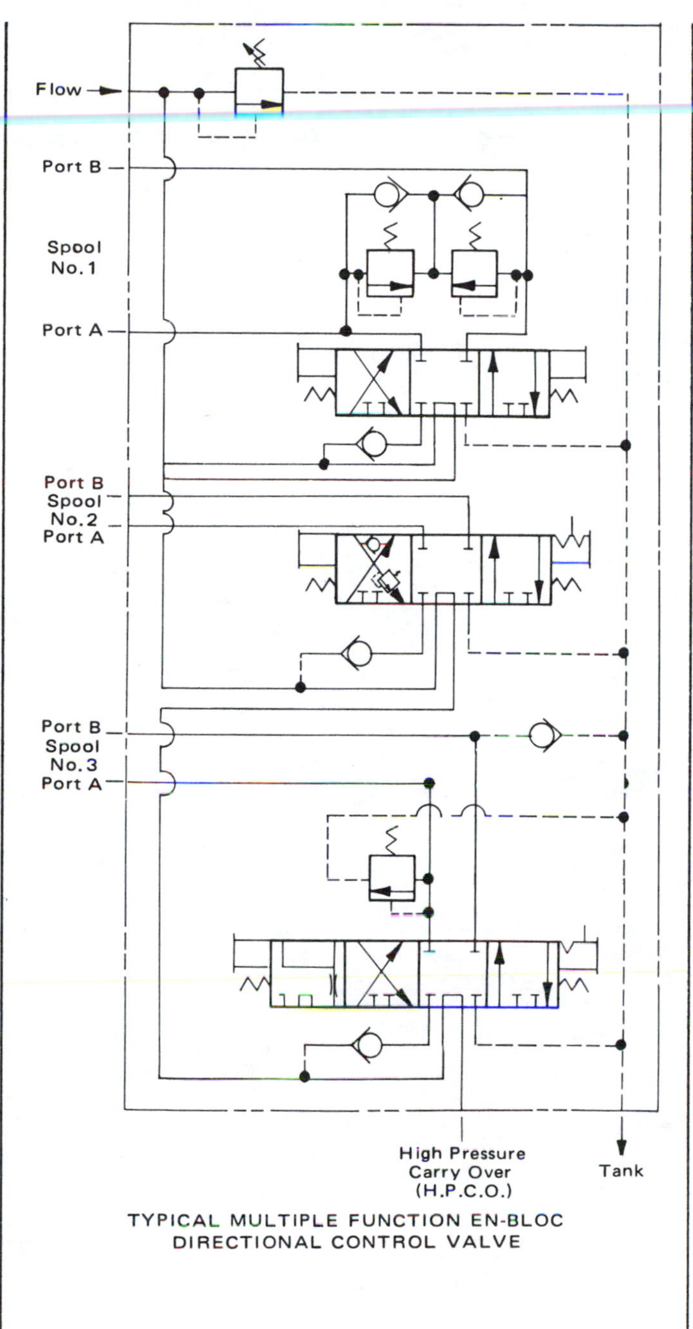

TYPICAL MULTIPLE FUNCTION EN-BLOC DIRECTIONAL CONTROL VALVE

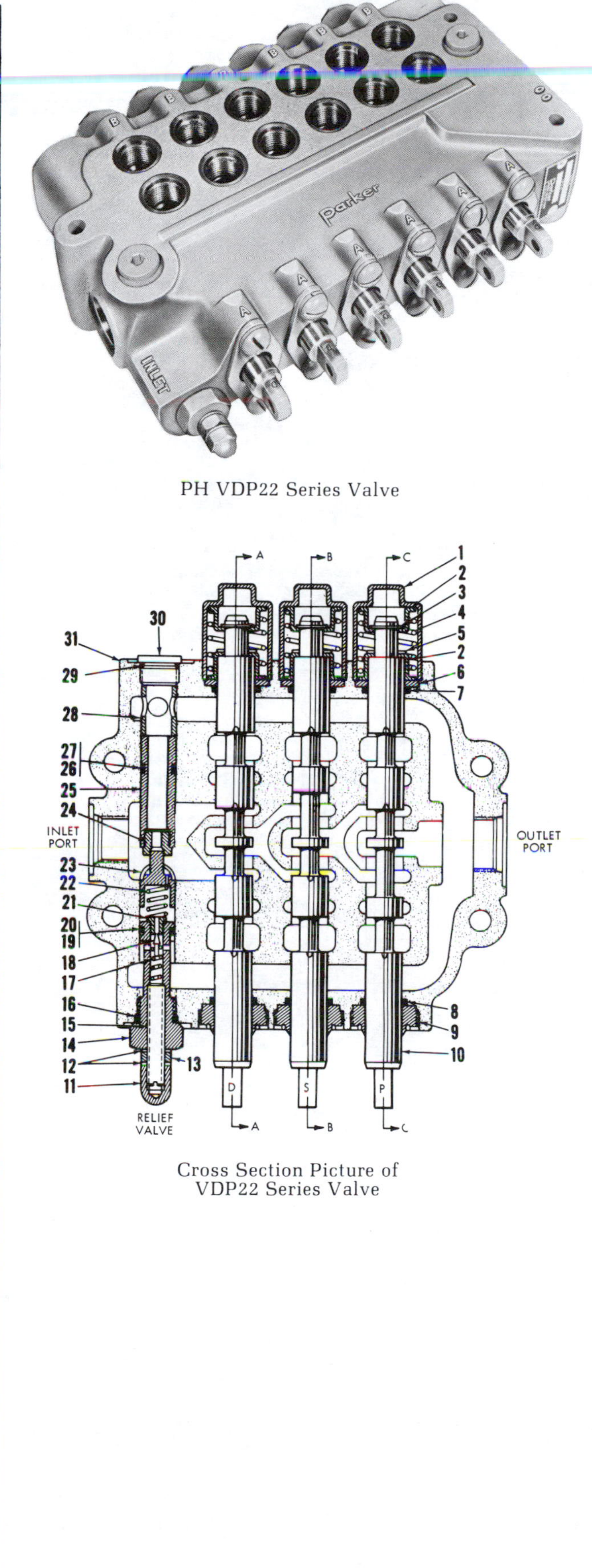

PH VDP22 Series Valve

Cross Section Picture of VDP22 Series Valve

PH. Multiple Spool Mobile Valves

Shown in cross sectional view is a typical Parker mobile multiple spool valve. Multiple ports are shown with the various spools available.

D = Double acting spool
P = Single acting spool, pull to power cylinder
S = Single acting spool, shove to power cylinder
C = Motor spool
F = Float spool

notes

ROTARY ACTUATORS

List of Symbols 2

Types of Rotary Actuators 3

Comparison Charts 7

Calculating Torque Requirements 8

Mass Moments of Inertia Table 10

Velocity & Acceleration Equations 10

Table of Torque Applications & Equations 11

Approximating Actuator Size 19

Calculating Required Pump Flow 23

Circuit Recommendations 26

Installation Instructions and Options 29

List of Symbols

A Area

d Diameter

F Force

I Mass mount of inertia

m Mass

P Pressure

Q Volumetric flow rate

r Radius

r_b Bearing radius

t Time

T Torque

T_α Angular acceleration torque

$T_{\alpha *}$ Angular deceleration torque

T_c Cushion torque

T_d Demand Torque

T_f Friction torque

V Volume displacement

V_s Specific Volume in terms of in^3 per radian

W Weight

x Distance or position

α Angular acceleration

μ Coefficient of friction

Θ Angular displacement or rotation

ω Angular velocity

ω_o Angular velocity at time = o

ω_t Angular velocity at time = t

Rotary Actuators

Introduction

A rotary actuator is the most compact device available for producing torque from hydraulic or pneumatic pressure. A self-contained unit, it is usually limited to one revolution or less and can provide oscillating motion as well as high and constant torque. Figure 1 shows the standard symbols for pneumatic and hydraulic rotary actuators.

There are many types of rotary actuators, each with design advantages as well as compromises. The three most commonly used are **rack and pinion, vane,** and **helical.** These type actuators are compared in Table 3 on page 7.

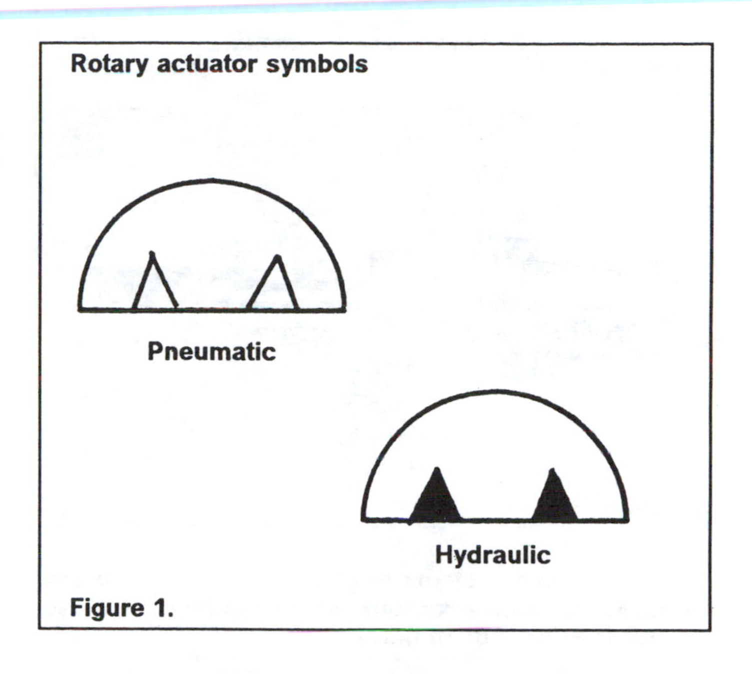

Figure 1.

Types of Rotary Actuators

Rack and Pinion: Rack and pinion actuators consist of a housing to support a pinion which is driven by a rack with cylinder pistons on the ends (See Figure 2). Theoretical torque output T, is the product of the cylinder piston area A, operating pressure P, and the pitch radius of the pinion r_p.

1) $$T = APr_p$$

Single, double, or multiple rack designs are possible and overall efficiencies for rack and pinion units average 85 to 90%. Because standard cylinder components can be used to drive the rack, many standard cylinder features can be incorporated into rack and pinion actuators, such as cushions, stroke adjusters, proximity switches, and special porting. Additionally, virtually leakproof seals will allow the actuator to be held in any position under the load.

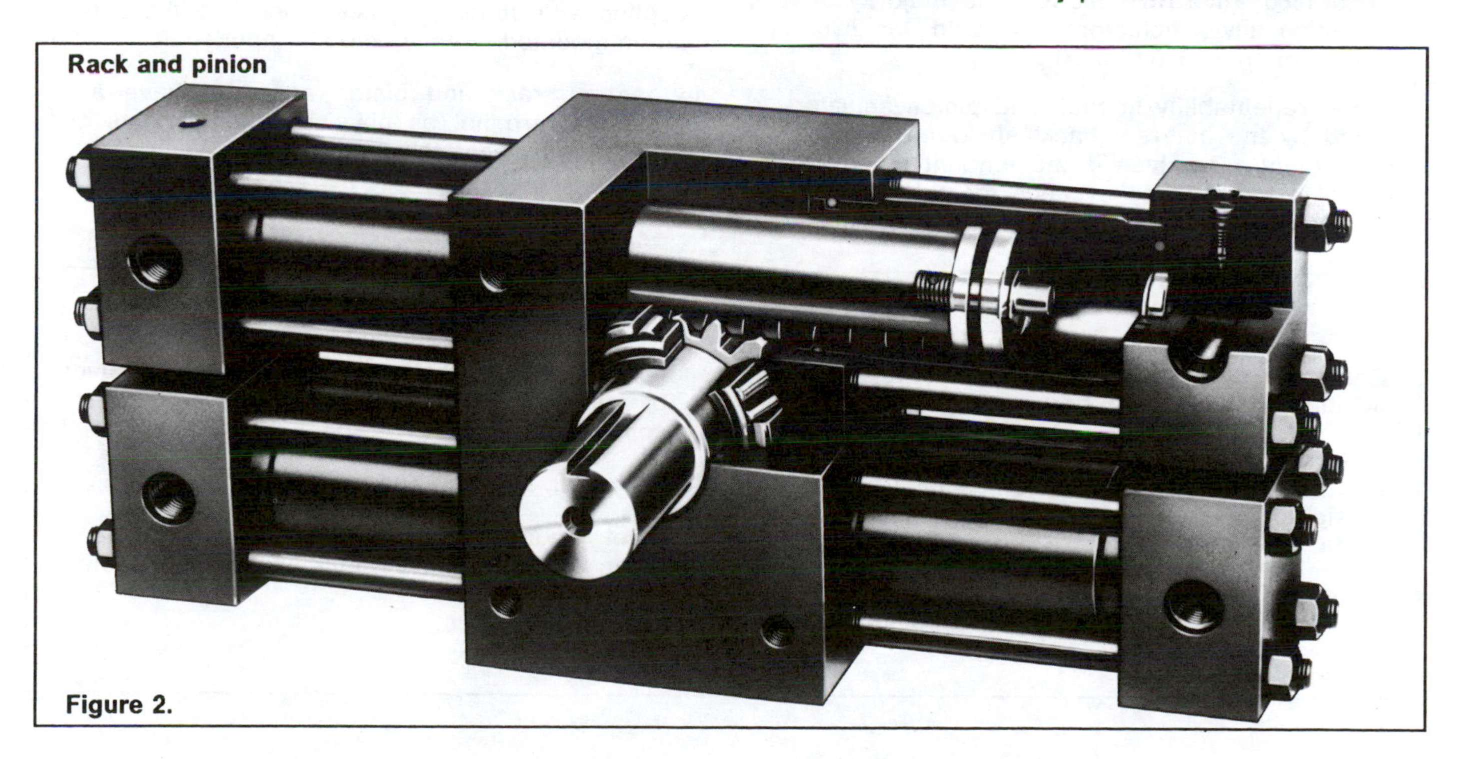

Figure 2.

Mill cylinder type construction

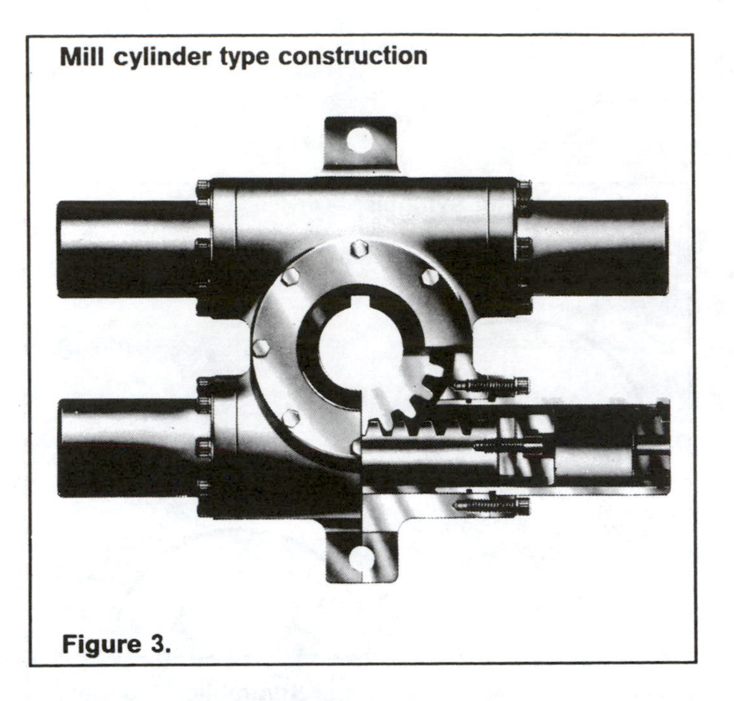

Figure 3.

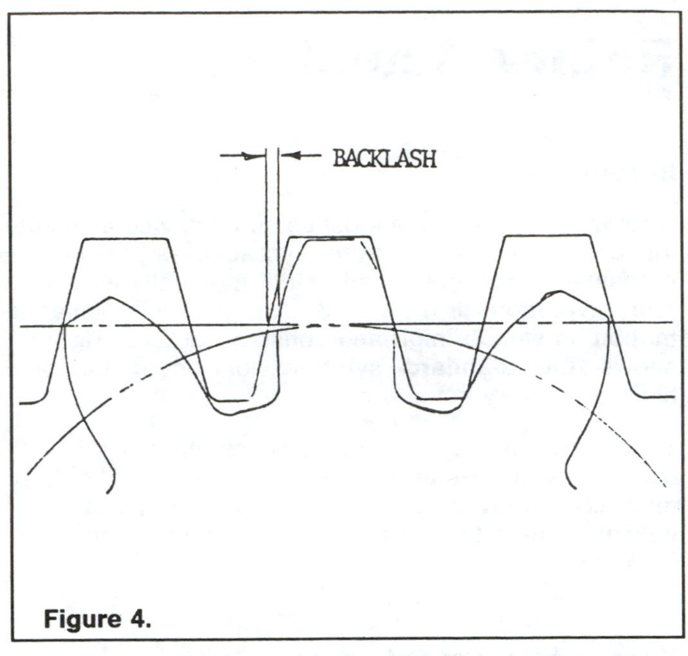

Figure 4.

NOTE: Some systems require a mechanical locking device for safety reasons, or for holding loads over extended periods of time.

Both tie-rod and mill cylinder type constructions are available (Figure 3), and most types allow for service of all pressure containing seals without removing the unit from its mounting.

Rack and pinion actuators cover the widest range of torque, from 6 lb-in pneumatic to over 50,000,000 lb-in in hydraulic units. Because of their construction, they are not limited to 360° of rotation and can easily be built to 1800° (five revolutions). The majority of rack and pinion style actuators are sold for hydraulic service, generally 500 to 3000 psi.

Position repeatability in rack and pinion actuators is affected by the inherent backlash found in any gear arrangement. **Backlash** is the amount by which the width of a tooth space exceeds the thickness of the mating tooth and can be as much as 0.5° on smaller size units (Figure 4). It should be noted, however, that this backlash can be reduced to almost nothing by pre-loading the rack into the pinion, but efficiencies will suffer to overcome the added friction.

Because the load ratings of the bearings used to support the pinion are large in comparison to the internal loading of the unit, external bearing capacity is usually available. This can eliminate the need for machine support bearings, or handle over hung and thrust loads which would be detrimental to other types of rotary actuators. In other applications, a hollow pinion is used, which eliminates the need for a coupling and support brackets because the actuator can be mounted directly onto the input shaft.

In general, rack and pinion actuators have a thin profile, but are not as physically compact as other styles of rotary actuators.

ROTARY ACTUATOR APPLICATIONS

General Industry: Camming, indexing, clamping, positioning, tensioning, braking, tilting, etc.

Material Handling: Switching conveyors, turning and positioning container clamps on lift trucks, tensioning and guiding, operating valves, braking, lifting.

Marine: Opening and closing hatches, swinging cargo handling gear, opening and closing fire and collision bulkhead doors, operating large valves of all types, positioning hydrofoils, steering control.

Robotics: Rotation and positioning.

Steel: Upending coils, turnstiles, rollover devices, tilting electric furnaces, indexing transfer tables, charging furnaces.

Table 1

Vane: Vane style actuators consist of one or two vanes attached to a shaft (called the **rotor**), which is assembled into a body, and then held in place by two heads (Figure 5). Rotation of single vane units is generally limited to 280° by a fluid barrier (called a **stator**). Double vane units are limited to 100° because two stators are required at opposite ends. The operating medium (air or oil) is ported across the shaft in double vane style actuators to eliminate the need for four ports. Fluid pressure acting on the exposed vane surface produces an output torque, or

2) $$T = LWP_r$$

Where the torque T is equal to the product of the vane length L, times the vane width W, times the system pressure P, times the radial distance r from the center of the rotor to the vane pressure center. Of course, a double vane style actuator will have twice the area of a single vane style actuator, and therefore twice the torque.

Available industry sizes range from a minimum of 1.5 lb-in with small pneumatic actuators to a maximum of 750,000 lb-in with high pressure hydraulic actuators.

Vane style actuators have no backlash, but because of the seal configuration, cannot hold position without pressure being applied. The vane seal typically has sharp corners to seal in the body/head interface. Since this corner cannot be sealed completely, there is always a slight bypass flow. There is additional bypass flow in the shoulder area of the vane, so even rounding the vane at the top does not completely eliminate leakage. Vane actuators require external stops, especially for high inertia and high speed applications, to prevent damage to the vane and stator.

Vane style actuator

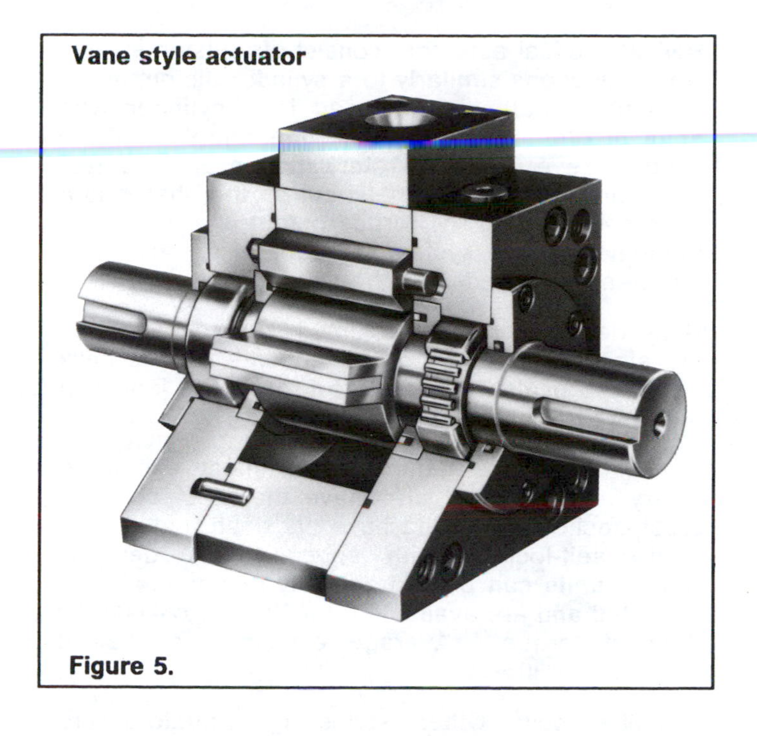

Figure 5.

Units are available with and without external shaft load capacity.

Vane actuators can be segmented into three general product lines:

1) Small pneumatic actuators for small parts handling, fixturing, etc.

2) 1000 psi hydraulic actuators for machine tool, automotive equipment, and transfer lines.

3) 3000 psi actuators for the mobile equipment industry. Efficiencies for hydraulic vane actuators are between 90-95% while pneumatic efficiencies are lower.

HYDRAULIC VS. PNEUMATIC ROTARY ACTUATORS

Hydraulic	Pneumatic
Can operate at high pressures for a better power to weight ratio and more torque for a given size.	Components are less expensive.
Better positioning and speed control are available due to the relative incompressibility of oil.	The smallest actuators available are pneumatic.
Smoother operation at low speeds. (No slip-start jerky movement).	Air is cleaner than oil.
	Pneumatic actuators can generally operate faster than hydraulic actuators.

Table 2

Helical: Helical actuators consist of a piston sleeve, which functions similarly to a cylinder piston, and a rotating output shaft encased in a cylinder type housing (See Figure 6). The linear motion of the piston sleeve produces rotary motion of the output shaft through the male helix cut on the shaft and a fixed helical nut. The torque output is proportional to the helix angle, system pressure, piston area, and the mean pitch radius of the helical shaft.

Helix designs provide maximum torque output for the smallest possible cross-section. Double helix designs are also available to reduce the length of the unit or double the torque output.

Helical units are generally the most expensive rotary actuators but also have the most compact cross-section. They do have backlash and can be made self-locking with special helix designs. Helical units can be hydraulically or pneumatically operated and are available from 20 lb-in to 4,000,000 lb-in of torque. Average efficiency for helical actuators is 80%.

Miscellaneous: Other styles of actuators are available. The majority of these simply enclose linkages and other devices in a self-contained package. One such device uses an air or hydraulic cylinder and a piston attached to a roller chain that drives a sprocket connected to an output shaft (Figure 7). The illustrated actuator requires maintained pressure at port C3 with pressure alternately applied at post C1 or C2 to cause rotation. These are generally light duty, low cost actuators available with rotations up to 1800° and torque up to 11,000 lb-in.

Another design encloses a cylinder and crank driving an output shaft. These units are limited by their configuration to a maximum of 100°, although stroke is easily adjustable (Figure 7). Note that these devices do not produce a constant torque throughout their rotation, instead torque varies with the sine of the resultant angle.

Helical type actuators

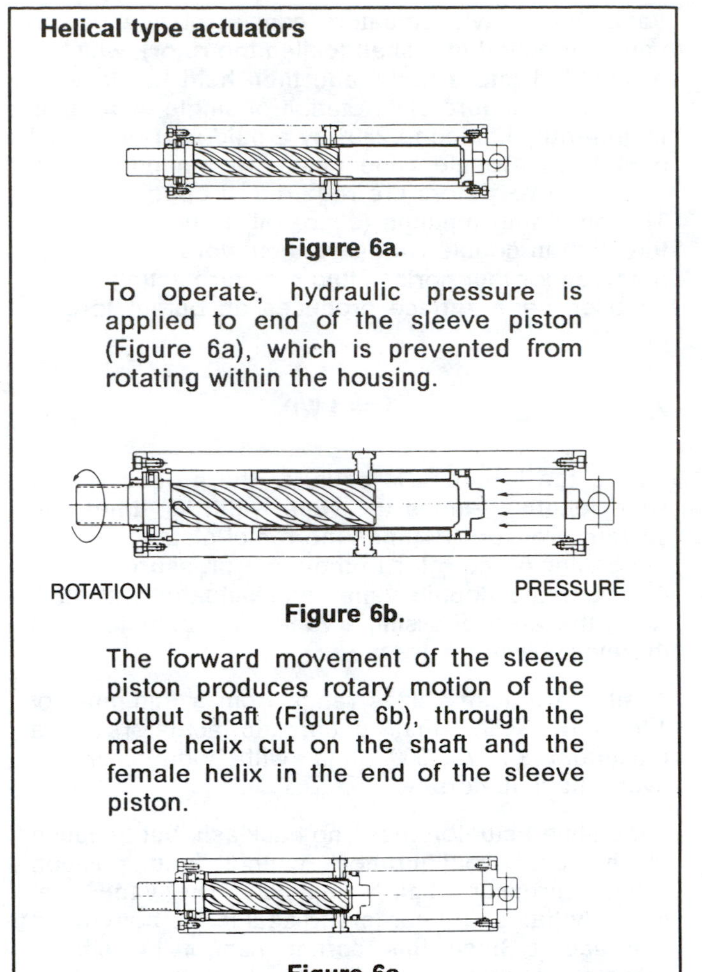

Figure 6a.

To operate, hydraulic pressure is applied to end of the sleeve piston (Figure 6a), which is prevented from rotating within the housing.

Figure 6b.

The forward movement of the sleeve piston produces rotary motion of the output shaft (Figure 6b), through the male helix cut on the shaft and the female helix in the end of the sleeve piston.

Figure 6c.

Reversing the application of the hydraulic pressure (Figure 6c), provides equal torque output in the opposite direction.

Figure 6.

Miscellaneous rotary actuators

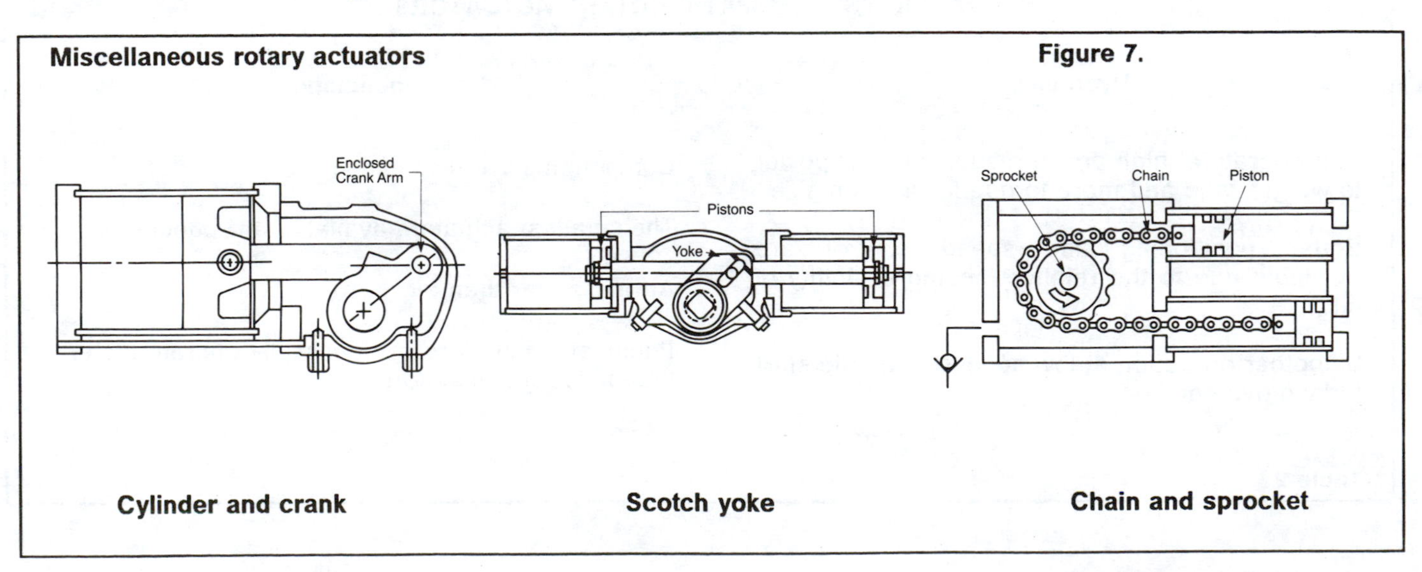

Cylinder and crank **Scotch yoke** **Chain and sprocket**

Figure 7.

ROTARY ACTUATOR COMPARISON CHART

FEATURE	RACK & PINION	VANE	HELICAL
1. Load Holding Ability	Leakproof cylinder seals allow holding of load in any position	Square vane seals and shoulder seals always have slight bypass flow.	Leakproof cylinder seals will allow holding of load in any position.
2. Positioning	Inherent backlash of rack and pinion cause position tolerance up to ½ °	Zero backlash allows for exact positioning anywhere in the rotation.	Some backlash, but can be made self-locking with special helix designs.
3. Efficiency (Hydraulic)	85 - 90% is average	90 - 95% is average	80% is average
4. Stops	External stops usually not required.	Internal stops available for some light duty applications, majority of applications require positive external stops.	External stops usually not required.
5. Cushions	Standard cylinder cushions can be used.	No cushions available.	Consult manufacturer.
6. Size	Thin profile but larger overall space and weight requirements.	Very compact, especially cross section.	The most compact cross section for a given torque.
7. Maintenance	Maintenance of pressure seals possible without complete disassembly of unit.	Maintenance of vane requires disassembly of unit.	Maintenance of seals requires disassembly of unit.
8. Mounting	Mounting styles include lug, foot, face, base, flange, or shaft mounting with hollow pinion.	Mounting styles include base, foot, face, or flange mounting.	Mounting styles include foot, flange, or body mounting.
9. Operating Medium	Air or hydraulic operation.	Air or hydraulic operation.	Air or hydraulic operation.
10. Available Rotation	90°, 180°, and 360° standard, specials to 1800°.	280° maximum single vane units. 100° maximum double vane units.	Consult manufacturer specials made to order.
11. Price	Generally more expensive than equivalent torque vane units.	Generally less expensive than equivalent torque rack and pinion units.	Generally much more expensive than equivalent torque rack and pinion units.

Table 3

A similar design, called a scotch yoke, functions in the same manner, except the yoke is a slot in which the cylinder rod actuates (Figure 7). These also have a variable output torque and are limited to a maximum rotation of 100°. This type of device is found primarily in the process valve industry.

Calculating Torque Requirements

DESIGN TORQUE represents the maximum torque that an actuator must supply in an application. This maximum is the greater of the **Demand Torque** or the **Cushion Torque**. If the demand torque exceeds what the actuator can supply, the actuator will either move too slowly or stall. If the cushion torque is too high, the actuator may become damaged due to excessive pressure. Demand torque and cushion torque are defined below in terms of load, friction, and acceleration torques.

Equations for calculating demand torque and cushion torque for some general applications are provided on the following pages.

T - TORQUE:

The amount of turning effort exerted by a rotary actuator.

T_D - DEMAND TORQUE:

This is the torque required from the actuator to do the job and is the sum of the load torque, friction torque, and acceleration torque, multiplied by an appropriate design factor. Design factors vary with the applications and the designers' knowledge.

T_L - Load torque:

This is the torque required to equal the weight or force of the load. For example, in Figure 8a; the load torque is 5000 lb-in; in Figure 8b the load torque is zero; in Figure 8c the load torque is 5000 lb-in.

T_f - Friction torque:

This is the torque required to overcome friction between any moving parts, especially bearing surfaces. In Figure 8a, the friction torque is zero for the hanging load; in Figure 8b the friction torque is 6880 lb-in for the sliding load; in Figure 8c the friction torque is zero for the clamp.

T_α - Acceleration Torque:

This is the torque required to overcome the inertia of the load in order to provide a required acceleration or deceleration. In Figure 8a the load is suspended motionless so there is no acceleration. In Figure 8b the load is accelerated from 0 to some specified angular velocity. If the mass moment of inertia about the axis of rotation is I and the angular acceleration is α, the acceleration torque is equal to $I\alpha$. In Figure 8c there is no acceleration.

Some values for mass moment of inertia are given in Table 4. Some useful equations for determining α are listed in Table 5.

T_c - CUSHION TORQUE:

This is the torque that the actuator must apply to provide a required deceleration. This torque is generated by restricting the flow out of the actuator (meter-out) so as to create a back pressure which decelerates the load. This back pressure (deceleration) often must overcome both the inertia of the load and the driving pressure (system pressure) from the pump. See applications.

WARNING: RAPID DECELERATION CAN CAUSE HIGH PRESSURE INTENSIFICATION AT THE OUTLET OF THE ACTUATOR. ALWAYS INSURE THAT CUSHION PRESSURE DOES NOT EXCEED THE MANUFACTURER'S PRESSURE RATING FOR THE ACTUATOR.

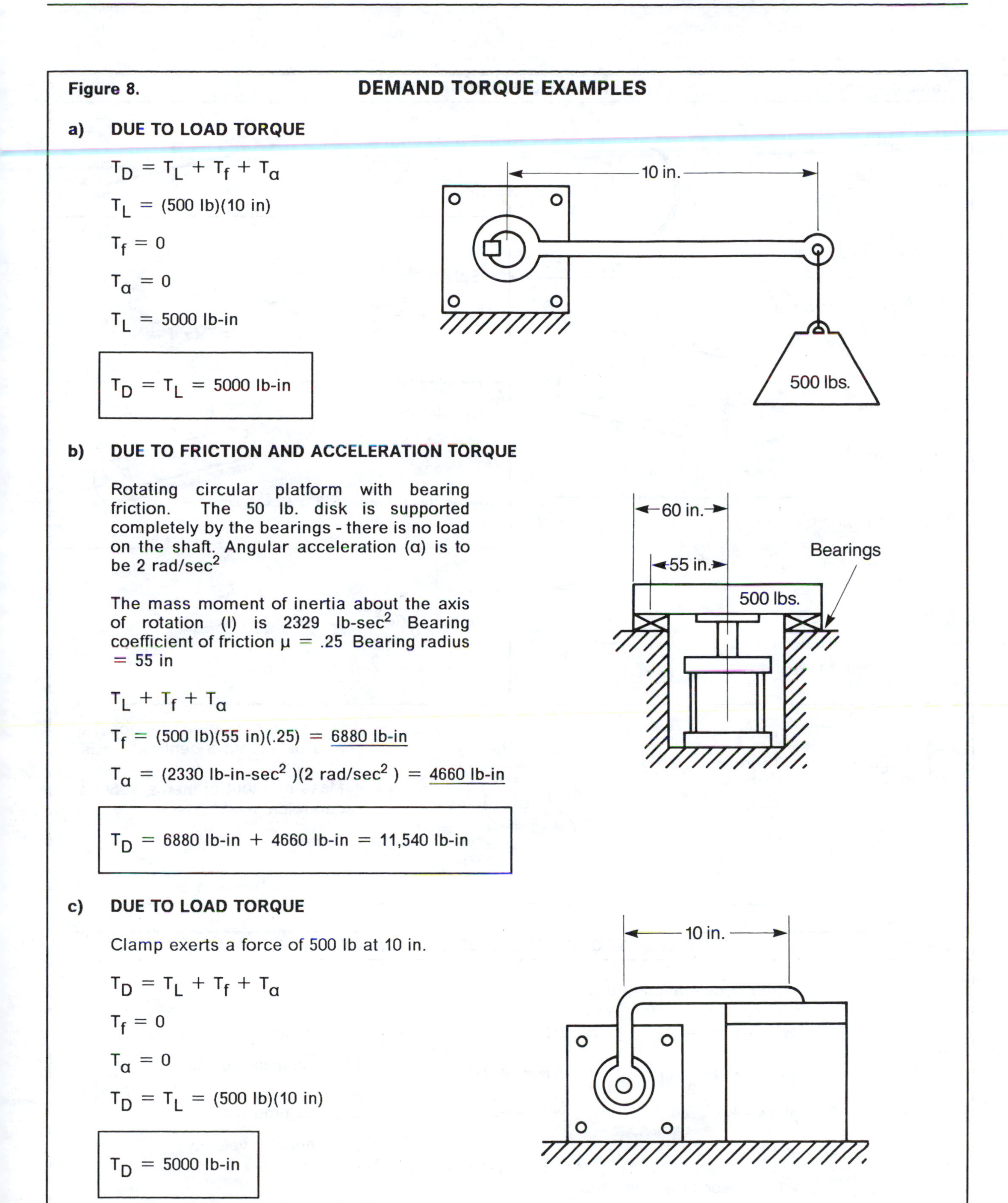

Figure 8. **DEMAND TORQUE EXAMPLES**

a) DUE TO LOAD TORQUE

$T_D = T_L + T_f + T_\alpha$

$T_L = (500\ \text{lb})(10\ \text{in})$

$T_f = 0$

$T_\alpha = 0$

$T_L = 5000\ \text{lb-in}$

$T_D = T_L = 5000\ \text{lb-in}$

b) DUE TO FRICTION AND ACCELERATION TORQUE

Rotating circular platform with bearing friction. The 50 lb. disk is supported completely by the bearings - there is no load on the shaft. Angular acceleration (α) is to be 2 rad/sec^2

The mass moment of inertia about the axis of rotation (I) is 2329 lb-sec^2 Bearing coefficient of friction μ = .25 Bearing radius = 55 in

$T_L + T_f + T_\alpha$

$T_f = (500\ \text{lb})(55\ \text{in})(.25) = \underline{6880\ \text{lb-in}}$

$T_\alpha = (2330\ \text{lb-in-sec}^2)(2\ \text{rad/sec}^2) = \underline{4660\ \text{lb-in}}$

$T_D = 6880\ \text{lb-in} + 4660\ \text{lb-in} = 11{,}540\ \text{lb-in}$

c) DUE TO LOAD TORQUE

Clamp exerts a force of 500 lb at 10 in.

$T_D = T_L + T_f + T_\alpha$

$T_f = 0$

$T_\alpha = 0$

$T_D = T_L = (500\ \text{lb})(10\ \text{in})$

$T_D = 5000\ \text{lb-in}$

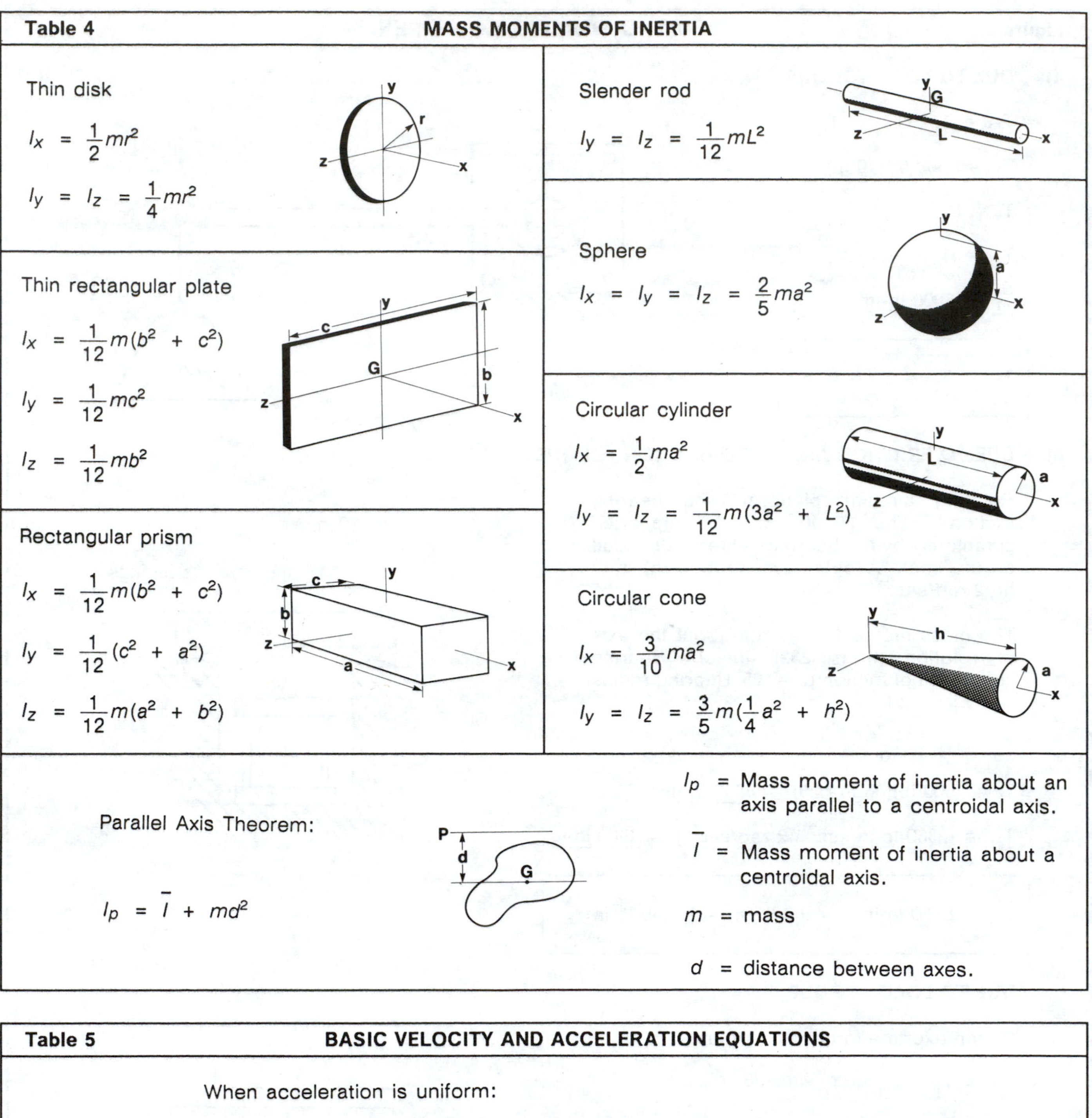

Table 4 **MASS MOMENTS OF INERTIA**

Thin disk

$$I_x = \frac{1}{2}mr^2$$

$$I_y = I_z = \frac{1}{4}mr^2$$

Slender rod

$$I_y = I_z = \frac{1}{12}mL^2$$

Sphere

$$I_x = I_y = I_z = \frac{2}{5}ma^2$$

Thin rectangular plate

$$I_x = \frac{1}{12}m(b^2 + c^2)$$

$$I_y = \frac{1}{12}mc^2$$

$$I_z = \frac{1}{12}mb^2$$

Circular cylinder

$$I_x = \frac{1}{2}ma^2$$

$$I_y = I_z = \frac{1}{12}m(3a^2 + L^2)$$

Rectangular prism

$$I_x = \frac{1}{12}m(b^2 + c^2)$$

$$I_y = \frac{1}{12}(c^2 + a^2)$$

$$I_z = \frac{1}{12}m(a^2 + b^2)$$

Circular cone

$$I_x = \frac{3}{10}ma^2$$

$$I_y = I_z = \frac{3}{5}m(\frac{1}{4}a^2 + h^2)$$

Parallel Axis Theorem:

$$I_P = \bar{I} + md^2$$

I_P = Mass moment of inertia about an axis parallel to a centroidal axis.

$\bar{I}$ = Mass moment of inertia about a centroidal axis.

m = mass

d = distance between axes.

Table 5 **BASIC VELOCITY AND ACCELERATION EQUATIONS**

When acceleration is uniform:

$$\Theta = \omega_o t + \frac{1}{2}at^2 \qquad a = 2\Theta/t^2$$

$$\Theta = \omega_o t + \frac{1}{2}\omega_t t \qquad a = (\omega_t - \omega_o)/t$$

$$\omega = \omega_o + at \qquad a = \frac{(\omega_t - \omega_o)^2}{2\Theta}$$

$$\omega = (\omega_o^2 + 2a\theta)^{\frac{1}{2}}$$

When velocity is constant: $\Theta = \omega t$

t = time

Θ = angular velocity

ω_t = angular velocity at time = t

ω_o = angular velocity at time = 0

Applied Torque Equations

1. **Round Index Table - No load**
2. **Rotary Index Table - Cylindrical load**
3. **Rotary Index Table - Rectangular load**
4. **Tube Bending**
5. **Screw Clamping**
6. **Simple Clamp**
7. **Linear Motion, Clamping**
8. **Modified Linear Motion, Clamping**
9. **Overcenter load**
10. **Harmonic drive**

NOTES:

1. The following equations are intended only as a guide. The design engineer should verify the accuracy of the equations and should be responsible for assuring that all performance, safety, and warning requirements of the application are met.

2. Unless specified otherwise, the following examples DO NOT take into account system or actuator efficiencies, and friction is neglected.

3. Deceleration torques are based upon the assumption that due to restrictor type flow controls, the actuator is subjected to relief valve pressure during deceleration.

(1) ROUND INDEX TABLE

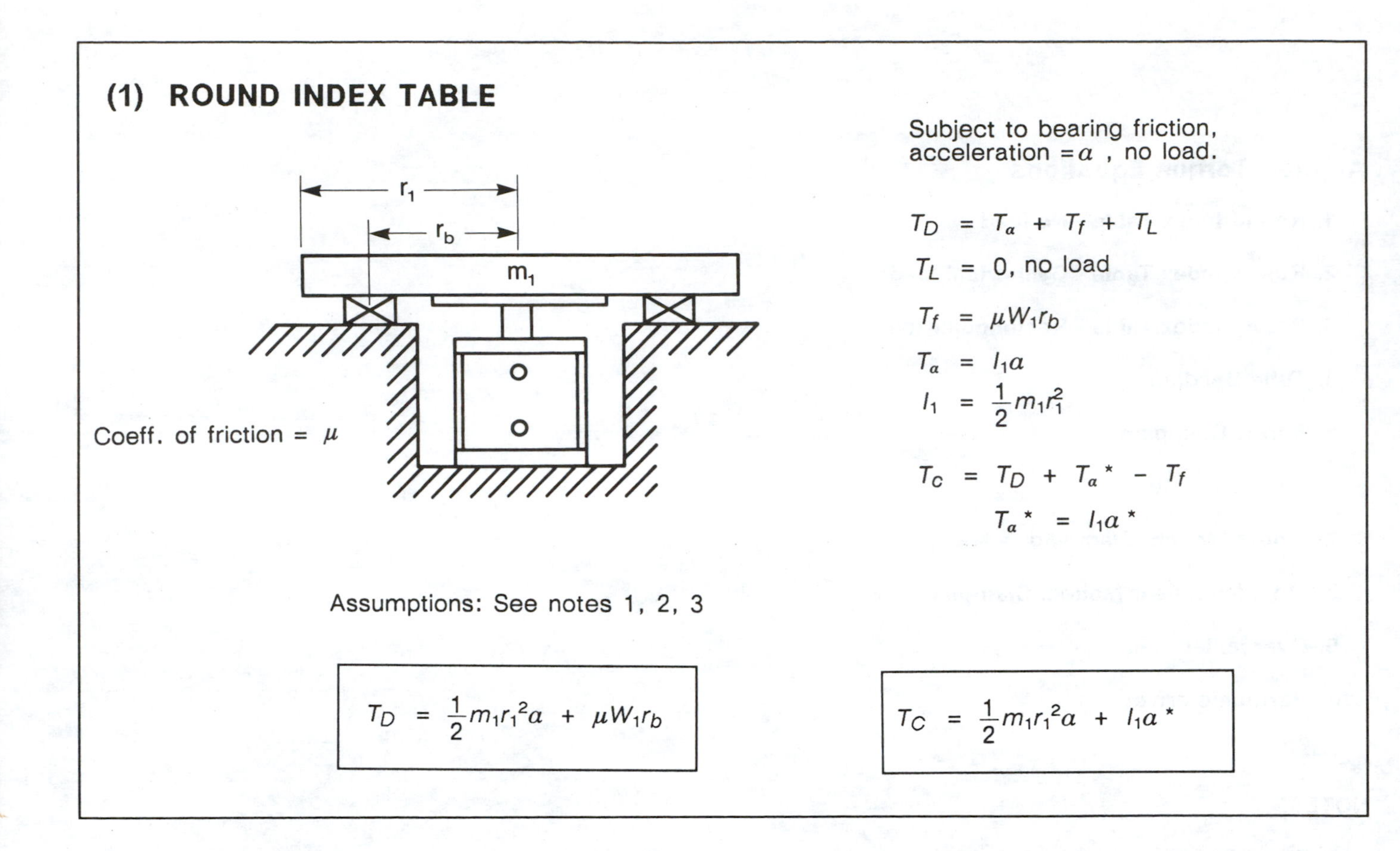

Coeff. of friction = μ

Subject to bearing friction, acceleration $= a$, no load.

$$T_D = T_a + T_f + T_L$$

$$T_L = 0, \text{ no load}$$

$$T_f = \mu W_1 r_b$$

$$T_a = I_1 a$$

$$I_1 = \frac{1}{2} m_1 r_1^2$$

$$T_C = T_D + T_a{}^* - T_f$$

$$T_a{}^* = I_1 a^*$$

Assumptions: See notes 1, 2, 3

$$T_D = \frac{1}{2} m_1 r_1{}^2 a + \mu W_1 r_b$$

$$T_C = \frac{1}{2} m_1 r_1{}^2 a + I_1 a^*$$

(2) ROTARY INDEX TABLE WITH CYLINDRICAL LOAD

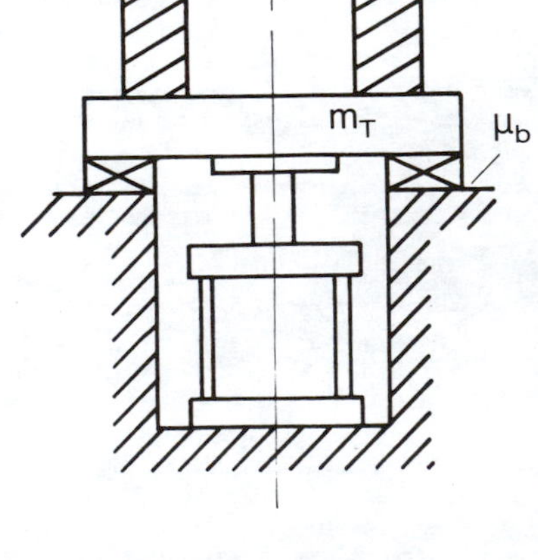

The index table rotates in a horizontal plane with a cylindrical load.

$$T_D = T_L + T_f + T_a$$

$$T_L = 0$$

$$T_f = (W_T + W_L) r_b \mu_b$$

$$T_a = (I_T + I_L) a$$

$$I_T = \frac{1}{2} m_T r_T{}^2$$

$$I_L = \frac{1}{2} m_L (a^2 - b^2)$$

$$T_C = T_D + T_a{}^* - T_f$$

$$T_a{}^* = (I_T + I_L) a^*$$

$$T_D = (W_T + W_L) r_b \mu_b + \frac{a}{2} \left[m_T r_T{}^2 + m_L (a^2 - b^2) \right]$$

$$T_C = \frac{1}{2} (a + a^*) \left[m_T r_T{}^2 + m_L (a^2 - b^2) \right]$$

Assumptions: See notes 1, 2, 3

(3) ROTARY INDEX TABLE WITH RECTANGULAR LOAD

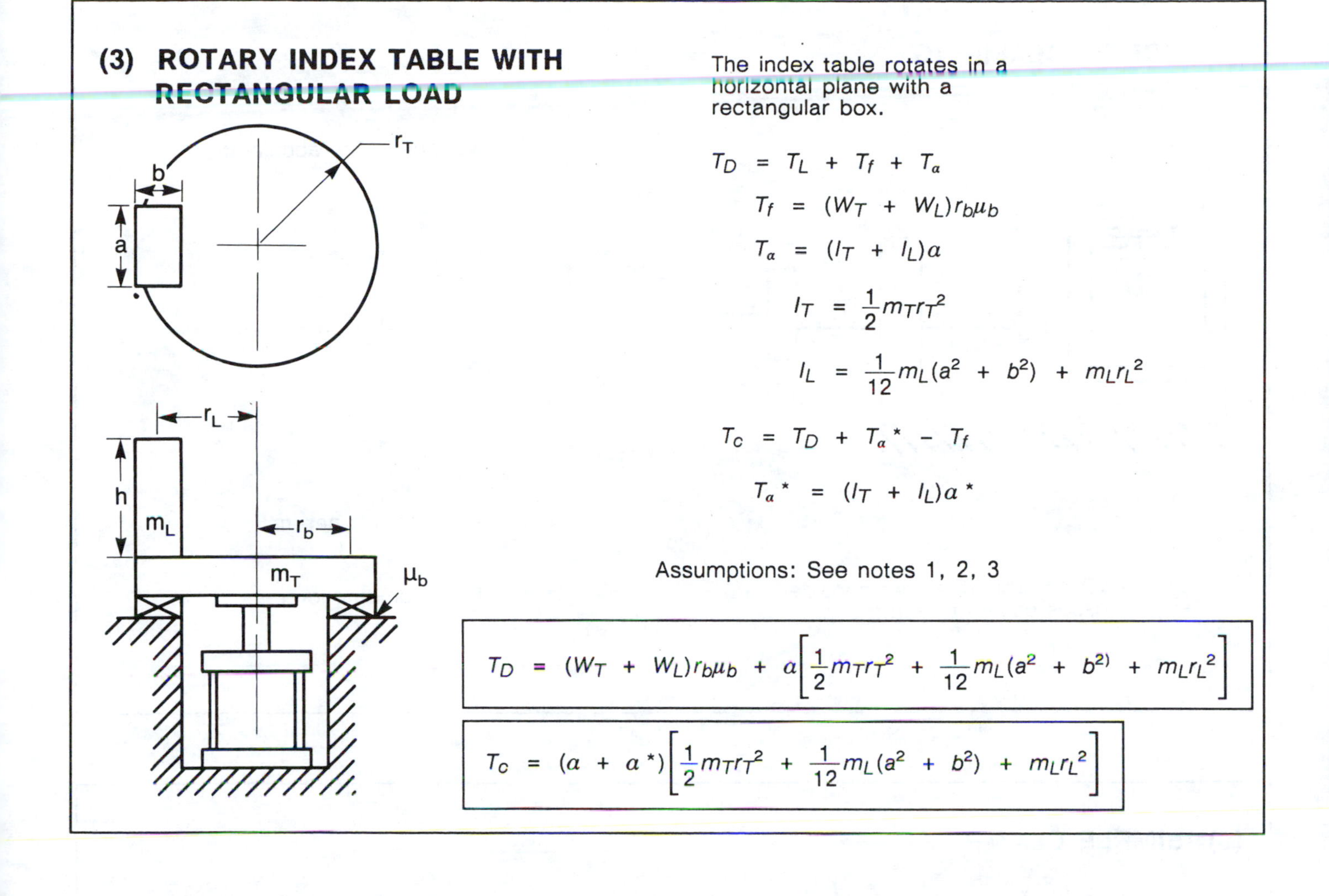

The index table rotates in a horizontal plane with a rectangular box.

$$T_D = T_L + T_f + T_\alpha$$

$$T_f = (W_T + W_L)r_b\mu_b$$

$$T_\alpha = (I_T + I_L)\alpha$$

$$I_T = \frac{1}{2}m_T r_T^2$$

$$I_L = \frac{1}{12}m_L(a^2 + b^2) + m_L r_L^2$$

$$T_c = T_D + T_\alpha^* - T_f$$

$$T_\alpha^* = (I_T + I_L)\alpha^*$$

Assumptions: See notes 1, 2, 3

$$T_D = (W_T + W_L)r_b\mu_b + \alpha\left[\frac{1}{2}m_T r_T^2 + \frac{1}{12}m_L(a^2 + b^2) + m_L r_L^2\right]$$

$$T_c = (\alpha + \alpha^*)\left[\frac{1}{2}m_T r_T^2 + \frac{1}{12}m_L(a^2 + b^2) + m_L r_L^2\right]$$

(4) WIRE OR ROUND TUBE BENDING

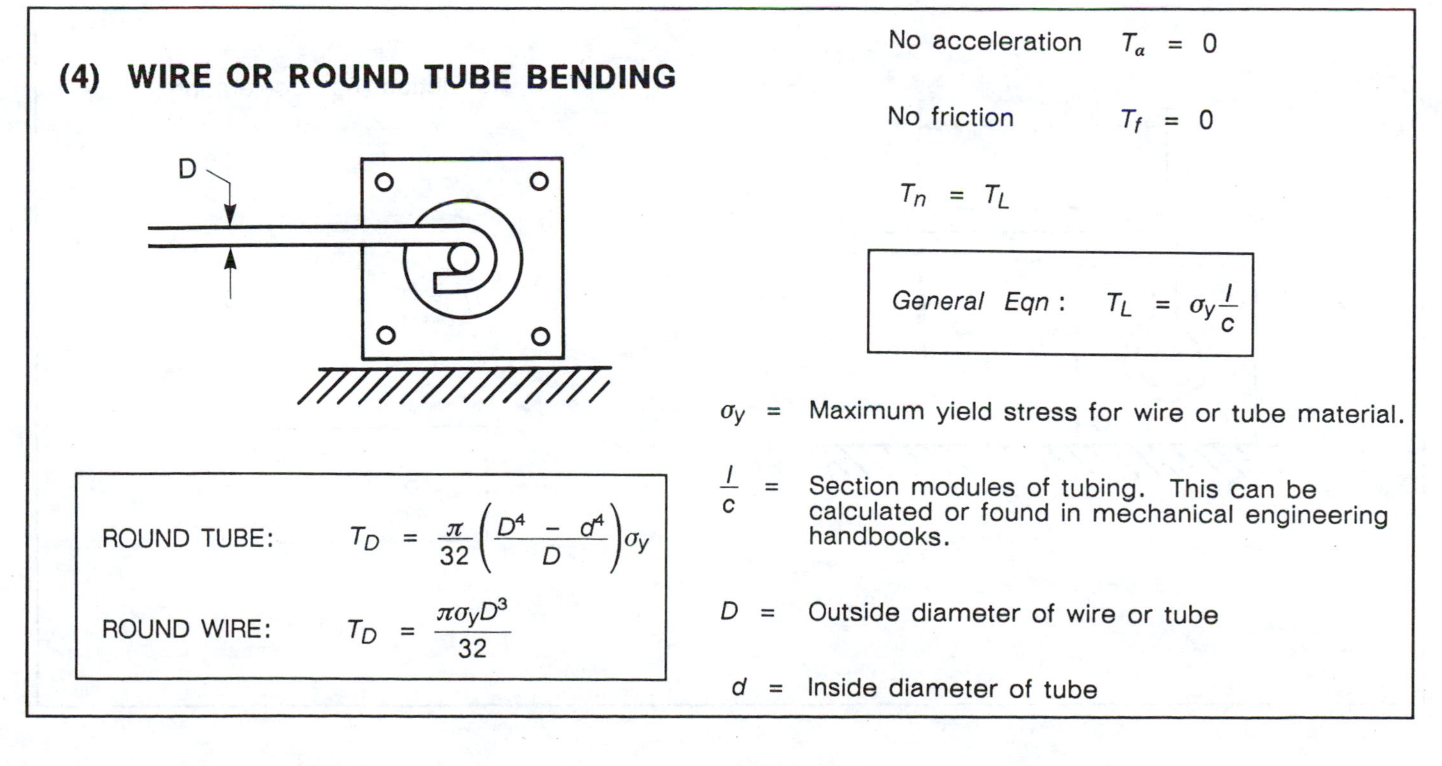

No acceleration $T_\alpha = 0$

No friction $T_f = 0$

$$T_n = T_L$$

General Eqn : $T_L = \sigma_y \frac{I}{c}$

ROUND TUBE: $T_D = \frac{\pi}{32}\left(\frac{D^4 - d^4}{D}\right)\sigma_y$

ROUND WIRE: $T_D = \frac{\pi\sigma_y D^3}{32}$

σ_y = Maximum yield stress for wire or tube material.

$\frac{I}{c}$ = Section modules of tubing. This can be calculated or found in mechanical engineering handbooks.

D = Outside diameter of wire or tube

d = Inside diameter of tube

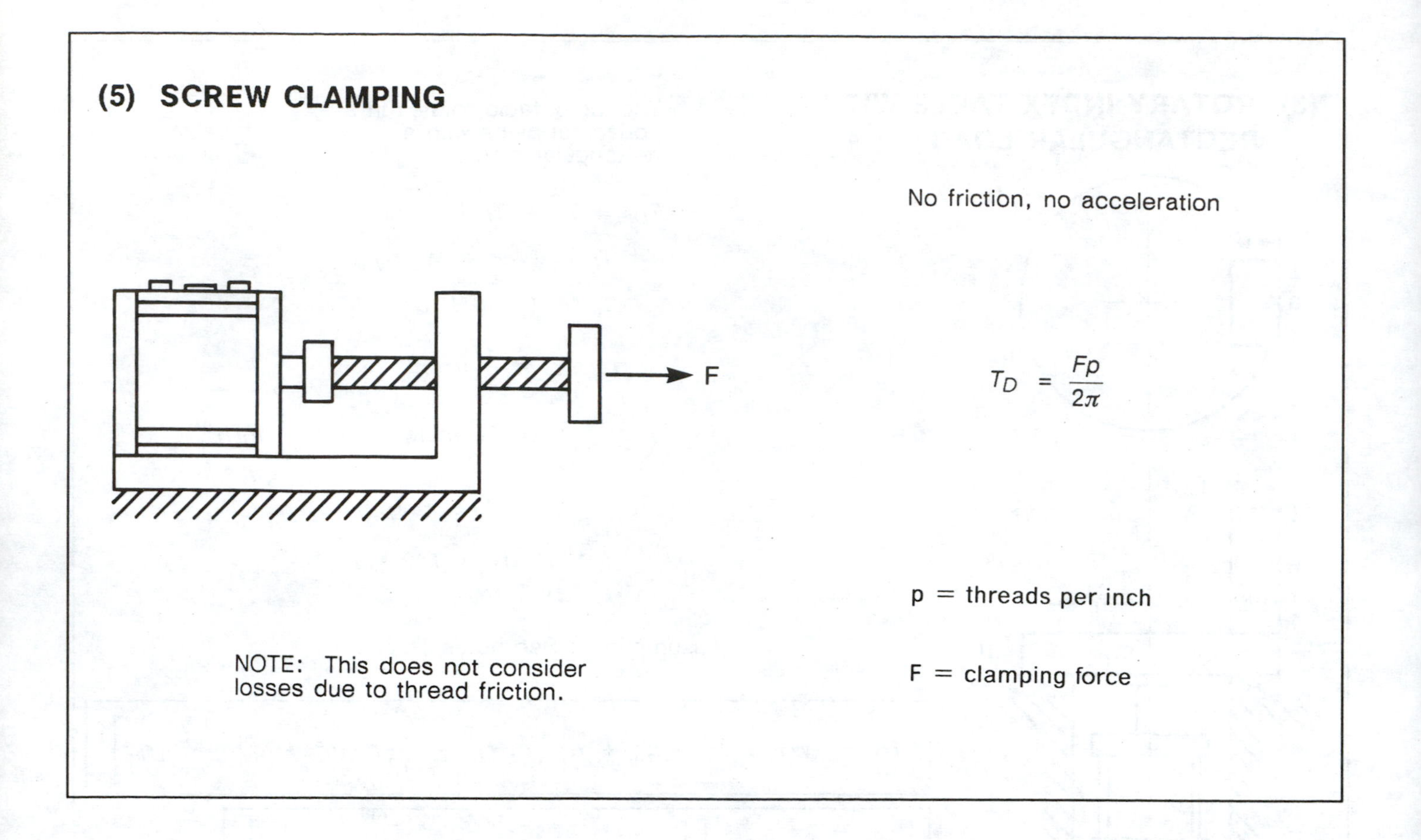

(5) SCREW CLAMPING

No friction, no acceleration

$$T_D = \frac{Fp}{2\pi}$$

p = threads per inch

F = clamping force

NOTE: This does not consider losses due to thread friction.

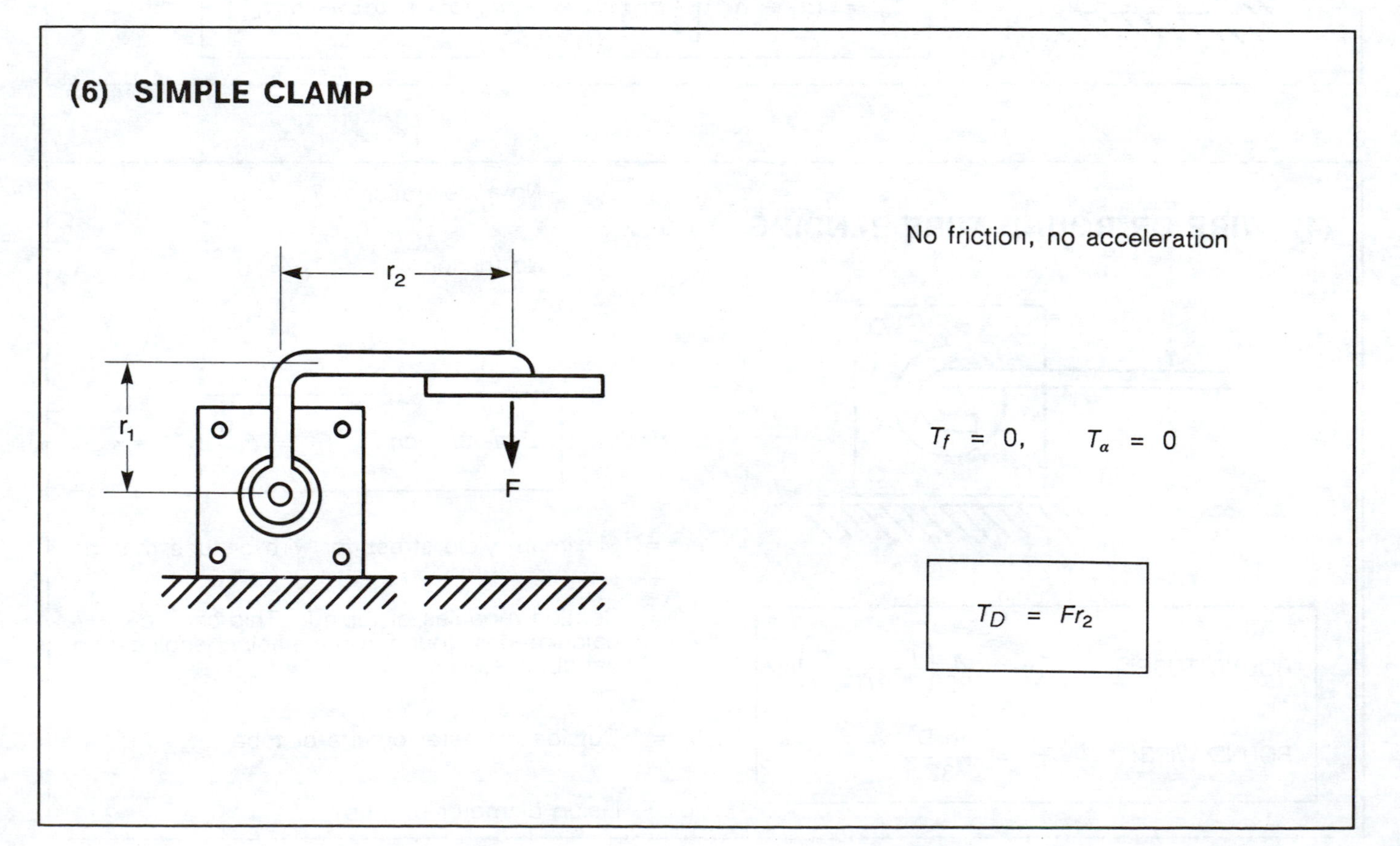

(6) SIMPLE CLAMP

No friction, no acceleration

$T_f = 0, \quad T_\alpha = 0$

$$T_D = Fr_2$$

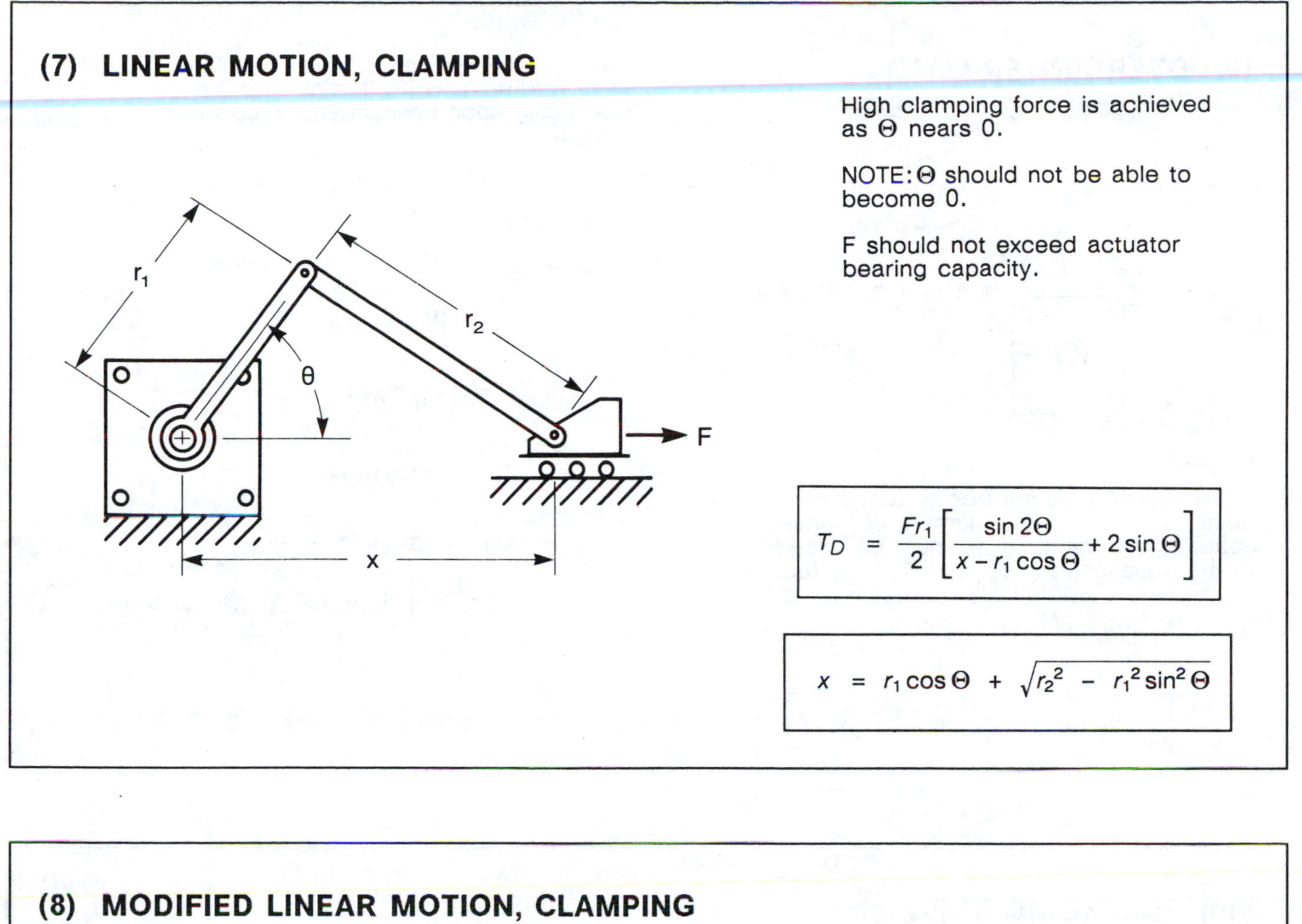

(7) LINEAR MOTION, CLAMPING

High clamping force is achieved as Θ nears 0.

NOTE: Θ should not be able to become 0.

F should not exceed actuator bearing capacity.

$$T_D = \frac{Fr_1}{2}\left[\frac{\sin 2\Theta}{x - r_1 \cos\Theta} + 2\sin\Theta\right]$$

$$x = r_1 \cos\Theta + \sqrt{r_2^2 - r_1^2 \sin^2\Theta}$$

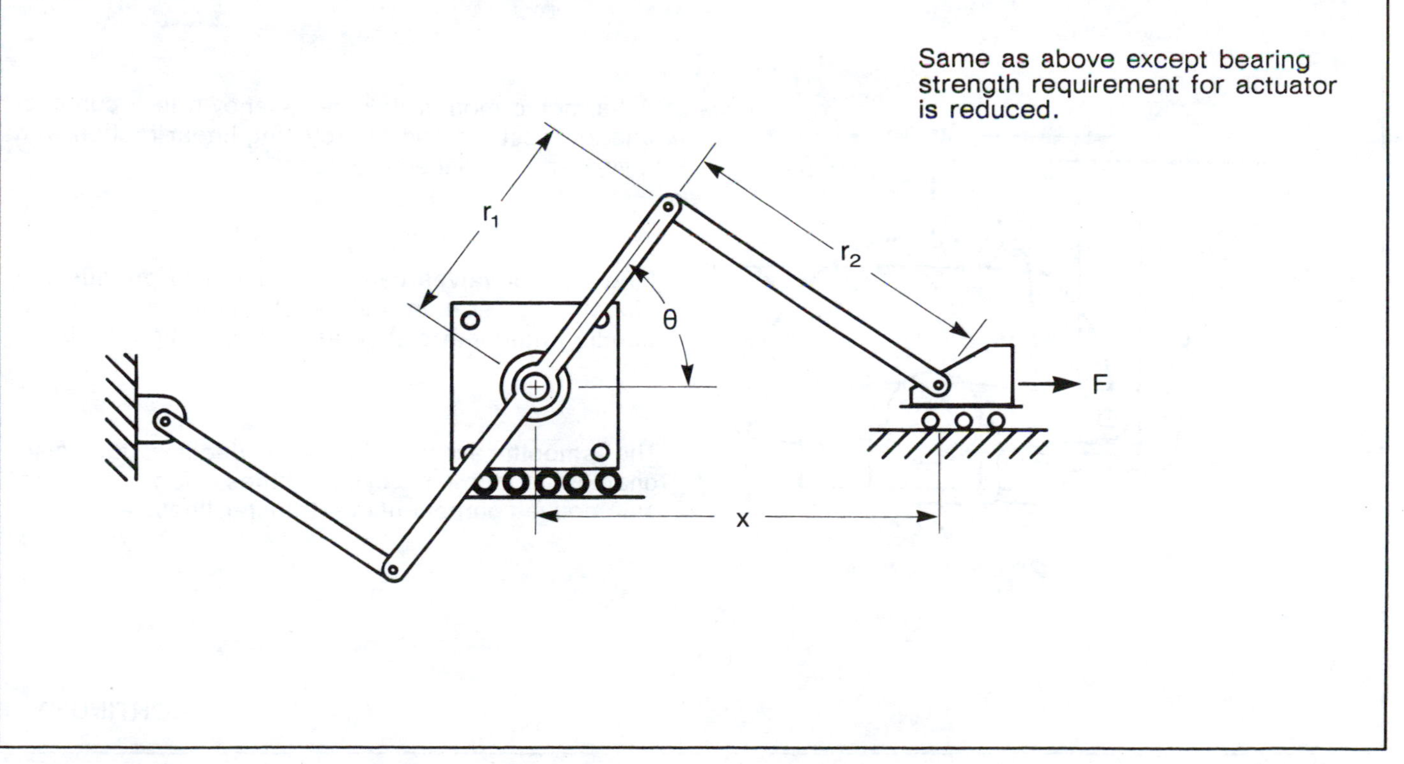

(8) MODIFIED LINEAR MOTION, CLAMPING

Same as above except bearing strength requirement for actuator is reduced.

(9) OVERCENTER LOAD

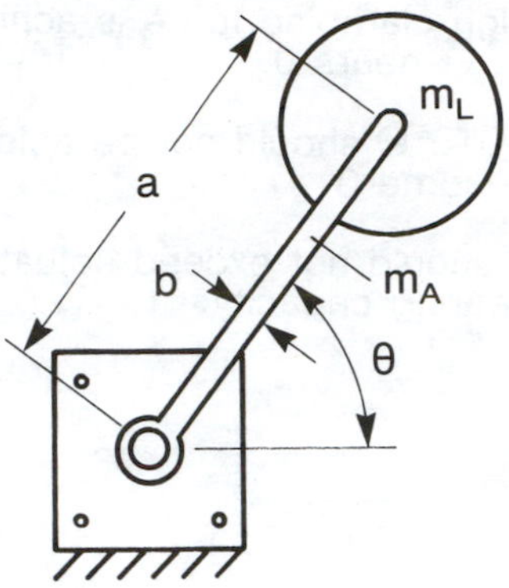

NOTE:

If the mass M_L is not free to turn about point p, then its mass moment of inertia about its own centroid I_L must be added to the equations for T_α and $T_\alpha{}^*$ as follows:

$$T_\alpha = \left[\frac{1}{12}m_A(a^2 + b^2) + I_L + m_L a^2\right]\alpha$$

$$T_\alpha{}^* = \left[\frac{1}{12}m_A(a^2 + b^2) + I_L + m_L a^2\right]\alpha^*$$

Load is rotated through a vertical plane. Load torque T_L is positve or negative depending upon position and direction of rotation.

$$T_D = T_L + T_f + T_\alpha$$

$$\pm T_L = (W_L + \frac{1}{2}W_A)a\cos\Theta$$

$$T_f = (W_L + W_A)r_b\mu_b$$

$$T_\alpha = \left[\frac{1}{12}m_A(a^2 + b^2) + m_L a^2\right]\alpha$$

$$T_{D_{MAX}} = (W_L \pm \frac{1}{2}W_A)a + T_f + T_\alpha$$

$$T_C = \pm T_L - T_f + T_\alpha{}^* + T_{D_{MAX}}$$

$$T_\alpha{}^* = \left[\frac{1}{12}m_A(a^2 + b^2) + m_L a^2\right]\alpha^*$$

Assumptions: See notes 1, 2, 3

(10) HARMONIC DRIVE

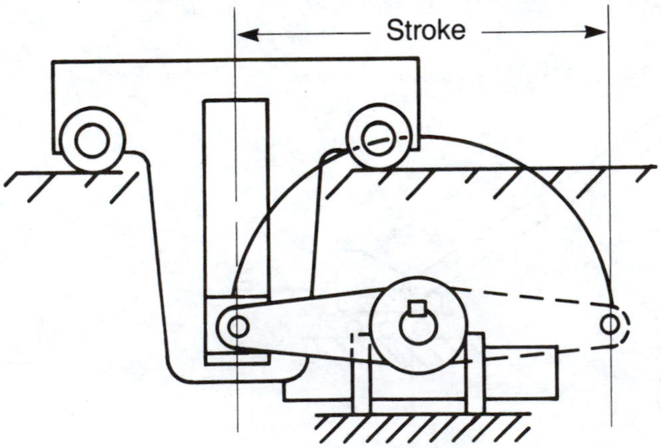

A harmonic motion linkage as shown is a compact and low cost method of providing linear motion with a very smooth acceleration.

Flow control valves can be adjusted to provide the smooth acceleration and deceleration necessary to handle fragile parts such as bottles or light bulbs.

The smooth acceleration and deceleration also enables optimum cycle times for handling automotive components on transfer lines.

CONTINUED

HARMONIC DRIVE (CONTINUED)

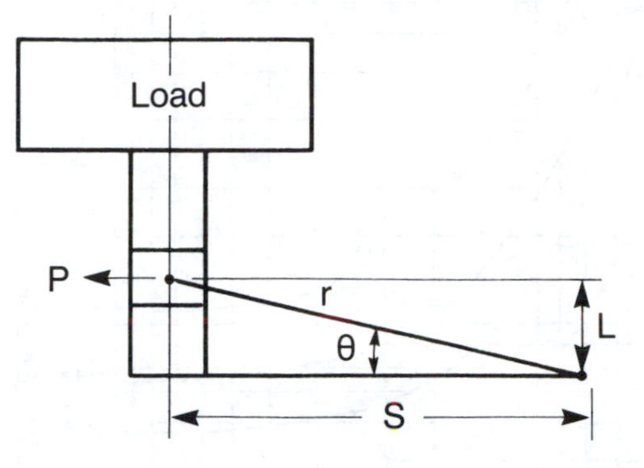

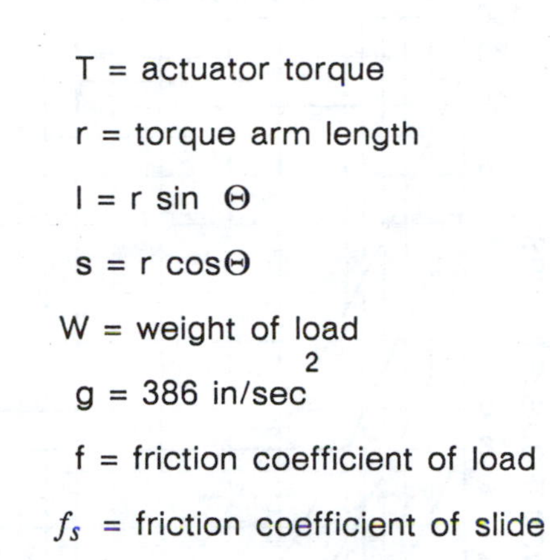

$$A)\quad \frac{T}{W} = \frac{r^2\omega^2}{g}\cos\omega t\ \sin\omega t + \frac{r^2\omega^2 f_s}{g}\cos^2\omega t\ \sin\omega t + rf\cos\omega t - rff_s\sin\omega t$$

$$B)\quad \frac{d(T/W)}{dt} = \frac{r\omega^2}{g}\ (2\cos^2\ \omega t + 1) - \frac{2r\omega^2 f_s}{g}\cos\omega t\ \sin\omega t + f\cos\omega t - ff_s\sin\omega t$$

By using the geometry of the linkage and including the friction of both the load and the slide, the equations A and B can be used to determine the required torque.

The curves (Figures 9 and 10) give solutions to equation A for various arm lengths and friction coefficients of .05 and .25 respectivity.

The preceding equations assume a constant actuator velocity; as the inertia from the moving load tends to drive the actuator during the decelerating phase, it is recommended that the circuit incorporates pressure compensated flow control valves. It is also recommend that, due to resistance of the flow control valves, plus the load deceleration requirements, figures obtained from the following graphs should be doubled to obtain maximum torque requirement.

Remember, the result must be doubled for deceleration using meter-out circuits, for high inertia loads. For high friction cases the following equation can be used:

$$\frac{T}{W}\text{(deceleration)} = 2\frac{T}{W}\text{(acceleration)} - \frac{T}{W}\text{(at 10 second throw time)}$$

HARMONIC DRIVE (CONTINUED)

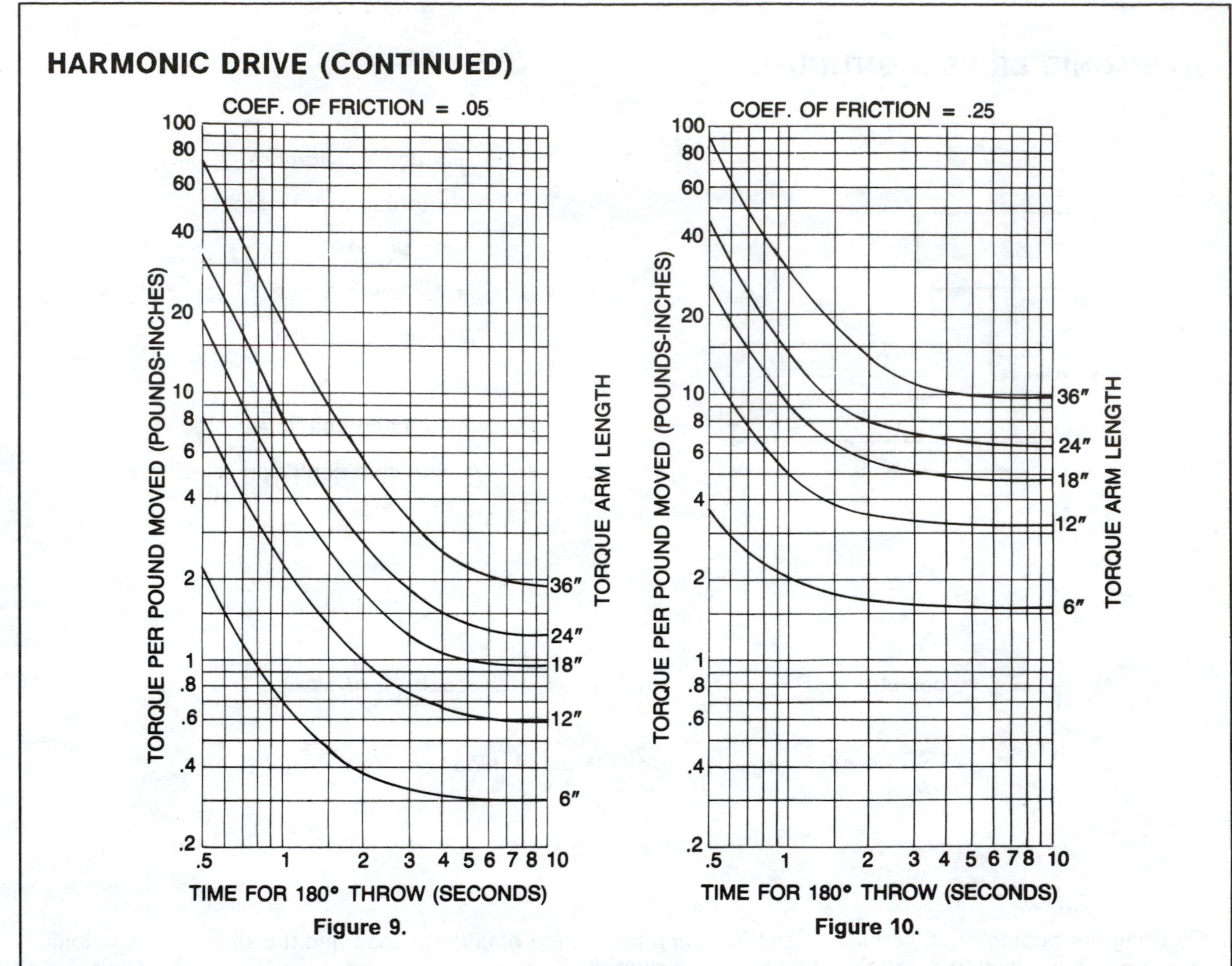

Figure 9. Figure 10.

EXAMPLE: A 400 lb. mass is to be moved a distance of 48 inches in four seconds. Assume that the load is supported on roller bearings with a coefficient of friction of .05.

SOLUTION: Using the curves for .05 coefficient of friction, draw a line from four seconds on the time scale to intersect with the 24 inch torque arm curve.

$$\frac{48'' \text{ travel}}{2} = 24'' \text{ torque arm length}$$

An intersecting horizontal line will show a requirement of 1.5 lb. of torque required per load, therefore:

$$T_D = 400 \times 1.5 = \underline{600 \text{ lb-in}}$$

Using meter-out circuit:

$$\begin{aligned} T_c &= 2T_\alpha - T_{10} \\ &= 2(600 \text{ lb-in}) - (1.2 \times 400)\text{lb-in} \\ &= \underline{720 \text{ lb-in}} \end{aligned}$$

Approximating Rotary Actuator Size

The smallest rotary actuator displacement that can be used in an application is that displacement which can both deliver sufficient torque to move the load and also withstand the pressure required to stop the load. (Recall that cushion torque is generated by back pressure that is often greater than system pressure.) A method for determining the smallest rotary actuator displacement is summarized in Figure 11 and outlined step-by-step below.

1. Determine the maximum allowable safe system pressure that the pump and components can tolerate. This is typically the highest pressure the pump can supply to the system; however, this is not the actual system working pressure. The actual working pressure is determined after an actuator is selected.

2. Calculate the demand torque required. The demand torque T_D is given by the equation:

3) $$T_D = T_L + T_f + T_\alpha$$

Definitions for the above torque components and examples for calculation of T_D were discussed previously under the heading "Calculating Torque Requirements."

3. Calculate a first value for V_s based upon T_D and the maximum system pressure chosen in Step 1. V_s represents the volume displacement per one radian of rotation for a rotary actuator. V_s can be calculated from the equation:

4) $$V_s = \frac{V}{\Theta} = \frac{T}{P}$$

Or by using the graph shown in Figure 12.

> **NOTE: The above equation for V_s is for constant torque rotary actuators such as vane, rack and pinion, or helical styles and does not consider actuator or system inefficiencies.**

4. Calculate the cushion torque T_c required. In any application where the actuator has cushions, a deceleration valve, or any form of meter-out flow control, the flow out of the actuator is restricted creating a back pressure on the outlet side of the actuator. This back pressure is what creates the cushion torque which acts to decelerate, or cushion the actuator as it approaches the end of its rotation. The cushion torque can be calculated by the methods presented under the heading "Calculating Torque Requirements."

5. Calculate a second value for V_s based upon T_c and the maximum rated pressure for the type of actuator selected. (This is not the same pressure used in Step 3; this pressure is found from the rotary actuator catalog.) The second value for V_s can be determined by using the same equation used in Step 3 or by using Figure 12.

6. If the second value for V_s is considered impractical for your application, some ways to reduce T_c are:

   a. Reduce system pressure, then recalculate steps 3 to 5.

   b. Increase the time for deceleration, then recalculate steps 4 and 5.

   c. Use an external shock absorber.

7. Choose the larger of the two V_s values calculated. Use Table 5 to convert V_s to V based upon the rotary actuator's degress of rotation. This value for V is the minimum ideal actuator displacement for the application.

Select a rotary actuator with a displacement equal to or larger than the V value determined.

8. Calculate system operating pressure based upon T_D , and the actual V_s value for the selected actuator. The relief valve setting must be less than the maximum pressure from Step 1, must not exceed the actuator's rated working pressure, and must be high enough to compensate for pressure drop through valves and lines.

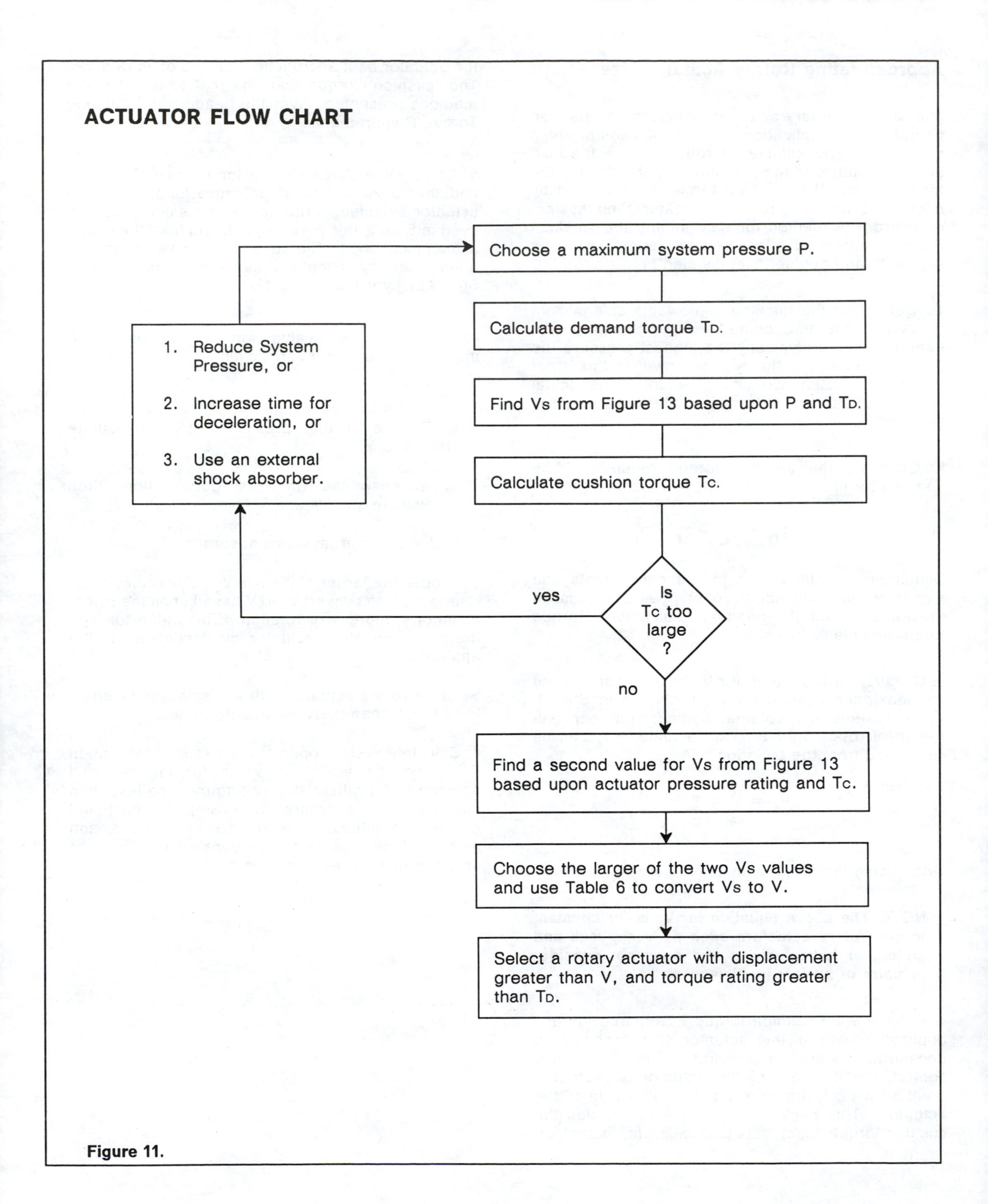
ACTUATOR FLOW CHART
Choose a maximum system pressure P.
Calculate demand torque TD.
Find Vs from Figure 13 based upon P and TD.
Calculate cushion torque Tc.
Is Tc too large ?
yes
no
1. Reduce System Pressure, or
2. Increase time for deceleration, or
3. Use an external shock absorber.
Find a second value for Vs from Figure 13 based upon actuator pressure rating and Tc.
Choose the larger of the two Vs values and use Table 6 to convert Vs to V.
Select a rotary actuator with displacement greater than V, and torque rating greater than TD.

Figure 11.

MINIMUM POSSIBLE ROTARY ACTUATOR VOLUME

This chart is based upon the equation $T = \frac{PV}{\Theta}$. Results from this table DO NOT account for actuator efficiency or any design factors.

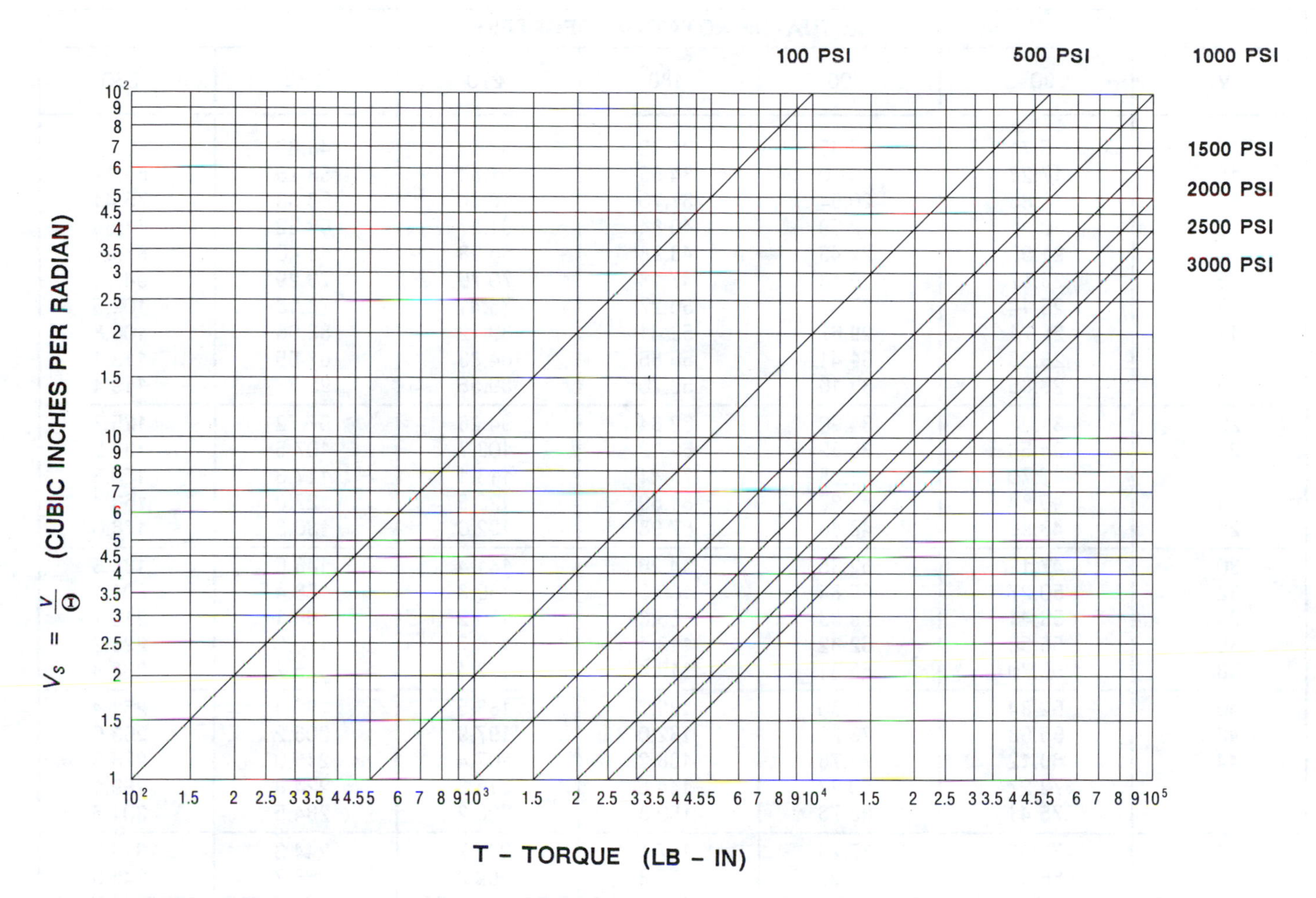

Figure 12.

TABLE 6: CONVERT V_S to V

Based upon equation: $V_S \theta = V$ for converting specific volume into volume.

ACTUATOR ROTATION VOLUME

	ACTUATOR ROTATION (DEGREES)					
V_S	90	100	180	270	280	360
10	15.71	17.45	31.42	47.13	48.86	62.84
11	17.28	19.20	34.56	51.84	53.75	69.12
12	18.85	20.94	37.70	56.56	58.63	75.41
13	20.42	22.69	40.84	61.27	63.52	81.69
14	21.99	24.43	43.98	65.98	68.40	87.98
15	23.57	26.18	47.13	70.70	73.29	94.26
16	25.14	27.92	50.27	75.41	78.18	100.5
17	26.71	29.67	53.41	80.12	83.06	106.8
18	28.28	31.41	56.55	84.83	87.95	113.1
19	29.85	33.16	59.70	89.55	92.83	119.4
20	31.42	34.90	62.84	94.26	97.72	125.7
22	34.56	38.39	69.12	103.7	107.5	138.2
24	37.70	41.88	75.41	113.1	117.3	150.8
26	40.84	45.37	81.69	122.5	127.0	163.4
28	43.99	48.86	87.98	132.0	136.8	176.0
30	47.13	52.35	94.26	141.4	146.6	188.5
32	50.27	55.84	100.5	150.8	156.4	201.1
34	53.41	59.33	106.8	160.2	166.1	213.7
36	56.56	62.82	113.1	169.7	175.9	226.2
38	59.70	66.31	119.4	179.1	185.7	238.8
40	62.84	69.80	125.7	188.5	195.4	251.4
42	65.98	73.29	132.0	197.9	205.2	263.9
44	69.12	76.78	138.2	207.4	215.0	276.5
46	72.27	80.27	144.5	216.8	224.8	289.1
48	75.41	83.76	150.8	226.2	234.5	301.6
50	78.55	87.25	157.1	235.7	244.3	314.2
55	86.41	95.98	172.8	259.2	268.7	345.6
60	94.26	104.7	188.5	282.8	293.2	377.0
65	102.12	113.4	204.2	306.3	317.6	408.5
70	109.97	122.2	219.9	329.9	342.0	439.9
75	117.83	130.9	235.7	353.5	366.5	471.3
80	125.68	139.6	251.4	377.0	390.9	502.7
85	133.54	148.3	267.1	400.6	415.3	534.1
90	141.39	157.1	282.8	424.2	439.7	565.6
95	149.25	165.8	298.5	447.7	464.2	597.0
100	157.1	174.5	314.2	471.3	488.6	628.4

Calculating Required Pump Flow

The flow rate required for a rotary actuator can be determined by the desired time for rotation and the rotary actuator's displacement. This is shown in Equation 5.

5) $$Q = \frac{V}{t}$$

where
Q = Flow rate
V = Rotary actuator displacement
t = Time to fill displacement

This equation is also plotted as Figure 14.

EXAMPLE: A 280° vane rotary actuator is chosen to provide a 194° rotation in 2 seconds. If the rotary actuator's displacement is 77.8 in^3 find what flow rate is required from the pump. Assume constant angular velocity.

SOLUTION: The actuator is only rotating 194° so the volume of oil required for this rotation is:

$$V = 77.8 \text{ in}^3 \frac{194°}{280°}$$

$V = 53.9$ in^3 for 194° rotation

Now using equation 4:

$$Q = \frac{V}{t}$$

$$Q = \frac{53.9 \text{ in}^3 \text{ (60 sec/min)}}{2 \text{ sec. (231 in}^3\text{/gal)}} = \textbf{7 GPM} \text{ ANSWER}$$

EXAMPLE: A 180° rack and pinion rotary actuator is to accelerate from 0 to some angular velocity ω during its first 10° of rotation, then remain at that angular velocity for the next 150° of rotation, then decelerate back to 0 radians/sec during the last 20°. The actuator is to rotate the total 180° in less than 2 seconds. If the actuator's displacement is 36 in^3 , find:

A. The angular velocity ω after the first 10° of rotation.

B. The pump flow rate required for the rotary actuator.

C. The pump flow required if the actuator traveled the entire 180° in 2 seconds at a constant angular velocity.

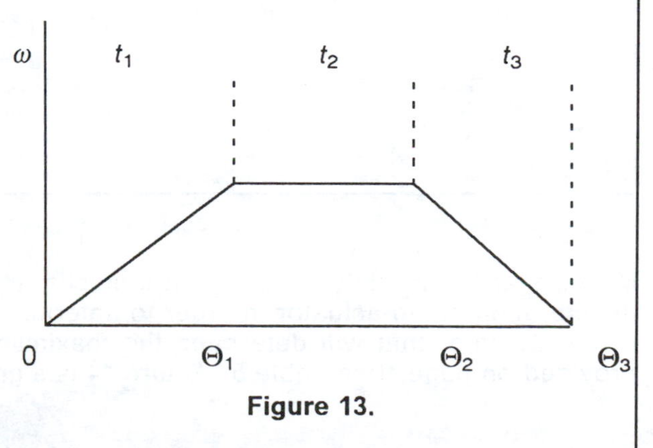

Figure 13.

SOLUTION:

A. Assume constant acceleration during the first 10° and constant deceleration during the last 20°.

$2 \text{ sec} = t_1 + t_2 + t_3$

CONTINUED ON NEXT PAGE

EXAMPLE CONTINUED

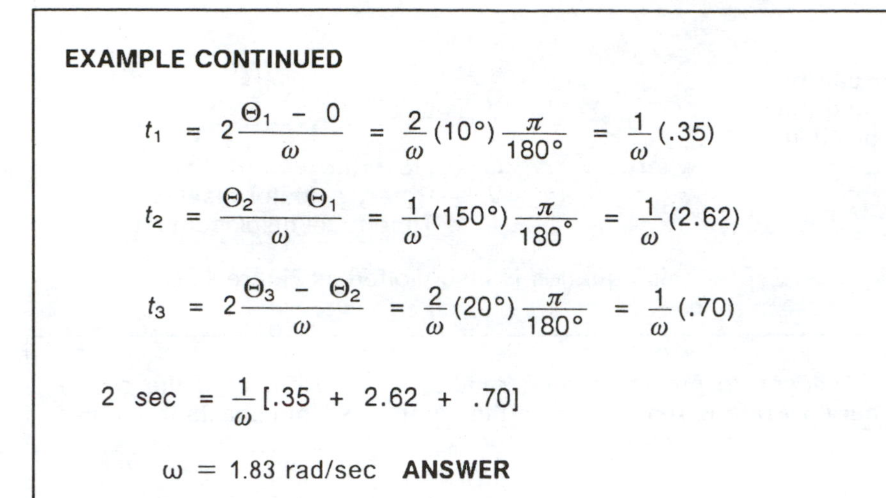

$$t_1 = 2\frac{\Theta_1 - 0}{\omega} = \frac{2}{\omega}(10°)\frac{\pi}{180°} = \frac{1}{\omega}(.35)$$

$$t_2 = \frac{\Theta_2 - \Theta_1}{\omega} = \frac{1}{\omega}(150°)\frac{\pi}{180°} = \frac{1}{\omega}(2.62)$$

$$t_3 = 2\frac{\Theta_3 - \Theta_2}{\omega} = \frac{2}{\omega}(20°)\frac{\pi}{180°} = \frac{1}{\omega}(.70)$$

$$2\ sec = \frac{1}{\omega}[.35 + 2.62 + .70]$$

ω = 1.83 rad/sec **ANSWER**

B. 180° rotary actuator has a volume displacement of 36 in^3
The cubic inches per radian can be expressed as:

$$V_S = \frac{V}{\Theta} = \frac{36 in^3}{\pi} radians$$

$$V_S = 11.5\ in^3/radian$$

NOTE: 180° = π radians

The actuator must be able to rotate at 1.83 rad/sec so the pump flow must be:

$$Q = V_S\omega$$

$$Q = (11.5\ in^3/radian)(1.83\ rad/sec)\frac{1\ gal}{231\ in^3}\ \frac{60\ sec}{min}$$

Q = 5.5 GPM **ANSWER**

C. If the entire 180° were traversed at constant speed in 2 seconds, the pump flow would be:

$$Q = \frac{V}{t}$$

$$Q = \frac{36\ in^3}{2\ sec}\ \frac{gal}{231\ in^3}\ \frac{60\ sec}{min}$$

Q = 4.7 GPM **ANSWER**

Notice that in the above example it is necessary to take into account the time required for acceleration and deceleration of the actuator in order to determine the maximum velocity required. It is the maximum velocity of the actuator that will determine the maximum flow required. Equations for velocity and acceleration are provided on page 10 in Table 5. Figure 14 is a graphic representation of Equation 5.

TIME / REVOLUTION – VS. – VOLUME

BASED UPON 100% EFFICIENCY AND THE EQUATION: $t = 26\ \forall/Q$

where t = time/rev. in seconds

$\forall$ = displacement in in^3

Q = oil flow in GPM

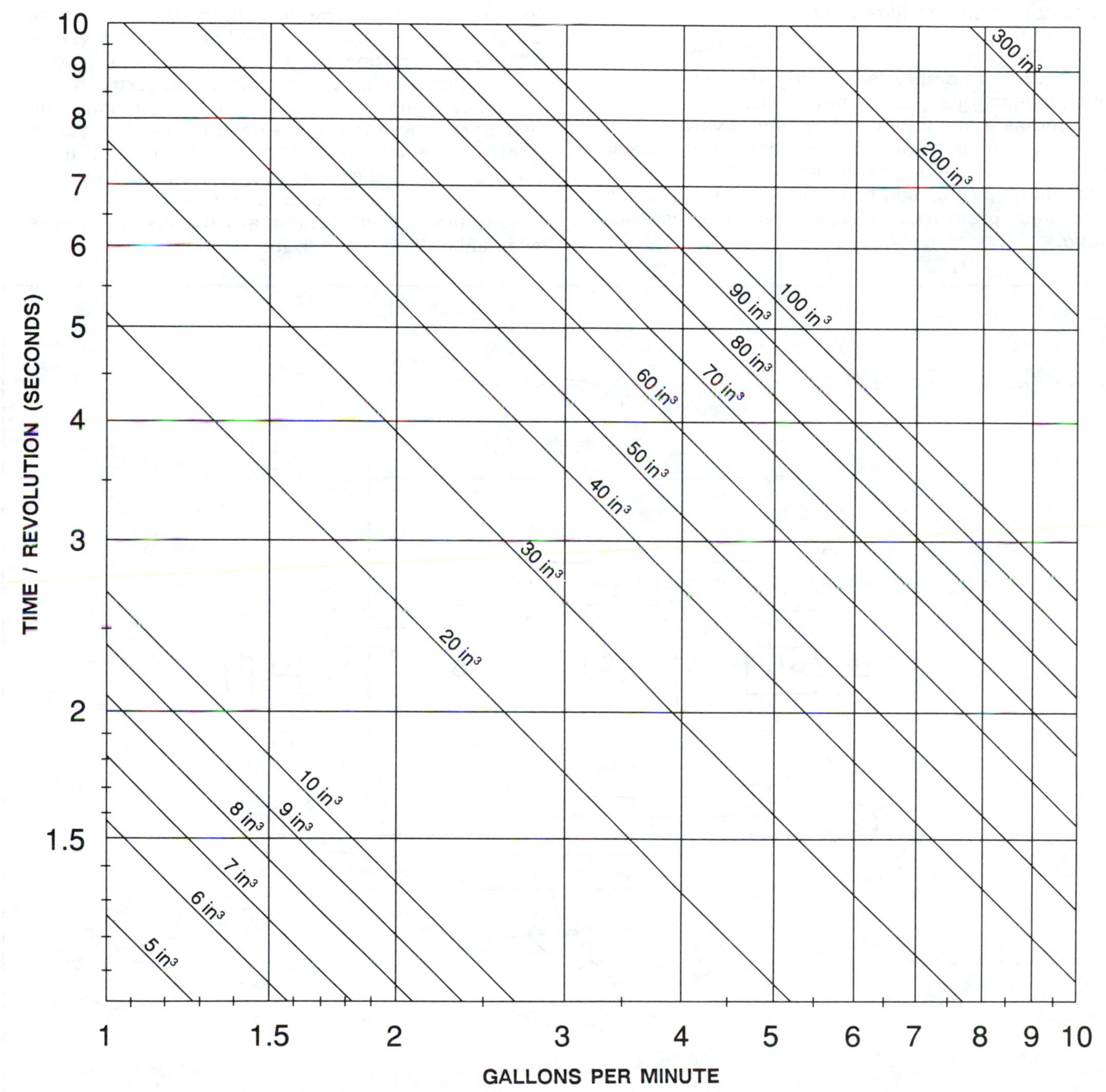

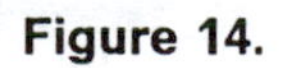
Figure 14.

Circuit Recommendations

I. Hydraulic Circuits

A. Composite Operating Circuit. When designing hydraulic operating circuits for rotary actuators, consideration should be given to the following criteria:

- actuator rotational velocity
- kinetic energy developed
- actuator holding requirement
- system filtration

Figure 15 is a composite drawing showing general recommendations for sample circuitry. It is intended as a guide only. Flow control valves (1) in the meter-out position provide controlled actuator velocity. Care should be taken if the load moves overcenter, as the combination of load and pump generated pressures may exceed the actuator rating.

To protect the actuator and other system components from shock pressures caused when the actuator is suddenly stopped in mid-stroke, cross-over relief valves (2) should be installed as close to the actuator as possible. These relief valves also protect the actuator and system if the load increases and "back-drives" the hydraulic system.

In applications involving high speeds or heavy loads, the built up kinetic energy may be too much for cushions to absorb during their 20° of operation. By using cam or lever operated deceleration valves (3) the deceleration arc can be increased beyond 20° so that kinetic energy can be absorbed more gradually and without over-pressurizing the actuator. Where there is a need to hold the load in intermediate positions for extended periods of time, pilot operated check valves (4) should be used. These must be used with leakproof actuator seals to hold the load in position; any bypass flow allows eventual drifting of the load.

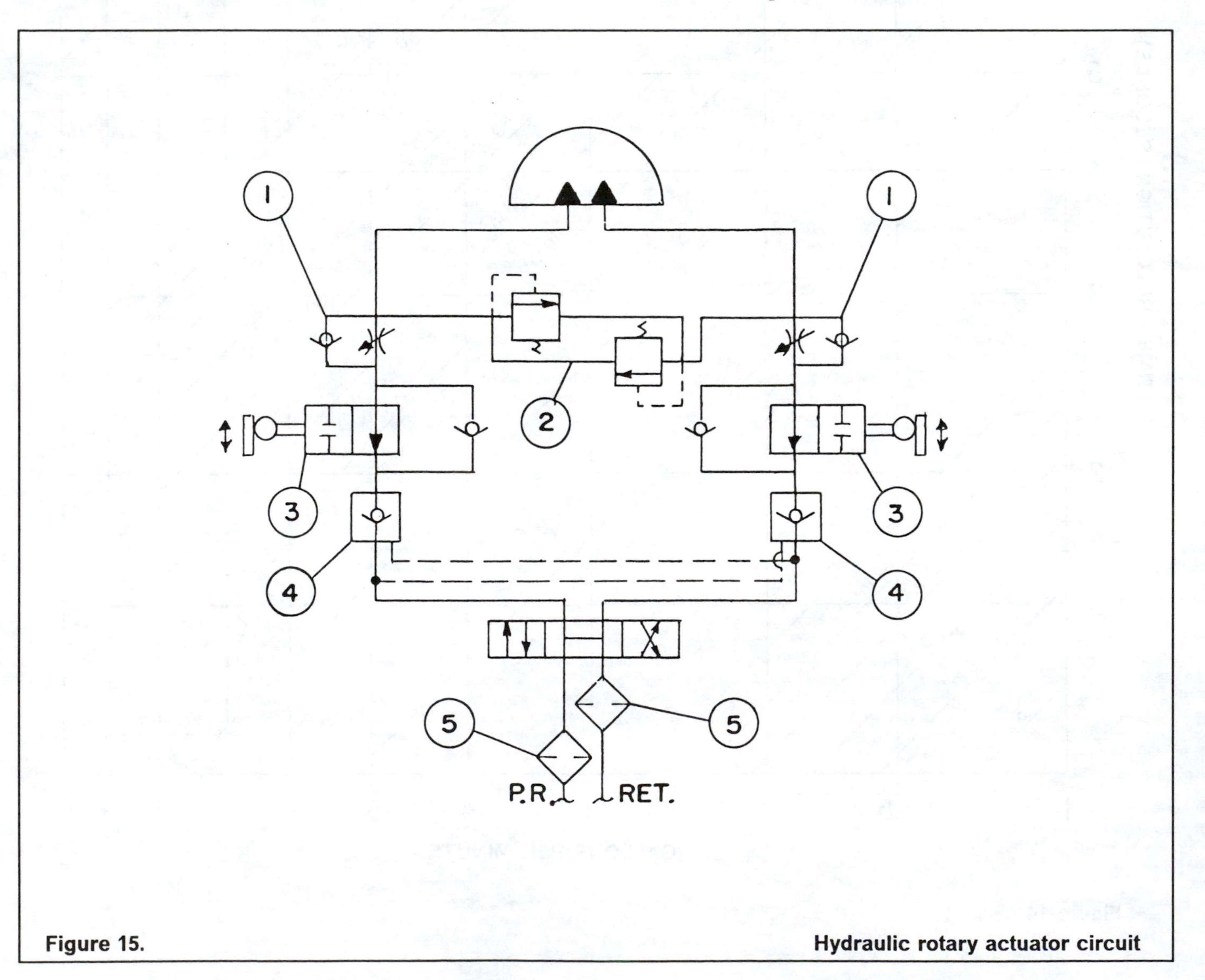

Figure 15. Hydraulic rotary actuator circuit

WARNING: FOR SAFETY REASONS, SOME APPLICATIONS REQUIRE A MECHANICAL LOCK ING DEVICE FOR HOLDING LOADS OVER AN EXTENDED PERIOD OF TIME.

As with most standard hydraulic circuits, rotary actuator applications should have filtration to provide a continuous cleanliness rating of no more than 390 particles greater than 10 micron per milliliter of fluid. This is an ISO 17/14 fluid cleanliness classification. Filters (5) should be fitted and maintained to ensure this minimum level.

B. Electrohydraulic Circuitry. The use of electrohydraulic components for rotary actuator applications can provide greater system flexibility. Figure 16 is a representative circuit showing some possible applications of electrohydraulic valves. Proportional or servo control valves (1) can provide precise position, velocity or acceleration control of loads, and "closing the loop" around a feedback device such as the ParkerJ option (2) can provide even greater control and velocity profiles for overcenter or varying loads. Even more precise position control is also possible with the use of vane actuators or anti-backlash devices on rack and pinion units.

Torque control can also be achieved with proportional pressure control valves (3) that can be used to vary the "stall torque" of an actuator or provide a torque profile for various machine processes or set-ups.

All of the considerations from part A are still relevant. Crossover relief valves should be installed if there are sudden stops in mid-stroke, and caution should be exercised when running overcenter, with high speeds, or high inertia loads. Filtration (5) is still a consideration, but the actuator requirements (ISO 17/14 class) are usually less demanding than the filtration requirements of today's electrohydraulics.

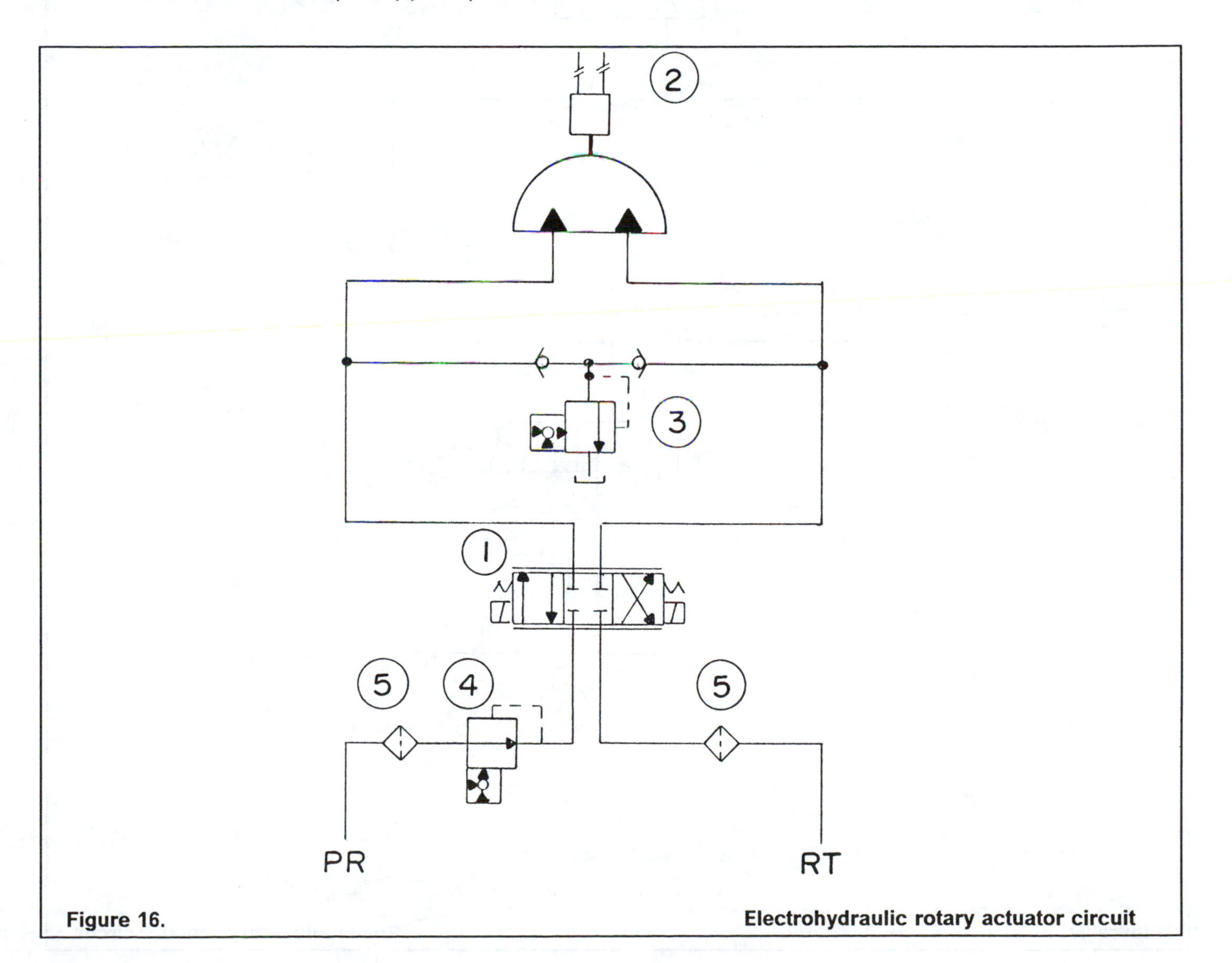

Figure 16. **Electrohydraulic rotary actuator circuit**

II. Pneumatic Circuits

All of the considerations for hydraulic actuator circuit design are also valid for pneumatic circuits, with the added caution that pneumatic systems generally run at higher speeds thus developing more kinetic energy. Caution is also required in that pneumatic cushion devices are not as effective as hydraulic cushions, as they cannot absorb as much kinetic energy. External shock absorbing devices may be necessary.

Figure 17 is a composite pneumatic circuit giving some general recommendations. It is intended as a guide only.

Meter-out flow control valves (1) should be used to control actuator velocity. Two speed circuits to reduce the speed before the end of stroke should be considered for fast rotation applications.

The use of two pressure circuits (2) should be considered where there are different loads in opposite directions to keep rotational speeds low and to conserve energy.

Compressed air should be filtered to at least 40 micron (3) and the use of air line lubricators (4) with a good quality lubricating oil is recommended.

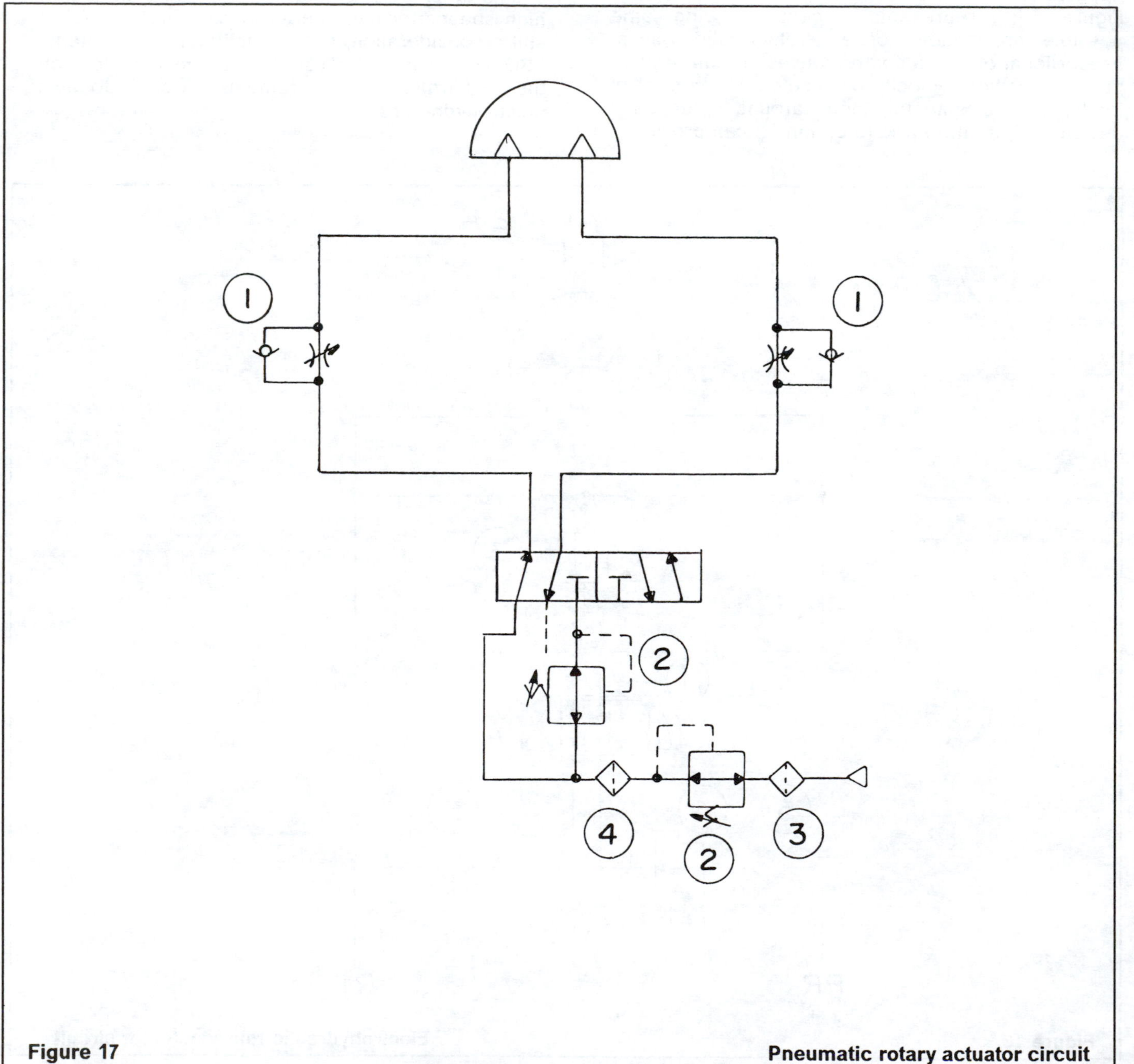

Figure 17 **Pneumatic rotary actuator circuit**

Installation Instructions and Options

A. Stops

Vane units should not use the vane and stator as a positive stop. For light to medium loads a taper lock stop (TLS) option is available, or external stops mounted to the machine framework may be used. For high inertia or high speed loads externally mounted valving or deceleration devices should be used to minimize system shocks.

Rack and pinion units can be stopped at the end of stroke provided the loads and speeds are not too high. Cushions can be used to decelerate the load to a gentle stop, providing the maximum actuator pressure rating is not exceeded by the cushions. Again, for high inertia or high speed loads externally mounted valving or deceleration devices should be used to minimize system shocks.

B. Surge Pressures

Surge or shock pressure in excess of the actuator rated pressure are detrimental to unit life and must be avoided. Crossover relief valves mounted adjacent to the actuator can help reduce these abnormal pressure peaks.

Pressure developed by cushion or deceleration valves should also be kept below rated pressure.

C. Angular Velocity

Angular velocity can be controlled by metering the flow into or out of the actuator ports. This can be accomplished by the use of flow control valves or if more sophisticated control is required; through the use of proportional or servo valves.

Care should be taken when using a meter-out circuit if the load moves over center, as the combination of load and pump generated pressures may exceed the actuator rating.

D. Drains

Some actuators are fitted with drain ports to minimize external leakage possibilities. These drain ports should be connected directly back to the oil reservoir with a minimum or back pressure (50 psi maximum).

E. Gear Chamber

Some rack and pinion actuators are supplied with the gear chambers filled with a molybdenum disulfide grease to better absorb gear stress and extend gear life. This chamber should be checked and filled periodically to ensure adequate gear lubrication. The housing is also fitted with a small relief valve that vents excess pressure in the gear chamber to the atmosphere. This is an indication of pressure seal wear, because high pressure oil is bypassing the piston seal and pressurizing the gear chamber. The piston seals should then be replaced.

F. Fluid Medium and Seals

For hydraulic usage, a clean, filtered, high-quality mineral-based hydraulic fluid with 150 to 500 S.U.S. Viscosity at 100°F is recommended for use with standard buna seals. Cleanliness should be maintained to an ISO code 17/14 level. The correct filters can be provided by your fluidpower distributor.

Air service units, including filter, regulator, and lubricator ensures correctly conditioned compressed air is available for pneumatically operated systems.

Standard seal compound is Buna N for mineral-based hydraulic fluid. Other seal materials can be provided for most operating fluids. To ensure correct seal compound, please provide the name and type of operating fluid to the factory.

G. Shaft Couplings

Couplings should engage the full length of the shaft keyway and pressure should only be applied after support has been provided on the opposite end of the shaft. Shafts should be within 0.005 TIR to ensure proper alignment.

NOTES

index

general index i
index of tables & graphs xix
index of circuits xxi
catalog reference directory xxii

general index

a

Abrasion
- Dirt Particles d-43
- Hose ..... j (c-9, c-11)

Absolute Rating d-39-40

Absorbing Shock
- Accumulators a-3
- Cylinders b-16
- Hydraulic Motors f-3

Absorptive Filter g-24

AC Motors, Dimensions c-3

AC Solenoid
- Current Flow c-6
- Failure g-24
- Wet Armature c-6

Acceleration Force, Cylinders b-6

Accumulator
- Adiabatic Performance a-2-3
- Application a-3-4
- Auxiliary Gas Bottles ..... a-4, k (b-2)
- Circuits a-5-6
- Cycle Time a-3
- Fluid Compatibility a-4
- General a-2-6
- Helpful Hints a-4
- Installation a-4
- Isothermal Performance a-2-3
- Operation a-3
- Piston a-2-3
- Precharging a-3-4
- Recharging a-5
- Safety Bleed a-5-6
- Seals a-4
- Sizing a-2-4
- System Design ..... h-9, k (b-2)
- Temperature Range a-4
- Unloading a-5
- Water Service a-4

Acoustic Baffling h-12

Acropor Element d-39

Actual Stroke b-5

Actuator
- Circuit Design ..... k (b-2)
- Controls ..... k (b-2)

Additives d-44, i-7

Adiabatic Performance a-2-3

Adjusters, Stroke b-16

Aeration d-42, e-6, h-13,

Aftercooler g-17

Aging, Solenoid c-7

Air Accumulator g-25

Air Bleed
- Cylinders b-16
- Pumps h-10
- Suction Line h-11

Air Brake Fittings ..... j (a-5)

Air Breather ..... h-3, i-5, k (b-4)

Air Circulation h-7, h-9, h-13

Air Cushioning g-13

Air Drop Lines g-10

Air Filters
- Absorptive g-24
- Application g-10
- Baffle g-10
- Bowl g-10
- Installation g-10
- Quiet Zone g-10
- Turbulence g-10

Air Formulae g-6-7

Air Flow
- Orifice g-6-7
- Pipe g-8

Air, Free b-12

Air Inclusion, Coupling ..... j (b-2)

Air Line
- Moisture g-10, g-17-18
- Slope g-18

Air Motors g-13

Air Mufflers g-25

Air-Oil System b-25

Air-Oil Tank e-2, e-6

Air Preparation g-10-12

Air Preparation Units g-10

Air Regulator
- Application g-10-11
- Flow Characteristics g-10
- High Pressure Pilot g-10
- Maintenance g-11
- Nonventing g-10
- Pressure Cracking Characteristics g-11
- Remote Controlled g-11
- Setting g-10-11
- Spring Controlled g-11

Air Requirement, Cylinder b-11-12

Air Valve
- Design g-5
- Icing g-25
- Overlubrication g-25
- Selection g-4-5
- Steer g-5
- Troubleshooting g-24-25

Alignment
- Coupling h-3, h-10
- Tubing ..... j (e-21, e-23)

Ambient
- Noise ..... k (b-5)
- Temperature h-7, h-13

Amplified Noise h-10-12

Antifoaming h-11

Analysis, Fluid d-43

index continued

Aniline Point d-44

Appliance Grade Solenoid c-7

Armature, Wet c-6

ASTM Standard Viscosity Temp Chart d-15

Automatic Drains, Air Systems g-17,g-19

Axial Piston Pump i-19

b

"B" Design Motor c-4

Back Flow Check Valve .... k (c-20)

Back Pressure b-11, b-16, b-38, f-3, h-10, i-14-15

Baffle
- Air Filter g-10
- Air-Oil Tank e-6
- Reservoir h-3, h-11-13

Baffling, Acoustic h-12

Ballistic Type Flaring Tool .... j (e-19)

Basic Cylinder Length b-5

Beam, Cylinder Application b-20

Bearing
- Failure b-33
- Length b-5
- Loads b-16
- Rod b-32

Bell Curve Filter Element d-39

Bent Axis Piston Pump i-19

Bi-Directional
- Hydraulic Motors f-6, f-8-9
- Pressure Filter d-42
- Pump f-9

Binding, Hydraulic Motors f-4

Bite Type Tube Fittings .... j (e-14-15, e-20)

Bladder Accumulators a-3

Bleed Off .... f-4, h-8-9, k (b-3)

Block and Pin Type Flaring Tool .... j (e-19)

Blowoff, Fitting .... j (c-9)

Blowout, Tubing .... j (e-15, e-20)

Bobbin, Solenoid c-6

Boost Pump f-9, h-10

Boyle's Law g-6

Braking Valve f-3, f-7

Brass Coupling .... j (b-2)

Breakaway
- Springs c-7
- Torque f-3

Breakdown Torque, Electric Motor c-4

Breather Cap .... h-3, i-5, k (b-4)

Buckling, Cylinder Rod b-33

Buna N (nbr) .... j (b-2)

Burn-out, Solenoid Coil c-7-10, g-24

Burst Pressure
- Couplings .... j (b-2)
- Hose .... j (c-7)
- Tubing .... j (e-12)

By-Pass Spring, Filter d-41

c

"C" Design Motor c-4

"C" Frame Press h-11

"C" Series Directional Air Valve g-22

Calculation Pressure Differential in Straight Line Lengths d-17

Calculation Pressure Differential in Fittings & Bends d-18

Calibration of Instruments, Noise Measurement .... k (b-6)

Cam
- Ring i-18-19
- Valve i-14

Cap, Cylinder b-32

Cap Cylinder Mounting
- Detachable Clevis b-3
- Fixed Clevis b-3
- Rectangular Flange b-3
- Square Flange b-3
- Trunnion b-3

Capacity Coefficient (Cv) .... b-13, g-4-5, k (b-16)

Cartridge Installation, Pump i-25

Case Drain
- Back Pressure h-10
- Variable Volume Pump h-10, h-13

Cavitation .... b-30, d-38, i-5, i-19, i-22, k (b-3), k (c-20)

Center of Gravity, Cranes and Beams b-20

Centerline Lugs, Cylinder Mounting b-3

Centrifugal Compressor g-17

Charles' Law g-6

Chattering, Valve h-11

Chemical Resistance, Hose .... j (c-7)

Chloroprene Rubber (cr) .... j (b-2)

Chord Factor Method b-19

Chuck, Vacuum g-22

Circuit Design .... k (b-2-4)

Clamp
- Hose .... j (c-9)
- Hose Fitting .... j (c-13)

Class
- "A" Electric Insulation c-7
- "H" Electric Insulation c-7

Cleanability, Filter Element d-40

Cleanliness Class d-44

index continued

Clevis Mounting, Cylinder b-3, b-18-19, b-25, b-33

Closed
- Center Valve b-38
- Loop Motor Circuits f-9

Closing Time, Solenoid c-6

Coefficient
- Capacity (Cv) .... b-13, g-4-5, k (b-16)
- Cv Calculator g-5
- Orifice g-6-7

Coils, Solenoid See Solenoids

Combination Gear Pump i-12

Compatibility
- Hose .... j (c-11)
- Seals .... j (b-2)

Compressed Air, Cost c-10, g-17-18

Compression
- Cylinder Failure b-34
- Fitting .... j (a-2-4)
- Loads b-4-5

Compressor
- Acid Formation g-17
- Centrifugal g-17
- Considerations g-17-19
- Indicator Card g-17
- Rating b-12
- Reciprocating g-17
- Single Stage g-17
- Two-Stage g-17

Conduit Sizes, Electrical c-5

Connectors, Installation .... j (e-22)

Constant
- Horsepower i-21
- Resistances, Suction Line d-41

Contamination
- Atmosphere .... j (e-16)
- Cylinder b-38
- Directional Valves c-9-10
- Fluid h-13-14, i-25
- Maintenance d-43
- Moving Components d-43
- New Systems d-43
- Pump i-5
- Solenoid c-9
- Tube Fitting .... j (e-20, e-22, e-24)

Cooling
- Heat Exchanger .... k (b-4)
- Reservoir h-7

Corrosion
- Filter Element d-40
- Hose .... j (c-8)
- Metal Parts h-8

Cost
- Filter d-38
- Filter Element d-40
- Hose .... j (c-7)
- System i-5

Counterbalance Valve .... k (b-3)

Coupler Body .... j (b-3)

Coupling, Drive h-3, h-5, h-10-11, i-4

Couplings, Hydraulic See Hydraulic Couplings

CPI Tube Fittings .... j (e-10)

Crane, Cylinder Application b-20, b-22

Creep
- Cylinder b-38
- Hydraulic Motors f-9
- Tube Fittings .... j (a-4)

Cross-Over Relief Valve f-3, f-6

Current
- Flow, AC Solenoid c-6

Cushion
- Adjustment Needle Location b-14
- Floating b-15
- Rating, Air Cylinder b-4
- Spear b-15

Cushioning
- Air g-13
- Accumulator a-3
- Cylinder b-7, b-10-12

Cv .... b-13, g-4-5, k (b-16)

Cv Calculator g-5

Cycle Time
- Accumulators a-3
- Machine .... i-22, k (b-2, b-6)
- Solenoid c-6-7

Cyclical Noise .... k (b-6, b-8)

Cylinders
- Acceleration Force b-6
- Actual Stroke b-5
- Air Bleed b-16
- Air Requirement b-11-12
- Applications b-18-26
- Back Pressure b-11, b-16
- Basic Length b-5
- Bearing Failure b-33
- Bearing Length b-5
- Bearing Loads b-16
- Cap b-31-32
- Cap Detachable Clevis Mounting b-3
- Cap Fixed Clevis Mounting b-3
- Cap Rectangular Flange Mounting b-3
- Cap Square Flange Mounting b-3
- Centerline Lugs Mounting b-3
- Circuits b-26-28
- Compression Failure b-34
- Compression Loads b-4-5
- Construction b-31-32
- Cup Packing b-32
- Cushions b-7, b-10-12, b-15, b-31
- Cushion Check Valve b-31
- Cushion Floating b-31
- Cushion Needle b-14, b-31
- Cushion Rating b-11
- Deceleration Force b-6-7, b-10-11
- Differential Lift b-25
- Double Acting b-32-33
- Dowel Pin Mounting b-35
- Drift b-38
- Duplex b-32-33

index continued

Dynamic Thrust b-12-13
Erratic Operation b-38
Failure b-33, b-38
First Class Lever b-18, b-22
Fluid Motor Coupled b-20
Force b-17
General b-2-38
Gland b-31-32
Gland Drain b-16
Head b-31-32
Head Rectangular Flange Mounting b-3
Head Square Flange Mounting b-3
Head Trunnion Mounting b-3
High Speed Air b-25
Hinged Lever b-25
Industrial b-4
Integral Key Mounting b-3
Intermediate Fixed Trunnion Mounting b-3
Jewel Gland b-16
Lifting b-25
Lipseal b-15-15, b-32
Manifold Ports b-14
Matched Flow Controls b-27
Mechanically Connected b-26
Metered-Out Air b-24-25
Minor Area b-7
Misalignment b-36
Motions b-18
Mounting Accessories b-4
Mounting Bolts b-3
Mounting Classes b-4
Mounting Selection b-4-5
Mounting Styles b-3
Net Stroke b-5
Non Shock Rating b-7
Operation b-31
Options b-16
Oversize Ports b-14
Piston b-31-32
Piston Construction b-16
Piston Rings b-31-32
Piston Ring Leakage b-16
Piston Rod Compared to Stroke b-5-6
Piston Speed b-8
Pivot Force Transfer b-4
Port Positions b-14
Port Sizes b-8, b-14
Power Factor b-19, b-24
Pressure Intensification b-30
Push and Pull Forces b-2
Rack and Pinion Connection b-26
Ram b-32-33
Regenerative Circuits b-29-31
Rod b-31-32
Rod Bearing b-32-33
Rod Buckling b-33
Rod End Connection b-5-6
Rod Extension b-5
Rod Gland b-31-32
Rod Gland Leak b-38
Rod Seals b-31-32
Rod Scraper b-32
Rod Speed b-17, b-29
Rod Wiper b-16, b-32
Rolling Load b-24
Rotary Motion b-21
SAE Straight Thread Ports b-14
Seals b-16, b-31-32
Seal Failure b-38
Second Class Lever b-18, b-21
Self-Adjusting Rod End b-36
Self-Bleeding b-16
Series Actuators b-27
Shear Key Mounting b-37
Short Non-Centerline Mounting b-34
Side End Angles Mounting b-3
Side End Lugs Mounting b-3
Side Loading b-33
Side Lugs Mounting b-3
Side Tapped Mounting b-3
Single Acting b-32-33
Sliding Loads b-24-25
Slippage h-8
Speed Controls b-13
Spreader Type Tie Rods b-35
Spring Return b-32-33
Step Cushion b-31
Stop Tube b-33
Straight Line Force Transfer b-4
Stroke Adjusters b-16
Stroke Considerations b-4-6
Tandem b-32-33
Tapered Pipe Thread Ports b-14
Telescoping b-32-33
Tension Failure b-34
Tension Loads b-4
Third Class Lever b-18
Tie Rods b-31-32
Toggle Force b-18
Toggle Mechanism b-18, b-24
Trouble Shooting b-38
Trunnion Mounting b-3, b-37
2:1 Ratio b-29-30
Uniform Acceleration Force Factor b-6-7
"V" Packing b-15, b-32
Vent Ports b-16
Vent Screws b-16
Vertical Lifting b-25
Water Service b-16
Wiperseal b-16
Working at an Angle b-19

Cylinder-to-Cylinder Intensifier e-4-5

d

"D" Design Motor c-4

Dampening
Noise h-11
Vibration .... j (e-16)

DC Solenoid
Failure c-9, g-24
Holding Current g-24

Deceleration Force
Air Cylinders b-10-11
Hydraulic Cylinders b-6

Decibel .... k (b-5-6, b-8-9)

Degree of Filtration d-38, d-40-41

Depth Type Filter Elements
Absolute Rating d-40
Advantages d-40
Bell Curve d-39

index continued

Disadvantages d-40
Mean Flow Pore Size d-39-40
Nominal Rating d-39
Types d-39

Design
"B" Motor c-4
"C" Motor c-4
"D" Motor c-4

Designing Hydraulic Circuits .... k (b-24)

Differential
Cylinder b-29-30
Lift b-25
Relief Valve .... k (b-4)
Unloading Relief Valve a-5-6

Directional Control Valve, Troubleshooting c-9-10

Dirt
Holding Capacity, Filter Element d-40
Removal d-39
Sources d-38, d-43

Double
-Acting Cylinder b-32-33
Extra Heavy Pipe .... j (e-10)
Rod Cylinder b-27, b-32-33

Dowel Pins, Cylinder Mounting b-35

Downtime, Machine d-43

Drain
Automatic, Air Systems g-17, g-19
Case h-10, h-13,
Cocks .... j (a-5)
Electric Power i-22-23
Gland b-16
Line, Installation h-3
Plug h-3
Top .... k (B-4)

Drift, Cylinder b-38

Drip Proof Motor c-4

Drive Speed, Pump i-5

Dryseal Pipe Threads .... j (a-5, e-18)

Dual
Air Exhaust Controls g-15
Frequency Solenoid Coils c-7
Pressure Intensifier e-2-5
Vane Pump i-11-12, i-14, i-19

Duplex
Cylinder b-32-33
Filter d-42

Duty Cycle, Solenoid c-6

e

Efficiency
Electric Motor i-22
Filter Element d-40
System .... j (e-13)
Volumetric i-18-19

810 Series Hydraulic Motor f-5

830 Series Hydraulic Motor f-5

Elbow
Suction Filter d-41
Return Line Filter d-42

Electric Current Abroad c-8

Electric Motors
Breakdown Torque c-4
Design "B" c-4
Design "C" c-4
Design "D" c-4
Drip Proof c-4
Efficiency i-22
Encapsulated c-4
Enclosures c-4
Explosion Proof c-4
Guarded Motor c-4
High Slip Motor c-4
Horsepower c-2
Locked Rotor Torque c-4
Open c-4
Series Wound i-22
Splash Proof c-4
Squirrel Cage c-4, i-22

Starter Sizes c-5
Three Phase c-4
Totally Enclosed c-4

Electrical
Conductivity of Hose .... j (c-7)
Conduit Sizes c-5
Controls .... k (b-4)
Signal Interlock c-9-10
Wire Sizes c-5

Elements, Hydraulic Filter See Filter Elements

En-Bloc Control Valve .... k (c-20-21)

Encapsulated Motor c-4

Enclosures, Electric Motors c-4

End Thrust, Pumps h-3, h-10

Environmental
Coupling Considerations .... j (b-2)
Stress Cracking, Fittings .... j (a-4)

Erosion d-43, i-5

Erratic Operation
Cylinder b-38
Solenoid c-9

Ethylene Propylene Rubber (EPM) .... j (b-2)

Exactol .... j (e-23)

Expansion Fittings .... j (a-4)

Explosion Proof Motor c-4

Extra Heavy Pipe .... j (3-10)

f

Fabrication Filter d-41

Fail Safe Conditions .... k (b-2)

Fast-Stor .... j (d-3)

Fast and Tite
Clear Vinyl Tubing .... j (d-2)

index continued

Fittings .... j (d-4)
Polyethylene Tubing .... j (d-2)

Fatigue, Filter Element d-40

Feed Lines, Solenoid c-9

Feedback, System i-21

Ferro-Magnetic Particles d-38

Ferrule .... j (e-20-21)

Ferulok Tube Fittings .... j (e-10)

Ferulube .... j (e-19)

Field, Solenoid c-7-8

50 Cycle Solenoids c-7

Filler Cap .... k (b-4)

Filling, Reservoir d-45

Filters, Air See Air Filters

Filters, Hydraulic See Hydraulic Filters

Filters, Noise Measurement .... k (b-6, b-8)

Fire Resistant Fluids d-38, h-13

First Class Lever b-18

Fitting
Blowoff .... j (c-9)
Make-Up .... j (e-20)

Fittings, Hose, See Hose Fittings

Fittings, Tube, See Tube Fittings

Fittings, Pipe .... j (a-5)

5000 PSI Pressure Filter d-42

Flange Mounting, Cylinder b-3

Flared Tube Fittings .... j (e-14-15)

Flareless Tube Fittings .... j (e-14-15, e-20-22)

Flaring Tools
Ballistic Type .... j (e-19)
Block and Pin Type .... j (e-19)
Power Actuated .... j (e-19)
Screw Type .... j (e-19)

Flaring Tube .... j (e-14)

Flat Response, Microphone .... k (b-6)

Flexible Hose .... k (b-4)

Flexing Component Housings d-43

Floating Cushion b-15

Flow
Air Through Pipes g-8
Clearance, Tube Fittings .... j (a-4)
Discharging Cylinders d-10
Dividing Valve b-28, f-8
Meter, Air g-9
Laminar d-16

Turbulent d-16

Flow Control Valves
Bleed Off .... f-4, h-8-9, k (b-3)
Heat Generation h-8-9
Matched b-27
Meter-In .... b-28, f-4, h-8, k (b-3)
Meter-Out .... b-24-25, b-28, f-3, h-8, k (b-3)
Motors b-27, f-3-5, f-7-8
Pneumatic g-8

Fluid
Additives d-44, i-7
Analysis d-11-12
Charts d-2

Contamination d-11, h-13-14
Identification h-13
Level b-30, d-41,
Molecular Structure .... j (e-15)
Monitoring d-44
Problems d-44
Sampling d-43
Supply c-10
Temperature c-10
Velocity b-8, b-14, d-41-42

Fluorocarbon Rubber (FPM) .... j (b-2)

Foam Depressants i-7

Foaming d-45, h-13

Force
Cylinder b-17
Solenoid c-7-8
Toggle b-18

Four-Bolt Flange Gear Pump i-9

450 Series Pump h-10

Friction
Air in Pipes g-8-9
Coefficient b-7, b-13
General b-6, b-13, i-10
Hydraulic Motors f-4
Liquids in Conductors d-16
Kinetic b-38
Solenoid c-7
Static b-38
Surface b-25

Free
Air b-12
Field .... k (b-6)
Wheeling, Hydraulic Motors f-5-9

FRL g-10

g

Gas
Bottles .... a-4, k (b-2)
Weight g-7

Gear Pumps
Advantages i-12
Combination i-12
Flow Divider i-12
General d-40, i-12, i-17-18
Operating Range i-18
Piggyback i-12
Spur i-18
Tandem i-12
Two-Bolt Flange i-21
Variable Delivery i-21,
Volumetric Efficiency i-18

index continued

Generated Noise h-10-11

Gland
- Drain b-16
- Jewel b-15
- Leakage b-38
- Rod b-31-32

Grades of Service Severity, Tubing .... j (e-10)

Gravemetric d-44

Gravity
- Assist Lubrication g-11
- Center b-20

Gross Cylinder Stroke b-5

Guarded Motor c-4

Guiding
- Cylinder b-5, b-33
- Load b-5, b-33

h

Head
- Cylinder b-32
- Rectangular Flange, Cylinder Mounting b-3
- Square Flange, Cylinder Mounting b-3
- Trunnion, Cylinder Mounting b-3

Heat
- Dissipation, Reservoir .... h-3, h-7-8, k (b-4)
- Dissipation, Solenoid c-6
- Exchanger h-8, h-13
- Generation, Solenoid c-6
- Generation, Sun h-8
- Generation, System h-7-8, i-14
- Reduction h-9, i-17
- Rise Estimation h-7

Heaters .... k (b-4)

Helical Gear Pump i-18

High
- Inertia Loads g-13
- Oil Temperature, Cause h-8
- Slip Motor c-4
- Temperature Oil h-8
- Transient Voltages, Solenoid g-24

Hi-Lo System a-6, i-11, i-15, i-21

Holding Current
- AC Solenoid c-6
- DC Solenoid g-24

Horsepower
- Constant i-21
- Electric Motor c-2
- Root Mean Square i-22

Hose
- Abrasion .... j (c-9, c-11)
- Ambient Temperature .... j (c-7)
- Application .... j (c-9-10)
- Bend Radius .... j (c-7)
- Burst Pressure .... j (c-7)
- Chemical Resistance .... j (c-7)
- Clamp .... j (c-9)
- Compatibility .... j (c-11)
- Construction .... j (c-11)
- Corrosion .... j (c-8)
- Cost .... j (c-7)
- Covers .... j (c-11, c-13)
- Electrical Conductivity .... j (c-7)
- Excessive Wear .... j (c-8)
- Fluid Velocity .... j (c-7)
- Inner Tube .... j (c-11)
- Kinking .... j (c-8-9)
- Operating Pressure .... j (c-7)
- Operating Temperature .... j (c-7, c-11)
- Premature Failure .... j (c-9)
- Pressure Drop .... j (c-4)
- Pressure Effects .... j (c-9)
- Reinforcement .... j (c-11)
- Safety Factor .... j (c-7-8)
- Spiral Wrap .... j (c-9)
- Spring Guards .... j (c-8-9)
- Swivel Adapters .... j (c-9)
- Turbulent Flow .... j (c-7)
- Vacuum Requirements .... j (c-9)
- Vibration .... j (c-8)
- Volumetric Expansion .... j (c-8)

Hose Fittings
- Clamp Type .... j (c-13)
- No Skive .... j (c-13)
- Permanently Attached .... j (c-13)
- Reusable .... j (c-13)
- Skive .... j (c-13)

Hum, AC Solenoid g-24-25

Hunting, Valve h-11

Hustler Air Valve g-4

Hydraulic Circuit Design .... k (b-2-4)

Hydraulic Couplers
- Air Inclusion .... j (b-2)
- Applications .... j (b-2)
- Burst Pressure .... j (b-2)
- Environmental Considerations .... j (b-2)
- Material Selection .... j (b-2)
- Proof Pressure .... j (b-2)
- Seal Material .... j (b-2)
- Vacuum Rating .... j (b-2)

Hydraulic Filter Elements
- Absolute Rating d-39-40
- Acropor d-39
- Bell Curve d-39
- Cleanability d-40
- Corrosion d-40
- Cost d-40
- Depth Type d-39-40
- Dirt Holding Capacity d-40
- Efficiency d-40
- Fatigue d-40
- Mean Flow Pore Size d-39-40
- Nominal Rating d-39
- Paper d-39
- Pore Size Control d-40
- Pressure Differential d-40
- Repairability d-40
- Shelf Life d-40
- Surface Type d-39-40
- Synthetic d-39
- Temperature d-40
- Viscosity d-40
- Water Base Fluid d-40
- Wire Mesh d-39-40

index continued

Hydraulic Filters
- Bi-Directional ... d-42
- Compatibility ... d-38
- Cost ... d-38
- Dirt Capacity ... d-38
- Duplex ... d-42
- Elbow Style ... d-41-43
- Element Cleanability ... d-38
- Element Replacement ... d-38
- Fabricated ... d-41-42
- Flow Rate ... d-38
- Maintenance ... d-38, d-40-45, h-13
- Never Stop ... d-41
- Pressure ... d-42, k (b-4)
- Pressure Drop ... d-38
- Return Line ... d-42
- Selection ... d-38, i-5-6
- Service ... d-45
- Submersible ... d-41
- Suction ... d-41, k (b-4)
- Tell Tale ... d-38
- Temperature ... d-38

Hydraulic Filtration
- Degree ... d-38
- Fire Resistant Fluid ... d-38
- General ... d-43
- Magnets ... d-38
- Membrane ... d-39
- NFPA Recommended Standard ... d-38
- Pressure Line ... d-42
- Suction Line ... d-41
- Return Line ... d-42

Hydraulic Motors
- Back Pressure ... f-3
- Binding ... f-4
- Bleed Off ... f-4
- Breakaway Torque ... f-3
- Circuits ... f-5-9
- Coupled ... b-28
- Creep ... f-9
- Cross-Over Relief ... f-3, f-6
- Displacement ... f-4
- Dual/Relief Braking ... f-6
- Efficiency ... f-4
- 810 Series ... f-5
- 830 Series ... f-5
- Flow Controls ... f-3-5, f-7-8
- Free Wheeling ... f-5-9
- Friction ... f-4
- General ... f-3-9
- Make Up Check ... f-5-6
- Meter-In ... f-4
- Meter-Out ... f-3
- Operating Speed ... f-3
- Overcenter Rotary Drive ... f-9
- Over-Run Limiter ... f-6
- Parallel ... f-8
- Piston ... f-4
- Running Torque ... f-3
- Selection ... f-3
- Series ... f-8
- Torque ... f-2-9

Hydro-Craft Clamping Hardware ... j (e-16)

Hy-Fer-Set ... j (e-23)

i

Icing, Air Valve ... g-25

Impulse Noise ... k (b-8)

Indicator
- Oil Delivery, Lubricator ... g-11
- Oil Level ... h-3
- Spool Position ... g-20
- Tell Tale ... d-45

Industrial
- Cylinder ... b-4
- Grade Solenoid ... c-7

Inefficiency, System ... h-8

Inertia
- Loads ... g-13
- Rotating Mass ... f-3

In-Ex Burring Tool ... j (e-23)

Inlet Conditions, Pump ... d-45

In-Line Piston Pump ... i-19

Inner Tube, Hose ... j (c-11)

Inrush Current, Solenoid ... c-6-7, c-9, g-24

Insta-Flare ... j (e-23)

Instrument Grade Tubing ... j (d-2)

Instrumentation Tube Fitting ... j (e-14, e-16, 3-21-22)

Intensifier
- Clamp Circuit ... e-3
- Cylinder-to-Cylinder Type ... e-4-5
- Dual Pressure ... e-2, e-4
- General ... e-2-6, h-9
- Operating Principle ... e-4
- Ram Size ... e-2
- Ratio ... e-2
- Selection ... e-2
- Single Pressure ... e-2, e-4
- Usage ... e-5-6

Intercooler ... g-17

Interlocks ... c-9-10, k (b-2)

Intermediate Fixed Trunnion, Cylinder Mounting ... b-3

Inverted Flared Fittings ... j (a-2)

Isolating Noise ... h-11-12

Isothermal Performance ... a-2-3

j

Jewel Gland ... b-16

JIC Hydraulic Standards ... j (e-10)

Jogging Pump ... h-10

k

Key Shaft Pump ... h-3

Kinetic
- Energy ... b-10
- Friction ... b-38

Kinking
- Hose ... j (c-8-9)
- Tubing ... j (e-23)

Kloskut ... j (e-23)

index continued

l

"L" Shaped Reservoir h-7
Laminations, Solenoid c-8
Large Molecule Media .... j (3-15)
Leakage
- Cylinder b-38
- Directional Control Valve c-10
- Internal b-28, h-13
- Mod-Logic Valves g-20
- Piping .... j (e-17)
- Piston Rings b-16
- Rod Gland b-38
- Seals b-38
- System a-3
- Tube Fittings .... j (e-14-15)
- Viscosity Change d-44

Level Gauge .... k (b-4)
Lever
- First Class b-18, b-22
- Hinged b-25
- Second Class b-18, b-21
- Third Class b-18

Life
- Solenoid c-7
- System d-38

Lift
- Suction d-41, d-44,
- Vacuum g-22

Lifting, Vertical b-25
Lipseal b-15-16, b-32
Liquid Level Control g-15
Load
- Acceleration b-6
- Deceleration b-6-7
- Friction Coefficient b-7, b-13
- Guiding b-5, b-33
- Misalignment b-38
- Rolling b-24
- Sliding b-24-25
- Solenoid c-6
- Surface Friction b-25

Lock Valve .... k (b-5)
Locked Rotor Torque, Electric Motor c-4
Locknut Fittings .... j (e-22)
Loop Air Distribution System g-18
Low
- Voltage, Solenoids c-9

Lubrication
- Excessive g-25
- Insufficient d-44,
- Mist g-12
- Micro Mist g-13
- Solenoid c-7
- Tube Fitting .... j (e-21)

Lubricators
- Adjustment g-12
- Application g-11-12
- Automatically Filled g-11
- Integral Reservoir g-11
- Manually Filled g-11

Lubriplate .... j (e-19, e-21)
Lug and Side Tapped Cylinder Mounting b-3

m

Machine
- Downtime d-43
- Envelope .... k (b-5-7)

Magnetic Aging, Solenoid c-7
Magnets and Filtration d-38
Main Header g-10
Maintenance
- Air Regulators g-11
- Filters d-6, d-45-46
- Source of Dirt d-43

Make Up Check f-5-6
Mandrel-Assembled Fittings .... j (c-4)
Manifold
- Cylinder Ports b-14
- Directional Valves .... i-15, k (b-2)

Manual Override g-25
Matched Flow Controls b-27
Mean Flow Pore Size d-39-40
Mechanical
- Damage Solenoid c-9
- Holding Power, Tube Fittings .... j (e-15)
- Life, Solenoid c-8
- Strain, Tubing .... j (e-10-12)

Mechanically Connected Cylinders b-26
Membrane, Filter d-39
Meter-In .... b-28, f-4, h-8, k (b-3)
Meter-Out .... b-24-25, b-28, f-3, h-8, k (b-3)
Metric
- Conversion .... k (b-19)
- Flow Factor .... k (b-16)

Microphones
- Frequency Response .... k (b-6)
- Noise Measurement .... k (b-6, b-8)

Minor Area, Cylinder b-7
Misalignment
- Cylinder b-36
- Load b-38

Mobile Directional Control Valve
- Back Flow Check Valve .... k (c-20-21)
- En-Bloc Control Valve .... k (c-20-21)
- Parallel Flow Path .... k (c-20-21)
- Regenerative Make-Up .... k (c-20-21)
- Series Flow Path .... k (c-20-21)

index continued

Mobile
- Power Unit c-4
- Pilot Operated Relief Valve i-14
- Tell Tale Filter d-38

Mod-Logic
- Base Markings g-21
- Clear View Spool Position Indicator g-20
- Cycle Life g-20
- Leakage g-20
- Servicing g-20

Moisture
- Air Lines g-18
- Oil h-13

Monitoring Fluid Porblems d-44

Motor Starter Sizes, Electric c-5

Motors, Hydraulic See Hydraulic Motors

Mounting
- Bolts, Cylinders b-3
- Bracket, Pump i-4
- Plate, Reservoir h-3-4, h-11
- Styles, Cylinders b-3

Muffler, Air g-25

Multipliers for Fittings & Bends d-19

n

National Machine Tool Builders Association (NMTBA) k (b-5)

National Sanitation Foundation Standard .... j (a-4)

Needle Valves .... j (a-5)

Net Cylinder Stroke b-5

Never-Stop Suction Filter d-41

Neutralization Number d-44

NFPA Recommended Standard For Filtration d-38

Nitrile (nbr) .... j (b-2)

Nitrogen Precharging a-3

No
- Delay Starting g-14
- Flow Pressure Setting g-10
- Skive Hose Fitting .... j (c-13)

Noise
- Ambient .... k (b-5)
- Dampening h-11
- Emanation, Machine Tools .... k (b-5)
- Isolating h-11-12
- Measurement Techniques .... k (b-5-15)
- Reduction h-11
- Regulations .... k (b-5)
- Reverberation .... k (b-5)
- Sources .... h-10-11, k (b-5)

Noise Measurement
- Cyclical Noise .... k (b-6, b-8)
- Decibel .... k (b-5-6, b-8-9)
- Fast Meter Response .... k (b-6)
- Filters .... k (b-6, b-8)
- Flat Response .... k (b-6)
- Free Filed .... k (b-6)
- Impulse Noise .... k (b-8)
- Instrument Calibration .... k (b-6)
- Instrumentation .... k (b-8-9)
- Machine Envelope .... k (b-5-7)
- Machine Operating Conditions .... k (b-5-6)
- Measurement Locations .... k (b-5)
- Meter Response .... k (b-6)
- Microphone Frequency Response .... k (b-6)
- Microphones .... k (b-6, b-8)
- Octave Band .... k (b-6)
- Octave Band Spectral Analysis .... k (b-6)
- Oscilloscope .... k (b-8-9)
- Reflective Surfaces .... k (b-5)
- Reporting Data Form .... k (b-9)
- Sound Pressure Level .... k (b-6)
- Steady State Noise .... k (b-6, b-8)
- Tape Recording .... k (b-6, b-8)
- Test Space .... k (b-5)
- Weighting Networks .... k (b-8)

Nominal
- Rating, Filter Element d-39
- Voltage c-7-8

Non-Repeat System g-5, g-16

Non-Shock Rating, Cylinders b-7

Non-Venting Air Regulator g-11

NPT Pipe Threads .... j (e-17-18)

NPTF
- Dryseal Pipe Threads .... j (e-18)
- Ports b-14

Nylon Tube Fittings .... j (a-4)

o

Octave Band .... k (b-6)

Octave Band Spectral Analysis .... k (b-6)

Oil
- Delivery Indicators, Lubricators g-11
- Mist g-12
- Micro-Mist g-13

Oil-Free Air g-17

Oldham Coupling i-22

101 Directional Control Valve c-8-9

111 Directional Control Valve c-9

Open
- Center Valve b-38
- Circuit, Solenoid c-9
- Circuit, Hydraulic h-10
- Motor c-4

Open-Dripproof-Motors h-11

Operating Pressure
- Tubing .... j (e-11)

Operating Temperature
- Hose .... j (c-7, c-11)
- System h-8, h-10-11
- Tubing .... j (e-12)

Orifice
- Air Flow g-6-7
- Air Flow Under Vacuum g-22
- Coefficient g-6-7
- Contamination c-9, h-8
- Shape g-7

index continued

Oscillating Motion g-16

Oscilloscope, Noise Measurement .... k (b-8-9)

Overcenter Rotary Drive f-9

Overheating
- System h-13, i-9
- Viscosity Change d-44

Overlubrication g-25

Override, Manual g-24

Over-Run Limiter f-6

Overtightening, NPT Thread .... b-14, j (e-18)

Oxidation, Fluid d-44-45

p

Paper Element d-39

Parallel Motors f-8

Particle
- Abrasive d-43
- Count d-44
- Ferro-Magnetic d-38
- Generation d-42
- Removal d-38
- Settling d-38
- Suspension d-38

Peak Pressure i-5

Pick Up Force, Vacuum g-22

Piggyback Gear Pump i-12

Pillow Blocks b-3

Pilot
- Controlled Air Regulator g-10
- Drain c-10
- Operated Check Valve .... b-27, b-29, b-38, k (b-3)
- Operated Relief Valve, Mobile i-14
- Orifice c-9
- Pressure c-10
- Spool c-9

Pintle, Piston Pump i-20-21

Pipe
- Air Flow g-8
- Double Extra Heavy .... j (e-10)
- Extra Heavy, Schedule 80 .... j (e-10)
- Fittings .... j (a-5)
- NPT Threads .... j (e-17-18)
- NPTF Dryseal Threads .... j (e-18)
- Schedule 160 .... j (e-10)
- Sealing Compounds .... j (e-22)
- Standard Schedule 40 .... j (e-10)

Piston
- Accumulators a-2-3
- Construction, Cylinder b-16
- Cylinder b-31-32
- Motors f-4
- Rings, Cylinder b-31-32
- Rod b-4-6, b-31-33
- Speed b-8

Piston Pumps
- Application i-13
- Axial i-19
- Barrel i-19
- Bent Axis i-19-20
- Cam Plate i-19
- Construction i-19-20
- Degree Filtration d-9
- In-Line i-19
- Operation i-19-20
- Radial i-20-21
- Reversing Flow i-20
- Shoe Plate i-19
- Swash Plate i-19
- Tilt Box i-19
- Tilting Blocks i-19
- Valve Plate i-19
- Variable Volume i-20
- Wobble Plate i-19

Pivot Force Transfer b-4

Plastic Tubing
- FDA Specification .... j (d-2)
- Instrument Grade .... j (d-2)
- Ultraviolet Inhibitor .... j (d-2)

Plug-In
- Mounting g-20
- Valves g-3

Plunger, Solenoid c-6-9

Pneumatic Circuitry g-13-16

PO Check Valve .... b-27, b-29, b-38, k (b-3)

Poly-Tite Fittings .... j (a-4)

Pore Size
- Control, Filter Element d-40
- Depth Filter Element d-39
- Surface Filter Element d-39

Port Plate i-18

Porta-Flare .... j (e-23)

Ports
- Overtightening NPT b-14
- Position, Cylinder b-14
- Sizes, Cylinder b-8, b-14
- Straight Thread b-14
- Tapered Pipe Thread b-14

Pour Point i-25

Power
- Actuated Flaring Tool .... j (e-19)
- Drain, Electric i-22
- Factor b-19, b-24

Power Unit
- Design h-3
- General h-2-14
- Mobile c-4
- Noise h-10-12
- Start Up h-10

Precharging, Accumulators a-3-4

Pre-Setting, Tools .... j (e-20, e-23)

Pressure
- Intensification .... b-30, k (b-3)
- Proof Hydraulic Couplings .... j (b-2)
- Rating, Pump i-5, i-9
- Reducing Valve .... h-8-9, i-10, k (b-3)
- Surge d-42

index continued

Pressure Compensated
- Dual Vane Pump i-12
- Flow Control Valves f-5, f-7-8
- Variable Volume Pumps i-12-13, i-17-21

Pressure Differential
- Control Spring d-41
- Filters d-40-41, d-44
- Suction Line d-41

Pressure Drop Charts d-20-37

Pressure Filters
- Bi-Directional d-42
- Duplex d-42
- 5000 PSI d-42
- Pressure Differential d-42
- System Design .... k (b-4)
- 3000 PSI d-42

Priority Flow Divider .... k (b-4)

Pull In Force, Solenoid c-8

Pumps
- Air Bleed h-10
- Alignment h-3, h-10,
- Calculations i-7
- Capacity .... k (b-2)
- Case Drain h-10, h-13
- Circuits i-14-16
- Coupling i-4
- Double i-11-12
- Drive Speed i-5
- Dual Vane i-11
- End Thrust h-3, h-10
- Environment i-5
- Gear i-12
- Inlet Conditions d-45, h-10
- Jogging h-10
- Key Shaft h-3
- Mounting Bracket i-4
- Noise i-5,
- Operation i-16-17
- Particles Generated d-42
- Port Connections h-10
- Pressure Compensated i-12-13, i-17-21
- Pressure Rating i-5, i-9
- Priming h-10
- Ratings i-17
- Replenishing f-9, h-10
- Risers h-6
- Safety Factor Application i-12
- Selection i-5, i-7
- Single i-11-12
- Sizing i-9
- Slippage h-8, i-17
- Speed h-11
- Suction Specification d-41
- Supercharging h-11, i-12
- Supplementing Delivery a-3
- Tang Shaft h-3
- Types i-17
- Wear h-12, i-5, i-14,

Push Pin, Solenoid c-9

q

Quick Couplings
- Double Shut-Off .... j (b-3)
- Single Shut-Off .... j (b-3)
- Straight Thru .... j (b-3)

Quiet Zone, Air Filter g-10

r

Rack and Pinion Connection, Cylinder b-26

Radial Piston Pump i-20-21

Ram, Single Acting b-32-33

Ratings
- Pumps i-17
- Solenoids c-7

Ratio, Intensifier e-2

Receiver Tank, Air g-17

Recharging, Accumulators a-5

Reciprocating Compressor g-17

Recommended Practice for the Use of Fire Resistant Fluids for Fluid Power Systems d-46

Reflective Surfaces, Noise Measurement .... k (b-5)

Regeneration
- Cylinders b-29-31
- Motors f-9

Regenerative Make-Up, Mobile Valve .... k (c-20-21)

Regulator, Air See Air Regulator

Relative Loss, Air Pressure g-8

Reliability, Fluid Power Components d-38

Relief Valves
- Accumulator Systems a-4-6
- Braking f-3
- Cross-Over f-3
- Differential Unloading a-5-6
- Heat Generation h-8
- Pressure Setting i-9
- Sealed i-5
- Solenoid Operated h-9
- Venting a-5-6, f-8, i-14-15

Remake, Tube Fitting .... j (e-21-22)

Removal, Dirt d-39

Repairability, Filter Element d-40

Replenishing Pump f-9, h-10

Reservoir
- Baffle h-3, h-11-13
- Breather Cap h-3
- Contamination d-43-44
- Cooling Capacity h-7
- Design i-5
- Drain Line Installation h-3
- Drain Plug h-3
- Filling d-45
- Fluid Level b-30, d-41
- General .... h-2, h-4, h-7-13, k (b-3)
- Heat Dissipation h-3, h-7-8
- "L" Shaped h-7
- Mounting Plate h-3-4, h-11
- Oil Level Indicator h-3
- Return Line Installation h-3
- Sizing h-3, h-8

index continued

Residual
- Magnetism, Solenoid c-7
- Stresses .... j (e-22)

Return Line
- Filters d-10
- Installation h-3
- Position .... i-25, k (b-4)
- Size d-10

Reverberation, Noise .... k (b-5)

Reversing Flow i-20

Rings, Cylinder Piston b-32

Rod, Cylinder
- Bearing b-32-33
- Buckling b-33
- End Connection b-5-6
- Extension b-5
- Gland b-31-32
- Gland Leak b-38
- Seals b-31-32
- Scraper b-32
- Speed b-17, b-29
- Wiper b-15, b-32

Rolling Load b-24

Rolo-Flair .... j (e-23)

Rotary Motion, Cylinders b-21

Rotor, Vane Pump i-18

Rules of Thumb, Filtration d-8-9

Running Torque f-3

S

SAE
- 45° Flared Fitting .... j (a-3)
- Pipe Fittings Standard .... j (a-5)
- Ports b-4
- Test Code i-17
- Tube Fittings Standard .... j (a-5)

Safety Factor
- Hose .... j (c-8-9)
- Pump Application i-12
- Tube Fittings .... j (e-15)
- Tubing .... j (e-10)

Safety Switches .... k (b-4)

Sampling, Fluid d-11

Saw - Skware, Saw Guide .... j (e-23)

Scale Layout Method, Cylinder Stroke b-25-26

Schedule
- 40 Pipe .... j (e-10)
- 80 Pipe .... j (e-10)
- 160 Pipe .... j (e-10)

Scraper, Rod b-32

Screw Type Flaring Tool .... j (e-19)

Sealed Relief Valve i-5

Seal, Vane i-19

Seals
- Accumulators a-4
- Buna N (NBR) .... j (b-2)
- Chloroprene Rubber (CR) .... j (b-2)
- Compatibility c-10, h-13
- Deterioration, Cylinders b-38
- Ethylene Propylene Rubber (EPM) .... j (b-2)
- Extrusion h-14
- Fluorocarbon Rubber (FPM) .... j (b-2)
- Nitrile (NBR) .... j (b-2)
- Wear b-38, h-8

Second Class Lever b-18, b-21

Self-Adjusting Rod End, Cylinder b-36

Self-Bleeding Cylinder b-16

Sequence of Operations .... k (b-2)

Series
- Actuators b-27
- Hydraulic Motors f-8
- Wound Motor i-22

Service
- Abuse, Tubing .... j (e-10, e-12)
- Air Preparation Units g-10
- Filter d-13
- Life, Fluid Power Systems d-6
- Mod-Logic Valves g-20

Setting
- Air Regulator g-10-11
- Relief Valve i-9

Settling, Dirt Particles d-6

Shear Key, Cylinder Mounting b-37

Shelf Life, Filter Element d-8

Shock
- Accumulator Absorbing a-3
- Cylinders b-16
- Hydraulic Motors f-3
- Tube Fittings .... j (e-14, e-16)
- Tubing .... j (e-10, e-13)

Shoe Plate, Piston Pump i-19

Short Circuit, Solenoid c-9

Shut-Off Valves .... k (b-3-4)

Side
- End Angles, Cylinder Mounting b-3
- End Lugs, Cylinder Mounting b-3
- Loading, Cylinder b-33
- Tapped, Cylinder Mounting b-3

Silt Index d-12

Silting d-12

Signal Lights .... k (b-4)

Single
- Acting Cylinder b-32-33
- Pressure Intensifier e-2, e-4
- Shut-Off Quick Coupling .... j (b-3)
- Stage Compressor g-17

60 Cycle Solenoids c-7

Skive Hose Fitting .... j (c-13)

Sliding Loads b-24-25

Slippage h-8, i-5, i-17

index continued

Sludge h-8, h-13

Small Molecule Media .... j (e-15)

Solenoid Operated Relief Valve h-9

Solenoid Valves, Troubleshooting c-9

Solenoids
- AC Current Flow c-6
- AC Hum g-24-25
- Appliance Grade c-7
- Bobbin c-6
- Breakaway Springs c-7
- Burn-Out c-7-10, g-24
- Closing Time c-6
- Coils g-24
- Corroded g-24
- Cycle Rate c-6-7
- Dual Frequency c-7
- Duty Cycle c-6
- Energized Simultaneously g-24
- Erratic Operation c-9
- Failure c-9, g-24
- Feed Lines c-9
- Field c-7-8
- Fifty (50) Cycle c-7
- Force c-7-8
- Friction c-7
- Heat Dissipation c-6
- Heat Generation c-6
- Heat Transient Voltages g-24
- Holding Current c-6
- Industrial Grade c-7
- Inrush Current c-6-7, c-9
- Interlock c-9-10
- Laminations, Coil c-8
- Life c-7
- Load c-6
- Low Voltage c-9
- Lubrication c-7
- Magnetic Aging c-7
- Manual Override g-25
- Mechanical Damage c-9
- Mechanical Life c-8
- Nominal Rating c-8
- Open Circuit c-9
- Plunger c-6-9
- Pull In Force c-8
- Push Pin c-9
- Ratings c-7
- Residual Magnetism c-7
- Short Circuit c-9
- Sixty (60) Cycle c-7
- Size Selection c-8
- Stroke c-6
- Tapped Dual Frequency c-7-8
- Temperature c-7-8
- Three Lead Dual Frequency c-7-8
- Troubleshooting c-7-9
- Voltage c-7-9, g-24
- Wet Armature c-6-7

Soluble Oil d-45

Sound Pressure Level .... k (b-6)

Sources of Dirt d-45

Spear, Cushion b-15

Specific Gravity d-44

Speed Controls b-13

Spiral Wrap Hose .... j (c-9)

Splash Proof Motor c-4

Spreader Type Tie Rods, Cylinders b-25

Spring Guards, Hose .... j (c-8-9)

Spring-Return Cylinders b-32-33

Springs, Directional Valves c-9

Spur Gear Pump i-18

Squirrel Cage Motor c-4, i-21

SSA Series Plug-In Air Valves g-3

SS 1200 Air Valve g-5

Stabilization, New Lubrication System g-12

Stack Valves h-8

Stainless Steel Coupling .... j (b-2)

Stall Point, Electric Motor i-23

Static Friction b-38

Steady State Noise .... k (b-6, b-8)

Steel Coupling .... j (b-2)

Stop Tube b-5-6, b-33

Straight
- Line Force Transfer b-4
- Threads .... j (e-18, e-22)

Straight Thread Boss
- Counterborer .... j (e-23)
- O-Ring Replacement .... j (e-22)
- Tap .... j (e-23)

Straight-Thru Quick Couplings .... j (b-3)

Stroke, Cylinder
- Adjusters b-16
- Considerations b-4-6, b-33
- Factor b-5
- Selection b-5

Submersible Suction Filter d-41

Sucker, Vacuum g-22

Suction Filters
- Application .... d-41-42, k (b-4)
- Collecting Dirt d-41
- Elbow Style d-41
- Fabricated Style d-41
- Never Stop Style d-41
- Pressure Differential Control Spring d-41
- Submersible Style d-41

Suction Lift d-41, d-44

Suction Line
- Constant Resistances d-41
- Filtration d-38, d-41
- Fluid Velocity d-41
- General .... i-5, k (b-3)
- Pressure Differential d-41

Suction Specification, Pump d-41

Sun, Reservoir Heat Dissipation h-8

Supercharge Pump i-12, h-11

index continued

Super-Lok Tube Fittings .... j (e-15)

Supply, Fluid c-10

Surface
- Finish, Tube Fittings .... j (e-15)
- Friction b-25

Surface Elements
- Absolute Rating d-39
- Advantages d-40
- Disadvantages d-40
- Pore Size d-39
- Wire Mesh d-39-40

Surge Pressures .... d-42, j (c-7-8)

Suspension, Dirt Particles d-38

Swash Plate, Piston Pump i-19

Swivel Adapter, Hose .... j (c-9)

Synthetic Element d-39

System
- Cost i-5
- Efficiency .... j (e-13)
- Feedback i-21
- Inefficiency h-8

t

Tandem
- Cylinders b-32-33
- Gear Pumps i-12

Tang Shaft Pump h-3

Tape Recording, Noise Measurement .... k (b-6, b-8)

Tapped Dual Frequency Solenoids c-7-8

Teflon Tape .... j (e-22-23)

Telescoping Cylinders b-32-33

Tell Tale Indicator d-45

Temperature
- Ambient h-7, h-13
- Compensation f-4
- Controls .... k (b-4)
- Filter Element d-40
- Fluid Sampling d-43
- Reservoir c-10
- Solenoids c-7-8, g-24
- System Operating h-8, h-10-11

Temperature Range
- Accumulators a-4
- Tube Fittings .... j (a-2-6)

Tension Failure, Cylinders b-34

Tension Loads b-4

Test Space, Noise Measurement .... k (b-5)

Thermal Expansion, System a-3-4

Thermoplastic Tubing .... j (d-2)

Thermostats .... k (b-4)

Third Class Lever b-18

37° Flared Fittings .... j (e-19)

Threaded Sleeve Fittings .... j (a-2-3)

Three Lead Dual Frequency Solenoids c-7-8

3 MD Directional Control Valve c-6

Three-Phase Motor Design c-4

3000 PSI Pressure Filter d-42

Three-Way Logic g-20

Thrust
- Air Cylinder b-12-13
- Cylinder at an Angle b-19
- Key Cylinder Mounting b-3

Tie Rod
- Cylinder Mounting b-3, b-32, b-35, b-38
- Spreader Type b-38

Tilt Box, Piston Pump i-19

Tilting Blocks, Piston Pump i-19

Toggle
- Force b-18
- Mechanism b-18, b-24

Totally-Enclosed Motors
- Fan Cooled (TEFC) c-4, h-1
- Nonventilated (TENV) c-4

Transient Voltage g-24

Transmitted Noise h-10-12

Triple-Lok Tube Fittings .... j (e-10)

Troubleshooting
- Air Valves g-24-25
- Cylinders b-38
- Directional Control Valves c-9
- Tube Fittings .... j (e-24)
- Solenoid c-7-9
- Solenoid Valves c-9
- System h-13-14

Trunnion Cylinder Mounting b-3, b-33, b-37

Tube Fittings
- Analyzing Requirements .... j (e-14)
- Bite Type .... j (e-14-15, e-20)
- Compression Type .... j (a-2-4)
- Creeping .... j (a-4)
- Dubl-Barb .... j (a-4)
- Environment .... j (e-14)
- Environmental Stress Cracking .... j (a-4)
- Fabrication .... j (e-16)
- Flared .... j (e-14-15)
- Flareless .... j (e-14-15, e-20-22)
- Flow Clearance .... j (a-4)
- Hazardous Liquid Approval .... j (a-2-3)
- Installation .... j (e-20)
- Instrumentation Type .... j (e-14, e-16, e-21-22)
- Inverted Flare .... j (a-2)
- Leakage .... j (e-14-15, e-24)
- Locknut .... j (e-22)
- Mechanical Holding Power .... j (e-15)
- Nylon Tube .... j (a-4)
- Poly-Tite .... j (a-4)
- Remake .... j (e-21-22)
- SAE 45° Flared .... j (a-3)
- Safety Factor .... j (e-15)
- Selection .... j (e-14, e-16)
- Shock .... j (e-14)

index continued

Super-Lok .... j (e-15)
Surface Finish .... j ((e-15)
Temperature Range .... j (a-2-6)
37° Flared .... j (e-19)
Threaded Sleeve .... j (a-2-3)
Troubleshooting .... j (e-24)
Value Factors .... j (e-16)
Vibration .... j (a-2-3, a-5, e-14-16)
Vibration Dampening .... j (e-16)
Weld Type .... j (e-16)

Tubing
Alignment .... j (e-21-23)
Benders .... j (e-23)
Blow-Out Force .... j (e-15, e-20)
Burst Pressure .... j (e-12)
Corrosion .... j (e-11)
Cracked .... j (e-24)
Deformation Capability .... j (e-14)
Flaring .... j (e-14)
Grade of Severity .... j (e-10)
Hardness Specification .... j (e-11)
Hoop Strength .... j (e-14)
Installation .... j (e-22)
Kinking .... j (e-23)
Mechanical Strain .... j (e-10-12)
Notching .... j (a-4)
Operating Pressure .... j (e-11)
Operating Temperature .... j (e-12)
Preparation .... j (e-19)
Safety Factor .... j (e-10)
Scratched .... j (e-24)
Selection .... j (e-10-13)
Service Abuse .... j (e-10, e-12)
System Advantages .... j (e-17)
Velocity .... j (e-11-12)
Wall Thickness .... j (e-10)

Turbulence
Air Filter g-10
Hose .... j (c-7)

Two Bolt Flange Gear Pumps i-21

2:1 Cylinder b-29-30

Two Stage Compressor g-17

Two-Way Valve i-14

u

"U" Tube g-9

Ultraviolet Inhibitor .... j (d-2)

Unidirectional Motor Applications f-5-7

Uniform Acceleration Force Factor b-6-7

Unipar .... j (e-22)

Unloading, System
Accumulators a-5
Cam Valve i-14
Low Pressure i-14
Tandem Center Valve i-15

Unloading Valve i-11-12

V

"V" Packing b-15, b-32

Vacuum
Chuck g-22
Lift g-22
Pick Up Force g-22
Rating, Couplings .... j (b-2)
Requirements, Hose .... j (c-9)
Sucker g-22
Tank Considerations g-22-23

Valve
Chattering h-11
Hunting h-11
Plate, Piston Pump i-19

Vane
Loading i-19
Seal i-19

Vane Pumps
Application i-12
Cam Ring i-18-19
Filtration d-8
Operation i-17-19
Port Plate i-18
Volumetric Efficiency i-19

Variable Volume Pumps
Dual Vane i-12
Gear i-21
General i-17
Piston i-13, i-20
Reversing Flow i-20
Vane i-19

Varnish d-44, g-24, h-8

VDP Directional Control Valve .... k (b-8)

Velocity, Recommended Fluid b-8, b-14, d-9-10, h-11

Vent
Ports, Cylinders b-16
Screws, Cylinders b-16

Venting
Brake Valve f-7
Differential Unloading Relief Valve a-6
Relief Valve a-5-6, f-8, i-14-15

Vibration
Dampening .... j (e-16)
Hose .... j (c-8)
Tube Fittings .... j (e-14, e-16)

Vertical Lifting b-25

Viscosity
Change d-44, i-5
Filter Elements d-8
Range, Start Up Fluid h-10

Voltage
Nominal c-7-8
Solenoid c-7-8, g-24
Transient g-24

Volumetric
Efficiency, Pumps i-18-19
Expansion, Hose .... j (c-8)
Output, Pumps i-17

index continued

W

Walsh-Healey Act .... k (b-5)

Water
- Base Fluid d-40
- Content, Petroleum Oil d-44

Water Service
- Accumulators a-4
- Modifications, Cylinders b-16
- Warranty, Cylinders b-16

Water Vapor, Air Line g-10

Wear
- Hose .... j (c-8)
- Plate, Gear Pump i-18
- Pump h-12, i-5, i-14

Weighted Filters, Noise Measurement .... k (b-8)

Weighting Networks, Noise Measurement .... k (b-8)

Weld Type Tube Fittings .... j (e-15)

Weldlok Tube Fittings .... j (e-10)

Wiper, Cylinder Rod b-15, b-32

Wiperseal b-15-16

Wire
- Mesh, Filter Elements d-49-40
- Sizes, Electrical c-5

Wobble Plate, Piston Pump i-19

Workmanship .... j (e-16, e-22)

Y

Yoke, Piston Pump i-20

Z

Zeroing, Bidirectional Pump f-9

index to graphs and tables

Title	Page
Absolute Viscosity Conversion Factors	k (b-24)
Accumulator Sizing Table	a-2
Air Brake Fittings - Turns to Seal	j (a-5)
Air Cylinder Cushion Ratings Table	b-11
Air-Oil Tank Rated Capacities	e-6
Aluminum Pipe	j (e-2)
Aluminum Tubing	j (e-8)
Annealed Low Carbon Steel Tubing	j (e-7)
Annealed Stainless Steel Tubing	j (e-6)
ASTM Graph Paper	d-15
Characteristics of the Major Valve Designs	g-5
Chord Factor	b-19
Combined Indicator Card Comparing Single-Stage and Two-Stage Compressors	g-17
Compression Factors and "A" Constants	g-4
Cost of Air Leaks	g-18
Conventional U.S. Conversion Factors	k (b-25-32)
Copper Pipe	j (e-2)
Cubic Feet Into Gallons Conversion Table	k (b-20)
Cubic Feet Per Minute — Linear Feet Per Minute With Air Flowing Through Tubes	g-3
Current Flow AC Solenoid	c-6
Current Flow AC Solenoid - Reduced Load	c-6
Current Flow AC Solenoid - Reduced Stroke	c-6
Cylinder Mounting Classes	b-4
Cylinder Port Position Table	b-19
Density and Viscosity of Air at Atmospheric Pressure	g-3
Drive Coupling Table	h-5
Dynamic Cylinder Thrust - Air Cylinders (80 PSIG)	b-13
Dynamic Cylinder Thrust - Air Cylinders (100 PSIG)	b-12
Effective Square Inch Areas for Standard Bore Size Cylinders	g-4
Electric Motor Horsepower	c-2
Extra Heavy Pipe	j (e-2)
Fast-Stor Tubing	j (d-3)
Feet Per Second Into Meters Per Second Conversion Table	k (b-21)
Fixed Displacement, P/C, Hi-Lo System Available Power	i-22
Flow Capacity of Parker Hose Assemblies at Recommended Flow Velocities	j (c-3)
Flow of Air Through Orifices Under Pressure	g-7
Flow of Water Under Pressure	g-2
Flow vs. Velocity vs. Tubing I.D.	j (e-3)
Fluids Charts	d-2
Foot-Mounted Motor Dimensions	c-3
Formula PV-1 Plastic Tubing	j (d-2)
Formula PV-2 Plastic Tubing	j (d-2)
Four-Bolt Mounting Flanges	i-2
Gallon Equivalents in Cubic Inches	i-6
Gallons into Cubic Feet Conversion Table	k (b-20)
Gallons Into Liters Conversion Table	k (b-20)
Gas Weight Per Cubic Foot	g-9
GPM and HP from Cu. In./Rev.	i-3
Heat Colors	k (b-23)
Holding Force Created by Vacuum on Suckers	g-22
Horsepower Into Kilowatts Conversion Table	k (b-21)
Horsepower - Speed - Torque Table	f-2
Hose Material Selection	j (c-11)
Hose Selection by Pressure	j (c-2)
Hydraulic Cylinder Port Sizes and Piston Speed	b-8-9
Hydraulic Filter Element Classifications	d-40
Inches Into Millimeters Conversion Table	k (b-20)
Intensifier Selector Chart	e-2
Inverted Flared Fittings - Working Pressure Ranges	j (a-2)
Kilograms Into Pounds Conversion Table	k (b-21)
Kilograms Per Square Centimeter Into Pounds Per Square Inch Conversion Table	k (b-21)
Kilowatts Into Horsepower Conversion Table	k (b-21)
Kinematic Viscosity Conversion Factors	k (b-24)
Kinetic Energy Graph - Air Cylinders	b-10
Liters Into Gallons Conversion Table	k (b-20)
Manufacturer's Oil No. 10	d-45
Manufacturer's Oil No. 10 Viscosity Change Due To Addition of Water	d-45
Mean Flow Pore Size	d-40
Measures of Volume (Compared)	i-6
Meters Per Second Into Feet Per Second Conversion Table	k (b-21)
Metric Conversion Table	k (b-19)
Metric Equivalents	k (b-22)
Metric Relationships	k (b-20)
Millimeters Into Inches Conversion Table	k (b-20)
Millimeters To Fractions To Decimals	k (b-20)
Motor Frame Size/Parker Pumps	i-4
Motor Pump Mounting Plate Data	h-4
Mounting Flanges For Pumps and Motors	i-2
Numerical Conversion Table	k (b-20)
Orifice Pressure Drop	k (b-17)
Piston Rod Stroke Selection Graph	b-6
Piston Rod Stroke Selection Table	b-5
Pounds Into Kilograms Conversion Table	k (b-21)
Power Factor Table	b-19

index of tables & graphs continued

Title	Page
Δ P (PSI)/ft Schedule 40 Pipe	d-20
Δ P (PSI)/ft Schedule 80 Pipe	d-22
Δ P (PSI)/ft Schedule 160 Pipe	d-24
Δ P (PSI)/ft Double Extra Heavy Pipe	d-25
Δ P (PSI)/ft Tube	d-26
Δ P (PSI)/ft Hose	d-32
Δ P (in. Hg)/ft Schedule 40 Pipe	d-34
Δ P (in. Hg)/ft Tube	d-36
Δ P (in. Hg)/ft Hose	d-37
Pressure Drop Conversion Factors For Chart Based on Texaco A (R & O) at 180°F	j (c-6)
Pressure Drop In Tubing For Various Air Flows	g-3
Pressure Drop — One Foot of Hose	j (c-5)
Pressure Drop — Two Full Flow Fittings	j (c-6)
Pressure Drop — Two Non-Mandrel Fittings	j (c-5)
Pump Risers	h-6
Pressure Loss Due To Air Friction In Pipes	g-2
Rated Full Load Torque Table	f-3
Recommended Min./Max. Tube Wall Thickness for Common Fitting Types	j (e-4)
Relative Sizes of Particles and Comparison of Dimensional Units	d-43
Reservoir Data	h-4
Reservoir Heat Dissipation	h-7
Riser Block Heights	h-6
SAE 45° Flared Fittings — Working Pressure Ranges	j (a-3)
Selected "SI" Units for General Purpose Fluid Power Usage	k (b-24)
Series Wound Motor Characteristics	i-22
Soft Copper Tubing	j (e-9)
Square Centimeter Into Square Inch Conversion Table	k (b-20)
Square Inch Into Square Centimeter Conversion Table	k (b-20)
Steel Pipe	j (e-2)
Temperature Conversion Table	k (b-22)
Theoretical Push and Pull Forces For Cylinders (Metric Units)	b-2
Theoretical Push and Pull Forces for Cylinders (U.S. Units)	b-2
Threaded Sleeve Fittings — Turns Required to Seal	j (a-3)
Threaded Sleeve Fittings — Working Pressure Ranges	j (a-2)
Thrust Developed — Air Cylinder	b-12-13
Two-Bolt Mounting Flanges	i-2
Type 304 Stainless Steel Seamless Pipe	j (e-2)
Uniform Acceleration and Deceleration Force For Hydraulic Cylinders	b-7
Velocity vs. Flow Table (Tube)	j (e-5)
Viscosity vs. Temperature	k (b-18)
Weight Table	b-10
Working Pressures for Fast and Tite Fittings	j (d-4)

index of circuits

Circuit No.	Title	Page
a-1	Basic Accumulator Circuit	a-5
a-2	Accumulator Unloading	a-5
a-3	Venting Relief With Small Unloading Valve	a-5
a-4	Differential Unloading Relief Circuit	a-6
a-5	Venting the Relief With Solenoid Controlled Two-Way Valve	a-6
a-6	Hi-Lo System Using An Accumulator	a-6
a-7	Venting A Differential Unloading Relief Valve	a-6
b-1	Mechanical Connection Cylinder Circuit	b-26
b-2	Rack and Pinion Connection Cylinder Circuit	b-26
b-3	Series Actuators Cylinder Circuit	b-27
b-4	Series Actuators Cylinder Circuit	b-27
b-5	Matched Flow Controls Cylinder Circuit	b-27
b-6	Coupled Motors Cylinder Circuit	b-28
b-7	Flow Dividing Valves Cylinder Circuit	b-28
b-8	Cylinder Regenerative Circuit	b-29
b-9	Cylinder Regenerative Circuit	b-30
b-10	Cylinder Regenerative Circuit	b-30
b-11	Cylinder Regenerative Circuit	b-30
b-12	Cylinder Regenerative Circuit	b-31
e-1	Intensifier Clamp Circuit	e-3
f-1	Unidirectional Motor Circuit	f-5
f-2	Unidirectional Motor Circuit	f-5
f-3	Unidirectional Motor Circuit	f-5
f-4	Unidirectional Motor Circuit	f-5
f-5	Braking Relief Unidirectional	f-6
f-6	Over-Run Limiter	f-6
f-7	Bidirectional Motors	f-6
f-8	Dual/Relief Braking	f-6
f-9	Cross-Over Relief	f-6
f-10	Optional Braking Or Free Wheeling	f-7
f-11	Relief Valve Controlling Braking Valve	f-7
f-12	Motor Control Unidirectional	f-7
f-13	Series Motors	f-8
f-14	Parallel Motors	f-8
f-15	Series Motor Circuit	f-8
f-16	Series/Parallel Motors Bidirectional	f-8
f-17	Multiple Motors in Series/Parallel With Free Wheeling	f-9
f-18	Closed Loop Conventional Overcenter Rotary Drive	f-9
f-19	Closed Loop Prevention of Creep	f-9
g-1	Adjustable Decelerating Air Cushioning	g-13
g-2	Remote Control of Reversible Air Motors	g-14
g-3	Rapidly Dropping Pressure	g-14
g-4	Liquid Level Control	g-15
g-6	Non-Repeat Circuit	g-15
g-7	Dual Air Exhaust Controls	g-15
g-8	Non-Repeat Circuit	g-16
g-9	Oscillating Circuit	g-16
i-1	Hi-Lo Circuit	i-11
i-2	Unloading The Pump With A Two-Way Valve	i-14
i-3	Low Pressure Unloading	i-14
i-4	Cam Valve Unloading	i-14
i-5	Venting The Relief Valve To Unload	i-15
i-6	Tandem Center Directional Control Valve Unloading	i-15
i-7	High Pressure Forward, Low Pressure Return	i-15
i-8	Hi-Lo System With Set-Up Control	i-15
i-9	Two Pressure System Unloading	i-16

catalog reference directory

Information about the Parker Hannifin Product Line, or the location of the nearest Parker Hannifin Distributor, can be obtained from any one of the regional sales offices throughout the United States. Each office is staffed by Technical Salesmen and a Chief Application Engineer who are trained to aid you in the application of the product.

Northeast Region — Parker Hannifin Corporation, 280 Midland Avenue, Saddle Brook, New Jersey 07662, (201) 791-2400

Middle Atlantic Region — Parker Hannifin Corporation, 28 Springdale Road, Cherry Hill, New Jersey 08003, (609) 424-3200

Fluidpower Sales Division — Parker Hannifin Corporation, 13103 Capital, Oak Park, Michigan 48237, (313) 398-6950

Southern Region — Parker Hannifin Corporation, 2837 Stasuma, Dallas, Texas 75229, (214) 241-7621

Midwest Region — Parker Hannifin Corporation, 500 S. Wolf Road, Des Plaines, Illinois 60016, (312) 298-2400

Western Region — Parker Hannifin Corporation, 18321 Jamboree Blvd., Irvine, California 92715, (714) 833-3000

Cleveland Region — Parker Hannifin Corporation, 17325 Euclid Avenue, Cleveland, Ohio 44112, (216) 531-3000

The most complete line of products or components required for designing fluid power systems can be chosen from the following Parker-Hannifin catalogs.

Catalog Name	Catalog Number
Mod-Logic Air Controls	0600
Compressed Air Preparation Units	0700
*Air Control Valves	0600
Hydraulic Components (Mobile)	1710
Power Units	2800
Air Cylinders	0910
**Hydraulic Cylinders	1110
Heavy-Duty Flanged Cylinders	0996
Hydraulic Accumulators	1630
Fluid Power Intensifiers	1010-B1
Fluid Filters	2300
Industrial Hydraulic Components (Manatrol)	2506
*Series "45" Air Control Valves	0610
Series "55" Air Control Valves	0630
Series "C & CC" Air Control Valves	0640
Series "GG200" Air Control Valves	0620-B1

Catalog Name	Catalog Number
**Hannifin Cylinder Selector Catalog	0800
Involvement Training	0200
Tube Fittings & Tube Fab. Equipt. (Book)	4300
MA6000 Heavy Duty Tube Fittings	4370
CPI Tube Fittings	4230
CPI Valves	4250
Adapter Fittings	4260
Weldlok Fittings	4270
Tube Fabricating Equipment	4290
Brass Fittings & Valves	3501
Thermoplastic Tubing and System Components	3600
Quick-Disconnect Couplers and Blow Guns	3804
Hose and Fittings	4403
Flexible Nylon Hose	4451
Hose of Teflon *DuPont registered trademark	4491
Hose Adapters	4490

Contact the nearest regional office or call Parker-Hannifin at 216-531-3000.

Write: Parker-Hannifin Corporation
17325 Euclid Avenue
Cleveland, Ohio 44112

Attention: Catalog Service Center

Telex: Parker-Hannifin Corporation
9-8474

notes

notes